DATA COMMUNICATIONS

WILLIAM L. SCHWEBER

McGRAW-HILL BOOK COMPANY

New York Atlanta Dallas St. Louis San Francisco Auckland Bogotá Guatemala
Hamburg Lisbon London Madrid Mexico Milan Montreal New Delhi Panama
Paris San Juan São Paulo Singapore Sydney Tokyo Toronto

Sponsoring Editor: John Beck
Editing Supervisor: Melonie Parnes
Design and Art Supervisor: Annette Mastrolia-Tynan
Production Supervisor: Catherine Bokman

Text Designer: Suzanne Bennett & Associates
Cover Designer: Renée Kilbride-Edelman

Library of Congress Cataloging-in-Publication Data

Schweber, William L.
    Data communications.

    Includes index.
    1. Data transmission systems.   I. Title.
TK5105.S386   1988   004.6   87-3974
ISBN 0-07-001097-8

The manuscript for this book was processed electronically.

DATA COMMUNICATIONS

Copyright © 1988 by McGraw-Hill, Inc. All rights reserved. Printed in the United States of America. Except as permitted under the United States Copyright Act of 1976, no part of this publication may be reproduced or distributed in any form or by any means, or stored in a data base or retrieval system, without the prior written permission of the publisher.

1 2 3 4 5 6 7 8 9 0   DOCDOC   8 9 4 3 2 1 0 9 8 7

ISBN 0-07-001097-8

*To my family,
whose patience
and understanding
make it all
possible*

Contents

Preface		ix
Chapter 1	**An Introduction to Communications**	**1**
1-1	What Is Communication?	1
1-2	Uses of Communications	3
1-3	The Structure and Types of Communications Systems	5
1-4	Communications Systems and Data Communications	8
Chapter 2	**Communications Channel Characteristics**	**13**
2-1	The Communications Channel	13
2-2	Electromagnetic Waves	16
2-3	Frequency and Wavelength	19
2-4	The Electromagnetic Spectrum	22
2-5	Bandwidth	26
2-6	Bandwidth and Channel Capacity	29
2-7	Bandwidth and Distance	34
Chapter 3	**Modulation**	**40**
3-1	Modulation and Demodulation	40
3-2	Fourier Analysis	42
3-3	Types of Modulation	53
3-4	Amplitude Modulation	54
3-5	Frequency Modulation	59
3-6	Phase Modulation	64
3-7	Analog versus Digital Modulation	67
3-8	Synchronous and Asynchronous Modulation	73

Chapter 4 Analog Communications and Multiplexing — 80
- 4-1 Analog Communications Systems in Today's World — 80
- 4-2 Functions Within an Analog Communications System — 81
- 4-3 Multiplexing — 85
- 4-4 Space-Division Multiplexing — 87
- 4-5 Frequency-Division Multiplexing — 89
- 4-6 Time-Division Multiplexing — 94
- 4-7 Combined Modulation Systems — 105
- 4-8 Shortcomings of Analog Communications and Multiplexing — 109

Chapter 5 Digital Communications — 118
- 5-1 Description of Digital Systems — 118
- 5-2 Advantages of Digital Systems — 122
- 5-3 Sampling Theory — 128
- 5-4 Analog to Digital Conversion — 133
- 5-5 Encoding of Digital Signals — 140
- 5-6 Multiplexing and Modulation of Digital Signals — 147

Chapter 6 Communications, Media, and Problems — 161
- 6-1 The Role of the Medium — 161
- 6-2 Wire and Cable — 164
- 6-3 Air and Vacuum — 167
- 6-4 Fiber Optics — 171
- 6-5 Noise — 174
- 6-6 Noise Measurements — 178
- 6-7 The Effects of Bandwidth Limitation (and Related Problems) — 184
- 6-8 Common Mode Voltage — 192

Chapter 7 Communications System Requirements — 205
- 7-1 Data Communications System Issues — 205
- 7-2 Codes and Formats — 210
- 7-3 Protocol — 215
- 7-4 Synchronous and Asynchronous Systems — 222
- 7-5 Data Rates and Serial and Parallel Communications — 228
- 7-6 Protocol Examples — 235
- 7-7 Hardware versus Software: Protocol Conversion — 241

Chapter 8 The RS-232 Interface Standard   251
- 8-1 Introduction to RS-232   251
- 8-2 RS-232 Voltages   252
- 8-3 Data Bits   257
- 8-4 RS-232 Signals   261
- 8-5 Some RS-232 Examples   266
- 8-6 Making RS-232 Interconnections Work   272
- 8-7 Integrated Circuits for RS-232   279

Chapter 9 Other Communications Interfaces   292
- 9-1 Additional Interface Needs   292
- 9-2 Multidrop Communications   293
- 9-3 Other Key EIA Standards   298
- 9-4 The Current Loop   305

Chapter 10 Telephone Systems and Modems   315
- 10-1 Basic Telephone Service   315
- 10-2 Dialing   323
- 10-3 Telephone Lines   329
- 10-4 Private Exchanges   336
- 10-5 The Role of Modems   339
- 10-6 Some Specific Modems   348
- 10-7 Other Specialized Modems   352

Chapter 11 Networks   361
- 11-1 What Is a Network?   361
- 11-2 Topology   364
- 11-3 Basic Network Protocols and Access   371
- 11-4 Media, Modulation, and Physical Interconnection 382
- 11-5 Local Area Networks   386
- 11-6 Some Network Examples   389
- 11-7 Wide Area Networks, Packet Switching, and Gateways 396
- 11-8 Network Layers   399
- 11-9 State Diagrams and Network ICs   407
- 11-10 Cellular Networks and Systems   412
- 11-11 The Integrated Services Digital Network 417

Chapter 12 Error Detection, Correction, and Data Security   **425**

- 12-1 The Nature of Errors   426
- 12-2 Parity   429
- 12-3 Cyclic Redundancy Codes   436
- 12-4 Dealing with Errors   444
- 12-5 Forward Error Correction   447
- 12-6 The Nature of Data Security   455
- 12-7 Random Sequences   461
- 12-8 Spread Spectrum Systems and PRSQ Encryption   463
- 12-9 The Data Encryption Standard   466

Chapter 13 Test Techniques and Instrumentation   **481**

- 13-1 Basic Tests   481
- 13-2 Breakout Box and Line Monitors   485
- 13-3 Loopbacks   490
- 13-4 Pattern Generators and Bit Error Rate Analyzers   495
- 13-5 Protocol Analyzers   499
- 13-6 Transmission Impairment Measurement Sets   501
- 13-7 Time Domain Reflectometry   504
- 13-8 Testing Fiber-Optic Systems   508

Appendix A   **518**

Appendix B   **519**

Appendix C   **524**

Glossary   **526**

Answers for Odd-Numbered Problems   **538**

Index   **558**

Preface

Data Communications covers the fast-moving dynamic field of communications. This area of study involves the sending and receiving of information in digital format over distances ranging from a few inches to thousands of miles. The text is intended for courses in which equipment and systems are emphasized. These courses are generally offered as part of an overall program in electronics technology. Prerequisites are a basic understanding of analog and digital circuits, signals, and concepts together with the mathematics necessary to understand these principles. However, a brief review of these basic areas is provided as a refresher and to put these topics in the proper context and setting for data communication.

The book covers both the essential and the fundamental topics in data communications and, at the same time, many of the newer and important areas of growing technical need. These include networks, error detection and correction, data security, integrated all-digital networks, fiber optics, and the test equipment and methods that are unique to data communications.

Each chapter is divided into sections, which are followed by questions and problems designed to reinforce the ideas presented in the section. Each chapter has a summary, which brings together all the key points of the chapter, and end-of-chapter questions and problems. Answers to all odd-numbered problems are in the back of the book. Appendixes contain critical information that the student will have to use on a frequent basis in data communications work. A glossary of terms in the book is also provided.

This textbook is divided into two parts. Chapters 1 through 6 provide the concepts and theory of communications, along with the constraints and problems that any communications system must handle. The second part, Chapters 7 through 13, covers the application of this theory to real-world situations and shows how the ideas of the first part are put into actual practice in modern systems.

Chapter 1 explains what a communications system is, the many ways that a system can be used, and the different appearances of a communications system. It provides the framework for all subsequent chapters. Chapter 2 discusses the technical characteristics of a communications channel and signal, such as bandwidth, frequency, and channel capacity. These define the ultimate performance and limitations of the system.

Modulation is a key aspect of a communications system and is the subject of Chapter 3. Modulation converts the original signal into one more compatible with the channel. The overall system may use several different types of modulation, depending on desired performance, cost, and other technical considerations.

In Chapter 4, analog communications systems are studied. Electronic communication has traditionally been performed with analog systems, since the original voice or picture signal is analog in nature. Multiplexing allows several analog signals to share the same channel. The weaknesses and limitations of analog systems lead into digital communications, discussed in Chapter 5, which inherently can provide superior performance. The digital system is designed to handle only a specific set of communications signal values, and so the overall electronics of the system can be optimized for these values. Digital signals are also more directly compatible with computers and can be used easily for communication of computer data. Even analog signals can be put into digital format, to take advantage of the superior performance that a digital system can provide.

The physical path of the communications signal is the subject of Chapter 6. The choice between copper wire, coaxial cable, fiber-optic cable, and radio links is made on the basis of performance, flexibility, and cost. The technical features of each media type are studied, along with the weaknesses of each type, including susceptibility to noise and distortion.

In Chapter 7, the supporting theory of communications systems and data communications is applied to typical, practical applications. A complete, functioning system requires data encoding, protocols, data rates (baud), handshaking, and many other important elements. The various ways—both circuitry and software—of implementing these are discussed.

Chapter 8 is dedicated to a single communications interface, called *RS-232*. This is the most common interface in use. By studying RS-232 in detail, students will understand clearly many of the technical issues and considerations of other interfaces. The chapter also shows the problems and performance limitations that can occur with RS-232 and ways to overcome them in practical installations. Standard integrated circuits (ICs) for RS-232 are shown.

Other interfaces are studied in Chapter 9. Some of these extend the capabilities of RS-232, while others are used where RS-232 is not the correct technical choice. These interfaces include RS-422, RS-423, and multidrop RS-485 standards. The current loop is often used in electrically harsh environments, but has some technical peculiarities that are examined in detail.

The telephone system is the basis for many data communications systems. It is available almost everywhere and can provide effective communications under many circumstances. In Chapter 10 the basic operation of the telephone (dialing, switching, tones, and pulses) is studied, along with the role of the telephone central office and system. This role is expanded to show how and why modems are used to interface data communications circuitry to the telephone system. Specific modem models and their performance are shown.

Chapter 11 examines networks, a very important topic in modern data communications. There are many network configurations in use, each with certain performance characteristics, costs, benefits, and drawbacks. International standards are used to define network operation and the many layers of message transfer that occur. The chapter details the major network topologies, protocols, and operational sequences, along with the ICs that support these. Cellular networks that allow many portable terminals to use a limited bandwidth are studied. An all-digital type of telephone network, called *integrated digital services network (ISDN)*, is now being developed, and the overall function and ICs of ISDN are shown.

The real world of data communications must also deal with errors caused by many channel and circuit problems. Chapter 12 shows how error detection circuits are used to determine that an error has occurred and how error correction circuitry (which is more complex) can actually correct many errors. The mathematics of error detection and correction is shown, along with the protocols, ICs, and circuitry which implement the mathematics. A natural consequence of the error detection and correction methods is the ability to make digital data secure from unauthorized persons who may eavesdrop, or even try to put false data on the channel. The chapter shows how data security can be achieved by using many of the principles of random-bit sequences and digital data transmission.

Anyone involved with data communications systems must understand how they are tested, what equipment is used, and what the various tests reveal. Highly specialized equipment is usually needed, and this is shown in Chapter 13. The instruments and tests studied cover lower-level interfaces and operations, complex systems such as networks, and fiber-optic systems.

My thanks and appreciation to all those whose comments helped in the preparation and review of this book, including Ernest G. Arney, Jr., Stanley P. Creitz, Ralph H. Green, and David H. Tyrrell.

William L. Schweber

1 An Introduction to Communications

This chapter provides a basic understanding of what a communications system is, the goal of communications, and the reasons why communications systems are vital in modern society. It begins with a brief historical background of the beginning of electrical communications and links this to modern electronic communications systems. The many different structures that a communications system may have are discussed, along with some obvious and less obvious roles of communications. The chapter also covers the changing nature of communications systems—from voice signals in analog form to digital signals, which can be either voice signals or computer data. Telecommunications, the transferring of large amounts of data in digital format from one location or computer to another, is explored.

1-1 What Is Communication?

The goal of communication is to send a message from one point to another and to ensure that the message is received properly and understood. Behind this simple concept is an extremely important and complicated subject. Communication, the ability to send and receive messages reliably and predictably, forms the nervous system of a civilization. Just as civilization has changed over the past thousands of years, the nature and practice of communications have also changed dramatically. The development of the computer and the modern information-oriented society has been both the incentive for improved communications and the result of communications capabilities.

Although communication may seem simple, it is not. In order to successfully send and receive a message, the sender and the receiver have to agree on many factors—the way the message is sent, the symbols used, the rate at which the message is transmitted, for example. Like a baby learning language, communication is a complex subject.

Ever since individuals learned to communicate with others, there has been a desire and a need to communicate faster, better, and more reliably. Data communications is the most recent development in this constant striving for improved communications. Regardless of what information the message represents—instructions, ideas, data (numbers), or reports—the concept of communication remains the same.

A society has a need to get two things from one place to another: information and physical objects, such as people or products. Before electrical and electronic communication, the only practical way to transfer information was to send it with someone. (Drums, smoke signals, and carrier pigeons have been used, too.) This, of course, limited the speed with which messages could be transported and also meant that in many cases delivery was very difficult or impossible, since travel could be dangerous and time-consuming. Electrical and electronic communications changed this; once the communications system was put in place, only the information had to be sent. This completely changed the way people did business, countries conducted foreign affairs and diplomacy, and individuals spread ideas.

Electronic communication of data began with the telegraph, perfected by Samuel Morse in 1854. This was a relatively simple communications system, which connected one end to the other in a predefined, dedicated way, with no possibility of different connections to other users. The telephone, first patented by Alexander Graham Bell in 1876, was the next major milestone. It spurred the development of communication systems because now the actual instrument of communication was located in the house or office of the user, who could choose to communicate with any other user who possessed a telephone (see Fig. 1-1). The development of the digital computer and improvements in computer performance and reduction in cost, which began in the 1950s, have increased the need for powerful communications and systems tremendously. The microprocessor integrated circuit, which puts many of the functions of a computer into a small, inexpensive box, has further increased the demand for communications since these boxes are capable of sending and receiving messages while performing other useful tasks.

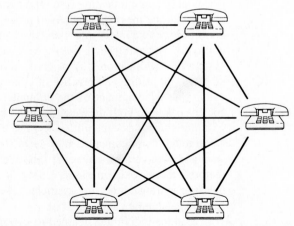

Fig. 1-1 A telephone system requires that any user be able to connect to any other user, as functionally shown above. Of course, this is not the way the phone company actually implements the connections.

Along with the increased need for communications spurred by the telephone and computer, the pathways for communications have also changed dramatically. When the telegraph and telephone were invented, the only practical method for transferring the message was copper wire, strung from one instrument to the other. The development of radio systems added another means of connecting the ends of the communications users and also gave a great deal of flexibility to the system, since the users were not tied down to the locations where the wires had been run. Worldwide communication became a reality, based on the construction of the appropriate radio transmitters and receivers. The space age has further improved on radio by using satellites in space to act as "relay" stations linking the various systems users, whether they be people on earth, astronauts in space, or space probes going to the other planets.

In the last decade, another method of transferring the message has grown in prominence. Fiber optics uses glass and plastic fibers as the pipe, or conduit, for the communications signals, which are sent by pulses of light through the fiber. This use of simple glass and fiber required the perfection of new technologies, such as ultrapure glass and plastic, easy-to-control sources of light, and sensitive light detectors. Now, fiber optics is used extensively for connecting the computers and phone systems of major cities and equipment centers and brings some important technical advantages that could not be achieved with copper wire, radio, or satellites.

QUESTIONS FOR SECTION 1-1

1. What are two of the many factors that have to be agreed on by the sender and receiver of a message for successful communication?
2. What is responsible for the tremendous increase in the need for better communications?
3. What are three of the many pathways available for electronic communications?
4. What is fiber optics? How is the message carried?

1-2 Uses of Communications

The use of electronic communications is much more extensive than many people realize. Besides the obvious examples of communications, such as the link between a central bank computer and an automated teller machine, there are many other types of electronic communications:

Person-to-person. The messages and information are sent by voice, usually by telephone or radio, so this is often called *voice communication*. People can also "converse" by typing at a terminal and seeing the other end of the conversation on a terminal screen. This can be a true two-way conversation, with give and take, or it can be a message typed in to be read at a later time by the recipient.

Computer system to peripheral device. In this system a computer system is sending messages, such as reports, to a printer, which is normally attached directly to the computer. The message is the data for the printer, and the computer needs to know whether the printer is available and ready for printing before it sends the data, or whether it becomes unavailable (someone shuts it off or it runs out of paper) during printing.

Computer to computer. An example would be a computer used in a national newspaper office. The main office computer is used to enter the articles, edit them, and prepare the layout of the pages. However, printing is done at printing facilities around the country, to ensure delivery the next morning to all readers. The main office computer transmits the contents of the newspaper to the computer in each local printing plant, which then uses this to generate an exact re-creation of the page as seen at the main office. This re-creation would be used to set up the printing press in the field plants.

Distributed systems. A *distributed system* is a system which uses individual intelligent boxes to preprocess some of the information; these boxes send only important information back to the central computer (Fig. 1-2). The intelligent boxes are called *local front ends,* or *controllers,* because they are closest to the actual source of the data. For example, a factory may use small microprocessor-based controllers to direct and monitor the operation of a large number of vats, where different batches of paint are being produced. At each vat there is a need to measure and control the flow of raw ingredients, regulate the temperature, and mix the ingredients according to the paint production process used. Associated with each vat is a local controller, which performs the actual measurement and control of the raw materials and temperature. The local controller uses a recipe that is sent to it from the central computer, which is managing the overall production plant and the use of each vat. It reports any unusual conditions to the central computer. If the controller is unable to get the temperature up to the desired value, it reports this to the main computer so that maintenance staff can be called and production temporarily stopped at that vat. The local controller in this way acts to reduce the amount of data collecting, processing, and decision making that the central computer has to perform. At

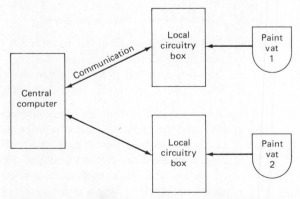

Fig. 1-2 A distributed system uses local intelligent circuitry to preprocess the data and communicate with the central computer.

the same time, the reliability of the overall system is increased, because the operation at each vat is not dependent on the central computer. Each local controller can manage the vat independently if communication to the central computer, or the central computer itself, breaks down. As the cost of microprocessor-based equipment comes down and its capabilities increase, distributed control systems will become more common.

Intracomputer communication. There is often a need to transfer data from one part of the system to another, even within a single chassis. Very often the designer of a system has divided the overall system into subparts, with circuitry (often microprocessor-based) to handle part of the total application. The various subsystems need to transfer data among themselves. For example, a computer may generate sophisticated color graphics to be used as part of a film production. One part of the computer may be performing the calculations needed to determine where the various pieces of the total image on the screen should go and what shapes to use in what places. It then passes the results of this analysis to another part of the system, which is actually responsible for drawing the picture on the screen by placing the proper shapes and colors in the right locations.

QUESTIONS FOR SECTION 1-2

1. Give two examples of computer-to-peripheral-device communications.
2. Give an example of the need for computer-to-computer communications.
3. What is a distributed system? Why does it need communications?
4. What are the advantages of a distributed system? Why is it such an important topic in modern system design?
5. Give two examples of communications within a single system.

1-3 The Structure and Types of Communications Systems

There are many ways to build a communications system that will successfully send a message from one point to another. Regardless of the system design, every communications system has the same basic functional blocks (Fig. 1-3). At one end, there is the sender of the message. This can be a person, computer, or other piece of equipment. The message then goes to a transmitter, which may be a radio transmitter, a computer designed to pass messages, or some simple circuitry within a system. The message goes over the actual communications link, which is the physical path between the sending end and the receiving end. At the receiving end, there is equipment which receives the message, extracts it from the link, and passes it on to the message user, which can also be a person, a computer, or another part of a circuit.

A properly designed and functioning communications system must integrate all the elements of the message path—from the sender through transmitter, link, and receiver equipment and to the actual user (receiver) of the message. The system can also be classified into one of three categories of ability: simplex, full-duplex, and half-duplex.

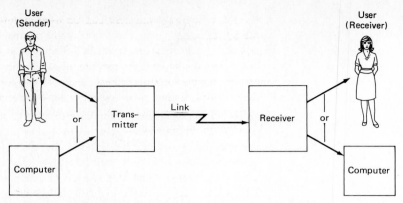

Fig. 1-3 All communications systems have the same functional blocks.

A *simplex system* is one in which the message can be sent in one direction only, from one end to the other (Fig. 1-4). There is only one transmitter and one receiver. An example of a simplex link would be a cable TV system, where the picture for the TV screen is sent from the central studio to the individual homes that are wired for cable. Another example would be a public address system, where a message can be broadcast to anyone in the listening area. (In a simplex system, it is common to have many listeners for a single transmitter.) A computer sending characters to a printer would also be a simplex system, since a printer does not send characters back.

A *full-duplex system* (or "duplex") is one in which the link is capable of transmitting in both directions at the same time. There is a transmitter and

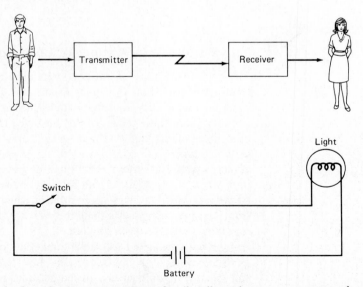

Fig. 1-4 In a simplex system, shown functionally at the top, messages can be sent in one direction only. A simple circuit for simplex communications is shown also.

receiver at each end, and they can be used simultaneously (Fig. 1-5). A telephone system is one example, since both parties on the phone can talk at the same time if they want to. (In many full-duplex systems, the capability for full-duplex operation is present but not used all the time, because it may cause confusion to have multiple talkers, just as in conversation.) Not only does a full-duplex system require more circuitry at each end and two links, but it also requires that the computer system or circuitry that is using the link be capable of both listening and talking simultaneously. Just as in conversation, this is often difficult to do in practice, even though the capability for it exists.

A third class of communications link is *half-duplex*. In a half-duplex link, each end may transmit, but only one at a time. This requires both transmitting and receiving circuitry at each end, but the actual link between the two ends may be shared (Fig. 1-6). Very often, this is the reason for a half-duplex link. For technical reasons it may be difficult to establish or build a link that can be used in two directions simultaneously, but much simpler to allow the link to be switched from one direction to the other. In other cases, the half-duplex link is used because the electronics system at each end cannot send and receive simultaneously but can only do one at a time. A microprocessor by itself can only perform one operation at a time, no matter how fast it is capable of operating. The microprocessor, therefore, would either be set to send messages or to receive them, but could not send and receive at the same instant of time. The half-duplex link is very common because it often provides the best choice for technical, performance, and cost reasons. Half-duplex is often used when one end does most of

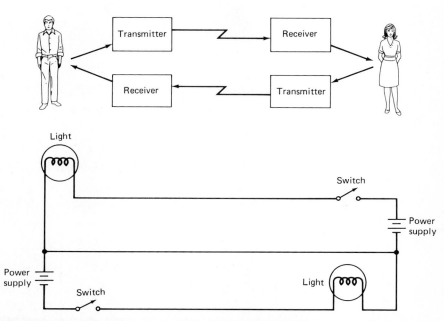

Fig. 1-5 In a full-duplex system, messages can go in both directions simultaneously as shown functionally and with a simple circuit schematic.

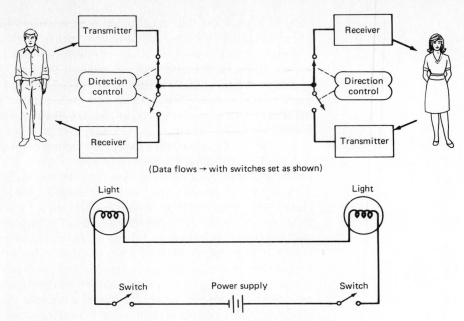

Fig. 1-6 A half-duplex system, shown functionally at the top, can send messages in only one direction at a time, usually using the same link for either direction. A simple circuit can provide basic half-duplex communications.

the sending, but it needs a brief response or acknowledgment from the other end. An example of a half-duplex link is a citizens band (CB) radio, where a frequency channel is shared and each party has to say ''over'' to switch the direction of communication.

QUESTIONS FOR SECTION 1-3

1. What are the basic parts of a communications system? What do they do?
2. What is a simplex system? Give two examples.
3. What is a full-duplex system? Give two examples.
4. What is a half-duplex system? Give an example.
5. What are the equipment and performance pros and cons of a half-duplex versus a full-duplex system?
6. Why is a half-duplex system often preferred to a full-duplex system?

1-4 Communications Systems and Data Communications

Up to this point, no mention has been made about the exact appearance of the message sent over the communications system. Traditionally, in telephone and radio systems, the message consisted of information conveyed by voice. The voice signal is an analog signal and so could take on any value within the overall range allowed (Fig. 1-7a). For example, if the telephone system were set up to handle voice signals which ranged from zero to one volt (V), the values transmitted at any instant could be 0.345 V, 0.179 V, and so on. This is called *analog signal communication* because the signal can be any value within the range.

The growth of computer technology and digital logic circuitry has also caused a tremendous change in the way that communication is performed. In a digital or computer system, all information is represented by a digital signal, which can take on only certain values (called *discrete values*) in the range. For example, a digital system may allow only values of 0.0, 0.025, 0.5, and 0.75 V to be used (Fig. 1-7b). Even more common is the *binary system*, in which only two values are allowed. Instead of referring to their actual voltage values, the two values are just named "0" and "1", or "low" and "high" (Fig. 1-7c). At any instant in the communications system, therefore, only one of these two values can exist on the link. (Ironically, the first form of electrical communications was the telegraph, a digital system.)

It may seem that by restricting the system to only two values, there is much less information that can be sent through the system. This is correct, in theory. However, a digital system has some overwhelming advantages that far outweigh this apparent limitation. In fact, these advantages actually allow the construction

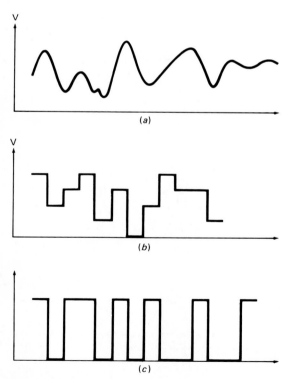

Fig. 1-7 (a) An analog signal can take on any value within the overall range. (b) A digital signal can take on only specific, predefined values within the range. A four-level digital signal is shown. (c) A binary signal can take on only one of two allowed values.

An Introduction to Communications

of data communications systems that can transmit as much information as an analog system, without any errors or corruption in the message. The reliability of the digital system in actual practice is higher than that of the analog system because the analog system is subject to many types of problems that a digital system does not have. Also, the digital data signal is more compatible with the computer systems and terminals that are often the source and receiver of the message that is being transferred. Data communications involves the transmitting and receiving of information in digital format. The message is called the data. The data can represent numbers, letters, or the digitization of an analog signal, such as someone's voice. The term "telecommunications" is also used when the distance between the end of the communications systems is long enough to require a special link. In effect, any data communications system (except one within a single box or chassis) is a telecommunications system. The link may be radio, wire, or fiber optics, but the concept of sending a large amount of digital data from one point to the other remains the same. A telecommunications system may use only binary representation, or a four- or eight-value digital representation, depending on several technical, performance, and cost considerations.

When computers were first invented, they were used for calculations and number processing, usually with an operator terminal right at the computer itself. As computers became more powerful, additional user terminals were added. Often these were not in the immediate area of the computer, and data communications was needed to connect the various users to the computer. Now, the wide use of computers and the many applications of computer systems have created a need for the computers to communicate directly with each other. It is data communications (and telecommunications) that makes this both possible and practical.

QUESTIONS FOR SECTION 1-4

1. What is an analog signal?
2. What is a digital signal? How does it differ from an analog signal?
3. Explain why a binary signal is a special case of a digital signal.
4. How is a binary signal often represented?
5. Is an analog signal more efficient than a digital one at sending information? Explain why or why not.
6. What is the advantage of using digital signals for messages?
7. What is telecommunications? How does it relate to the transmission of data from one point to another, when the two points are separated by miles? By a few inches of a circuit board?

SUMMARY

This chapter has discussed the function of a communications system, the importance of communications in modern society, and the various structures that a communications system can have.

The goal of a communications system is to transfer a message from one point to another efficiently and reliably. To this end, many types of systems can be used. All systems, however, have a message sender (source), transmitting circuitry, a link (path) for the actual message to travel on, receiving circuitry, and a message receiver. Depending on the application, performance requirements, cost, and other factors, each part of the system may be implemented in different ways. The communications system can allow messages to pass in one direction only, in both directions at the same time, or in both directions but only one at a time. These are called *simplex*, *full-duplex*, and *half-duplex systems*, and each has advantages and disadvantages.

The explosive increase in the power of computers and their use for more than just number processing, coupled with the decrease in their cost, has merged with the improvements in communications technology. The resulting data communications (or telecommunications) systems can transfer large amounts of data from one computer to another, or from one user to another, quickly and accurately.

END-OF-CHAPTER QUESTIONS

1. Explain why both sides in a conversation or communication system must agree to some rules and standards. What happens if they don't? Give an example of such a case.
2. Describe where a radio link would be preferred, and where a copper wire link would be more suitable. Is the choice always clear-cut?
3. Give two examples of the use of computer to computer communications that may not be obvious to the users of a system.
4. Describe a system for gathering data from a spacecraft carrying astronauts and analyzing it, using a central computer for data gathering, and using distributed data gathering. What equipment would be needed in each case? What types of communications systems?
5. Compare the capabilities of simplex, half-duplex, and full-duplex systems.
6. Compare the amount of equipment needed at each end for the three types of communications links of Question 5. Also compare the types of link required.
7. Why is full duplex difficult to actually implement and use? Give an example of an application in which it would be absolutely necessary.
8. Why is the half-duplex link very common? Give an example of an application in which it would be entirely adequate for two-way communication.
9. For each of the following, classify the communications as simplex, half duplex, or full duplex: telegraph, telephone, intercom, signaling by waving flags, talking, radio and television, computer-to-terminal (screen and keyboard).
10. Describe the basic parts of any communications system, from one end to the other.
11. Identify each of the following with the basic parts of Question 10: telephone, telephone wire on telephone poles, television, person listening to radio.
12. Explain the difference between an analog signal and a digital signal.
13. Explain how a digital signal can have more than two values. Give an example.

14. Explain why a digital signal is just a special case of an analog signal, and why a digital signal is easier to describe.
15. Explain why a binary signal is just a special case of the digital signal.
16. What are two common ways of describing binary signals? Can you think of any other ways?
17. Explain why an analog signal would carry more information than a digital one in a fixed amount of time. Why, then, are digital and binary signals used extensively?
18. What is telecommunications? What are the major applications of telecommunications?
19. What has been a major spur of the increased need for telecommunications?
20. How have electrical and electronic communications changed the way society can send information and messages? Why is this significant?

2 Communications Channel Characteristics

The communications channel is the path for the energy which carries the actual message from the sender to the receiver. The channel can take many forms, as a transmitted signal or as energy along a wire. The channel has many characteristics that affect its performance and capabilities, including frequency, speed, bandwidth, and capacity. All of these factors play a role in determining which kind of communications channel is selected for the communications system. The influence of each of these and the interrelation of the factors is discussed.

2-1 The Communications Channel

All communications systems and methods require a channel. This is because sending a message from one point to another involves the transmission of energy. All communications depend on the transfer of energy. The energy may be in various forms, such as light, electromagnetic waves, heat, sound, or mechanical motion. The channel is the path, or conduit, for this energy. It is important to understand the nature of channels and their characteristics because these characteristics have shaped many of the features of data communications systems.

It may seem difficult to understand that sending a message involves the transfer of energy. There are many examples of energy transfer that most people are already familiar with, although they may not have thought of them in terms of energy. Light, for example, is a specific type of electromagnetic energy and can be used for communications. A light beam from a flashlight can be used for signaling, in which case the air is the channel. If the light beam is sent through a glass pipe, then the glass pipe forms the channel.

The term "channel" as used in the communications industry includes both the path energy and the path for the energy, but it may also encompass other aspects of the overall link. These other aspects are not necessarily tangible. They relate to the characteristics and nature of the connection between the sender and the receiver. These aspects include the behavior of the link—does it vary with time, temperature, weather?—and the way the channel can be characterized. A channel may carry one signal, multiple signals in the same direction, or multiple signals in opposite directions.

For example, a red flashlight and a green flashlight can be used to send messages through air. Both flashlights can be used at the same end, and the color differences ensure that the signals will not be confused (Fig. 2-1a). The two flashlights can also be used at the opposite ends, to allow two completely separate messages to be sent at the same time in opposite directions (Fig. 2-1b). Again, the colors allow the messages to be sent at the same time without confusion.

Similarly, a source of sound energy such as a loudspeaker can send messages through the air. Two distinct tones can be sent from the same end, and the difference in pitch (frequency) will ensure that the signals can be separated (Fig. 2-1a). Two loudspeakers can also be used at opposite ends, to allow separate messages to be sent at the same time in opposite directions (Fig. 2-1b). Again, the different pitches allow the messages to be sent at the same time without confusion.

In some other cases, it is desirable for technical reasons to use a single pitch at the transmitter, but to send two independent messages. To do this, different and independent characteristics of the energy being transmitted must be varied, one for each message. Using sound energy again as an example, a single pitch tone could be used, with the exact pitch representing one message. At the same time, the intensity (amplitude) of the sound energy could be used to send the other message (Fig. 2-2). The equipment at the receiving end of the channel would have to contain circuitry that would determine both the frequency of the tone and its intensity. The two messages as sorted out would then go to the two users.

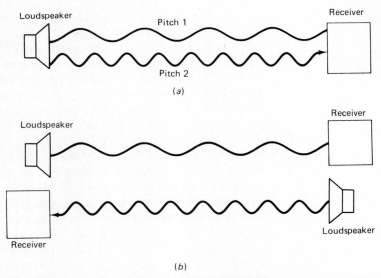

Fig. 2-1 Different tones allow the same channel path to carry two messages at the same time in either (a) the same direction or (b) opposite directions. The different tones do not interfere with each other, even over the identical path.

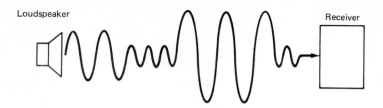

Frequency and amplitude variations

Fig. 2-2 A single signal can also be used to carry two messages over a single path, by varying two characteristics of the signal energy, in this case the frequency and the intensity.

Virtually all channels used for data communications and telecommunications use the energy of electromagnetic waves as the carriers of the message. These electromagnetic waves can span a wide range of frequencies, from extremely low—for example, 10 s of hertz (Hz, cycles per second)—to extremely high—for example, as in microwaves and even light at billions of hertz. Regardless of the frequency used, there are certain physical laws and characteristics that apply. At the same time, the nature of the channel that is employed is related to the frequency that is used, since certain channels can support only certain frequencies effectively. The choice of which type of path and which wavelength is decided by the designer of the system, depending on cost, reliability, performance, and other required factors. In many cases, first the path is determined by physical or cost issues, and then the appropriate frequency is selected.

The channel is the total connection path between the user sending information and the user at the receiving end. In most cases, the channel is built up by a series of individual links, like a chain. For example, a company may have a main office in New York and a sales office in London. Someone at the London office needs to get some pricing information from the computer at the New York headquarters. The overall channel might consist of these links (Fig. 2-3):

The computer terminal of the sales representative in London

A link from this terminal to the telephone system (which may also include some special interconnection circuitry)

The link from the telephone connection to the telephone central office

A connection from the telephone office to the satellite ground station (or the undersea cable, if that is the way the signal is routed)

The satellite link or undersea cable

The link from the receiving ground station (or undersea cable) to the telephone system in the United States

The connection from the telephone system to the computer in the New York office

This particular channel is just one of the many that could be constructed. Regardless of the specific details, the key issue is that a channel is a series of individual links. Like a chain, the channel is no better than its weakest link. The

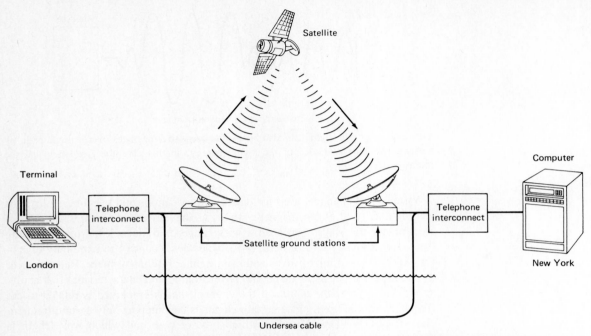

Fig. 2-3 A total communications channel usually encompasses many subsections, and within each subsection there may be alternate paths. Here, satellite stations and satellites may connect the U S and Europe, or an undersea cable may be used in place of the satellite link.

performance and reliability of the channel will be determined by the limitations of each piece. In some systems, these weakest links are identified, and either an improved link is put in place or backup links are available if the weak link fails.

QUESTIONS FOR SECTION 2-1

1. What is actually sent to transfer data?
2. What is a channel? What role does it have?
3. What are some of the key aspects of a channel?
4. Does a channel carry more than one signal? How can this be done?
5. What type of energy is used for nearly all modern communications channels?
6. What range of frequencies may be used?

2-2 Electromagnetic Waves

It is important to have a general understanding of the laws of physics that apply to electromagnetic waves, since these waves are the actual energy for carrying the communications message along the channel. The rules which govern the behavior of electromagnetic waves have been studied for over 100 years and are well understood. In some rare circumstances, data communications is achieved

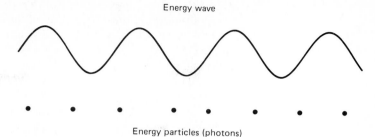

Fig. 2-4 Physics theory and experiments show that electromagnetic energy has characteristics of both pure waves and particles called *photons*. These two characteristics are not in conflict with each other; they are two perspectives.

by using nonelectromagnetic waves, such as sound, and in those cases different laws of physics apply to the communications channel. However, the same concepts of communications will apply to both the sending end and the receiving end of the link. For example, communication between undersea ships, such as submarines, is sometimes accomplished with acoustic waves because the seawater does not allow electromagnetic energy to pass for any more than a very short distance. The channel and the energy are different, but the way that the signal to be sent is organized, and the way it is handled, once received, uses the same principles and practices as electromagnetic channels.

Electromagnetic waves carry energy via the electric field and magnetic field that form the wave. From a physics perspective, the energy can be thought of both as a wave and as particles, or bundles, of energy called *photons* (Fig. 2-4). The fact that electromagnetic energy simultaneously has both a wave aspect and a particle aspect is used to explain many observed characteristics. The particle description is much more complicated than the wave description. Fortunately, most data communications situations are adequately covered, to the detail needed, by the wave description.

A single equation describes the most important property of electromagnetic waves, which is the relationship of the frequency, wavelength, and velocity of the wave:

$$\frac{\text{Velocity}}{\text{Wavelength}} = \text{frequency}$$

The wavelength is the distance between successive crests of the wave (Fig. 2-5). In a vacuum, such as in space, the value of velocity is 3×10^8 meters/second

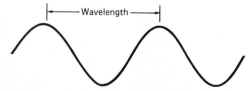

Fig. 2-5 The definition of "wavelength" is the distance between the same relative point on successive cycles, such as the crest or valley.

(m/s), or 186,000 miles/second (mi/s). This is often called c, the speed of light. The value of c in air is about 98 percent of the value in a vacuum. In a wire, the speed of the wave is from 50 to 85 percent of the speed in a vacuum, depending on the type of wire and its insulation. (Note: The SI system of units is used for most communications study.)

The speed of an electromagnetic wave, such as light, in a vacuum is the fastest velocity that can exist. (This was one of Einstein's points as he developed the theory of relativity.) Despite the fact that this is the fastest speed and is much faster than any speed we encounter in a car or an airplane, it is measurable with today's instruments. In fact, the effects of the speed are easily seen with a few simple calculations using the fundamental relationship

$$\text{Distance} = \text{Velocity} \times \text{time}$$

or

$$\text{Time} = \frac{\text{distance}}{\text{velocity}}$$

For example, a signal from a transmitter on the moon takes

$$\frac{240{,}000 \text{ mi}}{186{,}000 \text{ mi/s}} = 1.3 \text{ s}$$

to reach the Earth. A signal from a spacecraft approaching Venus (our nearest neighbor planet at a distance of 26×10^6 mi when closest) would take almost 2.5 minutes (min) to reach Earth. A conversation via satellite from the United States to Europe would have a delay of 0.24 s since the satellite is orbiting 22,000 mi above the Earth. (Why?) Although 0.24 s is not a long time, it can still have some undesired effects on the performance of the communications system, which will be discussed in later chapters.

The effect of propagation speed in various media is seen with a few calculations:

1. A radio signal from New York to Los Angeles (3000 mi) takes

$$\frac{3000 \text{ mi}}{186{,}000 \text{ mi/s}} = 0.016 \text{ s}$$

2. The same signal, sent through a cable with a propagation speed of 75 percent of c, takes

$$\frac{3000 \text{ mi}}{186{,}000 \times 0.75 \text{ mi/s}} = 0.0215 \text{ s}$$

QUESTIONS FOR SECTION 2-2

1. Why is an understanding of some of the concepts of electromagnetic waves important?
2. What actually carries the energy in an electromagnetic wave?
3. From a physics perspective, what are the two ways of characterizing electromagnetic energy? Which is more important for communications?
4. What is the meaning of frequency, wavelength, and velocity for a wave? What equation relates the three of them?

5. What is the speed of electromagnetic wave travel in a vacuum? In air?
6. Can data and energy travel from one point to another faster than *c*?
7. If *c* is so large, can it ever introduce significant delays in communications? Explain.

PROBLEMS FOR SECTION 2-2

1. How long does an electromagnetic wave take to travel 100,000 mi in space? To travel 3×10^6 m in space?
2. A certain type of cable allows the electromagnetic wave to travel at 65 percent of the speed in space. How far can the signal travel in 10 billionths (10×10^{-9}) of a second?
3. A system needs to send a message a distance of 1000 m with a time delay of less than 5 microseconds (μs) (5×10^{-6} s). Can this be done with air as the channel? What about using a wire with a propagation speed of 50 percent of vacuum?

2-3 Frequency and Wavelength

The velocity of an electromagnetic wave is determined by the medium. It only depends on whether the wave is traveling through vacuum, air, or some other material. As a result, for a given channel, there is a clear relationship between the wavelength and the frequency of the electromagnetic wave. Rearranging the previous section's equation shows this:

Velocity = frequency × wavelength

Therefore, in a given channel, as the frequency goes up, the wavelength goes down. Frequency is measured in cycles per second, or hertz (Hz).

The range of frequencies and wavelengths used for communications is enormous. Frequencies from 10 Hz through several hundred billion hertz are used, depending on various requirements of the channel. The corresponding wavelengths, in a vacuum, would be 30 million meters to less than centimeters. The total range of frequencies that can be used is called the *electromagnetic spectrum*. The spectrum has been divided into many groupings, or *bands,* and different bands are assigned for different uses. If the electromagnetic wave is traveling through the air or space, having many users within the same bands can cause interference with each other. An international commission meets to decide and assign which frequencies should be used by various countries and operations. For example, the range of frequencies from 540 to 1600 kilohertz (kHz) is assigned to the regular amplitude-modulated (AM) broadcast radio band of each country. Within the band, each country can assign individual frequencies to stations as it chooses. The velocity-wavelength-frequency equation is extremely important and useful. A radio station broadcasting at 1 megahertz (MHz), which is the middle of the AM broadcast band, is transmitting a wavelength of

$$\frac{3 \times 10^8}{1 \times 10^6} = 3 \times 10^2 = 300 \text{ m}$$

By contrast, a signal at 500 MHz (from an airplane to the control tower) has a wavelength of:

$$\frac{3 \times 10^8}{500 \times 10^6} = 0.6 \text{ m}$$

If the electromagnetic energy will not be going through free space, but is instead confined to a wire, then there is very little chance of interference with outsiders. Users can then choose to use whatever frequencies best suit the application itself. The situation is analogous to television, which can be sent by broadcasting from a main transmitting station and antenna, or by cable wired to each home. The broadcast TV stations are assigned frequencies of operation called *channels,* and two stations on the same frequency are separated by a large distance, typically 500 to 1000 mi. At this distance, the signals on the same frequency will not interfere, because the range of the signal is less than the separation distance. By contrast, cable TV systems use an actual wire, called a *cable,* to send the signal energy to each house. Cable systems in different towns can still use the same channels since the electromagnetic energy is confined to each town's cable and will not escape to cause interference.

There is technical importance associated with the frequency and wavelength of the electromagnetic wave. Consider this situation: The wavelength determines to a large extent whether the wave will stay confined to the wire or cable, or whether the wire will begin to act as an antenna and some of the electromagnetic energy will radiate out into the air or space around the wire. This occurs when the length of the wire is approximately the same as or longer than the wavelength. The energy will tend to radiate from the wire (Fig. 2-6). The wire thus acts as an antenna transmitting the energy into the space surrounding the wire. In some

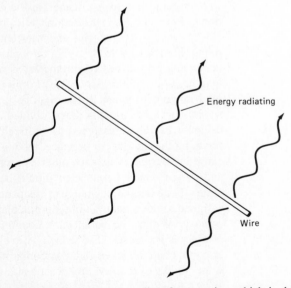

Fig. 2-6 Electromagnetic energy can radiate from a wire, which is then acting as an antenna.

cases, such as a transmitter, this is desirable. In other cases, such as a system which is sending data from one part of the circuit to another within the system, this is a bad feature. The energy that is radiated can cause electrical interference and malfunctions within the rest of the system, or with nearby equipment.

In many practical systems it is impossible to design everything so that the length of the wire is less than the wavelength. For example, a 1 m run of wire would require that only frequencies below 150 MHz be used if the wave speed were 50 percent of light. The solution is to use *shielding*, which is a metal enclosure around the wires. The shield acts as a "fence" and prevents the energy from passing beyond the shield. The shield itself must be connected into the system electrically, usually to the circuit or system ground. If the system is a mechanical chassis, the shield may be made of sheet metal. If the energy is radiating from a wire, a sheet metal enclosure would be too rigid. Instead, fine wires are spun into a fine mesh, called a *braid,* which then encloses the wire (Fig. 2-7). The braid allows the wire to flex as needed but still prevents the electromagnetic energy from escaping.

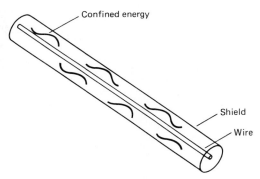

Fig. 2-7 Shielding the wire prevents this radiation from going past the shield and affecting other systems or circuitry.

Because the wavelength is closely related to the size of the wires and cables, antennas for transmitting signals are designed for a specific frequency or group of frequencies. In fact, very often one of the factors that goes into selection of channel and frequency is the amount of space available for an antenna. An antenna for the transmission of standard AM radio signals is several hundred meters long. A transmitting antenna for signals from a satellite, at several hundred megahertz, need only be several meters across for practical operation.

QUESTIONS FOR SECTION 2-3

1. What is the relationship between frequency and wavelength for a given, fixed velocity? What happens to one as the other increases? decreases?
2. Why is the use of the electromagnetic spectrum governed to a large extent by international agreements?
3. Do signals that are carried by wire normally create interference with other signals? Explain.

4. Why is the actual wavelength value important to know, with respect to possible interference? How can it affect the design of the system and its wire lengths?

5. How can signals radiated from a wire be reduced or limited?

6. How does wavelength affect the type of shielding needed?

7. How is antenna design related to the wavelength?

PROBLEMS FOR SECTION 2-3

1. For an electromagnetic wave of 100 MHz in space, what is the wavelength?

2. What is the wavelength for a 100 MHz wave in wire, with a velocity of 65 percent of space?

3. What is the ratio of wavelength for a frequency of 30 MHz to a frequency of 20 MHz in space? In the cable with a 55 percent velocity factor?

4. Will a wire mesh with a spacing of 0.1 millimeter (1 mm, 0.001 m) be effective at shielding electromagnetic waves at 1 MHz? at 10,000 Hz?

2-4 The Electromagnetic Spectrum

Different portions of the wide range of the electromagnetic spectrum are used in communications, depending on the technical requirements of the application. The electromagnetic spectrum has been divided into general bands, for convenience (Fig. 2-8). These bands are:

Frequency Band	Name
3–10 kHz	Extremely low frequency (ELF)
10–30 kHz	Very low frequency (VLF)
30–300 kHz	Low frequency (LF)
300–3000 kHz	Medium frequency (MF)
3–30 MHz	High frequency (HF) (also called "short wave")
30–300 MHz	Very high frequency (VHF)
300–3000 MHz	Ultra high frequency (UHF) (also called "microwaves")
3–30 GHz	Super high frequency (SHF)

[**Note:** For example, 3000 kHz is the same as 3 MHz.]
1000 kHz = 1 MHz
1000 MHz = 1 GHz

The spectrum of visible light is at even higher frequencies than the SHF band. Visible light has frequencies from 4300 to 7500 GHz. Light can be used for communications, but because of the extraordinarily high frequencies, systems using light must employ a completely different set of design schemes, even though light is an electromagnetic wave.

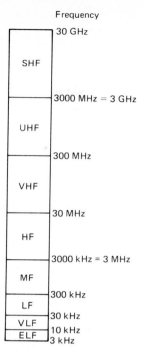

Fig. 2-8 The major divisions of the electromagnetic spectrum (not shown to scale).

All electromagnetic waves travel in a straight line, unless deliberately steered or reflected. This includes the entire range from ELF to light.

All the frequencies in the electromagnetic spectrum follow the same basic laws of physics. However, because of additional practical and real-world considerations, such as water vapor in the air, the energy of waves, their ability to penetrate solid objects, and the way they bounce and reflect, the performance of communication channels is greatly affected by the frequency which is used. Different communications systems require different characteristics or may have different guidelines. For example, the medium- and high-frequency signals are reflected back to Earth by various layers of the atmosphere, at a height of between 50 and 200 mi. This allows the signals to travel greater distances than the several hundred miles that simple "line of sight" would allow and lets the MF and HF bands be used for worldwide communications (Fig. 2-9). However, the exact amount of reflection, and the height of the atmosphere that causes it, varies with time of day, weather, and season, so that the repeatability and consistency of this "bounce" are erratic. They can be forecast with some accuracy, on the basis of past patterns and theories.

By contrast, the UHF frequencies have very short waves, which tend to penetrate the reflecting layers of the atmosphere. They therefore cannot be used directly for global communications but they can be used for communications with space vehicles and satellites. This ability to travel in a straight line, without being reflected by the atmosphere, allows the UHF waves to be used for global communications only by a satellite. The satellite is positioned 22,300 mi above

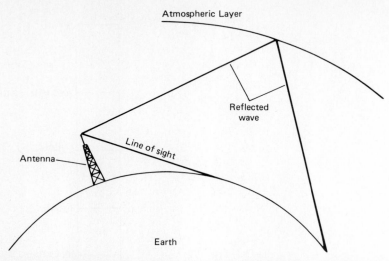

Fig. 2-9 Transmitted signals can travel by direct line of sight or by reflection from layers of the atmosphere.

the surface of the earth and takes exactly 24 hours (h) to revolve (Fig. 2-10). In this way the satellite is geocentric and has the appearance of being over the same spot on the globe all the time as the earth rotates in a 24-h period. The satellite can then be a fixed relay point in space. The UHF waves are broadcast from a

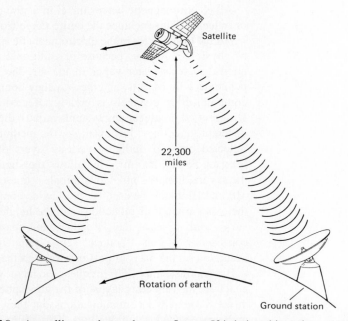

Fig. 2-10 A satellite can be used as a reflector. If it is in orbit at the correct altitude (approx. 22,300 mi) it rotates with the earth exactly once per 24 h, and thus appears stationary over some point on the ground. This is called a *geostationary orbit*.

ground-based transmitting antenna to the satellite, which receives them and then retransmits them to a ground-based receiving antenna. A satellite at this height can provide coverage for about one-third of the earth's surface in a more consistent and reliable way than relying on the reflection of waves from the atmosphere, which varies continuously.

The three most important characteristics of the travel, or *propagation,* are absorption, reflection, and noise:

Absorption refers to the way the wave energy is absorbed as it passes through anything that is not a vacuum. The UHF waves, for example, are absorbed by the water vapor in air, and so lose part of their strength as they travel through humidity or rain. (In fact, the home microwave oven uses UHF waves to heat the food by sending wave energy into it. The water in the food absorbs this energy and it becomes heat. Foods with low moisture content are not suitable for cooking in a microwave oven.) Lower-frequency waves, such as VLF and LF, are not absorbed by moisture to the same extent. The LF waves are used to communicate with submarines because the wave energy can penetrate the water for hundreds of meters.

Reflection refers to ability to bounce the waves, from a solid or semisolid surface, just as a mirror reflects light waves. Electromagnetic waves can also be reflected by metallic surfaces (and nonmetallic surfaces, to a lesser extent). The amount of signal that is actually reflected depends on the wavelength and the physical structure of the reflector. Longer wavelengths can reflect from relatively coarse surfaces, such as the electrically charged layer of the atmosphere that reflects low-, medium-, and high-frequency waves. This coarse layer is not smooth enough, though, for the much smaller waves of the UHF spectrum. The extremely short waves of UHF need reflectors of solid metal or very tight metal mesh.

Noise is an undesired electrical signal that is superimposed on the desired signal. The atmosphere of the earth and the vacuum of space have many other sources of electromagnetic signals besides the ones deliberately generated by the transmitter for the channel. These noise sources interfere with the desired signal and may corrupt the signal on the channel to the point where the signal cannot be retrieved properly. Noise includes static electricity "crackles"— from electrical discharges in the atmosphere (like lightning but smaller), steady low-level hissing from the many types of equipment that use electrical power (motors, generators, other radios), and even noise from other stars and objects in space. These various sources contribute noise in differing amounts to the many parts of the electromagnetic spectrum. Noise from outer space is usually significant primarily for VHF and higher frequencies, for example; noise from electrical discharges is much more of a factor in the low and medium frequencies. Noise is more than an annoyance. It is also a major factor in determining whether a signal and its data can be recovered without error at the receiving end. The type of noise present is a major factor in determining what type of channel is used. If there were no noise present, the design of an effective communications system and channel selection would be greatly simplified.

The propagation characteristics of electromagnetic waves, through air, space, and wire, are among the factors used in determining which part of the spectrum is used in the application.

QUESTIONS FOR SECTION 2-4

1. What is the ratio between the highest and lowest frequencies in each band of the spectrum?
2. How can medium- and high-frequency waves be used for long-distance communication?
3. Approximately how far can "line-of-sight" travel be? What determines the line-of-sight distance?
4. How can UHF waves be used for worldwide communications?
5. What are the relative disadvantages and advantages of using atmospheric reflection versus satellites for worldwide communications?
6. What is wave "propagation"?
7. What are absorption and reflection as related to wave propagation? What is noise as related to propagation?
8. Where can LF waves propagate that UHF cannot?
9. How can the reflective characteristic of UHF waves be used?
10. What are some synthetic noise sources? Naturally occurring ones?
11. What is the effect of noise on the channel?
12. What is a kilohertz? Megahertz? Gigahertz?

PROBLEMS FOR SECTION 2-4

1. Express 150 MHz as kilohertz and in gigahertz.
2. Express 150 kHz as megahertz and gigahertz.
3. Express 3 GHz as megahertz and kilohertz.
4. What is the range of wavelengths for the MF band in a vacuum?
5. What is the range of wavelengths for the UHF band in a vacuum? In cable with a velocity factor of 60 percent?

2-5 Bandwidth

A communications channel uses a specific frequency to transmit the electromagnetic energy which represents the data. This frequency may be 10.5 kHz, 30.325 MHz, or some other value. However, transmitting information takes more than a single frequency. A band of the spectrum around the nominal frequency is required. This is called the *bandwidth* of the signal.

Bandwidth is an extremely important concept in data communications. The communications channel must have sufficient bandwidth to handle the amount of data information that must be passed over it. If the bandwidth of the channel is too low, the rate of data transfer may be less than required. If the channel is to handle more than one signal, then the bandwidth of the channel must be equal to

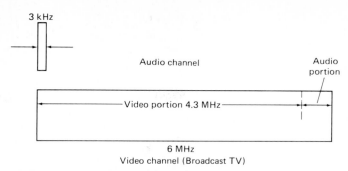

Fig. 2-11 The relative bandwidth used for audio signals versus the bandwidth of a TV video signal. Since the video signal has much more information, it requires a wider bandwidth.

the sum of the bandwidths of each signal. Bandwidth is a simple case of "you can't get something for nothing." The price paid for transmitting data at the desired rate is the bandwidth needed.

Some typical examples of bandwidth will illustrate the relationship between bandwidth and information rate (Fig. 2-11). A voice signal, transmitted over the telephone, uses a bandwidth of 3 kHz. A standard TV channel uses 6-MHz bandwidth, by contrast, of which 4.3 MHz is for the video information. The reason is that a video picture is transmitting information at a much higher rate. Data flowing from a computer to its printer typically can require a bandwidth of about 50 to 100 kHz. In each case, sending information at a higher rate requires proportionally more bandwidth.

The result of the need for bandwidth is that signals take up frequency space within the band that they are in. The high-frequency band of the spectrum is 3 to 30 MHz and so can hold fewer than five standard TV signals, but many thousands of voice signals. By comparison, the VHF band is almost 300 MHz wide and so can hold hundreds of TV signals. There is more total bandwidth available in the higher-frequency bands; that is one of the reasons they are often mandatory for channels that will be carrying large quantities of information at high transmission rates, such as many video channels from a remote sporting event such as the Olympics. Technically, it is difficult to design circuits and systems for the UHF and SHF bands, but the effort is worth it for the wide bandwidths that these bands offer. The fact that bandwidth is a fixed, finite resource means that it must be used very carefully. Steps must often be taken by users to make sure that their signals fit into the allocated bandwidth, or else they may interfere with another signal or the user will have to pay for a wider bandwidth channel.

Because the electromagnetic spectrum must be shared, various government and regulatory groups have been authorized to allocate parts of the spectrum. The band of frequencies from 88 to 108 MHz is assigned to standard frequency-modulated (FM) radio stations. Each station is allowed a bandwidth of 150 kHz. In the United States the Federal Communications Commission decides which station is assigned which frequency in a city. There are FM radio stations at 88.1, 88.3, 88.5, 88.7, 88.9 MHz, and so on, every 200 kHz, up to 107.9 MHz. The stations that are immediately above and below the frequency assignment are called the *adjacent channels*. In many cases, the assignment of the frequency and the bandwidth is made so that there is some unused band, or *guard band,* around the used bandwidth, to make sure that there is no interference between assigned stations. The space between adjacent station frequencies is made slightly larger than the required bandwidth. Of course, since spectrum bandwidth is such a precious commodity, these guard bands are kept as small as possible. For the FM broadcast band, the bandwidth is 150 kHz for each station, and the adjacent station channel-to-channel separation is 200 kHz, for a 50-kHz guard band (Fig. 2-12).

The different frequency bands have characteristics and applications that make them especially useful for many specific needs. For example, the HF band of 3 to 30 MHz is subdivided for shortwave international broadcasting, navigation signals, citizens band radio, amateur radio, and even radio astronomy, as shown in Fig. 2-13. Appendix A shows details of the division of the entire electromagnetic spectrum.

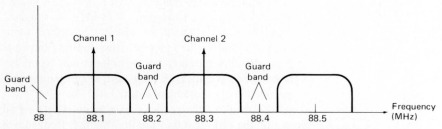

Fig. 2-12 The frequency allocations and guardbands for the standard FM radio broadcast band, shown from 88 MHz. The full band extends from 88 to 108 MHz.

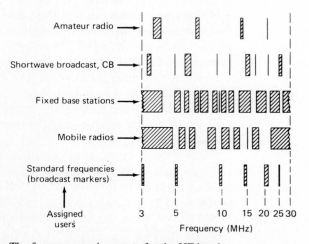

Fig. 2-13 The frequency assignments for the HF band.

QUESTIONS FOR SECTION 2-5

1. What is bandwidth? Why is it an important topic in understanding communications channels?
2. What is the meaning of a low-bandwidth channel? A high-bandwidth channel? What does each imply?
3. What is the bandwidth of a standard TV signal? Of a voice signal?
4. Why do the higher-frequency bands offer more bandwidth capacity? Why do system designers go to higher-frequency bands when their bandwidth needs are large?
5. How are the frequency bands divided? By whom?
6. What is an adjacent channel?
7. What is the guard band?
8. What are some of the applications of frequencies within the medium-frequency band?

PROBLEMS FOR SECTION 2-5

1. What are the specific FM station frequency assignments from 90 to 92 MHz?
2. Standard AM radio stations start at 540 kHz, with station assignments every 10 kHz. What are the frequencies assigned from 1250 to 1280 kHz?
3. What percentage of the FM band is "lost" to the guard bands?
4. How many 3-kHz voice signals can fit into the entire LF band? The VHF band?
5. How many 6-MHz TV signals can fit into the MF band? The SHF band?

2-6 Bandwidth and Channel Capacity

A wider bandwidth is needed to carry information at a higher rate. What is the specific relationship between the bandwidth needed and the data rate that can be achieved (called the *channel capacity*) with that bandwidth? In 1948, Claude Shannon showed by mathematical analysis that there was a specific MWh, simple formula that related bandwidth and capacity:

$$\text{Capacity} = \text{bandwidth} \times \log_2 \left(1 + \frac{\text{signal power}}{\text{noise power}}\right)$$

where the capacity is measured in bits/second (bits/s), bandwidth in hertz, and signal and noise powers must be in the same units.

Note: $\log_2$ is log to the base 2, and for any number X

$$\log_2(X) = \frac{\log_{10}(X)}{\log_{10}(2)} = \frac{\log_{10}(X)}{0.3}$$

Some examples show what Shannon's theorem indicates:

1. A system has 10 watts (W) of signal power, 1 W of noise power, and a bandwidth of 1 kHz. The capacity C is therefore:

Communications Channel Characteristics

$$C = 1000 \times \log_2\left(1 + \frac{10}{1}\right) = 1000 \times \log_2(11)$$

$$= 1000 \times \frac{\log_{10}(11)}{0.3} = 3.47 \text{ bits/s}$$

2. A system has 100 W of signal power, going into a channel with noise of 10 W. In order to send 10,000 bits/s, the bandwidth needed is:

$$\frac{10{,}000}{\log_2\left(1 + \frac{1000}{10}\right)} = \frac{10{,}000}{\log_2(101)}$$

$$= \frac{10{,}000}{\frac{\log_{10}(101)}{0.3}} = 1497 \text{ Hz}$$

Shannon's formula does not apply to all types of noise. The noise must be spread evenly throughout the channel bandwidth; fortunately, this is the most common situation in reality.

Although Shannon's formula is the basis for much of the theory of data communications, it does not indicate how much extra bandwidth is needed for different kinds of noise or when other channel problems exist. Therefore, the bandwidth required for a given application is usually determined by experience or tests. In many cases, the bandwidth available is predetermined and fixed, and the best that the system designer can do is try to get the maximum data rate within that bandwidth. If a video channel is set at a 6-MHz bandwidth, the designer will try to get the maximum amount of data into that bandwidth. There is no practical possibility of getting 7 MHz, since that extra 1 MHz is part of some other user's channel.

To illustrate how available bandwidth affects the rate at which useful information can be sent, consider the standard TV picture. It is transmitted on the allocated 6-MHz bandwidth channel in $\frac{1}{30}$ s. Sending the same amount of information over a telephone channel, which is limited to a 3 kHz bandwidth, requires about 30 s.

Sometimes the opposite situation applies. The channel has a wide bandwidth, but the actual rate of information is relatively low and needs only a much smaller bandwidth. Rather than let the unused bandwidth of the channel go to waste, it is common to combine several low-bandwidth channels into a higher-rate, high-bandwidth channel. This process is called *multiplexing* (Fig. 2-14). A telephone uses a 3 kHz bandwidth to the central office, for example, and this central office is linked to others by a wide bandwidth channel. The many individual telephone signals are therefore combined and multiplexed together before being transmitted to the next central office, over this wider-bandwidth channel. Ten users of 3 kHz each need a single channel of $10 \times 3 = 30$ kHz when multiplexed. The reason for using a wide-bandwidth channel between central offices is that it is cheaper in this application to install a single wide-bandwidth channel than many low-bandwidth channels with an equivalent total bandwidth.

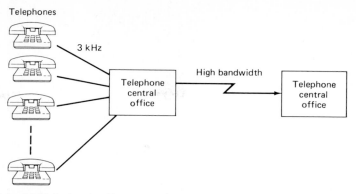

Fig. 2-14 Multiplexing combines several low-bandwidth channels into a single, higher-bandwidth channel.

The choice of whether to multiplex many low-bandwidth signals into one high-bandwidth signal, versus sending many low-bandwidth signals, is generally determined by the specific requirements of the application for cost, reliability, and performance.

In many cases, systems designers use special techniques to reduce the amount of bandwidth they will need. These techniques are called *bandwidth reduction* or *compression* and make use of the fact that often the data being sent has unnecessary elements or conveys no new information.

The technique of bandwidth reduction is used in sending a voice over a standard telephone line. The human voice typically uses a band of frequencies up to about 8 kHz, and so would require an 8-kHz bandwidth in the telephone channel. However, the actual understanding of the voice and the information from the voice are contained in a much smaller bandwidth, up to 3 kHz. The frequencies from 3 to 8 kHz do not carry any of the energy that makes the voice intelligible, but they carry the information that makes the voice distinguishable from other voices (Fig. 2-15). The telephone system is designed to conserve bandwidth and thus limits the available bandwidth per phone to 3 kHz. This is why voices on the phone are understandable, but sometimes the person speaking cannot be recognized with certainty.

Bandwidth compression is a more sophisticated method of reducing the band-

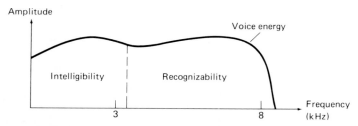

Fig. 2-15 Voice signals contain energy up to 8 kHz, but the energy that provides the intelligibility is only up to 3 kHz.

Communications Channel Characteristics

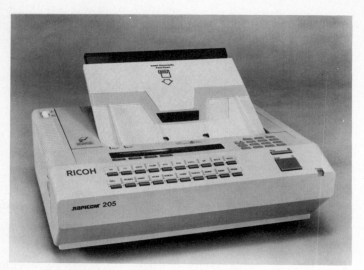

Fig. 2-16 A facsimile machine can send any drawing over a telephone line, by converting the visual information to energy in the audio band. (Photo courtesy of Ricoh Corp.)

width needed, by carefully examining the data to be sent and seeing whether there are areas where the data contains no new information and is redundant. An example is a facsimile (''fax'') machine which is designed to transmit a copy of a printed page over standard telephone lines. A facsimile machine is especially useful where the sender wants to send not just letters and numbers, but a drawing or sketch which would be too complicated to describe in words (Fig. 2-16). The facsimile machine divides the original sheet of paper into many small areas, and then goes through these areas in a fixed sequence. Where the paper is white, the machine sends one signal to the other end; where it is dark (from whatever is written on the page) it sends another signal. The receiving end can therefore reconstruct the original picture (Fig. 2-17).

The problem is that the bandwidth of the telephone system is limited to 3 kHz and yet many pieces of data have to be sent (a typical figure is 2×10^6 data

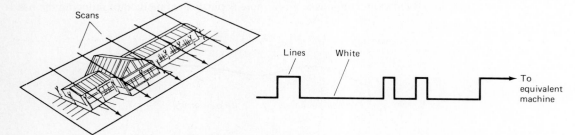

Fig. 2-17 The operation of a facsimile machine. The page is scanned with many repeated, narrow lines. Where there is blank white space, one signal is sent; where there is writing, another signal is sent.

points representing the dark and light areas). Since the cost of the telephone line is directly related to the total time it is used, having to send all these pieces of data at such a slow rate is costly, yet the rate cannot be increased because of the limited channel capacity. The solution is to have a facsimile machine that can look at the whole page and send information about the overall groupings of white and dark spots. Instead of sending data that indicates "spot 1 is white, spot 2 is white, spot 3 is white, spot 4 is dark," and so on, the facsimile machine can compress the data by saying "spots 1 through 3 are white; spot 4 is dark." A typical page has large areas of white, so simply describing where a white area begins and ends will reduce the quantity of data to be sent by a large amount, typically 80 percent. The facsimile machine at the receiving end would have to be able to reconstruct the picture from the description by area rather than on a spot-by-spot basis. The cost for the data compression and reduced telephone time is that more complicated and expensive facsimile machines are needed.

One of the advantages of sending information in computer digital format is that the user can examine the information by means of the computer before sending, to see what unnecessary data is in the group to be sent, and then use techniques of data reduction and compression to minimize the amount and thus the bandwidth required. In many cases, significant reductions in bandwidth can be achieved. The channel can be less wide, saving cost, or can handle other signals at the same time, resulting in higher efficiency. A user pays for the full bandwidth of the channel (in equipment or service charges), whether or not it is used to the fullest. Thus, it is in the user's interest to pick a channel with a bandwidth no larger than needed or to make maximum use of the channel bandwidth.

QUESTIONS FOR SECTION 2-6

1. What is channel capacity? How is it related to bandwidth?
2. What is the intent of the Shannon formula?
3. What rule is used to determine the bandwidth needed for a particular application?
4. How long does a standard TV screen take to transmit with its 6-MHz bandwidth? How long would the same information take with only a 2-kHz bandwidth? Why the difference?
5. What is multiplexing? Why is it used (give two reasons)?
6. What is bandwidth reduction? How does it differ from bandwidth compression? Give an example of each.
7. What is lost when the bandwidth of a voice channel is reduced from 8 to 3 kHz?
8. What is a facsimile machine? How is bandwidth compression used with one? Why?
9. What is the "cost" of using bandwidth compression in a facsimile machine?
10. What is the advantage of digital format for signals in bandwidth compression?

PROBLEMS FOR SECTION 2-6

1. A channel has a bandwidth of 10 MHz. How many 500-kHz signals can be multiplexed onto the channel?
2. It is desired to combine 32 low-bandwidth voice signals (3 kHz each) into a single wider-bandwidth signal, and then use a single cable from the multiplexer to the computer in the next building. What bandwidth is needed, at a minimum, on the single cable?
3. What percentage of available bandwidth is saved by reducing the telephone voice signal bandwidth to 3 kHz from 8 kHz? By what percentage does the number of voice signals that can be carried increase on a channel of a given bandwidth, when this reduction is made?
4. What is the channel capacity for signal power of 50 W, noise power of 3 W, and bandwidth of 15 kHz?
5. What is the channel capacity for the signal power equal to the noise power, both at 10 W, and bandwidth of 1 kHz? What if the signal and noise power are both 100 W?
6. What bandwidth is needed for a capacity of 10,000 bits/s, when the signal power is 8 times the noise power?
7. Repeat Prob. 6 when the signal power is one-half the noise power.

2-7 Bandwidth and Distance

The general need to send information has involved sending large amounts of data at high rates and for long distances. In some cases, the application requires more bandwidth than distance; other applications require the opposite. Some applications require both bandwidth and distance. Technically, it is most difficult to achieve both distance and bandwidth. A high cost must be paid, in equipment and channel needs. It is much easier to get one or the other, but not both at the same time. A review of some typical applications will show the relative amounts of bandwidth and distance needed:

Application	Bandwidth	Distance
Morse code signal (on/off)	Very low	Depends on locations
Local telephone call	Low	Short
Television signal from local station	High	Short
Television signal from another country	High	Long
Telephone call from another country	Low	Long
Computer signals to nearby printer	Medium	Short
Large amounts of data from one computer to another	Medium to high	Medium

(continued on next page)

(continued)

Application	BANDWIDTH	Distance
Data from one part of a computer to another	High	Short
Data from computer keyboard to computer	Low	Short
Signals from temperature probes in a manufacturing plant	Low to medium	Medium

The tradeoff between bandwidth and distance is illustrated by the use of a standard cable or wire as part of the channel. For physical reasons, the communications signal gets distorted and reduced as it travels along a wire (Fig. 2-18). This distortion alters the original shape and size of the signal and at some point prevents the information of the original signal from being retrieved without errors. The amount of distortion is directly related to the length of the cable. Therefore, the system designer must choose some combination of bandwidth and distance. Very often, the communications channel using this cable will have an associated chart, which shows what bandwidth can be achieved at what distances (Fig. 2-19).

Bandwidth and Digital Computer Data

So far, the relationship between bandwidth of a channel and the rate at which digital computer data (in the form of bits, or 1s and 0s) has not been defined. This is because there is no simple, single formula linking the two. Typically, the

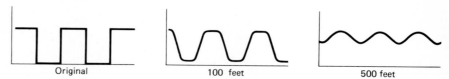

Fig. 2-18 A signal is both distorted and reduced in amplitude with distance.

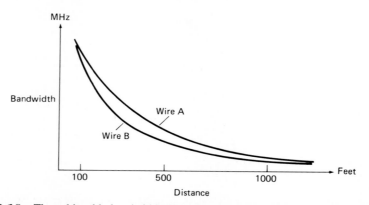

Fig. 2-19 The achievable bandwidth decreases with distance. The tradeoff also varies with the exact type of wire or channel used.

bandwidth needed to send a certain number of bits per second is 5 to 10 times the bit rate. Thus, sending 1000 bits/s would take a bandwidth of 5 to 10 kHz. The reason there is no simple formula is that there are many engineering design decisions and tradeoffs to be made. These are based on the way the signal will be sent, the nature of the channel, the kinds of noise expected on the channel, and the number of errors that can be accepted. One of the overall goals of data communications systems designers is to get the maximum rate of bits per second through a channel of a certain bandwidth, with an acceptably low rate of errors. A voice or TV channel can tolerate a few errors and still be usable, but a channel sending financial data must have no undetected errors.

New design and signal techniques, as well as new systems concepts, are often employed to get the maximum performance. The user decides whether the price is acceptable. For a system communicating with a satellite deep in space, the dollar cost is usually less of a concern than the guaranteed performance. However, a connection from a central computer to many dozens of terminals may be able to accept slower performance if the cost of 20 percent additional performance is twice as high.

QUESTIONS FOR SECTION 2-7

1. What is a general overall goal in a high-performance data communications system?
2. What is the meaning of the "distance and performance" tradeoff?
3. Characterize the bandwidth needs of the following as low, medium, or high: a telephone voice signal, a television signal (standard TV), a person using a terminal at home lined to a computer in an office, a computer connected to an automated bank teller machine.
4. What is the tradeoff of bandwidth and distance when using cable as compared to radio transmission? Why?
5. What bandwidth typically is needed for sending n bits per second? What prevents an exact formula?
6. Explain why different rates of error may be acceptable in different applications.

SUMMARY

A complete communications system requires not only a sender (transmitter) and receiver but also a path between the two. This path is called the *channel*, and it carries the electromagnetic energy that actually is the information being sent. The channel can be air, vacuum, or wire, and the physical properties of the channel play a large part in determining the performance of the system. A single channel can carry one or more signals, depending on how it is used.

The role of the channel is to carry the desired amount of information over the required distance, within an acceptable time period. The factors that affect the channel are:

The path medium, whether through air, vacuum, wire, or other medium

The number of subparts

The frequency (and wavelength) to be used

The bandwidth of the channel, which determines the rate at which data can be transferred

The noise that the channel adds to the signal

The way the channel absorbs and reflects energy at various frequencies

The amount of error that is acceptable in the application

A further complication is the fact that users who wish to broadcast a signal must use bands of frequencies that have been designated by international agreement so as not to cause interference with other users. The system designer must often consider many of the channel characteristics when making the design decisions and tradeoffs of which channel to use, in which way.

END-OF-CHAPTER QUESTIONS

1. Explain how actual information is sent from one point to another. What represents the information?
2. What is the channel, both physically and as a concept?
3. What is electromagnetic energy? Give some examples of electromagnetic energy.
4. Can different frequencies of electromagnetic energy share the same channel? Explain.
5. What is the relationship between electromagnetic energy wave velocity, frequency, and wavelength?
6. What determines the velocity of the wave? How can the velocity be changed?
7. What determines the frequency and wavelength of the wave? How can they be changed?
8. What is the speed of a wave in air, as compared to a vacuum? What is the speed in a wire or cable, as compared to a vacuum?
9. For communications applications, is the wave description or particle description of electromagnetic energy more useful?
10. How is the use of various frequencies of the spectrum governed? Why is this necessary?
11. How can electromagnetic energy be confined to a wire? Under what circumstances will it radiate from the wire?
12. How does electromagnetic energy get from a radio transmitter to the air? How does the antenna make the transition for the energy?
13. What are the designations of some of the bands of the spectrum? Why is the spectrum divided into these bands?
14. Do electromagnetic waves in the various bands have the same practical characteristics or do they vary? Explain.

15. What is line-of-sight communication? Give an example. What is the reason for the line-of-sight limitation?
16. Give two ways of communicating with electromagnetic waves beyond line of sight.
17. What are three key aspects of wave propagation? Explain and give examples of each.
18. What can noise do to communications? Is noise caused by artificial events only? Explain.
19. How many kilohertz equal 1 GHz? How many megahertz in 1 GHz?
20. What is bandwidth? Why does the transmission of information require bandwidth?
21. What is the general relation between the bandwidth needed and the rate at which data is to be sent?
22. What is channel capacity? What are the factors involved?
23. Does the standard voice signal need a low-, medium-, or high-bandwidth channel? How about a standard TV signal?
24. What is the total bandwidth available in the MF band? The VHF band?
25. What is the relationship between adjacent channels and guard bands?
26. What kinds of users share a typical band? How are the divisions made?
27. How can the bandwidth needed for a particular channel capacity be decided?
28. What is the true full bandwidth of a voice signal? What is it usually reduced to? What is lost and not lost by doing this?
29. What is bandwidth reduction? Compare it to bandwidth compression.
30. What is the tradeoff in bandwidth compression? Give an example.
31. Give an example, other than a facsimile machine, in which bandwidth reduction can be used.
32. What is the general relationship between distance and bandwidth capability in a cable?
33. Give three examples for which different amounts of error may be acceptable or for which no error is acceptable.

END-OF-CHAPTER PROBLEMS

1. Compute the time it takes for an electromagnetic wave to travel 1000 m in space. How about for 2000 m through a cable with a propagation velocity of 50 percent of light?
2. Modern circuits operate at very high speeds. The internal clock of a microprocessor system typically is at 10 MHz. How far on a wire (50 percent velocity of propagation) would a signal travel during one clock cycle?
3. A circuit board is 0.5 m square. The system clock circuit is located at one corner. How long does it take for this clock signal to reach the farthest corner of the board, if the signal travels at the speed of light in a vacuum?
4. A satellite is located in a stationary orbit 22,300 mi above the earth. A

speaker in New York is talking to a person in London, 4000 mi away. How much time is lost from wave travel when a question is asked and an answer received? How much time would be lost on a cable (50 percent velocity of propagation) between New York and London?

5. What is the ratio of the speed of light to an airplane traveling at the speed of sound, 700 mi/h?

6. What is the wavelength of a signal in space at 15.4 kHz? How about the same frequency in a cable, with a velocity of 60 percent of c?

7. A computer chassis is enclosed in metal, for shielding. The edges of the enclosure have small gaps, about ⅛ inch (in.) wide. What signal frequency radiating from the chassis will have a wavelength equal to this gap?

8. Express 1.5×10^6 Hz in kilohertz, megahertz, and gigahertz.

9. Express 25.5 kHz in megahertz and gigahertz.

10. How many megahertz in the VHF band? In the UHF band?

11. What is the longest wavelength in the LF band?

12. What is the shortest wavelength in the UHF band?

13. A band of frequencies from 100 up to 160 kHz is being allocated for channels. Each channel is 5 kHz wide, with a 1-kHz guard band. Sketch the channel assignments from 100 to 120 kHz.

14. A 30-kHz channel is used to carry a single 3-kHz voice signal. How much bandwidth is wasted? How many additional voice signals could be multiplexed?

15. What is the channel capacity with a bandwidth of 100 kHz, signal power equal to 1000 W, and noise power of 50 W?

16. Repeat Prob. 15 for noise power of 10 W.

17. Calculate the missing values in the table:

Signal Power	Noise Power	Bandwidth	Capacity
100	1	3 kHz	?
100	2	?	20 kbits/s
10	10	10 kHz	?
10	50	?	10 bits/s

3 Modulation

The process of modulation is required in any communications system. Modulation imposes the data signal on top of another, larger signal in a way that is compatible with the channel. The type of modulation used depends on many factors, including the type of channel, the available amount of signal energy, the system noise, and the performance requirements of the application. At the receiving end of the channel, demodulation is used to extract the desired signal information from the total received signal.

To better understand the concept of modulation and the different ways that it can be performed, it is necessary to understand how any signal can be studied both as a signal value versus time and as a signal value versus frequency. The mathematical principle of Fourier analysis links the two and shows that they are really just equally valid ways of looking at the same thing.

Modern data communications uses a specific form of modulation called digital modulation, *in which the modulating signal can only have certain predefined values. This type of signal is compatible with the data signals generated by computers and other digital systems and brings many performance advantages. It does, however, have some drawbacks that must be understood as well. The demodulation of a digital signal can be achieved in several ways, and it brings with it a new group of technical problems. For highest accuracy and performance, the receiving circuitry must have some synchronization with the transmitting circuitry, and various techniques are available to achieve this.*

3-1 Modulation and Demodulation

Whenever data is sent from one point to another via a channel, it is necessary to have the data vary some signal that is being sent through the channel. This process is called *modulation*. The energy that is being modulated by the data is called the *carrier*.

Modulation of a carrier is a concept that exists for other areas as well as data communications. In a car, the driver's foot modulates the gas pedal to control the gas flow to the engine in order to signal how fast the car should go. The carrier is the steady flow of gas to the engine. The driver tries to keep the car speed at a constant value of 55 miles per hour (mi/h) as the road goes up and down by applying more or less gas around the average value used to maintain 55 mi/h on level ground. The driver's foot conveys this information by modulating the gas flow.

Another example of modulation is a beam of light that is used to send a signal. If, for example, a white light is used over the channel, then the light can be filtered to show different colors. These colors can represent different messages or symbols. The white light is the carrier, and the color filters placed in the path of the light beam at the sending end modulate this white light beam. The modulated light beam then travels through the channel, which is air or a vacuum.

Why is modulation needed? Doesn't it add another layer of complexity to the communications system? There are several reasons why modulation is needed:

It allows the desired data (and its required bandwidth) to be sent at a desired frequency in the spectrum.

It allows the data to be converted to a carrier variation type that makes the best use of the particular channel and some other strong characteristics of the channel, such as noise and distortion.

It allows a relatively small signal, the modulating signal, to control a much more powerful signal, the carrier, for the cases in which the actual signal strength of the data signal as originally generated is not enough to reach the receiving end of the channel.

At the sending end of the channel, the desired data is used to modulate the carrier. At the receiving end, the opposite process must occur to recover the original data (Fig. 3-1). The receiving circuit and system must demodulate the received energy to extract the data and discard the carrier. (The carrier itself contains no information. It is a steady signal, and a constant signal conveys

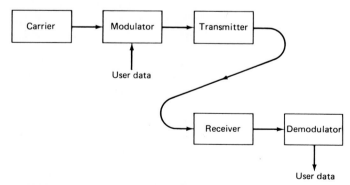

Fig. 3-1 A communications system requires that the sending end have the user data modulate a carrier which is then transmitted. The receiving end must take the received signal, demodulate it, and thereby extract the original data.

nothing new. It is the variations of the carrier, the modulation, that contain the information.)

Modulation is a useful and necessary process because it allows the data to be put into a signal form that makes the best use of both the channel and the energy that can be sent down the channel.

There are many ways that a signal can modulate a carrier. Mathematicians and engineers have studied the modulation process and ways to achieve it, and the modulation process can be analyzed precisely. The equations, however, can be complex. Nevertheless, it is important to understand some of the theory behind modulation in order to understand what modulation does, its consequences, and the advantages and disadvantages of the various types. The most important idea behind modulation is the concept of Fourier analysis.

QUESTIONS FOR SECTION 3-1

1. What is the role of modulation?
2. Give two "daily life" examples of modulations besides those discussed in the text.
3. What are three reasons modulation is needed?
4. What is demodulation?
5. What information is contained in the carrier?
6. What part of the modulated signal carries the information?

3-2 Fourier Analysis

Ordinarily, a signal is thought of as a voltage varying as time progresses. A drawing of the signal looks like Fig. 3-2, where the horizontal axis represents time and the vertical axis represents the signal level, usually voltage (but it may be signal energy or some other measure of the signal strength). A standard oscilloscope shows signals in this way on its screen.

However, there is an equally valid way of representing signals and the strength of the signal: using the horizontal axis to show the frequencies that are present in the signal. The vertical axis shows the strength of each frequency component. The concept of discussing signals in terms of their frequency is not really unfamiliar. Whenever a radio or TV station is identified, it is by its frequency.

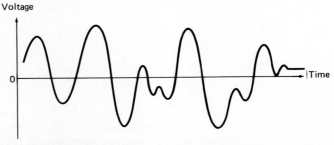

Fig. 3-2 A signal is usually thought of as a voltage that varies as a function of time.

An AM radio station is located at a specific frequency within the band, at 1440 kHz, for example. This means that the carrier of this station is at that frequency and does not identify the signal versus time, only signal versus frequency.

The discussion in Chapter 2 of spectrum and bandwidth also used the concept of signal versus frequency. A signal was described by the frequencies it occupied, not by its value in time.

The description of a signal versus time is called the *time domain description*. The description of a signal by its frequency components, and their values, is called the *frequency domain description*. Both descriptions are equally valid and legitimate; they just use a different perspective. Fourier analysis is named after the mathematician, Jean Baptiste Fourier, who developed in 1822 the mathematical theory now used to link the two domains. By using Fourier analysis, it is possible to take a mathematical equation that describes the signal in one domain and transform the equation to another equation that perfectly describes the signal in the other domain. Although the mathematics can become complicated, the results that the Fourier analysis provides are quite useful in understanding various communications concepts. Fortunately, it is not necessary to use the mathematics to gain the benefits of the result of the analysis.

Mathematically, this equation relates the time domain and the frequency domain:

$$S(f) = \int_{-\infty}^{\infty} f(t)\, e^{-j2\pi ft}\, dt$$

The equation states that $S(f)$, which describes the frequency spectrum, is equal to the integral of the time domain description $f(t)$ times the exponential factor, integrated from $t = -\infty$ to $t = +\infty$. It can be solved "on paper" for many waveforms $f(t)$, but not for all. $S(f)$ also shows the phase of each spectrum component. In practice, an instrument called a *spectrum analyzer* can produce the Fourier transform spectrum for any $f(t)$.

Spectrum analyzer circuits can be implemented several ways. One approach uses a large group of bandpass filters which span the entire bandwidth of the signal of interest. Figure 3-3 shows a block diagram of such a spectrum analyzer designed for analyzing low-bandwidth signals such as voice. A large number of filters are used in parallel. Each filter has a bandwidth of 50 Hz here, so the entire 5-kHz range of this particular instrument requires 5000/50 = 100 filters. The first filter has a bandpass of 0 to 50 Hz, the second is 50 to 100 Hz, the third is 100 to 150 Hz, and so on, up to 4950 to 5000 Hz for the last filter. The input signal $f(t)$ is applied to all filters simultaneously. Each filter then has an output amplitude that is directly proportional to the amount of energy that $f(t)$ had in the bandwidth range of that filter. The resulting filter outputs are displayed on a scale of frequency versus amplitude (Fig. 3-4). The resolution of the display (the "fineness" to which individual frequency components can be seen) is the *filter bandwidth*, in this case, 50 Hz. Finer resolution would use more filters and be more costly, but some applications require this additional information. Alternative designs of spectrum analyzers can provide greater resolution without many filters, but with other design complications.

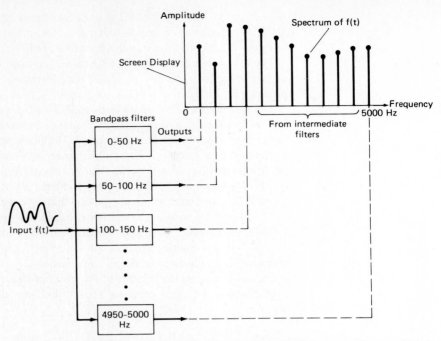

Fig. 3-3 The block diagram of one implementation of a spectrum analyzer, which is able to determine the frequency spectrum components of a signal.

Fourier Analysis Examples

The simplest signal is the sinusoidal wave (sine wave) of a fixed frequency (Fig. 3-5a). This oscillating signal has single frequency, and the amplitude versus time graph of the signal shows a sine wave. The frequency of the sine wave is determined by the simple equation of the wave, which states that the amplitude of the wave is related to the sine of the time variable. This same wave is shown in Fig. 3-5b as a frequency domain signal. It has a single component at frequency f with an amplitude equal to the peak value of the sine wave.

A sine wave with half the amplitude shown in the time and frequency domains would be as shown in Fig. 3-6. Similarly, the graph in both domains of a frequency equal to half the original wave, but with the same amplitude, would be as shown in Fig. 3-7.

A communications channel normally does not carry a single frequency, which is normally considered the carrier. The information of the channel is carried in the modulation that is imposed on the carrier. This modulation causes the signal to occupy some bandwidth around the carrier. For example, standard FM radio signals are assigned to the carrier frequencies such as 88.9, 90.1, 90.3 MHz, with the bandwidth of 150 kHz. The graph in the time domain of the waveform is extremely complex. Figure 3-8a shows it in greatly simplified form. However, the frequency domain graph is much simpler and often more useful (Fig. 3-8b).

The frequency domain analysis can also show what bandwidth is needed to carry a desired signal. A typical voice signal in the time domain may look like Fig. 3-9a. This does not show how much bandwidth is required to send the information contained in the voice. However, the frequency domain representa-

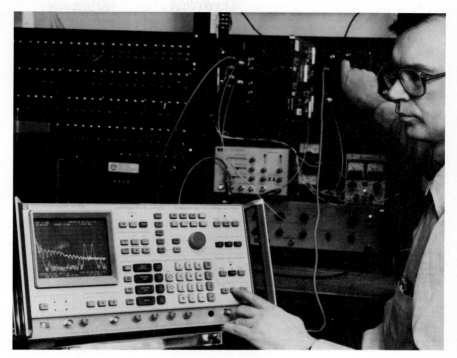

Fig. 3-4 The screen of a typical spectrum analyzer shows the amplitude of the various frequency components. The frequency values are on the horizontal axis, and the amplitude scale is the vertical axis. (Photo courtesy of Hewlett-Packard Co.)

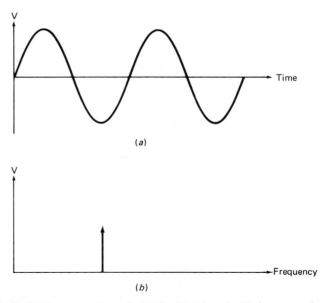

Fig. 3-5 The basic sine wave shown in the (*a*) time domain (*b*) frequency domain.

Modulation 45

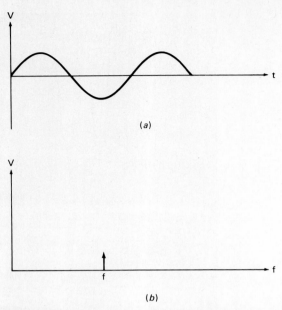

Fig. 3-6 The time and frequency domain representations of a sine wave of one-half the amplitude of the previous figure.

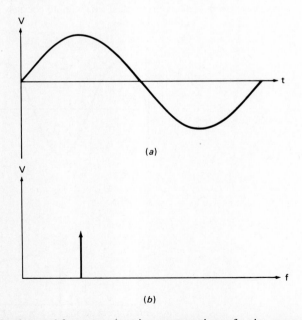

Fig. 3-7 The time and frequency domain representations of a sine wave with the same amplitude as the one is Fig. 3-5, but with one-half the frequency.

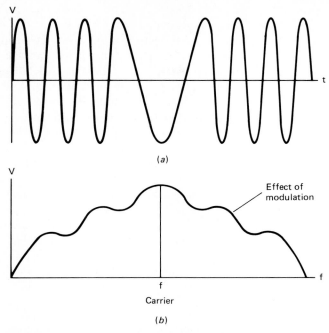

Fig. 3-8 The time and frequency domain representations of a standard FM broadcast channel signal.

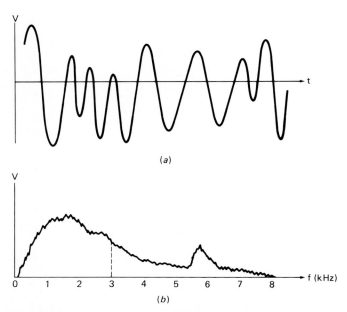

Fig. 3-9 A typical voice signal (*a*) time and (*b*) frequency domain representation. Note that most of the signal amplitude (and thus energy) is below 3 kHz, as seen in the frequency domain graph. This would not be evident from the time domain graph.

Modulation 47

tion of the same person speaking shows that most of the energy is in the range of frequencies up to about 3 kHz (Fig. 3-9b). Beyond that, the amount of energy, and thus information, is much less. Past about 8 kHz, the energy is almost zero. This means that a 3-kHz bandwidth can convey most of the signal, and 8 kHz can do a superior job. Bandwidth beyond 8 kHz for a voice is usually wasted.

A frequency domain analysis can also show the differences between signals that may seem to be similar but are actually very different. Musical instruments all play the same basic notes, or frequencies, and yet a flute playing a middle C sounds very different from a violin at the same note. The difference is not in the fundamental frequency of middle C (440 Hz) but in the amount of energy in the many harmonics of the fundamental frequency. (*Harmonics* are multiples of the fundamental: 880, 1320, 1760 Hz, and so on.) Figure 3-10 shows the relative amount of the various harmonics for a flute and violin which are both producing a steady note of middle C. The figure also shows that the bandwidth required to transmit accurately the sound of an instrument is much wider than needed for a voice signal. If the bandwidth of a channel carrying a music signal is very limited, then only the fundamental and one or two harmonics will be passed. The sound will be more like a dull, flat tone and the various instruments will be impossible to distinguish from each other, since it is the harmonics that make each instrument sound unique.

The frequency spectrum of a digital (on-off) signal is useful in understanding how much bandwidth is needed to convey the signal. For an idealized digital signal, with a perfect "right-angle corner" as shown in Fig. 3-11, the frequency spectrum is from 0 Hz to infinity! This is because the high frequencies contain the energy of the sharp corner, and the corner sharpness requires infinitely many of the high frequencies, although in lesser and lesser amounts. Sudden changes in time are the equivalent of the high frequencies.

A more realistic digital signal is shown in Fig. 3-12a. Here, the corner is somewhat rounded. The corresponding frequency domain representation, Fig. 3-

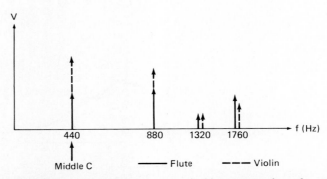

Fig. 3-10 The difference in sound between musical instruments depends on the number and type of harmonics, even at the same fundamental note. The flute and violin have differing harmonic amplitudes.

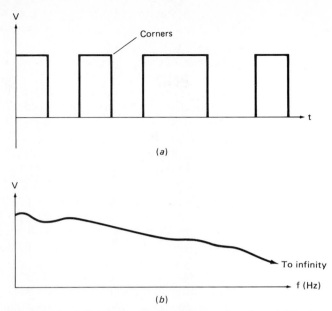

Fig. 3-11 An idealized digital signal (*a*) in the time domain and (*b*) with a very wide frequency spectrum in the frequency domain.

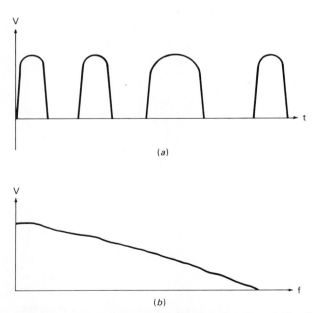

Fig. 3-12 A more realistic digital signal (*a*) in the time domain and (*b*) with a narrower spectrum in the frequency domain.

Modulation 49

12*b*, needs high frequencies, but these are of less importance. The rounding of the corner means that the frequency spectrum is not quite as wide, with less of the energy at the higher frequencies.

Another way of looking at this situation is to realize that, if the channel has a certain, fixed bandwidth, then there is a limit to how sharp the digital waveform can be. The reduction in sharpness means that the digital signal is not what would be considered ideal but is still very usable. However, if the bandwidth is too low for the digital signaling rate, then the digital signal will be severely distorted and begin to look like a sine wave (Fig. 3-13). This is why digital signals require bandwidth to be carried by the channel, and why the data rate that can be achieved for digital signals is related to the channel bandwidth.

It is also useful to examine the time and frequency domain graphs of two extreme signal cases. A signal which is constant with time (and therefore has a frequency of 0 Hz) has a very simple spectrum: a single line at the 0-Hz point (Fig. 3-14). On the other hand, a signal which is a very short pulse, or spike, in the time domain has a frequency spectrum that includes all frequencies from 0 to infinity in the frequency domain (Fig. 3-15). Note the corresponding nature of the two situations. The reason for this is that any sudden change in one domain requires a broad spectrum in the other, and anything which changes slowly in one domain requires a much narrower spectrum in the other. Thus, it is the sudden change in amplitude versus time for the spike that requires an infinite band of frequencies to represent that signal accurately. Signals which carry higher data rates have faster, sharper changes and require more bandwidth. Those that change more slowly require much less bandwidth. A limited bandwidth channel cannot pass sudden changes in amplitude.

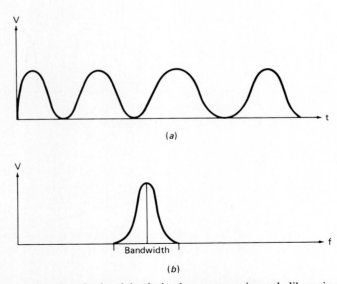

Fig. 3-13 A time domain signal that lacks sharp corners, is nearly like a sine wave, and uses little bandwidth in the frequency domain but also conveys very little information.

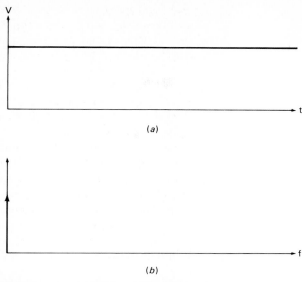

Fig. 3-14 The representation of a constant value signal (frequency of 0 Hz) in the (a) time and (b) frequency domains.

Superposition

There is another important concept associated with Fourier analysis, called the *superposition principle*. This principle shows, for example, that if a signal in the time domain is composed of the sum of several individual signals, then its representation in the frequency domain is the sum of the respective individual

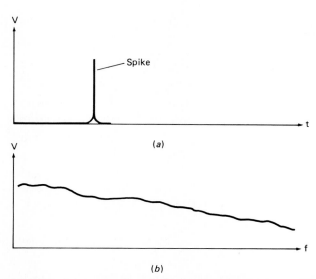

Fig. 3-15 A very short pulse has a narrow representation in the (a) time domain but occupies a wide spectrum of frequencies in the (b) frequency domain.

Modulation

frequency domain representations. This very powerful principle allows analysis to be performed by using the representation of signals already known, that comprise the total signal. A signal in the time domain with a steady, "dc" component (the term "dc" is often used to describe a 0-Hz wave, even though it stands for "direct current") and a 100-Hz sine wave twice as large will look in the frequency domain like a single spike at 0 Hz and a double-size spike at 100 Hz. The superposition principle also shows what happens in the other domain when two signals are added together in the time domain by the circuitry of the system, as is often done for technical reasons.

The principles of Fourier analysis, and time and frequency domains, apply to any waves, not just electromagnetic waves and energy. Mechanical engineers make use of the principles when examining structures such as machinery for unwanted vibrations and resonances which could cause the equipment to shake itself apart. The vibration of the machine is studied in the time domain, but it is also studied in the frequency domain. There may be a few frequencies at which the machine *resonates,* or has a natural vibration tendency. At these frequencies, the machine can actually get in tune with any small vibrations and begin to vibrate excessively. The frequency domain analysis shows the values of these resonant frequencies. The mechanical engineer will adjust the equipment, even adding more thickness or bracing, to eliminate these specific resonant frequencies. In some cases, the engineer "excites" the machine with a sharp tap from a hammer, and observes the results. The sharp tap in the time domain provides a very wide spectrum of excitation frequencies in the frequency domain, to show whether there are frequencies, if any, to which the equipment is naturally resonant.

QUESTIONS FOR SECTION 3-2

1. How is a signal usually drawn on most graphs?
2. What is an alternative way of representing the signal?
3. What is the time domain? The frequency domain? Explain.
4. Is one domain better than the other?
5. What is Fourier analysis? What does it show? What is the equation used to go from time to frequency domains?
6. What is the bandwidth of a typical voice signal? Where is the voice information in this bandwidth?
7. Why do instruments at the same note not sound the same? What is the effect of limited bandwidth on a voice or instrument signal?
8. What is the bandwidth of a digital signal? Explain.
9. What is the effect of a sudden change in a signal in one domain on its representation in the other domain? Give an example of each.
10. What is the superposition principle in Fourier analysis? Why is it useful?
11. What other types of engineering can use the principles of Fourier analysis? How?

12. What does a spectrum analyzer do? What is one straightforward implementation of a spectrum analyzer?

PROBLEMS FOR SECTION 3-2

1. Sketch a 1-V, 10-Hz sine wave, with 1 in of paper representing 1 V and 0.2 s, for the vertical and horizontal axes, respectively.
2. Sketch the same waveform in the frequency domain, with 1 in representing 5 Hz on the horizontal axis.
3. Repeat Probs. 1 and 2 for a 2-V, 5-Hz waveform.
4. Sketch the frequency domain representation of a 1-V dc waveform which has a 1-V amplitude, 10-Hz sine wave riding on top of it. Use the scale factors of Probs. 1 and 2.
5. Sketch the time domain representation of the signal in Prob. 4.

3-3 Types of Modulation

The purpose of modulation is to impose the desired data information onto the carrier. The carrier acts as the vehicle for the data, which is like the passenger. They both travel over the roadway, the channel.

The way that the information signal is imposed on the carrier is called the *type of modulation*. Different types of modulation are available. The choice of which type of modulation is used depends on many factors:

The types of noise that the channel may add to the signal

The amount of signal power available

The circuit complexity for modulation and demodulation, which varies with the type and affects cost, circuit size, and reliability

The amount of bandwidth available

The kinds of distortion that the channel may itself induce on the signal energy passing through

Modulation falls into three general categories. These are amplitude modulation, frequency modulation, and phase modulation. Each type of modulation primarily varies a single specific characteristic of the carrier. With *amplitude modulation*, the data causes the amplitude of the carrier to vary. *Frequency modulation* causes the data to vary the frequency of the carrier but leaves the amplitude unchanged. In *phase modulation*, the phase of the carrier signal is modulated, and the amplitude and frequency are unchanged.

Although the modulation directly varies only one of the three characteristics (amplitude, frequency, phase), the effect of modulation is to cause the original carrier to have some of the energy in its single initial frequency spread out, in differing amounts, to other frequencies. This is what causes the modulating data to require bandwidth in order to convey the data over the channel. The effect of any modulation is to make the single-frequency components, of various amounts, at other frequencies within the bandwidth of the channel. The specific way that the carrier is modified depends on the type of modulation used.

At the receiving end of the channel, the reverse process of *demodulation* must take place in order to recover the original data from the modulated carrier. The circuitry used for this demodulation must be designed for the particular type of modulation that is used. Although any modulation type—amplitude, frequency, or phase—causes changes in the frequency contents and energy variations of the carrier, these changes are very hard to interpret properly. The demodulation circuit must therefore look at what was done to the carrier by the modulation and try to reverse the process, ignoring the other lesser effects of modulation.

QUESTIONS FOR SECTION 3-3

1. What are three factors in the choice of modulation type?
2. What are the three types of modulation? What is the predominant feature of each?
3. What are the effects of modulation on the variable not being modulated?

3-4 Amplitude Modulation

Historically, amplitude modulation (AM) is the oldest and simplest form of modulation. All of the early work in radio transmission was done by using AM, since there are very simple ways to achieve the modulation and demodulation.

In AM, the amplitude of the constant-frequency carrier is varied, or modulated, by the amplitude of the information-bearing signal. As shown in Fig. 3-16a, there is a carrier of a relatively high frequency and a signal that contains the information to be sent, at a lower frequency. The modulating signal is not constant, but varies with the information. Typically, it could be a voice signal from someone speaking on a radio channel. The effect of the modulation is a signal which looks like Fig. 3-16b, where the amplitude of the carrier has been varied. The shape of the modulated carrier is called the *envelope*. It is this envelope that contains the information.

In the frequency domain, the carrier looks like a single line at the carrier frequency as shown in Fig. 3-17, and the modulating signal occupies a band of frequencies that correspond to its bandwidth. The result of the modulating process, in the frequency domain, is a new spectrum with the carrier unchanged in amplitude and frequency. There also are two new frequencies called *sidebands*. These are the result of the interaction of the modulating signal and the carrier. The sidebands contain the actual information energy.

What is the frequency of the sidebands? The AM process produces the sidebands as follows:

A lower sideband, called the *difference signal* or *lower sideband*, is
  carrier frequency − modulating frequency

An upper sideband, called the *sum signal* or *upper sideband*, is
  carrier frequency + modulating frequency

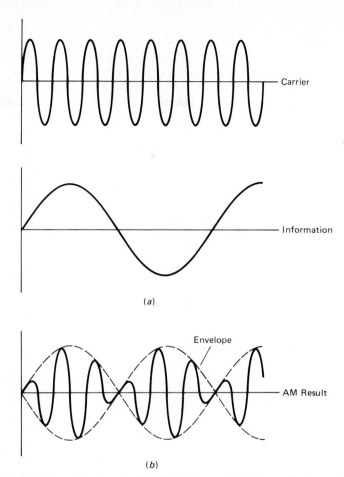

Fig. 3-16 (*a*) A modulating signal and carrier combine in amplitude modulation to produce (*b*) a modulated signal with information in the envelope.

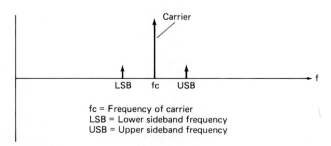

Fig. 3-17 The frequency domain result of the AM process is a carrier with associated upper and lower sidebands.

Modulation   55

Thus, a carrier at 100 kHz that is modulated by a 1-kHz signal would produce sidebands at 99 and 101 kHz. If the modulating signal itself were not constant, but occupied a bandwidth (which is the case in most situations), then the sidebands would occupy the same bandwidth, but at a frequency related to the carrier. A voice signal of 3-kHz bandwidth modulating a 100-kHz carrier would result in sidebands from 97 to 100 kHz and 100 to 103 kHz. (Why?)

The effect of the AM process is to move the frequency of the information signal to another part of the frequency spectrum, determined by the carrier. In this way, a large group of signals that originally are in the same frequency bands can be moved to different points in the frequency spectrum, so they do not interfere. In standard AM radio, the music and voice signal from each radio station is amplitude-modulated to carriers of different frequencies. Each station is assigned unique carrier frequency, so XYZ may be at 600 kHz, ABC at 650 kHz, DEF at 700 kHz. In this way, the broadcast spectrum can accommodate many users through the same channel.

The process of amplitude modulation is sometimes called *mixing*, because two frequencies are mixed to produce a result, the sidebands at the sum and difference frequencies. The modulating circuitry is often called a *mixer*. It is important to remember that the process of modulating the amplitude of the carrier results in a change in frequency band of the result, as shown by the frequency domain figures.

Advantages and Disadvantages of AM

Amplitude modulation is used extensively in data communications, but there are applications for which it is not the best choice. The advantages of AM include the following:

It is easy to implement and demodulate.

It provides a straightforward way of getting signals into different frequency bands.

Mathematical analysis of the AM process is relatively simple.

Testing and repair of AM can be done, in many cases, with standard electronic instrumentation (voltmeter, oscilloscope, and so on).

However, AM does have some drawbacks. The first drawback is that AM signals are easily corrupted by any electrical noise that is picked up by the channel, for whatever reason. When the signal is demodulated, the noise cannot be distinguished from the true modulating signal, and so the receiver recovers not the original signal, but a corrupted version. This would mean errors, or problems with the received information, and the receiver would have no way of determining this. This electrical noise comes from natural atmospheric sources, from the channel, and from the circuitry itself that is performing the modulation and transmission of the signal.

The second problem with AM is that it does not use power efficiently. Any communications system has to be concerned about power use. A circuit board can only supply a certain amount of power before it needs a larger, more costly power supply and larger amplifier components. A radio station using 50,000 W

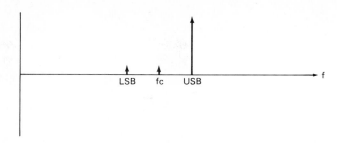

Fig. 3-18 A single-sideband AM process suppresses the carrier and one sideband, to minimize bandwidth used and concentrate transmitter power.

to broadcast a signal around the world wants to make sure that the extremely large transmitter and electric bill are going to get that signal to the receiver. Analysis of AM signals shows that after modulation a large percentage of the carrier power, about 60 percent, remains in the carrier, and each sideband has about 20 percent. Furthermore, each sideband has the same information as the other one, so the second sideband really is redundant. In summary, only about 20 percent of the original carrier power is used to transmit the actual information. (Ironically, it is the large amount of power that remains in the carrier that allows for very easy demodulation of AM signals.) In effect, the situation is like transporting two persons to work in a car. The car is the carrier, and one of the people is just going along for the ride. The goal is to get the one person to work, but the car and other passenger come along anyway.

One solution to the poor power efficiency of AM is to use a special form called a *single-sideband suppressed carrier*, usually called SSB. In this mode of AM, the modulation process is performed as discussed previously. However, before the signal goes to the main amplifier and transmitter, special filtering circuitry is used to suppress one sideband and the carrier (Fig. 3-18). This allows the full amplifier power to be used on a single sideband, and not wasted on the other sideband and carrier. The advantages of SSB are that it is more efficient and also frees up the spectrum that was occupied by the suppressed sideband for another signal. The disadvantages are the high cost and complexity of the suppression circuitry at the transmitting end, and the more complicated and difficult circuitry needed at the receiving end to demodulate in the absence of the carrier. Nevertheless, many special purpose AM systems, such as worldwide communications networks, use SSB. Lower-cost, lower-performance systems such as standard consumer AM radio do not, since the cost of the circuitry in the home receiver would be too high.

Baseband Systems

It may seem that the concepts of amplitude modulation are irrelevant to the simple but widely used method of varying voltages on a signal line, without actually having a carrier wave. This scheme is very simple and effective in many cases. It is used, for example, in the telephone system between the individual phone and the telephone company local office. The person speaking into the phone causes a voltage to vary in proportion to the loudness of the speech, and the rate of variation is determined by the frequencies in the voice.

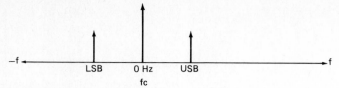

Fig. 3-19 Baseband AM is very common and uses a carrier of 0 Hz. The lower-sideband energy is effectively combined with the upper-sideband energy, since negative frequencies are the same as positive ones.

This is called a *baseband system*. The carrier frequency is a sine wave of 0 Hz, a nonchanging waveform. The analysis of AM systems with other carrier frequencies still applies, except that the carrier value is set to 0. The sidebands come at 0 minus the modulating frequency, and 0 plus the modulating frequency (Fig. 3-19). Note that the lower sideband thus has a "negative frequency," but the minus sign is meaningless and the sideband energy is in the same place as the upper sideband, at positive frequencies. The baseband system still requires bandwidth to carry the information and is very easy to demodulate. The bandwidth of the channel must be enough to handle the data rate flowing through the channel. Baseband systems are easy to build and reliable. The drawback is that the channel can handle only one baseband signal at a time, so the baseband system makes poor use of the potential capacity of the channel. A pair of wires within a building may have a bandwidth of 50 to 100 kHz, but a baseband system for voice uses only the first few kilohertz of that. Baseband systems are used for simple phone systems, links from a computer to another nearby computer, and transmission of data from a computer to a local printer, for example.

QUESTIONS FOR SECTION 3-4

1. What is amplitude modulation? What is the result of the AM process?
2. What is the signal envelope?
3. What are the sidebands? What do they contain? Where do they get their energy?
4. What are the frequencies of the sidebands?
5. How is the AM process used to put many voice signals into the same AM broadcast band?
6. What is mixing?
7. Does AM affect amplitude only, or does it cause frequency changes? Explain.
8. What are three advantages of AM?
9. What is the noise disadvantage of AM? The power usage disadvantage?
10. How much power is left in the carrier of a fully modulated AM signal?
11. What is SSB? How is it different from regular AM?
12. Why is SSB better but more difficult to use than regular AM?
13. What is a baseband signal? Why is it a special case of AM?
14. Where is baseband used?
15. What are the advantages and disadvantages of baseband AM?

PROBLEMS FOR SECTION 3-4

1. A 10-Hz sine wave is amplitude modulated by a 2-Hz sine wave of equal amplitude. Sketch the time domain waveforms for the individual waveforms and the result of the AM process.
2. Sketch the frequency domain representations of the waveforms of Prob. 1.
3. A carrier at 100 Hz is amplitude-modulated by a signal with a bandwidth that goes from 5 to 10 Hz. What are the resultant frequencies? Which is the upper sideband? The lower sideband?
4. Sketch the result of the AM process of Prob. 3.
5. A radio station at 100 kHz is carrying 0 to 5 kHz voice, through the AM process. Another station, at 110 kHz, is carrying a similar voice signal. Sketch the result in the frequency domain.
6. The carrier frequencies of the stations of Prob. 5 are moved closer, to allow more stations into the band. They are now at 100 and 108 kHz. Sketch the frequency domain of the result. Will this cause a problem?
7. The stations are back at 100 and 110 kHz, but one decides to broadcast music with greater fidelity to the original and allows a signal of 0 to 15 kHz to modulate its carrier. Sketch the result in the frequency domain.
8. A carrier is at 100 MHz. It is amplitude-modulated by the following signal bandwidths: 0 to 10 kHz, 5 to 50 kHz, 0 to 1 MHz, and 1 to 2 MHz. For each case, what are the frequencies of the resulting sidebands?
9. For the signals of Prob. 8, what is the resultant bandwidth of the carrier plus its sidebands?

3-5 Frequency Modulation

To overcome some of the drawbacks of AM, frequency modulation (FM) was developed by Major Edward H. Armstrong during the 1930s and 1940s. In an FM system, the modulating signal causes the frequency of the carrier to vary in proportion to the amplitude and frequency of the modulating signal. Figure 3-20 shows a carrier and a modulating signal, and the effect of the FM process in the time domain.

The frequency domain picture is very complex for an FM signal. Even if the modulating signal is a simple, single frequency, the effect on the carrier in the frequency domain is a wide spread of frequency components (Fig. 3-21). This points to one of the disadvantages of FM: The resultant signal requires a much wider bandwidth than the original modulating signal. This is in contrast to the AM method, in which the modulated result uses only twice the original bandwidth (for both sidebands) and can be reduced by SSB techniques. Typically, an FM signal requires 5 to 10 times the bandwidth of the modulating signal for intelligibility in practical systems. Standard broadcast FM radio uses 16 times the original 15-kHz bandwidth, or 240 kHz, for full fidelity and low distortion.

Another disadvantage of FM is that the circuitry for modulating and demodulating is much more complicated than for AM. The carrier must be very stable

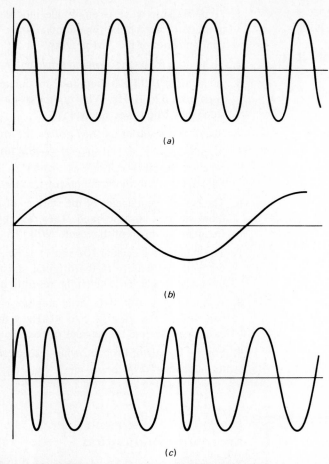

Fig. 3-20 In the FM process, the modulating signal causes the carrier frequency to vary: (*a*) carrier, (*b*) modulating signal, and (*c*) FM result.

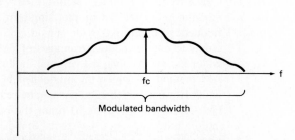

Fig. 3-21 The frequency domain result of the FM process is complex, and the resulting spectrum is wider than the modulating signal.

with respect to time and temperature, and the receiving circuitry must also be stable and precise. Historically, this need for signal stability at both ends restricted the practical use of FM for broadcast until new technologies such as advanced transistor circuits were developed.

Demodulation of FM signals can be achieved in several ways. The received waveform is sometimes passed through a "limiting" circuit, which clips all the amplitudes to the same value (Fig. 3-22). Since the FM system uses variations in frequency, the unintended variations in amplitude are of no interest. Then, the demodulation circuitry uses one of several methods to extract the variations in frequency and so recover the original modulating signal. One common way is to use the "zero crossings" of the received waveform, since these correspond exactly to the frequencies and variations. The frequency is measured between successive zero crossings and the variation from one cycle to the next in the original, modulating signal.

The Phase-Locked Loop

An alternative to limiting and zero crossing demodulation for FM is the *phase-locked loop* (PLL). This circuit scheme has been known since the 1930s, and it has been mathematically analyzed in many variations. The advanced integrated

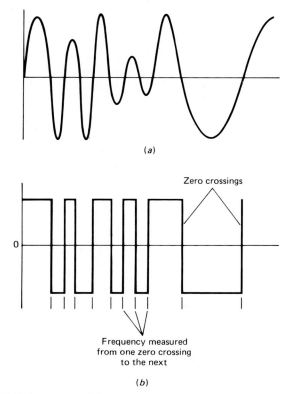

Fig. 3-22 FM demodulation with a limiting process eliminates the amplitude variations that may be present with the signal. The zero crossings of the limited waveform represent the original modulating signal (*a*) before limiting, (*b*) after limiting.

Modulation   61

circuits (ICs) of the last decade have made practical implementation of the PLL possible at low cost. A PLL is capable of providing high-accuracy demodulation under many difficult conditions and has many nondemodulation uses as well.

The block diagram of a PLL is shown in Fig. 3-23; it consists of three sections: a voltage-controlled oscillator (VCO), a multiplier, and a low-pass filter, arranged in a loop. The VCO is an oscillator whose output frequency is set by a controlling voltage, and the VCO output frequency can be varied as needed by this controlling voltage. A VCO may, for example, provide sine waves anywhere from 0 to 10 kHz in response to a 0- to 1-V controlling signal. (A 0.5-V controlling signal would cause a 5-kHz output in this case.) The multiplier is a circuit which takes two input signals and provides an output voltage which is equal to the product of the two inputs at any instant. This multiplication is the same as performing amplitude modulation, and the result of the multiplication of two sine waves is a sine wave at the sum frequency, and one at the difference. Finally, the low-pass filter is used to suppress the multiplier sum frequency output and allow only the difference output to reach the VCO.

The operation of the VCO is as follows: The input signal to be demodulated is fed to one input of the multiplier. The other input of the multiplier is the VCO output. The low-pass-filtered output of the multiplier then represents the difference or error between VCO output and the actual input. The error signal is fed back to the VCO to correct the VCO output so it more closely matches the input. The result of the feedback is that the PLL tends to lock on and track the input signal. As the input signal frequency varies with the modulation, the PLL VCO has a momentary difference, since it is putting out the last frequency value. These differences represent the original modulating signal. The error signal is the demodulated output and is also fed back to correct the VCO and have the loop lock onto the continuously changing input.

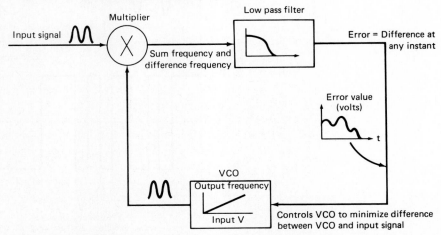

Fig. 3-23 The phase-locked loop uses a voltage-controlled oscillator, directed by the error output from a low pass filter, to try to match its output frequency and phase with the input signal. The input signal and VCO output are multiplied, and the low pass filtered result represents the difference between the two signals at any instant.

The PLL provides better performance than many other schemes because it separates the circuit element that determines the system center frequency from the one that determines the bandwidth. The VCO can be designed for any center frequency range, such as 88 to 108 MHz of the standard FM broadcast band. The low pass filter is designed for the bandwidth of the FM modulation, which would be 240 kHz for a channel of the FM broadcast band. Therefore, the performance of the system is selected separately for these two items, so that a change in one does not affect the design of the other. Some very extreme combinations are used in practice. A deep-space satellite may be transmitting at 400 MHz, but with a bandwidth of only 10 Hz. In theory, a bandpass filter of 10 Hz centered around 400 MHz could be used at the receiver, to filter out the desired signal and present it to the limiter and zero crossing detector. Such a filter is almost impossible to build in practice. Even if this filter could be built, it would have to be absolutely stable with time and temperature. Finally, the transmitter carrier may not be exactly the same as the filter frequency or may drift because of time and temperature, and even a perfect filter would soon be useless because the transmitted signal would soon be outside the passband of the filter.

The PLL design would break the problem into two parts. A VCO capable of operating around 400 MHz would be designed, with a range of perhaps 399 to 401 MHz. Separately, a low-pass filter with a 10-Hz bandwidth would be used. The output of the low-pass filter would be the demodulated signal. As the transmitter carrier drifted, the overall PLL would automatically track and lock in. The PLL would become a 10-Hz filter, initially centered around 400 MHz, yet it would be capable of automatically tracking and accommodating changes in the center frequency due to the feedback action of the low-pass filter error output which controls the VCO. The PLLs are capable, for these reasons, of capturing extremely weak signals such as spacecraft transmissions. They have many applications in regular data communications circuits as well.

Despite the bandwidth and signal stability limitations, frequency modulation is often used, especially where noise—atmospheric, circuitry, or channel—may be present in sufficient amounts to corrupt the desired signal. The reason is that the noise of most systems and channels is noise which adds to the amplitude of the signal being sent. But FM systems ignore the amplitude and look only for the variations in frequency. Therefore, as long as the noise is not so large that it causes false zero crossings, PLL misfunction, or other interference with proper operation of the demodulation circuitry, the noise-induced variations are of no concern. Only when the noise is very large does it create a problem, but the immunity to noise of an FM system is high.

There are also channels which induce distortions in frequency characteristics of the signal being sent. These channels are relatively rare. One example would be the use of sound waves to send data through water. As the sound waves (which are not electromagnetic energy) pass through the water, the movement of the water from waves, water currents, ripples, and so on, causes the actual characteristics of the channel to vary. The result is variations in the received frequency even though the transmitted frequency is constant. This type of distortion would cause an FM demodulator to make many mistakes. In general, though, channels such as air, vacuum, and wire do not have this problem or have it to a much smaller extent.

QUESTIONS FOR SECTION 3-5

1. What signal characteristic is varied in FM?
2. What are the advantages of FM in communications systems, as compared to AM?
3. What are the disadvantages of FM as compared to AM?
4. For what kinds of channels is FM not a good choice?
5. How is FM demodulation achieved with limiters and zero crossings?
6. What is a PLL? How does it work? Why does it provide good performance in many difficult circumstances?

PROBLEMS FOR SECTION 3-5

1. A carrier at 100 kHz is frequency-modulated by a 10-kHz signal. Assume the modulation results in a bandwidth 5 times the modulating signal. What frequency range is occupied as a result of the modulation?
2. A carrier is modulated by a signal which increases in amplitude as a steady ramp, from 0 to full scale. At full scale, the carrier frequency is double its unmodulated value. Sketch the unmodulated and modulated waveforms of this FM process.
3. An FM signal is received and limited to ± 1 V in the demodulation circuit. The demodulator looks for the zero crossings to decide what the frequency is. Sketch the limited waveform. How much noise can be tolerated before the demodulator is forced into an incorrect decision?

3-6 Phase Modulation

Any periodic signal, such as a sine wave of a carrier, has a phase. Two signals of identical frequency and amplitude can still differ in their time relationship to each other, as shown in Fig. 3-24. This difference is called the *phase*. It is normally measured in the number of degrees, out of the 360° of a complete cycle, that the two waveforms are offset from each other. Therefore, a signal that is offset by one quarter-cycle is 90° out of phase, and one offset by one half-cycle is 180° out of phase. It is important to note that although the frequency and amplitude of a wave depend only on that waveform itself, the measurement of the phase difference requires that one waveform be compared against another waveform, called the *reference*. It is meaningless to look at a single waveform and measure its phase. In some applications the two waveforms are two different signals. In other applications, the reference signal is the sine wave before it has gone through the circuit, and the second wave is the reference after it has passed through. The circuit may induce some delays in the wave that cause a phase difference, or *shift*, to appear. In some situations this has no meaning, but in other applications this phase shift can be useful.

In phase modulation, the modulating signal does not directly affect the frequency or amplitude of the carrier. Instead, it causes the phase of the carrier to vary, as shown in Fig. 3-25. The resultant frequency domain spectrum of the

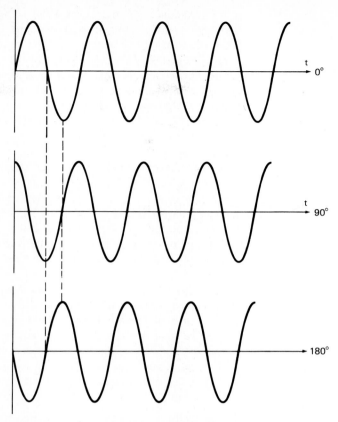

Fig. 3-24 Identical frequency signals can have different phases with reference to each other. The middle waveform has a phase difference of 90°, and the bottom waveform has a phase difference of 180°, compared to the top waveform.

phase-modulated (PM) signal is similar to that of an FM signal (Fig. 3-21). As with AM and FM, although one characteristic of the carrier is changed, there are effects in the frequency spectrum as well. The changes in carrier phase, although not varying the carrier frequency, do cause new frequency spectrum components to appear in the frequency domain.

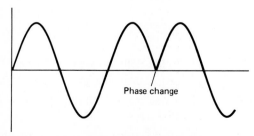

Fig. 3-25 In phase modulation, the modulating signal causes the phase of the carrier to vary. Shown is a sudden phase shift of 180°.

The demodulation process for a PM signal involves establishing a reference waveform and then comparing the modulated waveform phase against the reference, which has the same frequency but no phase changes. It may seem that this is fairly complex. Either the reference waveform must be sent along as another signal on the same channel or it must somehow be derived from the modulated signal. Both methods are used in practice, but the second one is much more common. Modern circuitry is available to implement at low cost a complex scheme that is able to reconstruct the exact frequency and phase of the reference signal from the modulated one and then use this reference to demodulate the received signal.

A PLL is often used to extract the reference signal. The *reference signal* is the lowest-frequency component of the received signal spectrum. The PLL VCO is centered around this frequency, and the low-pass filter has bandwidth to accommodate jitter, drift, and variations of the reference waveform. In operation, the PLL locks and tracks the reference as embedded in the received signal. The VCO output (not the error) becomes the rederived reference.

Since the frequency spectrum of a PM signal is similar to that of an FM signal, it may seem that there is no real advantage to using PM over FM. There are two reasons, however, that PM may be preferred. First, since the actual frequency of the received signal is fixed, and only its phase changes, it is possible to use frequency filters that are designed for a single frequency rather than the band of frequencies that an FM signal would require. This greatly simplifies the cost and design of the receiver, since most practical circuits require such filters in various parts of the system. Second, for the case in which the modulating signal can take on only a few distinct values (called a *digital system*), the phase-modulating and -demodulating circuitry can be greatly simplified. The wide use of digital systems and modern computers means that modulation and demodulation schemes must be chosen on the basis of their performance for digital signals as compared to continuous analog signals.

The importance of digital signals and their implications for modulation and demodulation will be studied in the next section.

QUESTIONS FOR SECTION 3-6

1. What is phase?
2. What is PM? What is varied? What other characteristics change as a result?
3. Why is a reference signal needed in PM?
4. How is a PM signal demodulated?
5. Compare FM and PM. How are they similar, and how are they different?
6. What is the advantage of PM over FM?

PROBLEMS FOR SECTION 3-6

1. Sketch a sine wave and another sine wave 90° out of phase with it.
2. Sketch a sine wave and a 45° shifted version.
3. A sine wave is shifted by 180° after 5 cycles. The shift occurs at the zero crossing. Sketch the reference waveform and the shifted waveform.
4. For the signal of the previous problem, the shift occurs at the peak of the waveform. Sketch the phase-modulated waveform.

3-7 Analog versus Digital Modulation

Up to this point, the discussion of modulation techniques has not made any special assumptions about the nature and shape of the modulating signal. When the modulating signal is allowed to take on any value within its total range of zero to maximum, the modulating signal is an analog signal. Historically, analog modulation was developed because that was the way that most of the information to be transmitted existed, and the circuitry that was available was really suitable only for analog signals.

Before discussing analog and digital modulation, it is useful to summarize the three types of modulation and the essential characteristics of each, as shown in Fig. 3-26.

The explosive growth of computers has changed the nature of the modulating signal. Computers use digital signals, which can have only one of several distinct values rather than any value within the total range. Along with the development of computers has come the development of digital circuitry which can use thousands of transistors to process digital data. Before the development of inexpensive and reliable circuitry for digital data, it was impractical to process and manipulate digital signals. Huge racks of vacuum tube equipment would be needed, as compared to far fewer for analog signals. Thus, the use of computers, made possible by advanced circuitry, has also encouraged the use of digital modulation techniques which are more suitable for computer data communications.

In a digital system, only certain specific values of the modulating signal can exist (Fig. 3-27). Therefore, whether AM, FM or PM is used, the resultant modulated signal will also have only certain shapes and characteristics. A digital modulation scheme is related to the binary logic and circuitry used in computer systems. In a binary system, only two values are allowed. They may be called "high and low," "one and zero," or "on and off," among other terms. A digital system is more general than a binary system and includes a binary system. In a digital system the number of allowed values may be two, but it can also be more. Only this specific, predefined group of values is allowed. A binary system is one type of digital system, with the group of allowed values limited to two.

Characteristics	AM	FM	PM
Bandwidth	Low	Wide	Wide
Spectrum complexing	Simple	Complex	Complex
Power efficiency	Poor	Good	Good
Ease of modulation/ demodulation	Simple	Difficult	Moderate
Noise resistance	Low	High	High

Fig. 3-26 The key characteristics of AM, FM, and PM compared.

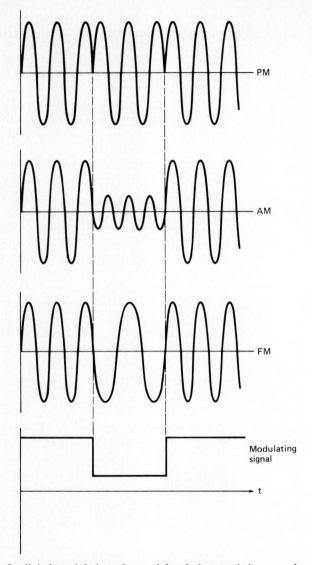

Fig. 3-27 In digital modulation, the modulated characteristic can only shift between specifically allowed values. Shown is a binary modulating signal causing PM, AM, and FM of a carrier.

Advantages of Digital Modulation

The advantages of digital modulation in practical systems are many: The major advantage of a digital scheme is that noise in the system or channel does not have the same detrimental effect on the received and demodulated signal. If an analog signal has some noise, however small, the demodulated signal is corrupted. This corruption due to noise is indistinguishable from the desired information signal. For example, if the modulating signal were in the range of 0 to 1 V, and the specific value sent were 0.56 V, a small amount of noise might make the demodulated value 0.58 V. The receiver would believe that the correct value was 0.58, instead of the actual 0.56.

In a digital system, only certain values are allowed, and these are separated in value by more than most noise values that the system will encounter. The values may be 0.0, 0.25, 0.5, and 0.75 V for a digital system with 4 values. A signal that is sent as 0.5 and received as 0.53 is examined by the receiver and restored to its correct value of 0.5 (Fig. 3-28). There would be potential for error only if the noise or distortion were sufficiently large to cause a 0.5-V signal to look closer to 0.75 or 0.25 than to 0.5, so that the receiving circuitry would decide incorrectly. Binary systems have the largest noise resistance, since the two values can be widely separated.

This incorrect decision making can occur if the noise or distortion is large, but it is very unlikely. Therefore, to a large extent, the digital method is resistant to noise and can tolerate quite a bit of it before errors in decision occur.

The second advantage of digital techniques is that it is possible to encode the messages in special ways so that the receiving system can determine whether an error has occurred or not. This is a further level of security, beyond the noise resistance that digital modulation provides, as discussed previously. Encoding details will be discussed in a later chapter. The basic idea is to send the desired data in digital format and send along with the data some type of error checking group of characters. The receiver then combines the data and the checking characters and determines whether there have been any errors in transmission. In this way, even noise that has caused 0.5 V to look like 0.75 V (and in turn caused

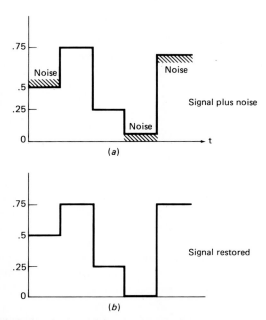

Fig. 3-28 Digital signals have inherent immunity to errors caused by noise and distortion, since the original value can be determined (if the noise is not large enough to cause the signal to look like another digital value: (*a*) signal + noise, (*b*) signal restored.

the receiver to make a judgment that 0.75 V was sent) can be handled. The receiver makes its best determination of what was sent and then is able to make a further check on the integrity of the entire message. It is important to understand that although digital modulation in itself is responsible for the ability to withstand noise, the ability to detect errors is a capability not in the modulation scheme. Instead, it provides a framework which allows the receiving system to implement error detection by using special encoding schemes, if the designer of the system decides it is necessary.

By contrast, it is practically impossible to provide a method for error correction using analog modulation. If the modulating signal can be anywhere in the 0- to 1-V range, there is no reasonable way to determine whether the last group of values received—0.13, 0.47, 0.59 V, for example—has any errors.

Third, digital modulation and demodulation circuitry may be easier to implement for certain types of modulation and types of data. If the transmitting modulating system has to send only certain values, the required circuit may be simpler. The receiving demodulator may also be simpler if the circuitry at that end has to be designed only to decide on certain specific values. This simplicity does not exist in all cases, but there are many in which it applies. For example, it is very easy to implement digital phase modulation because circuitry to flip the phase between two values is easy to build. Another example is demodulating an AM signal which has only a few possible values. A system could be designed for that specific situation which would be much more compact than one which must demodulate an AM signal which has analog modulation.

Finally, digital signals are easier to combine together so that they can then be modulated as a group and sent on a single carrier. This combining process is called *multiplexing*. It allows a single carrier and modulation method to be used for multiple signals traveling over the same channel. Analog signals are very difficult to multiplex and demultiplex (separate) at the receiving end.

Does digital modulation and demodulation have any disadvantages? There are some situations in which it does. Some types of information are inherently analog and must be converted to a digital format for digital modulation to occur. The cost of the conversion process may be too high for the application, especially if the signal has to be reconverted to analog format at the receiving end. Standard broadcast AM radio is one such case. The radio is used for sending voice and music, which are analog signals. It is too expensive to build AM receivers that can demodulate a digital signal and convert it to analog, at a price acceptable to radio listeners. This is especially true when the required accuracy and freedom from errors are not high. In practical terms, if a voice signal on the radio is slightly distorted or otherwise not quite perfect, the practical consequences are small. For example, if the voice is 0.01 percent louder than it should be for 1 millisecond (ms), the listener does not even notice it. However, if the signal represents numbers being passed between computers, even a slight error is unacceptable.

One other disadvantage to digital modulation is that it requires more bandwidth than amplitude modulation to carry a given amount of information at a desired rate. In other words, the efficiency in using the existing bandwidth is lower. This is because the digital scheme uses only certain possible values, while the analog scheme allows the use of any values within the total signal voltage range. Therefore, fewer points of different distinct information can be represented by a digital signal as compared to an analog signal. In order to have the digital signal represent the same amount of information, it must send more digital signals in a given amount of time than the analog signal. It is for this reason that many digital modulation systems use more than the two levels of a binary system. Practical digital systems for modulation often use 4, 8, or even 16 distinct levels, as a compromise between the noise resistance of the two-level binary system and the greater efficiency of the multiple-level system.

When digital modulation is used, the type of modulation (AM, FM, or PM) is combined with *shift keying*, since the modulating signal acts as a key to shift the amplitude, frequency, or phase of the carrier from one distinct value to another. Therefore, ASK is amplitude-shift keying, FSK is frequency-shift keying, and PSK is phase-shift keying.

Combined Modulation

In order to get the advantages of binary digital modulation and still have the channel efficiency of a channel which is supporting many discrete levels, some systems combine more than one modulation scheme. (Fig. 3-29) For example, AM and FM may both be used on the carrier. The binary AM modulation represents two values, and the binary FM modulation represents two others. As a result, there are four total possible values that the modulated signal represents at any given instant. These are often referred to by a pair of numbers in the format (X, Y)

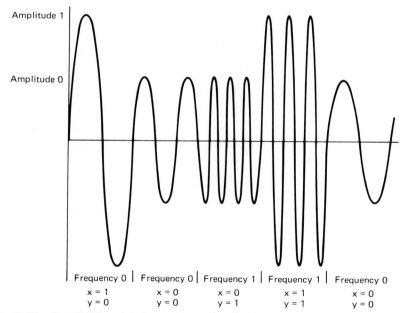

Fig. 3-29 Combined modulation varies two characteristics of the carrier, so that greater channel efficiency is achieved. At the same time, the noise immunity of digital signals that are widely separated is retained. The figure shows both the frequency and amplitude of the carrier changing to represent a pair of bits at each point.

where X can be a 0 or 1 as a result of the AM, and Y can be a 0 or 1 as a result of the FM. The modulated signal can thus represent (0,0), (0,1), (1,0), or (1,1) at any moment. The demodulation system really requires two separate demodulators working on the same signal. The AM demodulator is sensitive only to the AM process and has to look for only the two distinct amplitude values possible. The FM denominator is designed to separate and identify the two frequency values that are possible and is not affected by the AM. The four time domain representations of the combined AM and FM process are shown in Fig. 3-29. This is a four-level digital modulation scheme, with two levels provided by the AM and the other two by the FM.

QUESTIONS FOR SECTION 3-7

1. Historically, why was analog modulation used?
2. What is the difference between analog and digital modulation?
3. What is binary modulation? Compare it to digital modulation.
4. What is the effect of noise on an analog signal? Compare this to the effect of noise on a digital signal.
5. Why is a digital signal highly noise-resistant?
6. Why does a binary signal have the highest noise resistance?
7. What is the encoding advantage that digital can offer?
8. What is the role of digital modulation when looking for errors and correcting time?
9. What are the modulation and demodulation advantages of digital versus analog?
10. What are the disadvantages of digital versus analog?
11. Why does digital modulation require more bandwidth than analog modulation to carry the same amount of information in a practical system?
12. Why do many digital systems use multilevel, rather than binary modulation?
13. What is combined modulation? Why is it sometimes used?
14. What does a signal that has combined modulation represent at a single point in time?

PROBLEMS FOR SECTION 3-7

1. A 1-V, 10-Hz sine wave is amplitude-modulated by a 2-Hz on-off signal, of the same amplitude. Sketch each waveform and the result.
2. The same signals are used as in the previous problem, but the modulation is such that when the modulating waveform is on, the carrier shifts up in frequency by a factor of 2. Sketch the waveforms.
3. The same signals are used for phase modulation. The "on" modulating signal causes the phase of the carrier to shift by 180 degrees. Sketch the waveforms.
4. A sine wave at 10 Hz is phase-modulated by a 1-Hz square wave. This square wave causes the phase to shift 90 degrees when it is on. Sketch all the waveforms.

5. An AM system uses 10 levels for digital modulation, from 0 to 0.9 V. What is the amount of noise that can be tolerated before an error is made?

6. Many schemes are used in areas other than communications to detect errors. In accounting, not only are the individual numbers to be recorded sent, but also their sum and the total number of numbers. This can detect certain kinds of errors. A bookkeeping system is recording these numbers: 7, 4, 10, 3, 6, and 10. It also sends the sum, 40, and the number of entries, 6.
   a. Will any single error be detected? Explain by example.
   b. Will all multiple errors be detected? Explain.
   c. Which errors does the sum figure help detect?
   d. Which errors does the number of entries figure help detect?

7. A combined modulation scheme is being used with a 10-Hz sine wave. A 2-Hz square wave is causing amplitude modulation, while a 5-Hz square wave is causing 180 degree phase modulation. Sketch the carrier, the two modulating signals, and the result. What information is lost when the amplitude is 0?

8. For the signal of the previous problem, the 5-Hz signal is causing a frequency shift to 20 Hz when it is on, instead of PM. Sketch the waveform.

3-8 Synchronous and Asynchronous Modulation and Demodulation

When a channel is being used to carry data, the system at the receiving end must demodulate the received signal to separate the carrier from the modulating signal, which contains the data. In a digital communications system, this data consists of some representation of the discrete, specific values that the digital system was designed to use. There is a further problem that the receiving system must handle: at what instants of time should it look at the demodulated signal to determine what value was sent? For simplicity, assume a simple binary case. The modulating signal is a stream of 1s and 0s, at some fixed rate of bits per second. This is used to modulate the carrier and then is transmitted over the channel. At the receiving end, the signal is demodulated, and the receiving circuitry is left with a signal which varies in time, representing the 1s and 0s.

The problem is to determine when the receiving circuitry should look at the demodulated signal to decide whether the signal is 1 or 0. Although the data is sent at a known rate, and the receiver usually knows what this rate is, there is still the problem of how to start, or synchronize, the 1 and 0 decision making of the receiver with the transmitter timing. Also, the clock generating the timing signals for the transmitter and the one for the receiver will both differ by some small amount, since two clocks are never exactly the same. This slight difference can cause the receiving circuitry to look at the demodulated stream either a little too frequently or a little too infrequently and thus incorrectly decide whether 1s and 0s were sent, even if the signal was demodulated perfectly (Fig. 3-30).

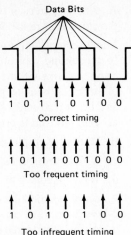

Fig. 3-30 In an asynchronous system, the receiver must use its own clock to determine where the data bits are changing. Even if the data is received perfectly, errors can result if the receiver looks at the data at the wrong instants of time, too often, or too infrequently when deciding whether a 1 or 0 was sent.

Asynchronous Systems

The problem of timing and synchronization can be solved by having the transmitted data stream contain some special characters, called *synchronization* or *sync characters*, which alert the receiving circuitry. These sync characters are used to indicate to the receiver that a new group of data characters is about to begin. The receiving circuitry can then look at the demodulated signal based on this sync signal occurring. The receiving circuitry must have its clock set to the same value as the transmitting circuitry, of course.

This system of recovering the original data from a demodulated signal is called *asynchronous*. This is because there is no inherent synchronization between the transmitter and receiver. The receiver must get a special sync signal in order to align itself in time with the transmitted data. Asynchronous systems are used often in data communications, because they are low in cost and simple in construction. However, they have two drawbacks. First, because the sync characters must be sent quite frequently, a considerable percentage of the overall communications time is spent with sync characters. This time, typically 10 to 20 percent, prevents the channel from being used for data. Therefore, the efficiency of the channel can never be 100 percent, since time is wasted with sync characters. Second, because the clocks of the two ends are not exactly at the same rate (a difference of 0.01 to 0.1 percent is typical), there is a practical limit to the clock rate after which it is almost impossible to keep the two clocks adequately aligned. In today's technology, this rate is about 20,000 to 50,000 bits/s. Below this rate, the error in the clocks is small enough that the data can be retrieved accurately. Above this rate, however, it is very error-prone, and it is almost impossible to establish sufficient accuracy. Also, any disturbances versus time in the channel, such as noise, will cause minute but important momentary shift in the received signal, called "jitter." This jitter prevents even a perfect receiver clock from accurately recovering the data, because the receiver clock and the data will be momentarily out of synchronization.

Synchronous Systems

An alternative to asynchronous communications is a synchronous system. In a synchronous system, the receiving circuitry provides a clock, but this clock is either derived from or synchronized to the incoming data itself (often using a PLL to "extract" the clock frequency). Any timing difference or jitter between the two systems is minimized because the transmitting clock is used to provide the receiving clock, or at least synchronize the receiving clock (Fig. 3-31). Synchronous communication is capable of sending and receiving data accurately at high rates (tens of megahertz) and with very high efficiency, since very few sync characters need to be sent. (A few are required, but there can be long streams of data after each sync character group.) The tradeoff is that synchronous systems are more expensive to build and implement.

How does the modulation choice affect the ability to use synchronous communications? For the three types of modulation discussed—AM, FM, and PM—there are different ways to extract the signal timing from the received signal. In practice, it is easiest to design a receiver that will extract the timing of the transmitted signal from a PM or FM signal. It is more difficult to extract the timing of the signal from an AM signal, although it can be done. The choice is made by the system designer on the basis of cost, complexity, channel characteristics (which can affect the corruption of the signal), and required performance. The receiving circuitry for a synchronous system actually locks on to the received signal to extract the clock and follow it (along with any jitter or other variations) within a reasonable range around the nominal value (typically a 2 to 5 percent swing). The output of the lock-on circuit is then used as the timing clock for removing the received data from the demodulated signal.

Synchronous systems also offer advantages in channels where there is considerable noise or only a very weak signal can be received. A good example is a deep-space probe or satellite. The transmitting power of such a satellite is very low, and the amount of signal that reaches the earth is incredibly small. It has often been described as trying to see someone light a match on the moon. The tiny signal that reaches earth is millions of times smaller than the noise of the channel, and it would seem impossible to recover this weak signal from all this

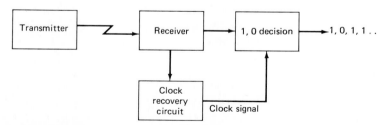

Fig. 3-31 A synchronous system overcomes this problem by developing the required receiver clock timing from the incoming data stream itself and so perfectly tracks any slight variations or jitter in the timing.

noise. However, by skillful design techniques and the use of synchronous communications, the signal from the space probe can be recovered. This is because the receiver on earth knows something about the signal it is looking for and attempts to find this "something," the frequency pattern, from within the random jumble of noise. When synchronous communications is used this way, it is often referred to as *coherent communications*.

QUESTIONS FOR SECTION 3-8

1. What problem does the receiving circuitry have after a digitally modulated signal has been demodulated?
2. Why is setting the receiving clock to the same rate as the transmitting clock not sufficient for proper reception of the signals?
3. How can the synchronization problem be solved in an asynchronous system?
4. What is jitter?
5. Why is an asynchronous system often inadequate? Where is it inadequate?
6. What is synchronous communications? Where does the receiver get its clock?
7. What are the pros and cons of synchronous communications?
8. Which type(s) of modulation are best suited to synchronous modulation?

SUMMARY

A channel can carry information energy only if the energy has been used to modulate a carrier, which acts as the vehicle. The modulation process has been extensively analyzed and is well understood. There are three types of modulation that can be used:

Amplitude modulation, in which the amplitude of the carrier is varied by the modulating signal

Frequency modulation, in which the carrier frequency is modulated

Phase modulation, in which the carrier phase is varied by the modulating signal

Although only one characteristic is varied by the modulating signal directly, other aspects of the signal are indirectly affected. Each of these types of modulation has specific advantages and disadvantages in the following areas:

Resistance to noise in the circuit and channel

Efficiency and use of transmitter power

Ease of implementation at the transmitting end and ease of demodulation at the receiving end

Use of bandwidth

Although normally a signal is thought of as a level that varies with time (the time domain), it is technically equivalent to think of signals in the frequency domain, in which the amplitude of the frequency components is studied. Fourier analysis shows that any signal characterized in one domain can be equivalently characterized in the other. Modulation and its effects are often better understood in the frequency domain.

In analog modulation, the AM, FM, or PM modulating signal can have any value within the entire allowed range. In digital modulation, only certain specific values are allowed. Digital modulation allows the building of systems that can provide better performance against noise and can even detect when errors have occurred. The tradeoff is the wider bandwidth needed for digital modulation systems. Some systems use the simplest form of digital modulation, called *binary*, in which only two signal values can be sent. Other systems use binary modulation of one type along with binary modulation of another type, combined to give the required performance and cost.

Lower-performance systems can use asynchronous communications, in which the receiving end has its own clock running at the same rate as the transmitting clock. Though low in cost, this sort of system is inefficient in using channel time and cannot perform at high signal rates. In synchronous communications, the receiving circuit actually derives its own clock from the transmitting clock, embedded in the modulated signal. Phase modulation and frequency modulation are especially suited to synchronous systems. A synchronous system can provide high performance and efficiency but requires more circuitry at either end.

Even the simplest form of communications between two systems, such as a computer's putting out a stream of two voltages when talking to its printer, is a form of baseband modulation for data communications. As such, even this kind of system involves the concerns and techniques of modulation.

END-OF-CHAPTER QUESTIONS

1. Why is modulation needed? What is the relation of the modulating signal, the carrier, and the channel?
2. What is the name for the process by which the original signal is recovered from the modulated signal?
3. Explain why the carrier contains no information. Does getting rid of the carrier have any advantages? Any disadvantages?
4. What is the time domain? Compare it to the frequency domain.
5. Which representation, time or frequency domain, is more correct? Explain why one or the other might be preferred.
6. How does the frequency domain show signal bandwidth? What information does extra bandwidth contain for a voice or instrument signal?
7. Explain how a rapid signal change in one domain appears in the other domain. What is the appearance of a slowly varying change signal representation in the other domain.
8. What are the three fundamental types of modulation? What is varied in each by the modulating signal?
9. Give two advantages and disadvantages of each of the three types of modulation.
10. What is a baseband signal? Give two examples of baseband modulation.
11. What are sidebands? How are they related to the modulating signal and carrier?

12. Compare the bandwidths of an AM and an FM signal carrying the same amount of information.
13. What is SSB? What problem does it overcome? What problems does it introduce?
14. Many early Morse code systems used a modulation scheme in which the carrier itself was turned on and off to represent a dot or dash (carrier on) or a space (carrier off). What type of modulation is this?
15. What is the overwhelming advantage of FM over AM? Why is this so?
16. How does a PLL work? How does it perform the demodulation of FM signals?
17. What is phase? Why is a reference phase needed to make PM meaningful?
18. Why is PM similar to FM in the frequency domain?
19. Why is PM often preferred in systems?
20. What are the advantages and disadvantages of digital versus analog modulation?
21. Why is multiple-level digital modulation often used instead of binary modulation, even though the data itself is binary?
22. Explain what combined modulation is, and why it is used.
23. Why does a receiver in a digital modulation system have to have some idea of the timing of the transmitting end?
24. Explain how asynchronous modulation satisfies this timing requirement. Explain the weakness of asynchronous modulation and its advantage.
25. What is synchronous modulation? Why is it used in place of asynchronous modulation in many cases?
26. Why don't all systems use synchronous modulation?
27. How does modulation allow many users to share the same general frequency band, each with its own bandwidth?
28. What are ASK, FSK, and PSK? Why are they so named?

END-OF-CHAPTER PROBLEMS

1. A system has two sine waves going through the channel. One is 1 V, 20 Hz, and the other is 1.5 V, 10 Hz. Sketch the frequency domain representation of the signals.
2. A 10 kHz carrier is modulated by 0 to 3-kHz voice signals. Sketch the frequency domain representation of the result. What are the frequencies of the sidebands?
3. A channel needs to carry 10 signals with a minimum total bandwidth. The allowable band to be used starts at 100 kHz. Each signal for this AM process has a bandwidth of 3 kHz. A guard band of 2 kHz is required between adjacent signals, but not at the edge of the band. What carrier frequencies should be used?
4. A triangular waveform at 1 Hz frequency-modulates a carrier at 10 Hz. When the modulating signal is at 0 V, the carrier is at its nominal value;

when it is at full scale, the carrier frequency is 50 percent above the assigned frequency. Sketch the modulated waveform in the time domain.

5. A 10-Hz sine wave is to be amplitude-modulated by a 4-level modulating waveform. The modulating waveform steps from 0 to 0.25, 0.50, and 0.75 V every 0.2 s. Sketch the modulated waveform in the time domain.

6. The 10-Hz sine wave is modulated by a square wave, which varies from 0 to full scale every 0.5 s. Sketch the time domain representation for these cases:

   a. AM

   b. FM, with the frequency shifting from 10 to 20 Hz

   c. PM, with a phase shift of 180°

7. Digital modulation is used to phase-modulate a 10-Hz sine wave. When the modulating signal is 0, the phase shift is 0. When the modulating waveform is 0.5 or 1 V, the shift is 90° or 180°, respectively. Sketch the modulating signal and the modulating result as the modulating signal steps from 0 to 0.5 to 1, then returns to 0 and repeats every 0.2 s.

8. An accounting system sends numbers, their sum, and the number of entries. The numbers are 23, 47, 10, 11, and 47.

   a. What should the sum value and the number of entries value be?

   b. The first number is received as 22. What changes?

   c. The first number is received as 22, and the second number as 48. What changes?

   d. The last entry is never received. What else changes?

   e. The sum is received as 100. What does this indicate about the validity of the received entries? Explain.

9. A combined modulation scheme is used to send a pair of numbers, represented by the pair (x,y).

   When x is 0, the amplitude is one-half the full-scale value; when it is 1, the amplitude is full-scale. When y is 0, the frequency is unchanged from nominal; when it is 1 the frequency doubles. For a 10-Hz carrier, sketch the modulated waveform when the data to be sent is (0,1), (1,0), (1,1), (0,0), and (1,0), changing every 0.2 s.

10. A transmitter and receiver start out perfectly synchronized. The clock of the receiver differs by 1 percent from the clock of the transmitter.

    a. After how much time will the two clocks differ by 1 ms?

    b. If the receiving clock is incorrectly synchronized by more than one-half the bit period, errors will occur. The bit rate is 1000 bits/s After how many bits in a row will the synchronization be unacceptable?

Modulation

Analog Communications and Multiplexing

Analog communications systems are used for sending inherently analog information, such as voice. They can also be used for sending analog information which has been converted to digital form or information from computers, which is inherently digital. An analog communications system often must accommodate many users at the same time, in order to get the most use and efficiency from the equipment and circuitry at each end and the communications link between them.

Multiplexing is used to allow many users to share the system. Three types of multiplexing are available: space division, frequency division, and time division. Each type has performance and cost benefits and disadvantages. A practical communications system used to carry information over long distances often has more than one type of multiplexing. The overall system consists of a series of links, and each link has a multiplexing scheme that is best suited to the needs of that particular link.

All analog communications systems suffer from the effects of noise. Noise corrupts the original signal value, making it difficult to send information with absolute accuracy. It can also cause system failures and problems with the multiplexing system.

4-1 Analog Communications Systems in Today's World

An analog communications system is designed to handle a signal that can have any value within the total allowable range. Many analog communications systems are in wide and successful use today. These include standard amplitude-modulated (AM) and frequency-modulated (FM) band radio, telephone lines between the telephone and the local phone company central office, and two-way citizens band (CB) radio, for example. These effective, inexpensive, and widely used analog communications systems will not change in the foreseeable future.

The growth of communications for computer and digital systems has created the need for systems designed to handle data (information), which generally exists not in analog form but in digital form. It may seem that it is unnecessary to understand analog communications systems, because the system is designed for digital communications. However, there are several reasons why analog communications systems must still be studied:

Digital signals are really just a special case of analog signals. Digital signals are allowed to have only specific values within the overall signal range; an analog signal can take on any value. This means that a system which can accommodate analog signals can also handle digital signals.

Even if the data to be sent is digital, the channel may be an analog channel, designed for handling analog signals such as telephone signals. A typical communications system consists of many interconnected links. Some of these may be analog and others digital, even though the signal being transmitted is a digital signal.

There are billions of dollars worth of analog systems already installed and in use for analog signals. These systems are used as part of many data communications systems, and they form the analog part.

Analog systems can be very inexpensive to implement and work with, and they provide adequate performance for many applications.

In the real world, the communications channel is an analog channel. The circuitry at either end may be designed to handle digital signals, but noise and distortion in the channel make it less than ideal. As soon as the channel corrupts the digital signal significantly, the channel is no longer handling an ideal digital signal but instead a signal that has values other than the specifically allowed digital values. Thus, the digital signals become somewhat analog.

Many of the technical issues that arise in digital communications systems really are related to the inherent analog nature of the channel or some other part of the system. Even as data communications grows in importance, and the number of data communications systems increases in the next decades, the analog aspects of the system play an important part in determining performance, operational characteristics, and potential shortcomings.

QUESTIONS FOR SECTION 4-1

1. How are digital signals related to analog signals? Explain why they are a subgroup of analog signals.
2. Give three reasons why it is important and necessary to study analog signals and modulation systems in the modern, digital technology environment.
3. Explain why noise makes digital signals and systems look somewhat like analog ones.

4-2 Functions Within an Analog Communications System

Any communications system consists of a series of links between the sender and the receiver. The communications signal is usually modified as it passes through each link and in order to be compatible with the circuitry of the link. The link may be the actual circuitry that generates the data, or it may be the channel which is the conduit for the energy over the distance between the sender and receiver. In most systems, there are several links between the source of the data and the actual transmitter which puts the energy onto the channel. Similarly, there are several links at the receiver end, between the physical receiving circuitry and the user of the received signal.

These additional links, or stages, are needed for two reasons: First, the signal in its original form may not be suitable for transmission. It may have the wrong voltage values, or signaling speeds, or frequencies, for example. The various stages convert or modify the signal generated by the source into those needed by the transmitting system. Second, the total communications system may handle many signals at once, and therefore the original signal must be combined in some predetermined manner with the other signals to be transmitted. The function of the communications system circuitry is to perform the necessary signal modifications at each stage of the transmission and then to perform the reverse operations at the receiver to undo these necessary modifications. Thus, the user at the receiving end has the original signal as the transmitter intended.

An example of such a system is a standard FM broadcast radio system, which broadcasts from a studio to many listeners. The total chain of the signal passage to the link is shown in Fig. 4-1. In this case, the signal comes from a prerecorded source, a phonograph record. (It could of course also come from someone speaking.) The original signal on the record is an analog voltage, which is recorded on the record in a base band AM mode. The needle of the turntable must get this analog signal, which is very small in value (hundredths of a volt, maximum), and send it to the amplifier. At the amplifier, the signal is boosted to a value of about 1 V. The signal is then used to frequency-modulate a carrier at the station's assigned frequency within the FM broadcast band. This modulated carrier goes through several more amplifiers, up to the final one, which is connected to the antenna. From the antenna, the electromagnetic energy is broadcast through the air, which is the channel in this example.

At the receiving end, another antenna extracts a minute amount of the signal energy from the channel. This energy is the modulated carrier, and the amount of signal extracted is extremely small, no more than 10 to 50 microvolts (μV). A preamplifier boosts this signal so that it can be effectively demodulated. The output of the demodulator circuit then is an AM signal on the record. Another

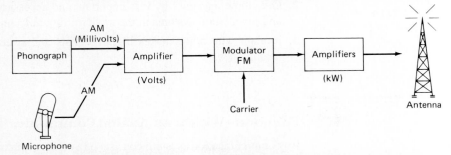

Fig. 4-1 A complete transmitting system, such as an FM broadcast band radio station, includes a signal source, amplifier, modulator, and power amplifier.

amplifier is used to take this 1-V signal and increase its value so that a loudspeaker can present the final signal to the ear.

The data being sent in this case are the music and voice of the record. The person listening recognizes the sounds and uses them for whatever purpose desired. If there were any noise, corruption, or distortion in the final demodulated and amplified signal, it would probably still be usable (unless these reached a high and irritating value). The music and voice would be recognizable even if the final value differed from the true value by 5 or 10 percent. Of course, for best and most enjoyable listening, it would be better if the differences were as small as possible, but the signal would be useful even with larger differences.

The same general type of system can be used for sending a signal representing numerical data. Suppose the system is to be used to report the temperature at the crater of a volcano to researchers located several kilometers away in safety. The signal source would not be the phonograph record and turntable needle. Instead, some sort of temperature-to-voltage transducer would be used. The most common transducer for this purpose is a *thermocouple*, which generates a predictable voltage at its terminals as a function of its temperature (Fig. 4-2). Just like the signal from the needle, the thermocouple signal is very small and must be amplified before being used to modulate the carrier.

At the receiving end, the modulated signal is demodulated and amplified to a more useful level of about 1 V. Instead of driving a loudspeaker, the signal is used to drive the pen of a chart recorder, which graphs the temperature versus time. The researcher can then look at this graph and perform additional analysis.

For this application, however, any distortions or noise cause the temperature value shown on the chart recorder graph to be in error. The amount of the error depends on the values of the noise and the type and amount of signal corruption that occur at any point in the overall system. Each link of the system may contribute some noise and distortion to the total. Of course, any inaccuracy affects the research that is being based on these results. The researcher prefers the error to be as small as possible, of course, so that the analysis of the data can be based on reliable numbers. If the errors are greater than even a few degrees, the data may be useless. This is in contrast to the system which sends music, in which large errors can be tolerated in many applications.

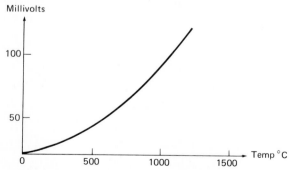

Fig. 4-2 A typical transducer for real-world signals has a very small voltage output level. The thermocouple generates a voltage proportional to temperature, with a typical voltage of 50 to 100 mV at 1000° (depending on thermocouple type).

Discovering the sources of the various distortions and noise is not always easy. The temperature transducer is not perfect and may put out a voltage for a given temperature that differs from the ideal value. The amplifier for this small signal may be distorting the signal as it amplifies it. This distortion is often called *nonlinearity*: the amplifier should be boosting all inputs by a certain factor but doesn't do so perfectly. The amplifier may be designed to amplify all signals by a factor of 100, but in reality amplifies some input values by more than 100, and some by less (Fig. 4-3). The FM modulator circuit is usually not ideal either, and its performance may be affected by the variations in its ambient temperature. (Recall that the unit is in a volcanic crater.)

The sources of potential problems do not end once the signal is transmitted to the air. The channel itself can have electrical noise which it adds to the electromagnetic energy which has been transmitted. At the receiving end of the system, the amplifier of the feeble antenna signal and the demodulator can have errors and distortions similar to those at the transmitting end.

In summary, there are many sources and types of errors which can creep into the original signal and build up to cause what may be a large error at the user's readout.

These problems can be overcome, or minimized, in the communications system by careful design to the specific needs of the application. The major goal of an analog communications system and its circuitry is to provide to the user a voltage which is based on the best *estimate* possible of what the true value is. For example, the receiver may be designed for a situation in which there is significant 60-Hz noise from nearby power lines. This noise can severely corrupt the signal as it comes into the receiver antenna, and within the receiver itself.

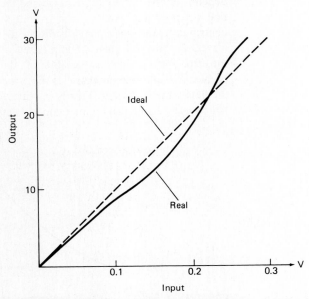

Fig. 4-3 The transducer signal amplifier has small nonlinearities and distortions, which make obtaining accurate measurements difficult.

However, the receiver can be designed with special filters that remove as much 60-Hz noise as possible from the signal presented to the user. This means that the receiver is providing a better estimate of the true value, based on the nature of the corrupting influence.

The quality of an analog communications system for data and numerical information is determined by the accuracy of the estimate it provides to the user. A system which is capable of consistently providing good estimates in the presence of noise and other interference is a useful one. A system which is not guaranteed to provide accurate estimates may not be useful to many users. A typical analog communications system specification may be an overall accuracy of about 1 percent under most conditions. It is difficult to do better than this value in most systems, and values far worse than 1 percent would probably not be useful for data that could be analyzed with confidence by a researcher.

QUESTIONS FOR SECTION 4-2

1. Why does a typical system involve several links at each end? Give an example.
2. What is the general function of the circuitry at each stage in a typical communications system that is carrying information from one point to another?
3. What are the stages in the transmission of prerecorded music from an FM radio station studio to the listener?
4. Compare distortion in a music or voice signal to that in a signal such as one from a temperature-measuring device. What are the differences in the consequences of this distortion?
5. What is nonlinearity? Give an example of a case in daily experience.
6. What are some of the possible sources of error in a system designed to acquire temperature and transmit it to a user for logging for scientific analysis?
7. What does it mean to say the analog receiver must provide a "best estimate"?
8. What is the typical accuracy of an analog communications and modulation system, from one end to the other?

4-3 Multiplexing

In most systems there is a need to have more than one user share the channel with other users. This means that various types of information from different sources need to have access to the channel. "Multiplexing" is the general name for various different ways of achieving this.

Multiplexing allows these various users to share the channel "simultaneously." Some multiplexing schemes allow full simultaneity, with the information from each source carried along the channel at the same time as the others. In other schemes, the channel is shared by the users, with the channel switching from user to user so quickly that all the necessary information is carried as fast as it is being generated. To the receiver, there is no significant time lag, although there are short moments of time when the channel is not directly connected for this user.

The choice of which multiplexing scheme is used depends on many factors. Some of these are technical, such as the nature of the channel, the amount of data to be transmitted, the number of users, and the bandwidth available. Other reasons are economic. Some multiplexing methods allow some of the circuitry at the transmitting and receiving ends of the system to be shared. This may mean less cost, and less physical space and power used. In some situations, physical space and power usage are critical issues—a space vehicle, for example, is very limited in both of these. Even when they are not a major concern, for example for a radio transmitting station on earth, cost considerations of sharing parts of the system can be important. Three types of multiplexing can be used. These are:

Space-division multiplexing (SDM), in which a separate wire is used as the channel for each user. The users therefore can be on the same frequencies at the same time.

Frequency-division multiplexing (FDM), in which the total channel bandwidth is divided into smaller bands. Each user is assigned one band.

Time-division multiplexing (TDM), in which the channel uses only a single frequency band, and the users share this, with each user having access at a different time.

Space-division multiplexing is the simplest and earliest form of multiplexing, since the idea of installing a separate transmitter, wire or cable, and receiver is fairly straightforward and logical. *Frequency-division multiplexing* was developed as the circuitry to modulate and demodulate carriers as specific frequencies became available. *Time-division multiplexing* is the most complex of the three forms, since it requires establishing precise time periods on both ends of the system. All of the three forms of multiplexing will be explored in more detail in the next sections, along with their advantages and disadvantages.

Multiplexing is a solution to the problem of transmitting the maximum amount of data from one point to another. It is similar to shipping more and more packages by truck between two cities, with packages from different factories intended for different stores. The space division approach is to build more roads between factory and store for more trucks, each road running alongside the others. In the equivalent of "frequency division," more width is made available by increasing the number of lanes in the highway, but many facilities, such as on and off ramps, are shared by all lanes. In the time division variation, trucks use the existing highway and lanes. Trucks from various factories go in sequence on the road and are directed to the proper store at the far end.

Many practical systems used for communications combine more than one multiplexing method. This is because the technical and performance needs, as well as cost constraints, are often different for different parts of an overall system. It is entirely practical and meaningful for one group of users to be combined by space-division multiplexing, for example, and then combined with another group via frequency-division or time-division multiplexing.

QUESTIONS FOR SECTION 4-3

1. What is the goal of any multiplexing system? What is the need for this?
2. What is the concept behind space-division multiplexing?
3. Repeat Ques. 2 for frequency-division multiplexing.
4. Repeat Ques. 2 for time-division multiplexing.
5. Which form of multiplexing is most complicated technically? Which form is least complicated?
6. What is the purpose of the demultiplexing process? Where is it done?

4-4 Space-Division Multiplexing

Space-division multiplexing (SDM) is used in many applications. In an SDM system, there is a physical link or path dedicated to each sender and receiver pair. A common example of an SDM system is the standard telephone and the local telephone office. Each phone is connected to the equipment at that local office by a pair of wires that no other phone shares. A standard central office serves a single exchange (the first three digits of the seven-digit phone number) and so has 10,000 wire pairs connected to it (for numbers -0000 through -9999) (Fig. 4-4).

In an SDM system, the signal is usually modulated as a base band AM signal. The frequency and time domain graphs of each signal are independent of each other and unrelated. Any variation in one has no effect on the others.

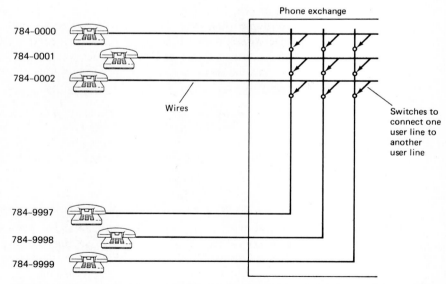

Fig. 4-4 A standard telephone exchange contains paths and switches so that any of the 10,000 users can be connected to any other, and many such pairs can be connected at any instant. Each user has a separate signal path.

Analog Communications and Multiplexing

There are advantages and disadvantages to an SDM system. On the positive side, consider these features:

An SDM system is simple to build since each link is independent of the others.

Any technical problem or failure in any link, such as a broken wire, does not affect other users. This makes SDM systems vital in critical applications, such as military control systems or industrial processes which must be fail-safe (control of a chemical plant, for example). The critical application might use several SDM links to carry the same message, in parallel at the same time, so no single failure will prevent the message from reaching the other end.

In concept, it is technically easy to add more users. Each user requires a transmitter, wire, and receiver as all preceding users do. Increasing the system capacity is straightforward and does not affect the other existing users. The only cost is the expense of the additional user parts.

The initial cost of the system is low in many cases. The only expense, not including cable, is for the equipment at the other end for the first user. There is no general cost for support equipment that must be provided regardless of the number of users the SDM system will support. Of course, the initial cable may require tunnels, telephone poles, and so on, which are expensive.

The performance of the system for each user is predictable and guaranteed. Since there are a path and circuitry dedicated to each user, there is no chance that a user will have to wait because someone else is using the system. Some applications require this absolute guarantee of service.

Full-duplex operation can be implemented by adding a pair of wires for each direction. Some applications require a pair of wires in each direction, for example, when specific frequencies are being used and these frequencies on the wire may interfere with the same frequencies from other users. In some cases, this may not be required. For example, the standard telephone is full-duplex but requires only one pair of wires (one link) between the users.

SDM systems have these advantages, but they also have drawbacks:

The initial cost of laying the cable bundle of wires (tunnels, telephone poles, and so on) means that the cabling should be able to handle many signals. In some cases, adding capacity may be very expensive because new wires have to be routed alongside the older, existing ones in hard-to-reach locations.

The cost of the cabling itself, and the thickness of the cable bundle, can become very high. Both of these factors mean that a practical cost and size limit are often reached well before enough lines have been installed.

Since all users have a transmitter system and a receiver system devoted to their use only, there is no possibility of sharing the cost of these systems among several users. This cost will increase in direct proportion to the number of users. For a large number of users, for example, in a phone system, this cost can be very high.

If the amount of use of the link is low, then much of the money and effort spent for the ends of the link and the link itself is wasted. In many applications, the user needs the actual connection to the other end for only a small percentage of the time. For the rest of the time, the link can be available to others, but it is unused because of the SDM system.

The decision on whether SDM is the best choice for multiplexing, as compared to FDM and TDM, is made by the system designers on the basis of the required performance reliability, expandability, and cost factors. In most cases, SDM is used only for shorter distances (a few miles) such as those between two parts of a system that are in the same overall enclosure, or in systems in which the users need to be independent of each other, such as the connections to the local phone exchange. However, the practical limitations of SDM in terms of cost and use of the link between users often necessitate that FDM and TDM be chosen. In many systems, SDM is used for part of the overall application, and another modulation scheme is used for another part of the system. The standard telephone system is one such case. All the local phones are linked via lines to the local office, and any one of these 10,000 phones can be connected to any other one via the switching system in the office. However, it is not practical to run 10,000 wires to the office of the adjacent exchange, and it is usually not necessary. In fact, it would soon become impossible, since all these wires would have to go to all the exchanges with which callers might wish to connect. Fortunately, there are equally effective alternative choices that the systems designers can use: FDM and TDM.

QUESTIONS FOR SECTION 4-4

1. Give two examples of space-division multiplexing systems.
2. What do the time and frequency domain graphs look like for a two-user SDM system, before and after the multiplexing? Explain.
3. What are three advantages of SDM? Three disadvantages?
4. How is system capacity (number of users) increased in an SDM system?
5. Where is SDM typically used? For what distance, in general? Why?

PROBLEMS FOR SECTION 4-4

1. A simple system connects user A to user B. The system must be expanded, via SDM, to connect user C (next to user A) to user D (next to user B). Sketch the original system and the expanded system. How many additional transmitters, links, and receivers are needed?
2. Ten more pairs of users are going to be added, above the original A-B and C-D pairs. How many additional sets of circuitry will be needed?

4-5 Frequency-Division Multiplexing

Frequency-division multiplexing (FDM) is used with both wire links and radio links. The idea of FDM is to modulate the signals of the various users onto carriers at different frequencies. These modulated signals can then share the same path (wire, air, or vacuum), and they occupy different parts of the overall electromagnetic spectrum. Before multiplexing, they have the same shape and bandwidth, but each user signal is centered around the carrier that is used for that signal (Fig. 4-5).

In FDM the type of modulation chosen can be amplitude, frequency, or phase. The use of FDM does not mean that frequency modulation of the carrier is the only type permitted. The modulation chosen depends on the overall requirements of the application. Standard broadcast radio is a common example of this. The AM band uses amplitude modulation in an FDM scheme to allow multiple stations in the 550- to 1600-kHz band; the FM band uses frequency modulation to allow for many stations in the 88- to 108-MHz band.

In FDM the total bandwidth required is equal to the sum of the individual bandwidths, plus any guard bands that are required. Since the link—air, vacuum, wire, or assigned frequency band—usually has a bandwidth much wider than the bandwidth of any individual user signal, the link's capacity is used more efficiently. A link made up of 22 gauge copper wire has a bandwidth of about 100 kHz, depending on wire spacing, insulation, and other factors. Therefore, sending a single voice signal of several kilohertz bandwidth using this sort of link is very wasteful.

An FDM system requires equipment as shown in Fig. 4-6. For each user, there are modulating circuitry, a transmitter, a receiver, and a demodulator. The channel is common to all users. Since each transmitter is using a carrier of a different frequency, there is no interference unless the sidebands or carriers are incorrectly assigned and therefore overlap. In some designs, it is practical to share a single transmitter or receiver.

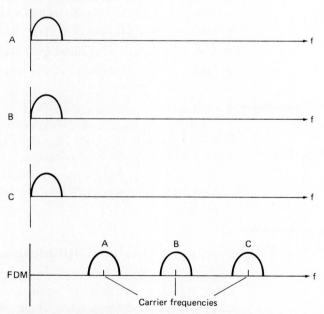

Fig. 4-5 In FDM, each modulating signal is shifted to a different part of the frequency spectrum by using a different carrier for each signal.

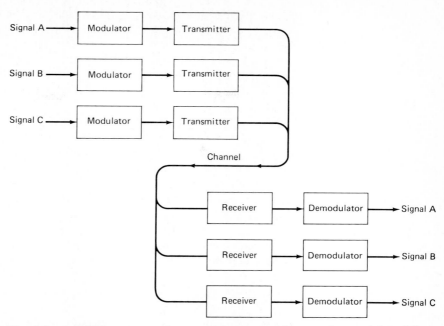

Fig. 4-6 An FDM system requires a modulator, transmitter, receiver, and demodulator for each user signal. The channel is common to all users.

In terms of advantages and disadvantages, there are many similarities between SDM and FDM. The main difference is that an FDM system uses a common pathway, and an SDM system requires a separate, unique path for each user. In an FDM system, the only initial expense is the cost, both in technical difficulties and in dollars, of setting up the path. This may include the cable between the two ends and the associated connectors for the cable, or it may be the sending and receiving antennas or satellite dishes. After the path is established, the additional cost for each user is mainly the expense of the transmitter and receiver for that particular user or the modulator and demodulator pair. (Depending on system design, it may be possible to share a single transmitter.)

In an FDM system, a problem for one user can sometimes affect others. If the modulator or transmitter of a user malfunctions, the specific type of failure may cause the equipment to begin putting out signals with the wrong carrier value, or with improper modulation that spills into other user sidebands. A problem with the path (antenna, cable, connectors) affects all users at the same time.

With an FDM system users can be added to the system by simply adding another pair of transmitter modulators and receiver demodulators. The maximum number of users is determined by the amount of bandwidth that each user is assigned and by the total bandwidth of the channel. In most systems, the carriers are spaced at even, preassigned intervals, so that various technical and design computations are easier. However, in some cases there may be some users who require low bandwidth, and others who require high bandwidth. For these applications, the bandwidths are assigned and the carriers spaced to get the maximum use of the total bandwidth. One example would be signals from a space vehicle carrying astronauts. There are several voice signals, and these need only a small bandwidth, typically 3 kHz. However, the signals carrying the video (television) from the space vehicle to earth-based stations need a much higher bandwidth, about 6 MHz. In this case, the system designers divide the overall bandwidth assigned to the space vehicle to calculate the exact bandwidth needed for each function (Fig. 4-7). In this way, the best use is made of the available resource.

For many other applications, however, each user is sending approximately the same signal. In these cases it is much easier simply to divide the overall spectrum evenly, since the installation and calibration of equipment will be easier. A system designed to carry many conversations between telephone offices is an example of this situation. The path between the offices may be designed to carry 20, 50, or even 100 individual conversations, each spaced with a carrier about 4 kHz from the next.

Many applications require full-duplex information flow. An FDM system is often chosen because it allows the same physical link to be used while user A talks to user B and user B talks to user A at the same time. A regular voice link telephone is an example of an application for which this is needed, since conversations between people are interactive, give-and-take, two-way communications. It is awkward to require that each talker speak in turn, and then say "over," without interruption or comments from the other talker.

The FDM systems are commonly used to connect computers or computer terminals to other computers using standard telephone lines. The telephone line has a bandwidth of about 3 kHz, and this overall band is divided into two subbands. One subband is used for data flowing from the remote computer or terminal to the central computer, and the other is used for the opposite direction. Full-duplex communication is therefore possible. The system at each end which implements this communication is called a *modem* (contraction of "modulator-demodulator"); it will be studied in greater detail in subsequent chapters.

One major drawback to the FDM system is that each user requires a precise carrier frequency, and this carrier must remain at its assigned frequency as the temperature of the system varies, and as the characteristics of components in the system change with age. These changes in frequency are called *drift*, and they

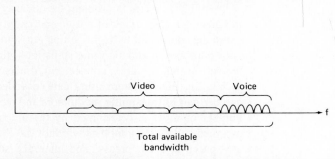

Fig. 4-7 The carriers of an FDM system do not have to be evenly spaced. Uneven spacing is used to obtain the maximum use of the available bandwidth, when the bandwidths of the individual signals (such as a wideband video signal and a relatively narrow band audio signal) differ by a large amount.

can become a problem under difficult operating conditions. They actually create a double problem since the receiving circuitry must be stable in order to receive the signal properly, although some available circuit techniques let a receiver "tune around" automatically and lock onto a signal that is not at its precise, assigned frequency. If the carrier frequency drifts too far, it may cause interference in an adjacent channel or be too difficult to capture and lock onto at the receiving end.

The stability and drift problem is not so severe in a standard radio broadcast, which is a form of FDM, since the user (the radio listener) is only listening to one radio station at a time, and the listener can manually adjust the radio when the station is first tuned in. However, imagine having a roomful of radio receivers, each tuned to a different station. The stability of each carrier and the ability of each carrier and receiver to work together properly are then much more difficult and important.

QUESTIONS FOR SECTION 4-5

1. What is frequency-division multiplexing?
2. What does the FDM process do to the time and frequency graphs of user signals?
3. Is FDM related to the AM, FM, and PM process? Explain.
4. What is the total bandwidth required in an FDM system for all the users?
5. What is the key difference between FDM and SDM with regard to adding another user?
6. What are the effects of equipment problems in an FDM system with multiple users?
7. How are carrier assignments made? What are the advantages of even versus uneven spacing?
8. Give two examples of application in which full-duplex communication is needed. Why is FDM often chosen?
9. What is drift? Why is it a concern with an FDM system? What problems can it cause?

PROBLEMS FOR SECTION 4-5

1. A system has to support three users through FDM. Each user has a 0- to 3-kHz bandwidth. Carriers are assigned at 10, 20, and 30 kHz, and AM modulation is used. Sketch the three signals, in the frequency domain, before and after the FDM process.
2. A single pair of users is supported by an FDM communications system. The system has to be expanded to handle a total of five pairs of users. Sketch the original system, showing the user, modulator, transmitter, link, receiver, demodulator, and other user. Sketch the system when expanded for the five pairs. What additional circuitry is needed?
3. A communications system is required to handle signals from a spacecraft to the ground station. It must handle two 6-MHz video signals, ten 200-kHz data signals from sensors on the equipment and astronauts, and five 3-kHz voice channels.

a. If all channels were evenly spaced at the width of the widest bandwidth channel, what total bandwidth would be needed?

b. If uneven spacing is used, how much total bandwidth is needed, assuming there are no guard bands?

c. Sketch a frequency domain graph of the signals, with uneven spacing, when the FDM process is used to combine the signals for a single link.

4-6 Time-Division Multiplexing

The last form of multiplexing to come into wide use was *time-division multiplexing* (TDM). In a TDM system, a single path and carrier frequency is used. Each user is assigned a unique time slot for his or her signal. A central switch, or multiplexer, goes from one user to the next in a specific, predictable sequence and time. When the switch is pointing to user A, for example, user A's signal is put onto the link. Then the switch goes to user B, user C, and so on, and then back to the first user (Fig. 4-8). At the receiving end, the demultiplexer reverses this process and sorts each signal received, in time sequence, to the correct user.

A TDM system is a *serial system*, because the signal from each user follows, in time, the signal from another user. The time domain graph of signals before and after the time-division multiplexing is shown in Fig. 4-9. The frequency domain representation is of course dependent on the specific signal that each channel has. The overall bandwidth of the TDM result is much wider than the bandwidth of any individual user signal. This is because the TDM output is carrying much more information, and information requires bandwidth. Typically, the final bandwidth is approximately equal to the sum of the bandwidths of the individual signals.

In a TDM system, the carrier frequency is often 0 Hz—a baseband system. This is the simplest system, in which each user, in effect, has exclusive use of the line for a specific time window. This allows many users to make use of the link. TDM is used within computer systems and integrated circuits (ICs) too. Many microprocessors and their associated buses multiplex address and data bits to reduce the number of line drivers and receivers and IC pins needed.

In TDM, each of the user signals gets its turn at being sent over the communications link. Consider three signals shown, versus time, in Fig. 4-10a. The TDM process passes each signal, one at a time. Therefore, the value of the user signal

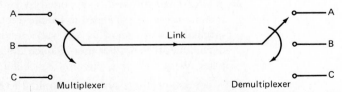

Fig. 4-8 In TDM, a switch at the transmitter passes each signal, in turn, onto the link; at the receiver, the incoming signal is sorted out to the corresponding user by a similar switch.

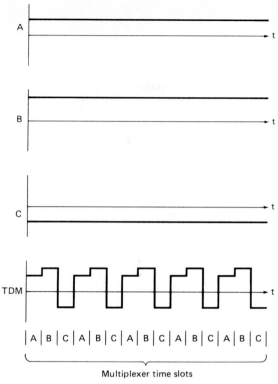

Fig. 4-9 The result of three constant but different signals A, B, and C after TDM. Each has a time slot in the multiplexed scheme.

at whose time slot the multiplexer is "pointing" appears at the output of the multiplexer (Fig. 4-10*b*).

A typical application for TDM is a system which is monitoring many temperature points in a building as part of the heating and cooling management. Each room or office may have a temperature sensor, which provides an analog electrical signal related to the air temperature. The building energy control system needs to know the temperature in each location so it can send the proper amount of hot or cool air to the various locations, while minimizing total energy use. If this were implemented with a space division multiplexing system, wires would have to run from each temperature sensor to the central computer (Fig. 4-11*a*). This would involve high costs for the signal wires and their installation around the building, as well as the physical space for the wires. Adding a new temperature sensor would also be expensive.

Instead, a TDM system would be used. For every floor or zone of the building, a multiplexing box would be installed. All the sensors for that floor or zone would go to this box, which in turn would have a single set of wires carrying the signal to the computer (Fig. 4-11*b*). The multiplexer would scan, in sequence, each sensor and put the signal from that sensor onto the wire leading back to the computer. In this way, only one wire would have to be run back to the computer.

Analog Communications and Multiplexing

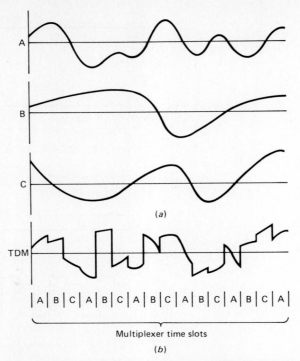

Fig. 4-10 The three signals A, B, and C normally vary. The result of the TDM process is shown in (b). The result is a scanning through, in fixed sequence, of each signal at the instant it has its time slot.

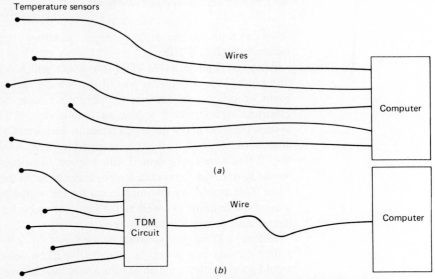

Fig. 4-11 A TDM system can be used to drastically reduce the number of wires needed in a building energy management system, in which many points of temperature readings must be reported to the central building computer.

96   Chapter 4

At the receiving end, the computer would have to receive these various signals in the sequence in which they arrive and match up each signal with the place in the building it represents. This brings up a vital issue with TDM systems: how does the receiving end of the system know what signal it is receiving? There are several ways that this channel identification is performed.

Channel Identification

One method is to have the multiplexer add a *channel number,* or identifying code, to each signal as it performs the sequence operation and sends the signal to the other end of the link (Fig. 4-12). This method is simple and effective. The receiving circuit simply has to check each signal as it comes in for this code. The drawback to this method is that sending these codes occupies time on the signal link, and time is a valuable commodity on any link. For a building heating and cooling system, in which the temperatures of each location are reported back very slowly, this may not be a problem. There is enough unused or slack time in the information flow to allow identifiers.

However, in many higher-performance applications, especially those between computers, there is a need to get maximum use from the system and its link. For these cases, the percentage of link time used just to send channel identifiers may be unacceptable. Instead, there has to be some way to coordinate the sending and receiving circuitry so that these identifiers do not have to be sent. This is called *synchronization.*

Most high-speed, high-performance systems are designed to synchronize the transmitted time slots and the receiver demodulation of these time slots. This is what happens: a starting signal, or *sync signal,* is sent from the transmitter to tell the receiver that the next group of time-multiplexed signals is about to begin. The rate of signaling at the two ends of the link is set in advance and is the same for both (and can be as low as one channel/second and as high as several hundred thousand channels/second). The receiver starts at the sync pulse and knows that each time slot after the sync signal represents the next user signal in succession from the transmitting end (Fig. 4-13).

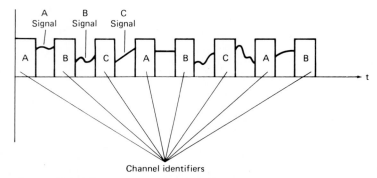

Fig. 4-12 A TDM system with an identifier for each channel. This allows the receiver of the TDM signal to know which channel is on the data line at any instant.

Analog Communications and Multiplexing

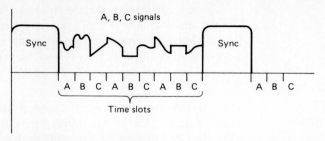

Fig. 4-13 A single synchronization signal can be used before every group of signals, to show that a new group starts at this point in time.

A few sample calculations show the difference between the two methods of coordinating the sending and receiving time multiplexing. Suppose the system has four temperature inputs and is required to send one temperature reading per second, and the actual temperature data takes 0.2 s to send. This means that each 1 s time slot has 0.8 s unused; the 0.8 s can therefore be used for a 0.2-s channel identifier without affecting the desired performance (Fig. 4-14a). Next, suppose the system has to send a new reading every 0.25 s. This means that out of each second, 0.2 s is occupied with the data from the sensor, leaving only 0.05 s for the channel identification (Fig. 4-14b). This is not enough. If the channel identifiers take as much time as the actual signal information, the channel time is half for the data and half for the identifiers. This channel has a 50 percent efficiency.

Next, change the situation so that a sync signal is sent before each group of four channels. This sync signal is 0.2 s long. After each sync signal, the four channels send their information, in 0.2 s time slots each (Fig. 4-15). In this case,

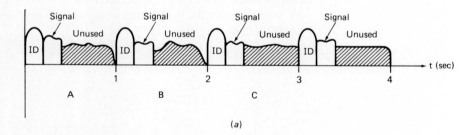

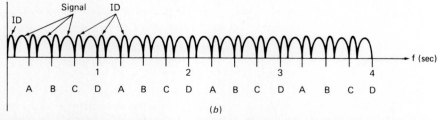

Fig. 4-14 Channel identifiers take up time. In (a) there is enough time for an identifier and the channel signal; in (b) there is not enough time for all the signals.

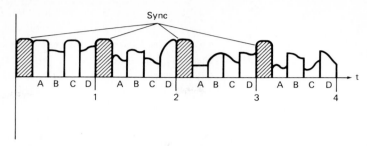

Fig. 4-15 A sync signal for every group of channels uses much less overall time and is more efficient, but it requires more circuitry at the transmitter and receiver.

the sync signal uses only 20 percent of the channel, and the real information has use of 80 percent. The result is a much more efficient system.

Both types of synchronization are used in TDM systems. The "identifier with channel" method is used when the total amount of data to be sent is low, the rates are slow, and full use of the link is not needed. The sync signal followed by a stream of channels is used when higher information rates are needed and the link must be more efficiently utilized. The lower-performance method is often cheaper and easier to implement, and that is why it is often still used.

TDM and Synchronous Communications

It is important to understand that the sort of synchronization in an FDM system is different from the concept of synchronous communications discussed in Chapter 3. In synchronous communications, some type of synchronization is needed to perform the demodulation operation. The synchronous demodulation is not related to the contents of the signal that result from the demodulation. The function of the demodulator is to provide, to the next stage in the receiving system, the analog signal value which was imposed on the carrier at the sending end. By contrast, synchronization at the receiving end of a TDM system is needed to begin to unwind the meaning of the stream of demodulated signal values. The TDM process does not know what type of modulation is being used. Synchronous modulation and TDM synchronization are independent processes in the overall system.

As an analogy, suppose a person is passing written, coded messages through a slot to another person. The messages consist of a long string of characters, with the starting character not clearly specified. In synchronous communications the transmitting person would put the messages into the slot repeatedly over a regular, predefined time period. The receiver would take messages out of the slot over the same period. The result would be that messages would never pile up in the slot, and the receiver would never reach for a message and find out that there were none waiting for pickup. However, in the type of synchronization that is similar to TDM the receiver would need some sort of marker on the message to show where to start reading the received string of characters before deciphering them. Synchronous demodulation produces a stream of signal values. Sync signals are used to point to the starting point when interpreting and using that stream.

It may seem that the method of sending only fragments of each channel, by sequencing in time, is a fraud. Is it possible to send only parts of a signal and still have a usable signal? What is lost by dedicating only a small percentage of the link time to each user? In some high-performance systems, dozens or hundreds of users share the link and thus have direct communication with the receiving end for only a very small percentage of time. Is this meaningful?

The answer is that it is very meaningful. Many signals change at a relatively slow rate, and devoting a full, continuous link (through SDM or FDM) to each one is unnecessary. In the case of the temperature sensors in the building energy management system, this is clearly true. Temperature changes occur relatively slowly, no more than 5° per hour, so sending the temperature value once every few seconds provides all there is to know.

Many other signals have the same characteristic. There is no need to continuously send the signal. Even though the signal is changing in value, there is a rate at which a TDM system can "revisit" that signal channel, send the present value, and then go on to the next channel without losing any new information. The rate at which any individual channel is sent is called the *sampling rate*. The next chapter discusses the relationship between the information signal and the required sampling rate. There is a mathematical analysis which shows that any signal can be fully characterized by samples only, if the sampling rate is high enough. The sampling rate needed can be calculated in advance, if the signal characteristics are known.

TDM Advantages and Disadvantages

Like the space and frequency multiplexing methods, TDM has certain characteristics which give it practical advantages and disadvantages. In general, the advantages of TDM are very significant and often make it the preferred multiplexing method.

One key advantage of TDM is that it uses a single link, like FDM. However, unlike FDM, it does not require precise carrier matching at both ends of the link. Only one carrier is needed for the complete signal. If the system is a base band signal, then no carrier at all is needed. Therefore, many users can share a single link, but without any carrier. This makes it relatively easy to expand the number of users on a system simply and at low cost.

The tradeoff for a TDM system is the need for precise division of time and synchronization, instead of precise carrier frequencies and stability. Before computer systems and digital circuits were developed and made available at low cost, dividing 1 s into thousands of parts, or millions of parts, was difficult. It was even more difficult to do this at two ends of a link. However, modern computer systems and circuitry are very good at precise time interval activities. This means that a technical obstacle to TDM no longer exists.

Another advantage of TDM is that adding users, up to the 100 percent use point, is relatively easy. The system can be designed, for example, to provide for a maximum of 10 users, each with 100 ms from a total 1 s time period. Not all users have to be connected at the same time. Initially, the system can provide for connections to only 5 users. As needs increase, the additional users can be brought into the system simply by having them connect at their respective time slots. No new circuitry is needed at the transmitter or the receiver, and the link does not have to be changed.

There is another advantage to a TDM system. The inherent nature of computer signals is compatible with TDM operation. Computer systems have precise internal clocks to drive their circuitry and to keep all the ICs properly in step with each other. It is relatively easy to use these ICs to perform the multiplexing operation of switching around in precise sequence and time to many inputs. The corresponding operation of sorting out the serial stream of information into many individual signals is also simple for computer ICs.

TDM systems do have some weaknesses. Since there is only a single transmitter, link, and receiver, all users suffer if there is any problem with these. Also, there may be practical limits to the TDM rate that is achievable. These limits can come from the link, which may not be able to handle the desired rate, or from noise and distortion problems in the link which corrupt the TDM signal so that proper synchronization cannot be achieved by the receiver. Finally, keeping the clocks of two systems in synchronization at high rates can be difficult, so the techniques of synchronous modulation and demodulation (discussed in Chapter 3) must often be used.

The trend in many modern systems is to use TDM whenever possible. This is because the advantages of TDM, both in theory and in actual implementation, are very large as compared to those of SDM and FDM in many cases. Some of the early disadvantages of TDM, such as the need to have channel identifiers or a synchronization signal, are overcome at very low cost by modern ICs. The advantages of TDM in systems required to serve many users become even greater when digital modulation, in which the signal being sent by each user can have only certain specific, predefined values, is used.

A TDM Circuit Using ICs

A TDM circuit for analog signals can be built from digital ICs and analog switch ICs. The block diagram of the circuit is shown in Fig. 4-16a. The circuit uses the following functional blocks:

A clock circuit: This circuit generates clock pulses to drive the entire circuit and set the timing of the TDM operation.

A 2-bit counter: This counter counts 00, 01, 10, 11, 00, 01, . . . , repeatedly. This indicates, in binary, which one of the four input signals should be multiplexed at any instant.

A 2-line to 4-line decoder: This IC takes the 2-bit count and decodes it to one of four distinct output lines. The output of the decoder steps through output positions 0, 1, 2, and 3 as it is driven by the counter value.

An analog switch IC: This IC contains four individual on/off switches which can pass any analog signal in the -10- to $+10$-V range, at frequencies up to 1 MHz. Each switch is turned from the off position to the "on" position by a control line. The control line for each switch is connected to the output of the decoder.

The overall timing of this circuit is shown by the timing diagram of Fig. 4-16b. The clock puts out a continuous stream of pulses. These are counted by the counter, which has two output bits to represent the four possible values. With each clock pulse, the counter increments its output value by 1, and the counter

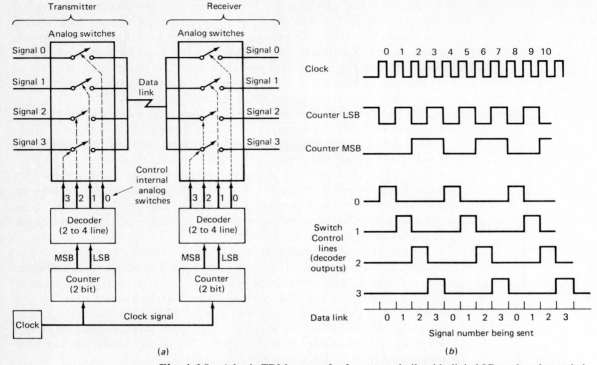

Fig. 4-16 A basic TDM system for four users, built with digital ICs and analog switch ICs. The users all share one data line, and a separate line sends the clock signal: (a) block diagram, (b) timing.

"wraps around" back to 00 after it counts to 11. The MSB and LSB of the counter drive the 2 line to 4 line decoder. This decoder steps through its 4 outputs, putting a logic 1 on the output corresponding to the count value. The result, at this point, is that the decoder is sequencing through the four outputs. With each clock pulse, the decoder puts a logic 1 level on the next one of its outputs and leaves the other three at logic 0.

The logic outputs of the decoder then control the four analog switches. As the switch control line is set to logic 1, the switch is turned on and passes the analog input signal of that user channel to the communications link. All other inputs remain blocked, since their switches are off. With each clock pulse, the next switch in sequence is turned on and the signal passed through to the link.

The receiving circuitry is very similar to the transmitting circuitry, and it uses the same ICs wired the same way. The clock line from the transmitter is used to increment the counter at the receiver, which in turn drives the decoder and analog switches. The analog switch takes the multiplexed signal from the link and parcels it out to one of the four user signal outputs. The four outputs each get a turn in sequence, in the same order as the circuit does at the transmitter. The only difference is that the transmitter selects one of four signals, in sequence, to put onto the data link, whereas the receiver takes that analog data link signal and puts it, in turn, onto the four output points.

The number of users and rate at which they can be served depend on the clock rate. If the clock is set to 100 Hz, then there are 100 time slots per second. Each user will have 100/4 = 25 time slots per second, each with a 1/100 = 10 ms period. If the clock rate is increased to 200 Hz, each user gets 50 time slots/second, but each slot is only 5 ms. More users can be handled by expanding the circuitry, but each user will be affected. Consider changing the number of user signals from 4 to 8, while maintaining the 100-Hz clock. Each user now gets 100/8 = 12.5 time slots/second, and each slot remains at the clock period of 10 ms.

In this circuit, the clock must be sent on a separate wire. More complicated circuitry, such as a phase-locked loop, could be used at the receiver to extract the clock rate from the data itself. Modern ICs also combine the counter and decoder function into one IC, which may also be part of the analog switch IC. More channels of user signals can be handled by using a larger counter, larger decoder, and more analog switches.

QUESTIONS FOR SECTION 4-6

1. What is time-division multiplexing? What is the concept behind it?
2. How does TDM allow a single link and transmitter to serve many users?
3. What is the bandwidth needed by a TDM link as compared to the bandwidth of the individual signals it is multiplexing?
4. Explain applications in which TDM is a better choice than SDM. Why?
5. Why is synchronization needed in a TDM system?
6. What are two ways of performing this synchronization? What are the pros and cons of each?
7. In a TDM system, what is the time slot?
8. Why is the single value to be sent via a TDM system "sampled"? Give two reasons why it is valid to send only these sample values, rather than the continuously varying value.
9. Give three advantages of TDM over other forms of multiplexing.
10. Compare the technical requirements for a TDM system to those of an FDM system. What equipment is required on each end?
11. Why are TDM systems now more practical, technically, than they were 10 years ago?
12. What is the difference between synchronous communications and synchronizing characters?
13. Explain how digital ICs and analog switches can be used to build a TDM system easily. What is the role of the clock signal? How can the clock line from transmitter to receiver be eliminated, if necessary?
14. The analog TDM circuit of Fig. 4-16 must be expanded to serve eight users. How many bits are needed in the counter? How many lines must the decoder have?
15. For the expansion of the TDM circuit of Fig. 4-16 from four users to eight, assume that the clock rate remains the same. What is the difference in the

frequency with which the single user gets a turn on the data link? If the clock rate is doubled, how frequently does each user get a turn, as compared to the four-user system with the slower rate?

PROBLEMS FOR SECTION 4-6

1. A TDM system is used to transmit five signals. Each signal has a 1-s time slot. The signals have the following constant values: 3, 6, 2, 7, and 9 V, within a 0- to 10-V range. Sketch the signal that comes out of the TDM process for 10 s if the modulation process is baseband AM.

2. A TDM system is used for three users, and each has a 1-ms time slot. The modulation used is FM, with a user signal of 0 to 10 V corresponding to a modulated frequency of 0 to 10 kHz. Sketch the first 5 ms of the TDM process output if the three users have constant signal values of 2, 4, and 3 V.

3. A TDM system uses an identifying code for each user of a five-user system. Each user is connected by the time division multiplexer for 20 ms, and the identifier takes 20 ms.

    a. What is the fastest rate at which each user can have access to the link, if all five users must be served?

    b. The identifier time is reduced to 10 ms. Can each user have access to the link at a rate of 10 times/s?

    c. Sketch the output of the TDM process for the first 200 ms, identifying each user A through E and the user identifier time periods, for case **a**.

4. A TDM system uses a single synchronization signal for the entire group of users. The system supports 10 users, each of whom has a 10-ms time slot. The sync signal itself is 20 ms.

    a. What is the total time needed to send the sync signal and all 10 user signals?

    b. What percentage of the time is occupied by a sync signal? What is the efficiency of the system?

    c. Compare the result in case **b** to the case in which each user has its own identifying code, only 5 ms long. What is the efficiency of this scheme?

    d. Sketch the user time slots for the 10 users A through J and the sync signal, for 150 ms, for part **a**.

5. The analog TDM circuit of Fig. 4-16 must be expanded to handle eight user analog signals. Draw the timing diagram, showing the clock, counter output, decoder output, and data link.

6. The clock is set to 1 kHz for four users. How many times per second does each user put its signal on the data link? For what period of time?

7. The clock rate is doubled to 2 kHz. How many times per second, and for what period of time, does each user have the data link?

8. Repeat Probs. 6 and 7 for the TDM system increased to eight users.

4-7 Combined Modulation Systems

The wide range of user needs and the various physical separations of communications users often necessitate that the overall user-to-user system be composed of several independent links. At each link, the incoming signals are demodulated and then combined with those of other users. The newly combined group is then modulated again, often in another form.

The purpose of this process is to provide the most efficient total system for the various numbers and groups of users. This efficiency means best technical performance, lowest cost, and maximum flexibility to handle differing numbers of users.

Once again, a good example is the telephone system. Since the telephone was the first general-purpose communications system and a telephone system requires maximum performance and flexibility to satisfy the large number of users, the telephone system has been a pioneer in many advanced communications system designs. It utilizes numerous forms of modulation and can choose the best link to suit the needs of the particular users and distances.

Suppose a telephone user in New York is speaking with a user in Los Angeles. There are several stages in the overall connection (Fig. 4-17):

An SDM path from the New York phone to the New York local central office for that phone

An FDM link combining the many long-distance users of this central office to the next level, the regional office

A TDM, very high performance and distance link, via microwave links, from the East Coast regional office to the regional office on the West Coast

Another TDM link, connecting the long-distance users from the West Coast regional office to the central office for the Los Angeles user

An SDM connection from the local office to the user in Los Angeles

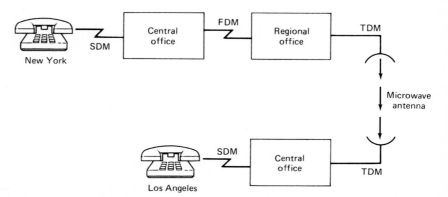

Fig. 4-17 A long-distance link usually involves many types of modulation from one stage to the next.

Analog Communications and Multiplexing

Note that different users can be combined or split apart at various points along the way. The regional office in New York, for example, groups many users from the New York area who are connecting to the West Coast. The West Coast regional office then sends the numerous signals it receives to the proper local offices in its region. The role of the various links is to allow the overall system to gather together many users who are going to the same destination, modulate them as a group of users, and send them down the link. At each stage, new users can be combined into the group, and other users split off to go down a different path (Fig. 4-18).

There are many ways that the same end-to-end connection can be achieved. The links between the regional office on the West Coast and the various local offices may be FDM or TDM (it is unlikely that SDM would be used, because the number of wires needed would make it very unwieldy). The choice would be made by the telephone system designers. The user message passes from one link, through the demodulator, and onto the next modulator without the user realizing it.

Combined modulation systems can also use the same type of modulation for successive links, but implemented with some different specific frequencies or time slots. In a system which must accommodate many users with great flexibility, such as the telephone system, this is often the case. For example, two independent groups of users may be multiplexed together by using the identical multiplexing method and technical specifications, and then these two groups in turn may be multiplexed together. The phone system does this, in some cases, by using an FDM system at each local office to connect many users to a regional office. Then, several regional offices go to a larger regional office and are combined via FDM and different carrier frequency into a larger group, for transmission across the country (Fig. 4-19). In this way, the identical multiplexing equipment can be installed in the various local offices, a great advantage from an engineering and maintenance perspective.

This is an important point in a modulation and communications system. Even though the original user signal is modified by the modulator as it passes through

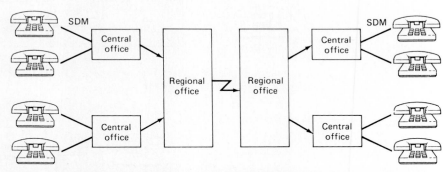

Fig. 4-18 In large communications system often various users and user groups add on and drop off along the way. The overall goal is to make maximum use of the channel capacity while preserving flexibility.

106   Chapter 4

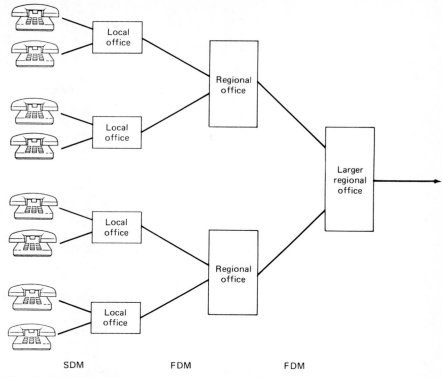

Fig. 4-19 The telephone system uses several levels of multiplexing from smaller groups into ever larger groups. The result is to make the best use of the office equipment and have all local offices and all regional offices have identical equipment.

the FDM or TDM system, it is restored to original shape and form by the demodulator. This means that the link is *transparent* to the user signal. So long as the user signal fits within the general specification for amplitude range and bandwidth, it is carried by the system without regard to its specific contents. The link is a transparent window that the message can pass through. The modulation and demodulation system does not actually examine the contents of the message to see what it contains. It simply applies the modulation operation in a fixed, constant way to the message and signal. This transparency means that the message can be plain English, or it can be in some sort of code that is compatible with computers, yet it makes no difference.

As communications systems become more complex, the concept of transparency becomes even more important. A transparent link allows users to have great flexibility, because they can put the message into a form that is most convenient, without worrying that the communications link will change that message. If users want to send a list of numbers, followed by their total, the link will handle that. Conversely, if they want to send the total first, and then the actual list of numbers, the link will pass the message through in that form.

Generally, the modulation and demodulation circuitry is designed to be transparent. This does not mean that every part of a total communications system must be transparent. As part of the system design, some parts of the overall system may modify or encode the message to make it more suitable to the kind of performance needed. This process of encoding is discussed further in the next chapter.

QUESTIONS FOR SECTION 4-7

1. Why are combined modulation systems needed? Give an example.
2. What happens to the various system users at each stage of the demodulation and remodulation process?
3. What does the term "transparent" mean?
4. Why is transparency important to users of communications systems?
5. Does every section of a total communications system have to be transparent? Explain.
6. How does the process of demodulating the incoming signal and then modulating again accommodate a complicated system that has many users who want to go in the same general direction, but with slightly different starting points and destinations?
7. Explain how and why groups of users are identically multiplexed, then combined with additional multiplexing.

PROBLEMS FOR SECTION 4-7

1. A multiple link system uses SDM to connect three users (A, B, and C) to a central point, and then these users are connected by FDM to a central receiving point where they are passed to a single computer.
    a. Sketch, with a block diagram, what this system looks like.
    b. Show the time domain graph of the SDM link.
    c. Show the frequency domain graph of the FDM link, assuming each user has a 5-kHz bandwidth and carriers of 10, 20, and 30 kHz are used.
2. Using the preceding situation, the single computer distributes the received signals from users A, B, and C to users D, E, and F, respectively, using TDM.
    a. Sketch the block diagram of this new part of the system.
    b. Show the time domain graph of the new part of the system for the signals going to users D, E, and F.
3. A communications system for long-distance transmission combines three users at a local office, by using FDM with carriers at 10, 20, and 30 kHz. Another local office, nearby, also combines three additional users with the same carrier frequencies. These two groups then go to a regional office, where they are combined into wider-bandwidth, longer-distance FDM links using a carrier at 1 MHz for the first group and 1.1 MHz for the second group.
    a. Sketch the block diagram of the system.
    b. Show a frequency domain graph for this system, after the initial grouping of the three users and the overall combination of the two groups.

4-8 Shortcomings of Analog Communications and Multiplexing

The weaknesses of analog communications and multiplexing are a logical result of the weaknesses of using analog signals to convey information. The greatest problem is noise. Noise can corrupt the desired signal and reduce its accuracy. It can also cause problems for the receiving circuitry if the demodulator has to synchronize to the received signal.

The noise problem for analog communications has a greater effect on TDM systems, a lesser effect on FDM, and least effect on an SDM system. In a TDM system, the desired signal is available only for a short time period, so that there is really only one chance for the receiver to get it correctly. If noise corrupts the signal at that instant, there is no second chance, no looking at the signal a short instant of time later (Fig. 4-20a). The noise may also affect the synchronization scheme, and in some cases cause the sync signal to be so corrupted that the receiver cannot synchronize. In that case, all data between the corrupted sync signal and the next one is lost.

In SDM and FDM systems, noise has an effect only on one or a few users in most cases. This is because the SDM system has a separate link for each user. Noise often affects only a small range of frequencies, so FDM systems are less affected. For both SDM and FDM systems, the user can check the received values several times in rapid succession, take the average value, and thus minimize the effect of the noise (Fig. 4-20b).

Averaging signal values is a common technique. For example, if the original value were 7.3, and the receiving system were concerned about noise, the system might take 10 readings of the received value. Suppose the received values were 7.0, 7.5, 7.2, 7.4, 7.3, 7.3, 7.7, 7.1, 7.0, and 7.4. If the receiver looked at any one value, it could be inaccurate by as much as the difference between 7.3 and the noisiest value received, if the noisy one just happened to be the one chosen. By averaging, the largest-value noise signals are smoothed out numerically. For the numbers given, the average value is 7.29, very close to the correct value of 7.3.

It is even difficult to define the amount of the error. Should the error be the largest deviation from the true value? Or should it be the average deviation? The "correct" definition of the error depends on the application—no single definition is inherently better than another. For the numbers in the previous paragraph,

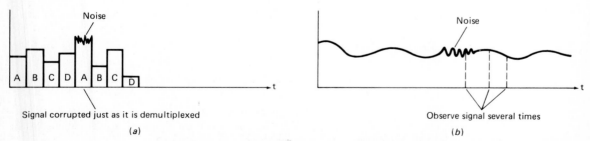

Fig. 4-20 The effects of noise in (a) a TDM system and (b) a SDM or FDM system.

the maximum error is 7.7 − 7.3, or 0.4, although the average error is 7.3 − 7.29, or 0.01. For some users, the concern is the worst error that occurs, but others only care about the average error since individual values may not be so important to their needs.

An analog communications system must also be designed to handle properly the full range of analog signal that is expected, from minimum value (usually 0) to maximum value. The multiplexing system must not introduce any distortions or nonlinearities as it combines the various user signals (whether by SDM, FDM, or TDM), and the demultiplexer at the receiving end must also be very accurate. Any imperfections in the multiplexer, transmitting circuit, demultiplexer, or receiving circuit are immediately passed on to the user as the actual received signal, and these errors and inaccuracies cannot be separated out by the user. The errors and inaccuracies become a part of the overall received signal and are indistinguishable from the correct values that would have been received if the system had been perfect.

Some applications can tolerate these problems, and others cannot. An audio signal usually can accept distortion, since the sound (voice, music, or message) is at least understandable (although not so clear as desired). However, an analog system carrying a video signal is much more sensitive to distortion since visual problems are easily seen by the user. For example, if the picture being transmitted shows straight lines as part of the total picture, system distortion problems may cause some of these lines to be curved, wiggly, or not so straight as originally intended. The eye of the user can easily see this problem and be confused by it—should there be a straight line there? Or was the system really sending a slightly curved line (Fig. 4-21)?

Analog systems do not offer opportunities for accommodating and even identifying where the received signal has been corrupted. In the case of SDM and FDM, the system is sending a continuous flow of analog values, which can take on any value within the total range. Perhaps there is some way to put into the system a special analog value which could be used as a check on the correctness of the received values. It turns out that this procedure is quite difficult to implement in practical systems. For critical applications, in which the message

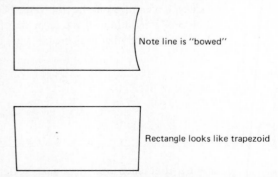

Fig. 4-21 Some types of information are more sensitive to noise and distortions. Video problems are easily noticed by the viewer.

must be received correctly. two users are sometimes assigned to carry the same message. However, this is expensive and doesn't really guarantee that errors are identified. Both signals could suffer from the same or different distortions that would affect their signals; using two signals would not ensure that one was perfect.

Two channels can be used to give some additional confidence in the received signal value, but not assurance. Perhaps the correct value is 5.4. If there are no problems, both channel A and channel B have 5.4. But both channels may be identically affected by noise, and receive 5.2, possibly leading the user to conclude that 5.2 is the correct value. What if one channel has 5.4, and the other 5.3? Which is correct? And what if one channel has 5.3, and the other is 5.6? Then neither is correct, and the correct signal value lies somewhere in between.

If the signal values represent some sort of text, it is often possible to determine what is wrong from the context: "I had a cit" should be "I had a cat," but couldn't "had" also be "hid"? Some errors can be determined from context, many cannot. For signal values that represent numerical data, of course, it is virtually impossible to decide from the adjacent numbers.

As an example of the difficulties in making an error determination, imagine an analog system that sends a 1-s message. Within the 1-s period, the signal voltage varies continuously from 0 to 1 V. What methods are there of ensuring that the message is received properly, or that errors are recognized?

Send the message again, after the first time, or send it in the multiplex system as another user on the system, in a parallel fashion. This takes system time or uses another user link.

Send a special signal after the message signal. The value of this signal could be equal to the average of the message signal values, but an average does not indicate what the exact previous values were—it is only a summary. (Why?)

Send another pair of signals equal to the largest value of the message, and then the smallest. These could be used to "calibrate" the communications system, to see how it performs at either extreme of its capabilities. However, the system problems and noise may vary with time, and thus the calibration factors should be different for the period when the actual message is sent.

A well designed, properly installed analog system can give good results in many applications, if the circumstances of its use are within the design goals. If the noise or distortions are greater or more frequent than the system was designed to handle, the performance can suffer as the results become severely degraded. For some applications, such as voice transmission, this is an acceptable situation since the listener can ask the speaker to repeat the message and speech contains many built-in redundancies that allow the listener to understand the speech that is not clear or noisy. However, many applications cannot accept this kind of performance. If the message is critical, involves exact numbers, or contains information (such as temperature values) such that the receiver cannot tell from the message context what the right interpretation should be then an analog system may fall short of the need.

QUESTIONS FOR SECTION 4-8

1. What is the general weakness of analog communications? How is SDM affected by noise? How about FDM?
2. What are the two ways that TDM can be affected by noise? Why is this serious?
3. How much can be lost if sync signals are not properly received in a TDM system?
4. What is signal averaging? How is it used to minimize the effects of noise?
5. What signal range must an analog system handle? Why is this difficult?
6. What types of signals and users can accept some distortion? What types cannot?
7. How can distortion and error be measured? Why is there no simple, single answer?
8. What is gained by sending another signal value, equal to the average of the previous values? What is the weakness of this method?
9. What are the pros and cons of sending the same information redundantly?
10. What are calibration signals? What are their good and bad points?
11. Give two examples of information that cannot be derived from the surrounding context of information. How does this relate to communications errors and problems?

PROBLEMS FOR SECTION 4-8

1. An analog system receives the following signal values for a signal that is a constant 4.2 V: 3.8, 3.9, 4.7, 4.1, 4.2, 4.3, 4.5, 3.8, 3.8, and 4.0 V.
   a. What is the maximum error in volts?
   b. What is the average error, if the 10 readings are averaged?
   c. Sketch the received signal and sketch the actual signal value on the same graph.
2. An analog system uses three parallel channels to redundantly transmit the signal from a single user to another user. The transmitted signal voltage is 5.0 V.
   a. Sketch the system.
   b. The received signal values are 5.0, 5.9, and 4.9 V.

   If the receiving users employ redundancy to estimate the correct value, what do they determine to be the correct value of the transmitted signal?

   c. The receiving user instead averages the three readings. What is the value that is estimated this way?
3. A user requires accuracy to 1 percent. How much error can be tolerated if the actual signal value is 5 V? If the actual signal value is 0.1 V? If the system provides a guaranteed error of less than 100 millivolts (mV), will this user be satisfied over the analog range of 1 to 10 V? Explain.

SUMMARY

Analog communications systems carry signals which can take on any value within the overall range allowed. They are more general than digital systems, which are designed to handle only specific signal values. Analog systems are often used to carry digital signals as part of an overall communications system.

In practical systems, technical and cost reasons require that more than one user share a system. This may be in an application in which all users are going from the same starting location to the same destination, or it may be one in which users are gathered at one end to use a common path to the other end, and then are dispersed at the receiving end (the telephone long-distance network, for example). Multiplexing allows this sharing to be done.

In space-division multiplexing, a separate wire or cable is run for each user. This is technically straightforward, but very costly. Frequency-division multiplexing assigns each user a specific carrier frequency. The users are then modulated onto their carriers (by AM, FM, or PM) and are able to share the path from the transmitter to the receiver without interfering with each other. In recent years, time-division multiplexing has become practical. In TDM, each user is assigned a specific time slot, and is sent, in sequential order, over the link. This allows many users to share the same transmitter, link, and receiver. TDM can be technically and economically attractive and is compatible with modern computer circuitry.

The key to the validity of TDM is that a signal does not have to be sent continuously in order to convey all of its information. Just sending samples, at a frequent enough rate, can provide all the information that the signal contains. The rate that is required is determined by the nature of the signal.

Most communications systems are built up from a series of subsections which are joined together. Each subsection may use a different form of modulation in order to provide the best performance for that particular part of the total system. Joining various subsections allows the total communications system to provide the flexibility and versatility that are needed in the application. SDM, FDM, and TDM may be used in various combinations in the completed communications system. Of course, many simpler systems only need one form of modulation and carry the signals of a group of users from one point to another without combinations of modulation or multiple subsections.

The greatest technical problem with an analog communications system is noise. Noise that comes from any part of the system—the modulator, the transmitter, the communications link, the receiver, or the demodulator—can corrupt the signal that the user sees and prevent the user from accurately estimating the value of the original signal sent. In an analog system, it is difficult to have accuracy greater than about 1 percent, which may be acceptable for applications such as voice and music but may be too great an error for scientific data and numerical information. There are techniques that can be used to determine whether noise has corrupted the signal. These include signal averaging, sending signals redundantly, and sending a checking signal. None of these methods is practical or effective for analog communications systems.

END-OF-CHAPTER QUESTIONS

1. Give three examples of analog systems in use today.
2. Explain why studying analog signals and modulation involves a more general case than studying only digital systems and modulation.
3. How does a digital signal, with noise added, come to look like an analog signal?
4. In a total, end-to-end communication system, why are multiple stages, or links, needed at each end? What happens to the signal at each stage?
5. What is signal distortion? Is it meaningful for music and voice? Is it noticeable? Give an example of a system distortion and the problems it can cause.
6. A system is designed to measure the motion of a sliding part of a machine shop lathe, and to display this to a central operators' room several hundred feet away. The lathe and room are linked by wire. What are the typical stages of a communications system designed to display this single quantity?
7. The system in Ques. 6 is to be upgraded to show two additional physical variables from the lathe. Should SDM, FDM, or TDM be used? Discuss three factors in making the decision.
8. The readout of the lathe position needs to be accurate to 2 percent. Can an analog system do the job? What if the readout had to be accurate to 0.02 percent?
9. Explain the key difference between SDM and FDM, and the similarities.
10. Explain the differences and similarities of FDM and TDM.
11. Explain the the differences and similarities of SDM and TDM.
12. Why was SDM the first form of multiplexing? Why was FDM next? Why was TDM the last form to come into widespread use?
13. What does the demultiplexing system do for SDM systems? For FDM systems? For TDM systems?
14. What is a very common system which uses SDM? Why?
15. Explain why using FDM does not restrict the actual modulation to FM.
16. How would you add another user to an existing SDM system? To an FDM system? To a TDM system? What new equipment is needed at each end?
17. What is the general relationship between the bandwidth of a single-user signal and the bandwidth of many signals that have been multiplexed? Why doesn't the final bandwidth remain the same?
18. Explain where uneven FDM carrier spacing would be used. Where would even spacing be used?
19. How does a TDM system manage to convey all the necessary information from a single user, even though the user signal is only sampled?
20. Why is the concept of time slot important in TDM? What does the slot represent?
21. What is the need for synchronization in TDM? What are the two common ways of achieving this?

22. Explain why TDM and modern computer systems are good matchups for each other.
23. Explain the need, in many communications systems, for multiple links along the way from one end to the other.
24. Why do systems, such as those discussed in the previous question, use different forms of modulation for different links?
25. What is a transparent link?
26. Is a communications system which transmits all numbers except those less than 0 transparent? Explain why or why not.
27. Explain the effect on a signal value being transmitted by an SDM system when noise occurs. What about in an FDM system? A TDM system?
28. What is the effect on the user at the receiving end if the signal received is corrupted by noise? Is it always a problem? Explain.
29. What are two of the techniques used to minimize the effects of noise in an analog system? How well do these techniques work?
30. When can the correct value of a signal be determined from its context? Give an example of where it can and cannot.
31. Explain why even defining the amount of error in an analog signal can be difficult.
32. What is the difference between synchronous demodulation and sync signals? Are they related? Give an example different from the one in the text, to illustrate this difference.

END-OF-CHAPTER PROBLEMS

1. A system is being designed to handle five pairs of users. Sketch the system, using block diagrams, for implementation with:
   a. SDM
   b. FDM
   c. TDM
2. Assume the user signals in Prob. 1 are essentially similar. Sketch:
   a. The time domain representation of the SDM signals
   b. The frequency domain output of the FDM process
   c. The time domain output of the TDM process
3. Standard television signals occupy 6 MHz, of which the major part is for the video signal and a small part is for audio signal for that particular video signal. If the video occupies 5.9 MHz, and the audio has the remaining 100 kHz, sketch the frequency domain representation for three adjacent stations.
4. A TDM system is designed to handle five user signals as shown in Fig. 4-22, with the time slots shown.
   a. Sketch the output of the TDM process for these signals.

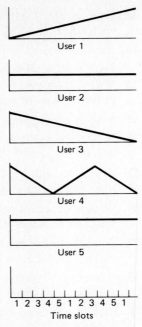

Fig. 4-22 Figure for End-of-chapter Prob. 4.

    b. Repeat for the case in which the time slots are half as wide. What can be determined about the value of user A's signal when it is not user A's time slot?

5. A system is designed to handle 25 users via TDM. Each user signal takes a 1-ms time slot. The user identifier takes 2 ms.

    a. Sketch the output of the multiplexer system for the first 10 users.

    b. What percentage of the time is devoted to user identifiers? Does it vary with the number of users?

    c. How many times per second can any given user be transmitted?

6. A system is designed to handle 25 users with TDM, and each user has a 1-ms time slot. A complex sync signal precedes each group of user signals, and it is 25 ms long.

    a. Sketch the output of the multiplexer system for the first 10 users.

    b. What percentage of the time is devoted to the sync signal? Does this percentage change if the number of users is increased?

    c. How many times per second can any given user be transmitted?

7. Complete this table, using general terms like "high," "low," "simple," and "difficult," for SDM, FDM, and TDM.

		SDM	FDM	TDM
a.	Use of link capacity			
b.	Initial cost			
c.	Complexity for adding users			
d.	Cost for additional users			
e.	Sensitivity to link problems			
f.	Sensitivity to other user problems			
g.	Technical complexity			

8. Three telephone central offices each take five signals and combine them, using FDM carriers of 100, 120, 140, 160, and 180 kHz. These combined user signals then go to a regional office, where the three groups are multiplexed, using FDM and carriers at 1, 1.25, and 1.5 MHz. Each user signal has a 5-kHz bandwidth.

  a. Sketch the frequency domain representation of the signal leaving each central office.

  b. Sketch the frequency domain representation of the signal after the regional office multiplexing.

9. A signal of 5.0 V is sampled at the receiver to minimize the effects of noise. Five samples are taken, with values 5.1, 4.8, 5.2, 5.0, and 4.9 V.

  a. What is the error if the average value is used to estimate the correct value?

  b. What is the maximum error if only one sample is used?

5 Digital Communications

This chapter discusses the transition from an analog communications system to a digital one. A digital communications system offers many advantages to the user that cannot be achieved with an analog system. It also has some important differences in design that simplify the job of the system designer. Present-day integrated circuits can be used to implement the requirements of digital systems effectively and at low cost.

A digital system that is handling signals which were originally analog must first sample the analog signal and convert it to digital format. There are theories and guidelines for ways this sampling should be done, and ways the analog-to-digital conversion process should be implemented. Digital signals may also have to be encoded, so that the stream of data being sent maintains some desired general properties. Multiplexing and modulation of digital signals are related to the same operations on analog signals, but with some differences and simplifications.

5-1 Description of Digital Systems

Digital communications systems are used to meet the needs of most new communications applications. A digital communications system can be as short as the distance between two integrated circuits (ICs) of a single computer chassis; it can be as long as the distance around the world or to a space research satellite. The enormous growth in the use of computers has been a large part of the need for digital systems, since computers are inherently digital. At the same time, new technologies and capabilities that computers and computer circuits provide have made digital communications systems more practical and inexpensive than even 10 or 15 years ago.

Digital communications systems may make use of analog links and concepts. However, they are a specific subgroup of the general-purpose analog system. Any time a system is designed to handle a wide range of signals, there are tremendous technical demands on every piece of the system. With a digital system, the system design can be specifically oriented to getting the best performance possible with a certain type of signal. Digital signals provide this opportunity because they are simpler than analog signals in many respects. Besides this, using digital communications opens new possibilities in overcoming some of the weaknesses of analog systems as discussed in the previous chapter.

Exactly what is a digital system? A *digital system* is designed to handle only *digital signals*, electrical signals which can take on only certain predefined values, instead of any value within a whole range. The entire communications system can be optimized to deal with these values and not be concerned about other values. The following sections of this chapter discuss in more detail all the distinct benefits that this provides, as compared to analog signals, in terms of errors, error detection, noise, accuracy, and essentially ensuring that the information that the user at the receiving end gets is exactly the same as the information the sender transmitted.

A digital system is a more general case of a binary system. In a *binary system*, only two signal values can exist. These are often called *0* and *1*, *low* and *high*, or *off* and *on*, but these names represent specific voltages (or currents). It is easier to call them by their names than the actual voltage levels. All modern computers use the binary system, because it allows certain technical advantages to be used to provide high performance at low cost.

A digital system allows more than just two values of the binary system. A digital system may use 4, 8, or even 16 unique values (Fig. 5-1). In effect, the digital system is somewhere between the binary system and an analog system. Why are multilevel digital systems, instead of the binary system of computers, used in communications? The reason is that the channel efficiency that can be achieved may be too low with a simpler binary (2-level digital) system. A binary system does not make use of as much of the channel potential as a 4-, 8-, or 16-level system can provide. Since getting the maximum channel efficiency (*efficiency* is the amount of data that can be sent through a channel in a fixed amount of time) and low-error performance are major goals of most systems, a communications system may use binary in some places and multilevel digital in others.

Digital communications systems use various types of signals and modulation to convey the information they are carrying. Various voltage levels (for amplitide modulation [AM]), frequencies (for frequency modulation [FM]), and phases (for phase modulation [PM]) are used and even mixed to provide the best combination of signals for the channel. Some examples are as follows:

1 3, 5, 7 V	four-level amplitude
0, 1, 2, 3, 4, 5, 6, 7 V	eight-level amplitude
1000, 2000 Hz	two-level frequency
1, 2 V; 1000, 2000 Hz	two-level amplitude and two-level frequency
$+90°, -90°$	two-level phase
$+45°, -45°$ 1, 3 V	two-level phase and two-level amplitude

Special circuitry is used in the modulating system to convert the initial digital

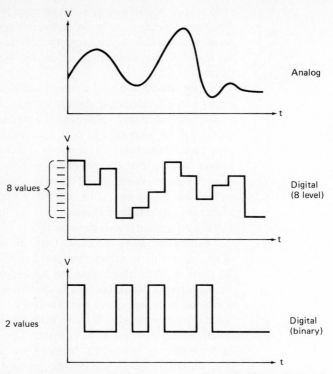

Fig. 5-1 An analog signal can take on any value within the range; a digital signal takes on only certain specific allowed values. A binary signal is a digital signal which has only two allowed values.

form to the type and combination needed for the digital modulation process. At the demodulator, circuitry performs the reverse operation and converts the digital amplitude, frequency, and phase operation to signals (usually voltages) representing the original digital levels (Fig. 5-2). For example, a computer which uses binary signals (two amplitude levels) may be sending messages with a system that is designed for two-level phase. The circuitry converts the amplitude values to the corresponding phase values.

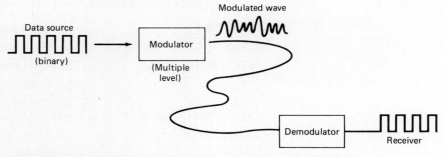

Fig. 5-2 The modulator and demodulator combine to encode the binary data and to retrieve it.

The term "data" is commonly used in digital communications systems. *Data* is any form of information that has been put into digital form so that it can be handled by a digital system. Data has some clear features: it is easy to measure, quantify, analyze, and manipulate. It can represent numerical values, alphabet letters, or special symbols. In an analog system, information is conveyed, but it is hard to measure and analyze the amount of information carried by an analog system and to determine the quality of the received information as compared to what was sent. In a digital system, the information or data is in digital form, which can be precisely and easily dealt with by the user.

The data itself is measured as bits ("bits" is a contraction of the term "binary digit"). A *bit* is a single value that can occur in a binary system. If the symbols 0 and 1 are used to represent the voltage values of a binary digital system, then a typical group of data information would be 01101001. This is an 8-bit group, which is called a *byte*. Even multilevel digital communications which are not binary use the term "bit," because it is possible to translate the information from one form to another. Since all computers are binary, the overwhelming proportion of data sent starts out as binary data, even if it is converted to some other number of levels as part of the data communications system. Nearly all measures and discussions of data communications system performance involve the use of bits.

What a digital system does is change the nature of the problem at the receiving end of the system. Recall that in an analog system, the system goal is to provide the best estimate possible of what the sender actually transmits, based on the value of the received signal, noise, and other disturbances or channel problems. In a digital system, the problem is changed from an estimation problem to a much simpler one, that of detection. In a detection situation, the receiver only has to detect that a signal has arrived and put the signal into the correct "bin" of allowable received values. For a binary system, the receiver has to detect only one of the two possibilities, for example. Suppose a binary system is designed to handle signals that are either 0 or 10 V. When a signal is received, the receiver does not have to estimate whether the value is 7.24 or 7.26. It only has to detect the signal and easily determine that the signal is closer to 10 V than it is to 0 V. Certainly, this is an easier activity. Because of this, digital systems can be simpler than analog ones, and there are fewer pieces of temperamental equipment that have to be fine-tuned for high-accuracy, low-distortion performance.

QUESTIONS FOR SECTION 5-1

1. What is a digital system?
2. Compare a digital signal to an analog one: what is each, and how are they used?
3. Compare a digital system to a binary one: what are the similarities? The differences? How does a digital system compare with a binary one and with an analog one?
4. What are two goals of most communications systems?
5. Give examples of four-level AM, two-level FM plus two-level AM, two-level PM digital systems.

6. What does the term "data" mean? What does data represent?
7. What is a bit? A byte?
8. What does a digital communications receiver have to do, as compared to an analog signal receiver?

5-2 Advantages of Digital Systems

Digital communications systems are used because they provide many distinct advantages to users and many choices for system performance improvement. Many systems designed for data communications do not need to make use of all these possibilities, but instead use the ones that fit the specific communications situation. For example, a system that is being used to send voice signals as digital data may not need a lot of error detection circuitry, since a small error in the data does not cause a problem for the listener. However, if the same system is to carry numerical information from a bank, then any error is a concern, and the data communications system uses more extensive error checking.

The greatest single advantage of digital communications is the resistance to system noise that the digital signals provide. Most communications systems, with very few exceptions, must be concerned about the effects of noise. For some systems, this is a minor concern, because of either the type of information the system is carrying (voice) or the fact that the designer knows that the system will experience very little noise. However, a data communications system designed to work over long distance, over regular telephone lines, by using radio links, or even within a single chassis, must assume that at some time a large amount of electrical noise will be present, even for a short moment. The designer must study what the effect of this noise will be and make sure the system performance is still acceptable and especially that the communications system does not "crash" (functionally fail) and not resume operation.

The sources of the noise can be internal to the circuitry at either the sending end or receiving end, or it can be caused by the nature of the link between the transmitter and receiver. In general, noise within the circuitry is much less a problem than link noise. The circuitry noise can be controlled with proper system design by the engineer who provided the circuitry. However, the link itself may be subject to noise that is beyond the control of anyone using or designing the system. Link noise comes from nearby electrical interference from motors, overhead lights, and machinery. It can also result from electrical discharges such as lightning and static electricity, and from unrelated sources of electrical energy such as TV and radio transmitters. The system powerline itself is also electrically noisy and can induce noise in the system. In summary, noise is an unpredictable, chaotic signal that is imposed on the desired signal and makes system design and performance difficult.

Digital systems provide noise resistance because the noise that is added to the digital signal is usually not enough to make the original signal level look like another one of the allowed signal levels. If the system is a binary digital system using levels of 3 and 7 V, for example, the receiver sets a boundary at the 5-V

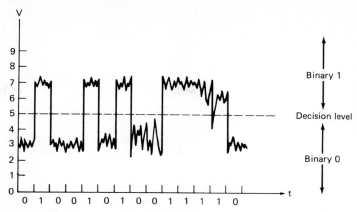

Fig. 5-3 A decision level is used to decide whether a binary 1 or binary 0 has been received, since the original signal voltage is corrupted by noise.

level. Any signal below 5 V is considered a 3-V signal, and any above 5 V is considered a 7-V signal. Therefore, noise of 2 V or more has to be present before the decision-making circuit makes a wrong decision on the value of the transmitted signal (Fig. 5-3).

Digital signals can defeat the effects of noise by using signal *regeneration*. As a signal, analog or digital, goes over a wire or any link, two things happen: the signal picks up noise from various sources, and the signal loses some of its strength, or voltage, as a result of resistance in the wire, the nature of radio waves, and other factors (Fig. 5-4). In a communications system that uses many links, amplifiers are usually used at each node between links to boost the signal back up to the original value. For analog signals, this process of amplifying also boosts the noise, which cannot be distinguished from the true signal. In a digital

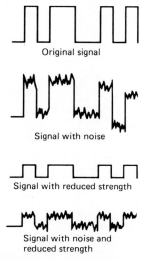

Fig. 5-4 Even digital signals are corrupted in the real world, by noise added to the signal and by attenuation from wire resistance, signal fading, and other causes.

system, the circuitry is designed to handle only the voltage values that the system uses. Therefore, a digital system can use "regenerative amplifiers" or signal buffers, which restore the signal to the correct original digital value. The effects of noise and signal loss are therefore nullified by each amplifier stage. The result is that noise and signal loss never become so large that the original digital signal value cannot be recovered.

Consider the example in Fig. 5-5, which shows a cable that carries phone signals across the United States. Because of signal loss, amplifiers are placed every 1000 mi, so there are two amplifier stages between the transmitter and receiver systems. The digital signals used in this case are $+10$ and -10 V. As the signal goes through the cable and approaches the 1000-mi amplifier, the voltages have been attenuated, or reduced, to only $+5$ and -5 V, and there is ± 2 V of noise on the signal as well. This means that the signal which was originally $+10$ V is down to $+5 \pm 2$ V, or between $+3$ and $+7$ V. The signal which was -10 V is reduced to -5 ± 2 V, or -5 to -3 V. At the amplifier, the circuitry is designed to restore all signals between 0 and $+10$ V back to 10 V and to restore all signals between 0 and -10 V to -10 V. This has the effect of making the signal coming out of the amplifier and going on the next 1000-mi link as good as the original signal. These amplifiers act as repeaters which correct and signal problems that have occurred, before they become too severe. This technique, of course, can only be used with digital signals, for which the circuitry can decide that any signal in a certain range really had a specific known value at the transmitter.

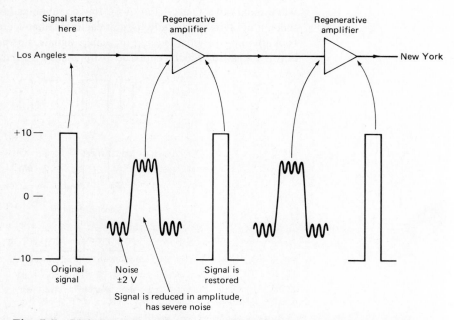

Fig. 5-5 Digital signals can be easily restored to their original noise-free values by regenerative amplifiers. Here, a binary signal is corrupted with distance but restored before the corruption makes the signal useless.

This process of restoring signals with amplifiers does not apply only to long-distance communications systems. Even within a single computer installation similar problems can occur. A printer connected to a nearby computer may have some of the data bits corrupted by noise and attenuation. However, special integrated circuits called *buffers* perform the regeneration process and completely restore the computer signal before the printer sees it.

Digital systems are clearly noise-resistant, but if the noise is great enough, they too can make errors. This leads to the next advantage of digital systems, the ability to allow the implementation of a scheme that can determine that errors have indeed occurred. With a digital system, the information is sent as data: a long string of numbers. It is technically possible to add special checking numbers to this string. Various types of checking numbers can be added, depending on how carefully the system needs to check for errors. Using these checking numbers, the receiver can determine whether it has correctly determined what was sent. If the receiver decides that it has made a mistake, the user can be informed that the received data has an error. The communications system can also be designed to automatically inform the sending circuitry that an error has occurred and that the last group of data values should be sent again. Some very advanced methods of determining whether an error has occurred can also detect where and what the error is and actually correct it automatically at the receiving end. This error detection and correction technique is very powerful for systems that need absolute data integrity.

The use of digital signals to carry the data also overcomes the measurement problem of analog systems. Recall that one problem in an analog system is evaluating how good the system performance is. Is the inaccuracy 1 percent, 0.1 percent, or some other percentage? How is this measured—by the average inaccuracy, the peak inaccuracy, or some other method? With a digital system, it is much simpler to evaluate. The value that is determined at the receiver is either right or wrong. Therefore, the system performance can be measured and specified. Suppose the sender wants to send 10 bits of data, and 1 bit is received in error. Then the error rate is 10 percent. A common measurement in data communications is the *bit error rate*, which defines what percentage of bits are in error from the point where the data originates to the point where the user at the receiving end gets it. A good data communications system can provide error rates of less than 1 or 2 in 1 million, or 0.0001 to 0.0002 percent. Not only is the bit error rate low, but it is measurable and definable without confusion and ambiguity.

Additional Advantages

Once the information to be transmitted is in digital form, it is also easier to manipulate and process than purely analog signals. The system may want to check, for example, for certain received signal patterns that indicate problems either at the transmitting side or at a link of the channel. To do this, the receiver circuitry may have to perform a series of numerical calculations with the received data. This is practical with digital systems, but not with analog ones. In addition, it is easier to perform additional manipulations on the data to improve the received signal. Techniques such as averaging are much more straightforward to implement in the digital situation.

A data communications system may be sending temperature data from inside a furnace to an operator in a control room. The receiving circuitry may be designed to check the successive temperature readings to see whether there are any sudden changes, since the design of this furnace is such that the temperature cannot change more than 10° per minute. If the receiver sees that the data, as received, indicates a greater change, it can inform the furnace operator. Either the data communications system or the temperature sensor is having problems, or the oven is grossly malfunctioning.

There is another advantage to data communications as compared to analog systems. By its very nature, the communications system is designed to handle only digital signals. This means that the system designer can analyze the performance of the system by using relatively simple equations, since the signal at any instant of time has only one of the allowed digital values. For the simplest case of a binary system, the allowed values are usually represented by symbols 0 and 1, regardless of the actual voltages. It is very practical to set up a model of the system, which will characterize and predict the system performance under many conditions, if the model has to accommodate only the values of 0 and 1. At any point in the system, the circuitry either leaves the value as it is or changes it to the other value, depending on the function of the circuit and the data. This model can be used on paper, or with a computer, as a "simulation" of the overall data communications system for various numbers and types of users, under many conditions. By contrast, an analog system must accommodate an infinite number of signal values within the allowable range. Therefore, a signal in an analog system would have to be described by an equation, such as $V = A \sin(2\pi ft)$ where V is the voltage at time t, A is the maximum amplitude of the signal, and t is the present time. Clearly it is much more manageable to use simple 1s and 0s than trigonometric equations.

As discussed in the previous chapter, time-division multiplexing (TDM) is often preferred because it gives a good combination of high performance and low cost for handling many users. Digital signals fit together well with TDM. This is because the computer circuitry which is generating the digital signals also is especially suited, from a circuit design view, to performing the TDM process. There is almost a "natural fit" between the operation of the computer circuitry and the technical requirements of a TDM system.

Digital signals from computers are better carried by a digital communications system. It is possible to convert the digital signal to analog form, for use in an analog communications system, but this is costly and the benefits of an all-digital system are lost. In practice, the opposite operation is the one most often performed: if the original signal is analog in nature (voice, temperature values, and so on), analog to digital converters are used to convert the analog format to digital format so that the preferred digital system can be used. For signals which are inherently digital in nature, such as those in a computer, an all-digital system is much more compatible.

One other essential difference between an analog system and a digital system are the types of test and diagnostic tools that are needed. Many of the traditional tools of electronic test, such as the voltmeter or oscilloscope, are useful only for some limited areas of test, and they are essentially of no use for many areas. New tools and techniques are used to test digital systems, but with these tools, much better and more meaningful testing can be done. For example, a data communications test instrument may send a specific known pattern of thousands of 0s and 1s over the channel. A test unit working in cooperation at the receiving end can see (1) how many of these bits were received incorrectly, (2) whether there is any pattern to the errors, and (3) whether the errors were caused by noise or by some other factor (such as some subsection of the total system that incorrectly changed the 1 to a 0, or vice versa). This level of detail and completeness is not possible with an analog system, because the test equipment must handle too many signal possibilities over the entire range, and analog signals and systems cannot be measured as precisely and easily as digital ones.

The many advantages of digital and binary communications dictate that for most new applications, in which large amounts of information must be sent or high accuracy and reliability are needed, digital communications is the preferred method. New communications systems that are installed are now designed expressly to handle digital signals only, and cannot even handle analog signals. These systems provide even better overall performance than an analog system which is carrying digital signals.

QUESTIONS FOR SECTION 5-2

1. What is noise? What problems does it cause for most systems? Where is it most critical?
2. What are sources of noise external to the system? Internal to the system?
3. What happens to electrical signals as they travel down a wire?
4. What do amplifiers do for analog signals?
5. How does regeneration work? How does it help overcome noise and attenuation?
6. What is the advantage of intermediate buffers?
7. Why are amplifiers needed even on short distances?
8. What happens to the system performance if the value of the noise becomes too large? How can a digital system usually overcome this inevitable problem?
9. What can a system do if an error is detected?
10. Why can the performance of a digital system be more easily analyzed than that of an analog system? How is this done?
11. How are digital error rates specified? What are typical values?
12. Why would systems in some cases need to process the digital data as it was received?
13. Why are the design and analysis of system performance easier for digital systems than analog ones? How would digital systems be described? Analog signals?
14. Why are digital signals and TDM a good combination?
15. Why are computers and digital systems a good match? What is done to convert analog signals into digital form?

16. What test equipment is used in digital systems? What traditional equipment is less useful?

PROBLEMS FOR SECTION 5-2

1. A digital system uses levels of +1, +3, +5, and +7 V for its four levels. Digital signals are received with the following amounts of noise: +0.5, −0.75, +1.3, +2, −0.95, −1.2 V. Which amounts of noise will cause errors in the receiver?
2. A digital system uses values of −3, −1, +1, and +3 V. For each of these four levels, what would be the most likely ranges of acceptable signals that would cause no error?
3. A binary system is designed to tolerate up to 5 V of noise before having an error. If one signal level is +10 V, what can the other signal level be?
4. Signals of +1, +3, +5, and +7 are used, and the noise values are +1.2, −0.5, −1.6, and +0.8 V, respectively. Sketch these signals before and after regeneration. Which are restored to their original value correctly?
5. A data communications system must provide a bit error rate of less than 1 in 1 million. What percentage is this? What about a rate of 3 errors per 100,000 bits received?

5-3 Sampling Theory

In the previous chapter's coverage of TDM, the issue of sending only samples of a signal instead of the entire signal was briefly discussed. Any TDM system, whether analog or digital, sends only the value of the user signal at certain selected times. The signal value is not sent continuously, as it would be with a space-division multiplexing (SDM) or frequency-division multiplexing (FDM) system. For the input signal shown in Fig. 5-6a, an analog system would send samples as shown in Fig. 5-6b, and a digital system would use the nearest digital value that was allowed (Fig. 5-6c). In both cases, however, only "snapshots" of the original signal would be transmitted.

At the receiving end, these sampled values are used to determine what the original signal looked like. The issue is, how can samples honestly represent a continuous and smooth signal? How many samples does it take to do this?

The answer was determined in 1924 by H. Nyquist working at Bell Telephone Labs. He was able to show, by mathematical proof, that in order to convey faithfully all the information in a signal with a bandwidth of N hertz, only $2N$ samples per second need be sent. This value of twice the bandwidth is called the *Nyquist rate* (Fig. 5-7). Therefore, if a sampling rate greater than 6 kHz is used for a voice signal with a bandwidth of 3 kHz, then more additional samples than theoretically needed are sent, but no new information is obtained. If a lower rate is used, then some of the original information is lost and not recoverable at the receiver.

This Nyquist sampling theory and the sampling rate needed for a specific signal bandwidth are the basis for much of the design of modern data communi-

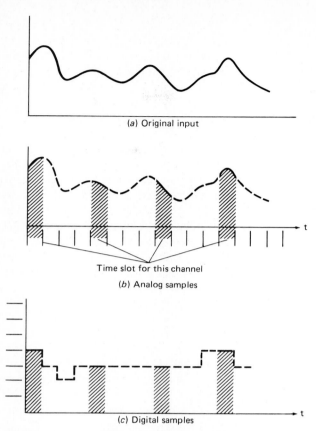

Fig. 5-6 TDM for analog and digital signals. For any varying signal (*a*), the four-channel TDM system transmits the variation that occurs during that channel's time slot, (*b*). For a digital signal, the transmitted value is constant during the time slot, (*c*).

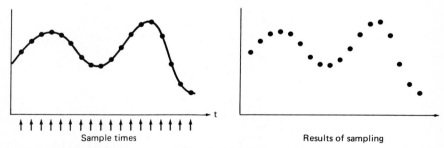

Fig. 5-7 Sampling of an analog signal. Only the values that occur at the sampling times are used. The signal values between sample times are not needed.

cations systems. They determine how many samples must be taken, at what rate, from each user, and therefore how many users can be served by a single TDM system.

For example, suppose a system is handling voice signals of 5-kHz bandwidth each. The system is designed to provide 100,000 samples/second. Since the voice signal bandwidth is 5 kHz, the Nyquist rate is 10 kHz per user. Therefore, a total of $100,000/10,000 = 10$ users can be handled by the overall system with these rates, and no signal information from any user is lost.

The Nyquist theory does not assume that the samples are taken at evenly spaced intervals. It only requires that the overall rate be twice the bandwidth, and that the sampling times be precisely known to the receiver, which reconstructs the original signal from the sampled values. In practical systems, it is always easier to use evenly spaced sampling times. This makes it much more straightforward for the receiver to know when the samples are taken, since the receiver can be synchronized to the sampling times, which are at constant intervals. If uneven sample times are used, then somehow information about the sampling times also has to be sent. This means that additional information has to be sent through the system, and this is undesirable in terms of channel efficiency.

The need to know precisely when the samples were taken points out another characteristic of Nyquist sampling. The theory assumes that the samples are taken at precisely known times. In a real physical system, there is always some error, however small, in the time when the samples are taken. Small amounts of back and forth variations in the timing, called *jitter*, occur since real systems are never 100 percent perfect. This jitter means that the actual samples are not taken at the supposed instants (Fig. 5-8). As a result, the reconstructed signal, which is developed at the receiver on the assumption of perfect sampling times, has some error in it, even if the sample values themselves arrive at the receiver without any error.

The solution to this problem is to sample at a rate higher than the Nyquist rate, as a form of "insurance." In typical systems, the sampling rate is anywhere from 50 percent greater than the Nyquist rate, up to 10 times greater. The amount of insurance depends on the accuracy needed in the final signal. Telephone systems commonly sample the 3-kHz-bandwidth voice signal at 8 kHz, compared to the 6 kHz needed in theory. Signals which represent some type of scientific informa-

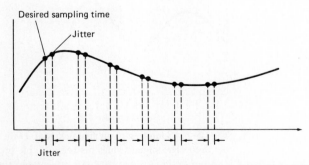

Fig. 5-8 Sampling jitter means that the sample times are not exactly where they should be, and actually vary slightly around the desired value. This causes inaccuracies in the reconstructed signal. Amount of jitter may vary from sample to sample. It is not a constant value.

tion, such as pressure and speed information from a car in a test laboratory crash, are sampled at the higher rate of 10 times the signal bandwidth to ensure the desired accuracy.

Aliasing and Filtering

What is the effect if the sampling is done less often than the absolute minimum of the Nyquist rate? If the Nyquist rate or a greater rate is not used, then serious inaccuracies can occur. The errors resulting from undersampling are called *aliasing errors*, because sample values are masquerading as representing something which they are not. If the sampling rate is lower than the Nyquist rate, these errors completely ruin any chance of recovering the original signal accurately. The nature of undersampling and aliasing errors is that there is no direct relationship between the percentage of undersampling and the percentage of error that occurs. For example, if the required Nyquist rate is 5 kHz, and the sampling rate is 10 percent low, 4500 Hz, it does not necessarily follow that the error is about 10 percent. Depending on the nature of the signal being sampled and the exact times the sampling occurred, the error could be 10 percent or it could be much, much higher. It is therefore crucial to the system designer to ensure that the sampling rate is high enough and aliasing is avoided.

The error due to undersampling is very different from the error that can occur from sample time jitter. Error due to jitter is usually very small, typically only a fraction of a percent.

Since it is very important to make sure that the sampling rate is above the Nyquist value, the system designer must carefully study the bandwidth of the signal being sent over the system. This is because the sampling rate is usually fixed in the system design, and the same sampling rate is used throughout the system. Therefore, the designer must ensure that the signal bandwidth is less than one-half the sampling rate. There are two aspects of this. For some signals, which have low bandwidths because of the nature of the information they are carrying, the information signal can be fed directly to the sampling circuit. A typical example would be a signal from a temperature sensor. Temperature is a slowly changing physical variable, and a bandwidth of 10 to 100 Hz is sufficient for almost any temperature application.

Other signals have wider bandwidth than the sampling rate allows, and so the bandwidth must be reduced by filtering before the sampling process. A voice is a good example of this. A voice has a bandwidth of up to about 10 kHz, although most of the necessary information is in the lower frequencies of up to about 3 kHz. Therefore, filters which bandwidth-limit the signal must be placed between the signal source (the voice and microphone) and the sampling circuit (Fig. 5-9). These antialiasing filters are essential to proper overall operation. In fact, the concept of limiting the bandwidth by filters and guaranteeing that it is limited is so important that most systems use some sort of filter, even when the signal itself is low bandwidth. For the preceding temperature case, the actual bandwidth is low and does not need antialiasing filters. However, some high-bandwidth electrical noise, from nearby power lines, static electricity, motors, and lights, may be picked up in the signal wires from the temperature sensor and add to the temperature signal (Fig. 5-10). This noise not only would corrupt the signal

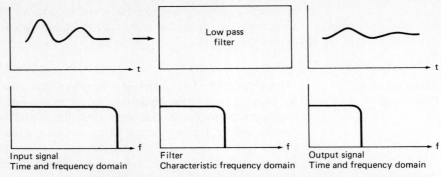

Fig. 5-9 Low pass filters are usually used between the signal and the sampling circuit to limit the bandwidth of the input to prevent aliasing.

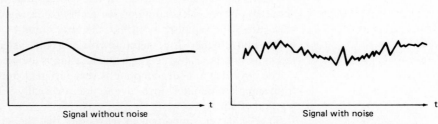

Fig. 5-10 The input filters also eliminate the effects of sampling on high-frequency noise corrupting a low-bandwidth signal. The noise causes overall aliasing errors if it is bandwidth-limited.

(because it is noise) but it could severely disturb the sampling process and cause even larger errors. Therefore, it is common practice to use bandwidth-limiting filters on every signal input, even if the bandwidth of the desired signal is lower than the value required by Nyquist sampling.

The concept of sampling an analog signal is crucial to almost any communication system. For analog systems, it allows many analog signals to share the system via TDM, if the sampling rate is properly chosen. For digital systems, in which the goal is to send all information in digital format, it allows the original analog signal to take the first step to conversion into digital format for use by the system. The next step is to convert the sampled value to a digital equivalent, which will be discussed in the next section. Fortunately, the theoretical relationship of the Nyquist rule makes it easy to determine what kind of sampling is needed to take one or more analog signals and sample them properly and accurately.

QUESTIONS FOR SECTION 5-3

1. What is the Nyquist rate? What rule does it state for the sampling of a continuous signal?
2. How does the Nyquist rate affect the number of users that a TDM system can handle?

3. What does Nyquist theory say about sampling time exactness? About sampling time evenness? Why are evenly spaced times usually used?
4. What is jitter? What is its effect in practical systems?
5. Why is sampling usually performed at a rate higher than Nyquist theory requires? By what range of percentages?
6. What is the effect of sampling at a rate lower than the Nyquist rate? Does a 10 percent shortfall cause 10 percent inaccuracy? Explain.
7. What is aliasing? Compare aliasing error to jitter error.
8. What is done in most systems to prevent the possibility of aliasing errors for a preset sampling rate?
9. Why do even low-bandwidth systems have filters for prevention of aliasing?

PROBLEMS FOR SECTION 5-3

1. What is the Nyquist rate for signals with these bandwidths: 100 kHz, 55 kHz, 10 Hz, 1 MHz, 6 MHz, 3 kHz?
2. What is the maximum bandwidth that could be handled without aliasing, in theory, by sampling at these rates: 100 kHz, 250 kHz, 40 kHz, 9.6 kHz?
3. A system is using sampling at 50 percent greater than the Nyquist rate. What sampling rates are required for the signals of Prob. 1? Repeat if the sampling rate is three times the Nyquist rate.
4. A TDM system is designed for digital signals. The system can send 100,000 samples/second. How many users can it handle if each user is being sampled at 15 kHz?
5. A system must be designed to support 18 users. Each user signal has a bandwidth of 3 kHz. The Nyquist sampling in this application needs 50 percent oversampling. How many samples/second must the system provide?

5-4 Analog to Digital Conversion

Signals which are in digital form, such as those within a computer, can be used directly with a data communications system. However, the signals which come from voice, telephone, television, or various sensors representing real physical quantities (temperature, pressure, speed) must be converted first to digital format. This is done by *analog to digital converters*. At the receiving end of the overall system, the signal may have to be converted back to analog form by a digital to analog converter, if the original analog form is needed by the user. In many cases, the data is needed only in the digital format, so no reconversion is needed. Data from sensors is often converted to digital form and then used by a computer for analysis, reports, and charts, so converting the data to digital format serves the needs of both the communications system and the researcher for data that is in computer-compatible format.

There are several important factors that must be considered when this analog to digital conversion process is performed. These factors are number of conversion bits, conversion time, and error sources. The system designer considers these factors and makes some technical tradeoffs when deciding on the combination of circuit and system characteristics needed to make sure the system meets the user requirements.

Operation of the Analog to Digital Converter

The *analog to digital converter* (ADC) converts any analog signal value within the overall input range into its digital equivalent. The output of the ADC is a binary value. The number of bits of the ADC conversion determines into how many groups, or subranges, the total input range is divided. A 2-bit converter divides the input into 2^2 unique values: 00, 01, 10, and 11, in the overall range, as shown in Fig. 5-11, with an input range of 0 to 1 V. A 3-bit converter divides the range into $2^3 = 8$ subgroups. The function of the converter is to determine which digital value is closest to the analog input value. For the 2-bit converter, the relationship between the analog input value and the associated digital value is fairly coarse, because there are only 2^2 groups into which the analog input is divided. The resolution of this converter is 2 bits, or 1 part in 4. The converter

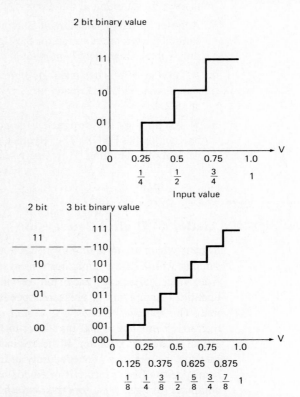

Fig. 5-11 A 2-bit A/D converter divides the analog signal range into four unique values, a 3-bit converter divides the range into eight values, providing greater resolution of the analog signal value.

may therefore cause large apparent errors in the conversion process, even if it is working perfectly.

Consider an analog input over the 0 to 1 V range. For a 2-bit converter, each digital value corresponds to a range of 0.25 V. Any analog value within the 0.25 V window causes the same digital value to appear. Inputs of 0.53 V and 0.6 V have the same digital output. The poor resolution of this converter means that it is itself a major source of inaccuracy, even if the rest of the data communications system is perfect.

Practical systems use many more bits to achieve the resolution that applications really need. A converter of 8 bits provides $2^8 = 256$ digital values and divides the analog input into 256 groups. The resolution is therefore 1/256, or about 4 percent. This may be acceptable for digitizing a voice signal. For better performance, or in applications in which the analog system represents data from temperature, for example, 10-, 12-, 14-, or even 16-bit converters are used. They provide this performance:

Number of Bits	Number of Digital Values	Approximate Resolution (%)
10	1,024	0.1
12	4,096	0.025
14	16,384	0.00625
16	65,536	0.0016

It would seem that a system should always use a higher-resolution converter for best performance. There are reasons why a data communications system would use the lowest resolution acceptable to the user and application:

Higher-resolution converters cost more.

Higher-resolution converters provide more bits of digital output. Since the number of bits/second that a system can send is usually set in advance, sending a higher-resolution digital value takes more time (and bandwidth) since there are more bits to be sent. Time and bandwidth are resources to be used carefully.

The error that an ADC causes because it has only a certain amount of resolution is called the *quantizing error*. To quantize is to divide a continuous value into small, individual packets, and the process of digitizing divides the overall range of value into many small ranges. The ADC then picks the subrange that is closest to the original analog value (Fig. 5-12). Since exact analog value is digitized to the closest value, there is always an uncertainty about what the original analog value was. (It has no relation to the efficiency of a practical converter as compared to a theoretical one.) This is because any analog value within the smaller range could have caused the same digital output value. This uncertainty is the *quantization error*.

For a 3-bit converter operating over a 0- to 1-V range, the analog and corresponding digital values are:

Analog	Digital
0–0.125	000
0.0125–0.250	001
0.250–0.375	010
0.375–0.5	011
0.5–0.625	100
0.625–0.750	101
0.750–0.875	110
0.875–1.00	111

Any signal value within the range from 0.5 to 0.625 V is digitized to 100. Therefore, the user has no way of looking at the digital value and determining the exact value of the original analog input. The user can only say that the analog input was between 0.5 and 0.625 V. This is a quantizing error. After the analog signal has been digitized, the digital output of the ADC is used as the binary data for the data communications system. In most systems, this is sent as a *serial bit stream*, meaning that the bits are sent in sequence, 1 bit at a time.

An ideal, theoretical ADC would perform the conversion process instantaneously. A real ADC takes some amount of time. Depending on the converter used, this conversion process can be relatively fast (microseconds) or much slower (milliseconds). The conversion speed must be matched to the bandwidth of the analog signal. If the converter is too slow to perform conversions at the Nyquist rate or better, then the system is not able to sample fast enough. Aliasing errors will result. The converter must be selected properly. High-speed converters cost more than low-speed ones, so designers choose a converter that is just fast enough. High-speed, high-resolution converters can be extremely expensive (several hundred dollars) so they are used only when absolutely necessary. An 8-bit, 10-ms converter costs only $5, in contrast.

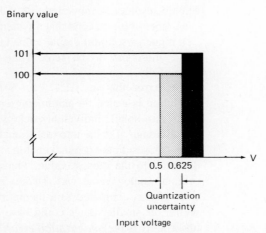

Fig. 5-12 Even a perfect A/D converter contributes some error to the system, because it quantizes the analog signal to the nearest digital value.

An Example with a 3-Bit Converter

Consider the analog signal shown in Fig. 5-13. This signal is varying over the range from 0 to 1 V. It goes to a 3-bit ADC, which therefore provides a converted value somewhere between 000 and 111, where 000 represents 0 V and 111 represents the full-scale value of 1 V. Since a 3-bit converter provides 2^3 values, the 1-V range is divided into eight subzones. Binary 000 represents 0 to 0.125 V, binary 001 represents 0.125 to 0.25 V, and so on, up to 111 representing 0.875 to 1 V. (A 2-bit converter provides only four subzones: 0 to 0.25 V, 0.25 to 0.5 V, 0.5 to 0.75 V, and 0.75 to 1.0 V.)

The ADC is instructed to perform the conversion at time intervals selected by the system designer. Circuitry in the system tells the ADC to perform this conversion. For the signal shown, at the times shown, the output of the conversion process for the first 10 conversions is the following:

Conversion Number	3-Bit Digital Value	2-Bit Digital Value
0	000	00
1	011	01
2	110	11
3	111	11
4	111	11
5	011	01
6	000	00
7	001	00
8	011	01
9	011	01

(The 2-bit conversion is shown for comparison.)

The ADC has provided the digital value at the instant the conversion is performed. Any changes in the analog value between conversion times are not seen by the ADC and the rest of the communications system. This is not a problem if the sampling rate is higher than the Nyquist rate, since all the information present in the analog signal can be reconstructed just from the samples.

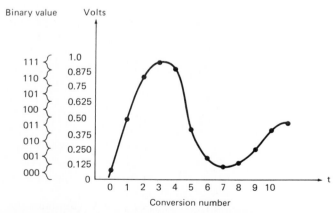

Fig. 5-13 The analog signal shown is digitized by a 3-bit A/D converter.

Digitization and Modulation

The digitized values of the original analog signal is next used to control the modulation of the carrier, if any. The patterns of 1s and 0s cause the modulator, whether AM, FM, or PM, to vary the amplitude, frequency, or phase of the carrier between two values. One value corresponds to binary 1, and the other to binary 0. Some typical schemes in use today for binary modulation are shown in the following table.

	Binary 1	Binary 0
AM	Full scale	10 percent of full scale
FM	1000 Hz	2000 Hz
PM	$-45°$	$+45°$

Many other methods are used, depending on the application and capabilities of the communications system.

The modulator receives the binary values as a serial stream, with a new bit provided at the system clock rate. For example, the preceding 3-bit conversion may start with the rightmost, least significant bit (LSB) of each conversion and move to the leftmost, most significant bit (MSB). Thus, the serial stream of bits and the corresponding FM frequency for a 1000/2000-Hz system would look like the following for the first three conversions:

Conversion Number	Bits	FM Frequency
0	000	2000, 2000, 2000 Hz
1	011	2000, 1000, 1000 Hz
2	110	1000, 1000, 2000 Hz

and so on. Each converted value must be modulated and transmitted before the next conversion occurs, or else data will "pile up" and new data will override the old. This is why a system must transmit the data bits at a rate fast enough to support the conversion rate and the number of bits per conversion.

To illustrate: suppose a system is providing 4-bit conversions every 1 ms. Then the system is producing 4000 bits per second (bits/s), and the remaining pieces of the communications systems must be able to support that rate. A system which is digitizing voice to 8 bits, at a Nyquist rate of 6 kHz, is generating 48 kilobits per second (kbits/s) (1 kbit = 1000 bits). The bit rate of data communications systems varies widely, and rates in use today range from 10 kbits/s to several megabits per second (Mbits/s) (1 Mbit = 1 million bits). From these simple calculations, it is easy to see how a system which has several users, each at moderate Nyquist rates and number of bits of resolution, can require relatively high bit rates.

An alternative to sending all the bits at the high rate is to store some of them in a special computer memory circuit, called a *buffer*. This is a feature that only digital systems can provide, since storing 1s and 0s is simple, whereas storing analog values is quite difficult. A small memory buffer can be used to hold the digitized values for transmission at a slower rate or later time. In some systems, this memory is used for the opposite purpose: to store bits that result from slow digitization, so they can be sent at a higher rate that is compatible with the rest of the system. If the overall system is designed, for example, to perform at several hundred kilobits/second, and the ADC process is producing only 64 kbits/s, either the entire system has to be slowed down, or the system capability is not being used to maximum. By using a buffer, bits from the slower ADC can be stored, then sent in a high-speed burst at the system rate. This allows the system to be used at higher efficiency.

The process of taking an analog signal and developing the digital equivalents in order to allow the analog signal to be handled in a digital system is called *pulse-code modulation* (PCM). With a PCM circuit, any analog signal can be converted into digital format, making it compatible with data communications systems. The analog signal can then be treated as a signal that was originally digital in nature, such as that from a computer, and can make use of all the advantages of digital systems for communications.

QUESTIONS FOR SECTION 5-4

1. What types of signals need to be digitized? What circuit element does this?
2. For what types of applications does the digitized signal not have to be reconverted to analog voltages? For what type does it have to be reconverted?
3. How many subranges does a 2-bit converter provide? A 3-bit converter? An 8-bit converter? What does the term "resolution" mean?
4. How does low resolution cause apparent accuracy problems? Is this error due to the performance problems in the conversion? Explain.
5. By what factor does the resolution improve as the conversion is done to another bit (i.e. 5 bits versus 4 bits)?
6. What is the idea of quantizing error? What does it mean with respect to determining the exact analog value that results in a certain digital value?
7. What determines the conversion time for an ADC? What determines the rate at which bits are sent?
8. Why do communications systems very quickly require high bit rates?
9. What is a buffer? Give two reasons it may be used in a data communications system.
10. What is PCM? What overall function does a PCM system perform?

PROBLEMS FOR SECTION 5-4

1. Assume an ADC is designed for signals that range from 0 to 5 V. What is the resolution in volts provided by converters with these numbers of bits: 2, 4, 6, 8, and 10?
2. Into how many subgroups does a 7-bit converter divide an input signal? A 13-bit converter?
3. What is the quantizing error, in volts, that can occur when a 10-bit converter is used with a 0- to 4-V signal?

Digital Communications

4. A 2-bit converter is used with a 0- to 1-V signal. What is the binary digital value that will occur for these values of input: 0.4, 0.3, 0.78, and 0.9 V? Repeat for a 3-bit converter.

5. A 3-bit converter is used for 0- to 5-V signals. What is the binary digital value that will occur for 1.2-, 2.7-, 3.2-, and 4.5-V signals?

6. A system is designed to sample analog signals, convert them to digital form with a 4-bit converter, and transmit them. What bit rate is required if the sampling rate is 10 kHz? 20 kHz?

7. A system is designed to sample at a rate 50 percent greater than the Nyquist rate for 5-kHz bandwidth signals and convert them with a 6-bit converter. What is the system bit rate required?

8. What bit rate would be needed to support five users identical to the preceding one, assuming the sampling and conversion circuitry were not a limitation?

5-5 Encoding of Digital Signals

The digital bits, either from the ADC process or from within a computer, may not be directly suitable for modulating the carrier of the data communications system. Even if there is no carrier, but simply a baseband system, there may be technical reasons why the bits may not be used without more manipulation. The reason for this problem is that the communications channel itself has some aspects that must be considered in the design of the way the digital information is carried. The simplest situation is the binary system, in which only two values must be handled by the system. On the average, in a long string of bits (1s and 0s) half are 0s and half are 1s. However, over any small group of bits, there may be many bits in a row in which there are only 1s or only 0s. The situation is similar to flipping a coin. Eventually, there will be an equal number of head flips and tail flips, but there will also be smaller periods with three, four, five, or perhaps more heads (or tails) in a row.

What is the implication of this in data communications? Suppose a system is sending a string of bits, at a rate of 100 kbits/s. If many bits in a row have the same voltage value (representing a group of 1s or a group of 0s) then for a long period of time the voltage does not change. Instead of having a new voltage representing a 1 or 0 every microsecond, there will be stretches of many microseconds when the voltage remains constant. This causes two types of problems.

One problem is that the average value of the voltage representing the bits changes, and actually shifts up and down. This problem is called *dc shift* because the average value is the "steady value" or direct current (dc) value that would be observed on a dc voltage meter. For the entire string of data bits, the average value is 1/2, midway between 0 and 1. However, if there is a stretch of many 0s, for example, the average signal value during that stretch drifts toward whatever voltage represents 0 (Fig. 5-14).

Why is this a problem? Many communications systems require that there be no "dc component" in the frequency spectrum of the stream of data bits for proper

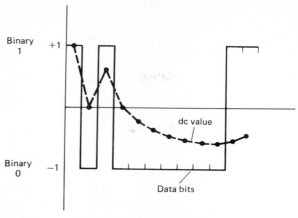

Fig. 5-14 The average, or dc value, of a stream of bits varies as the bit pattern contains runs of 1s or 0s.

operation. The reason is that many of the electronic components used in circuits prevent the passage of steady dc signals. Transformers and capacitors used for isolation between parts of circuits do this—they allow alternating circuit (ac) signals to pass, but by their very nature block dc signals. What the systems designer does, therefore, is use a negative voltage for one binary value (for example, -1 V for binary 0) and a positive voltage for the other symbol ($+1$ V for binary 1). The average value is then 0 V.

Therefore, if the dc component in the data stream is not zero, that component is blocked. The consequence of this is that the signal, after it passes this blocking component, looks very different from the way it would if the entire signal were passed (Fig. 5-15). It would then be impossible to receive the data properly.

The dc shift is not always a problem. When the data communication is between two subsections of a system there are usually no transformers, capacitors, or other dc blocking components in the signal path. However, many communications systems need some way to synchronize the clock of the receiving circuit with the clock that is used for timing the bits at the transmitter which is the second problem. Otherwise, if the clocks differ by even a small amount, many data errors result. This synchronization is performed by a circuit which derives the clock of the transmitter from the actual received data. However, if the data has a long run of 1s or 0s, then there is no change in the incoming signal and it is nearly impossible to derive the original clock. The best signal, from the viewpoint of the synchronization circuit, is one which has many changes from 1 to 0 and vice versa. These transitions act as markers to show where the original clock was.

Encoding Schemes   The solution to both the dc blocking situation and the clock synchronization problem is a technique called *encoding*. Encoding takes the original stream of data bits and manipulates them in some predictable way so there is a constantly changing 1 and 0 pattern. At the receiving end, the encoded signals are put through a circuit which performs the reverse operation and restores the original

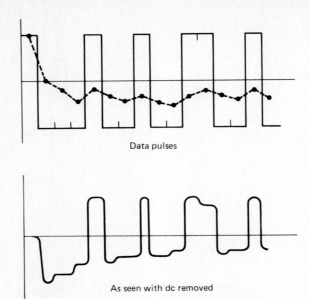

Fig. 5-15 If the receiver circuitry removes the dc value, the resulting data pulses look very different from the way they did when they originally were sent. The difference is related to the runs of 1s and 0s that previously occurred.

stream of 1s and 0s, as they came from the ADC or computer at the transmitting end. Then, the average value is always close to 0 and there is a signal transition to be used for 1 clock derivation.

Encoding is analogous to having a system which is to be used to transmit letters of the alphabet, but unfortunately over a channel whose circuitry can only send numbers. The solution is to assign a number to each letter, send the numbers, and then reconvert back to letters at the receiving end. In this way, the original data is transformed into a code the system can handle, and yet the user can get the desired information across. Many types of encoding techniques are in use today. Different encoding methods are designed to best serve the specific needs of certain applications. Two of these techniques are return to zero (RZ) and Manchester encoding.

The similarities and differences among the various coding schemes are seen if the same bit pattern is used as the example of data (Fig. 5-16a). Regular binary data consists of a stream of 1s and 0s. (Assume a voltage $+V$ represents a 1, and $-V$ represents a 0.) The level of the voltage stays at the same value throughout the time period assigned to the bit. This is called *nonreturn to zero* (NRZ). For the 8-bit data stream shown, 11000101, the voltage stays at the level ($+V$) for the first 2 bits, then goes to the 0 level ($-V$) for the next 3 bits, then back up to $+V$, then down to $-V$, and finally up to $+V$. The average value over the 8 bits is 0 V, since there are as many 1s as 0s. However, after the first 2 bits have been sent, the average is $+V$, since both bits were 1s. After 4 bits, the average value is back down to 0 V. After 5 bits, the average begins to become negative, since the next bit is a 0. During a long string of either 1s or 0s, there is no

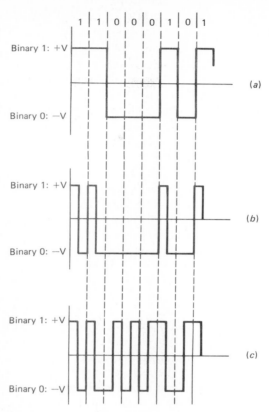

Fig. 5-16 Different types of encoding have characteristics that are crucial in different applications: (*a*) binary data in the common NRZ format; (*b*) the same bits in RZ; (*c*) Manchester encoding of the bits.

information present in the data stream that might help synchronize the receiver.

A modification of the NRZ method is *return to zero* (RZ) (Fig. 5-16*b*). In RZ, the data pulses that represent binary 1 return back to the binary 0 level in the middle of the time period allowed for the bit. This means that for every 1, there is a signal transition, which helps the receiver develop synchronization to the incoming data stream. A long series of 1s is seen as an alternating signal. A long series of 0s still looks like a single, constant value. The RZ method offers some improvement, but it does not solve the dc shift problem and does not completely solve the synchronization problem. The improvement may be enough for the particular circuit and system, however.

Manchester encoding solves both the problems. In *Manchester encoding*, each bit period has both the high- and low-voltage values (Fig. 5-16*c*). If the data is a 1, the first half of the bit time period is sent at the $+V$ level, and the second half of the period is at the $-V$ level. For a data bit of 0, the reverse is sent—first a $-V$ signal, then a $+V$ signal. The result is that for every bit, whether 0 or 1, there is a transition which can be used for a synchronization, and the average value for each bit sent is precisely between the 1 and 0 voltage values, regardless of the sequence of data bits. Manchester encoding is sometimes referred to as a *self-clocking encoding method*, since the data stream in effect carries its own clock.

Regardless of the encoding method used, the receiver must then perform the mirror-image operation to restore the data to the original sequence of binary data. To the user who is sending data, regular binary (NRZ), RZ, and Manchester encoding are transparent. The circuit which performs the encoding does this for any bit sequence and does not study the bits to see what information they represent. The user simply sends data from its source to an encoder circuit, which performs the operation, and then uses the result to modulate the carrier by whatever method is to be used. The user never sees the encoding process.

However, someone working with a system that may be malfunctioning sometimes has to study the signal patterns at various points in the circuitry. A technician, either working with a design engineer in the lab or fixing equipment at a user location, very often has to probe key points in the system with voltmeter and oscilloscope. The waveform that the technician sees at the output of the encoder looks different from the data that is being generated by the data source, whether the data is from a computer or the digitized values of a voice or temperature signal. This changed stream of 1s and 0s is the encoded version and has to be checked to make sure that it is correct and that the encoder is working properly. Otherwise, the total system may be providing bad data to the user at the receiving end and the problem will not be obvious.

Besides the encoding schemes discussed previously, there are others in general use. The decision about which one to use is based on specific requirements of the application, and whether there is some standard method of encoding that many users in a certain type of application have generally agreed to for commonality. Variations on the RZ and Manchester encoding can have some subtle advantages and disadvantages for certain sources of data used with some types of channels. For example, Manchester encoding has a transition for each data bit being sent. It therefore requires more bandwidth than NRZ, which only has transitions for one-half the bits, since in 50 percent of the cases the value of the next bit is the same as that of the previous bit. In this case, the choice is the self-clocking feature and guaranteed dc value of 0 for Manchester, versus the lower bandwidth requirements for NRZ.

Circuitry for RZ and Manchester Encoding

Digital logic gates can be used in simple circuits to encode regular NRZ binary data into RZ and Manchester data streams. Figure 5-17 shows a series of data bits that are in NRZ format, along with a clock that will be used to synchronize the encoding. The encoding is performed by a simple AND gate, which combines the clock signal and the data bits. When the data bit is 0, the output of the AND gate is 0; when the data bit is 1, the output of the AND gate is equal to the clock signal. This is exactly the desired RZ encoding. The timing diagram shows this.

The circuitry for Manchester encoding is more complicated than a simple AND gate, but still relatively simple. The rule that must be implemented for Manchester encoding is, if the data bit is a 1, transmit the clock signal; if the data bit is a 0, transmit the inverse of the clock signal (called *CLOCK'*). Figure 5-18*a*

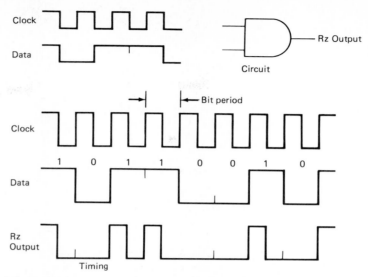

Fig. 5-17 A circuit to encode NRZ data bits into RZ format requires only a clock and an AND gate. The timing of this circuit for a typical bit pattern is shown.

on page 146 shows one circuit which implements this rule. The clock signal is passed through an inverter so that both CLOCK and CLOCK' are available. The data bits control two AND gates, which allow either CLOCK or CLOCK' to be passed through, but not both at the same time. The outputs of the two AND gates are combined in an OR gate, so that the desired signal reaches the modulator and link.

The timing diagram for this circuit is shown in Fig. 5-18b for key points in the circuit. A commercial Manchester encoding circuit is somewhat more complicated, because extra gates are added to ensure that no inadvertent glitches or transitional signals occur as a result of internal gate delays.

QUESTIONS FOR SECTION 5-5

1. What are two problems that a long string of 0s or 1s can cause?
2. Why is dc drift a problem? What circuit components can block dc components?
3. Why does a steady stream of 0s and 1s make synchronization difficult?
4. What part of the signal provides information for the synchronization to the bit clock?
5. What is encoding? What does it do to overcome problems?
6. What is NRZ? What is RZ? Compare them.
7. What is Manchester encoding? What are its advantages?
8. Why is Manchester encoding called "self-clocking"?
9. What does it mean to say that the encoding process is "transparent"?
10. Compare the bandwidth of a binary stream of bits that uses NRZ, RZ, and Manchester encoding.

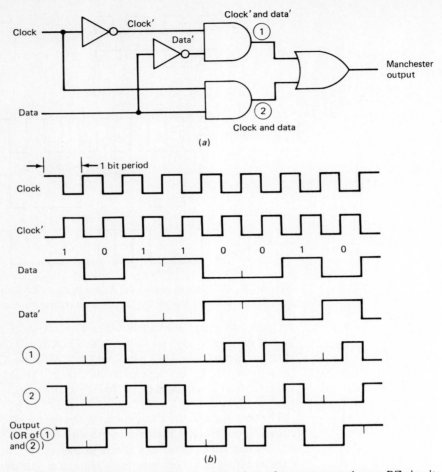

Fig. 5-18 A Manchester encoding circuit requires a few more gates than an RZ circuit: (a) conversion of NRZ signals to Manchester encoded signals; (b) timing and signal waveforms at key points in the circuit.

PROBLEMS FOR SECTION 5-5

1. A binary system is sending out the following pattern, starting with the bit on the left: 0011 1110 1001 (spaces are for clarity only). What is the average value after each bit is sent?
2. Repeat Prob. 1 for this pattern: 0111 1000 1010.
3. A system is using $+1$ V and -1 V to represent binary signals. What is the average value for this pattern: $-1, -1, +1, -1, +1, +1, +1, +1, +1, +1, +1, -1$?
4. For the pattern of Prob. 1, sketch the waveform as an NRZ, RZ, and Manchester encoded signal.
5. Repeat for the bits of Prob. 2.
6. Repeat for the values of Prob. 3.

7. For the circuit of Fig. 5-18, show the timing for the key circuit points when the data bit pattern is 1000 1111 0101.
8. Repeat Prob. 7 for the data pattern 0011 1010 0010.

5-6 Multiplexing and Modulation of Digital Signals

Among the major advantages of digital signals are the ease with which they can be multiplexed and the simple modulation circuitry that can be used. Modern digital integrated circuits and systems can operate both at low cost and with few components. This makes digital signals even more attractive in terms of overall performance and expense.

Digital multiplexing, like analog multiplexing, combines the signals of several users into one signal that can then share the link or channel. These users can be individuals, such as telephone system users, or can represent different signal sources within a single system. For example, a computer may need to send data bits from several points in one subsection to another. For various reasons, it may be desirable to have only a single, high-bandwidth wire link between the two parts of the system. The various digital signals can be multiplexed and the resultant signal used in base band mode for signal transmission.

Technically, digital signals can be multiplexed with SDM, FDM, or TDM. There are no major technical differences between multiplexing digital signals versus analog signals by using SDM or FDM. The signals that are being multiplexed are just a more limited group than the analog signals that an analog system must handle. The main difference is that the multiplexer can be designed for optimum performance on digital signals, without having to worry about its ability to handle signals that can take on any value within the total range.

There are some aspects of using TDM with digital signals that make TDM a very attractive scheme. A TDM system for digital signals uses a clock that is at higher speed than any of the digital signals, samples the various user signals briefly, and puts the sampled value onto the channel. This effect of TDM multiplexing of digital signals is really very straightforward. We can examine it more closely for the simplest case of binary signals. Suppose there are five users, each providing bits at a rate of 1 bit/s. These bits may be computer data, or they may be the digitized value of some analog signal. The digital signals of the five users are shown in Fig. 5-19a, along with the timing. Next, have a clock signal at five times the rate of any single user, that is, at 5 Hz. Use this clock to sample each digital signal, in sequence, starting with the first user, and make the output of the sampling circuit the resultant signal that goes to the modulator and communications system. In the first 1-s period, the first bit of each user is sampled and transmitted; in the second 1-s period the second bit of each user is sampled and transmitted; and so on. By rapid sampling from channel to channel, the bits of all users can be combined through this multiplexing process and sent over a single link (Fig. 5-19b).

At the receiving end, the opposite process occurs. A clock operating at the multiplexer rate of 5 Hz is used to sample the incoming bit stream. At the first

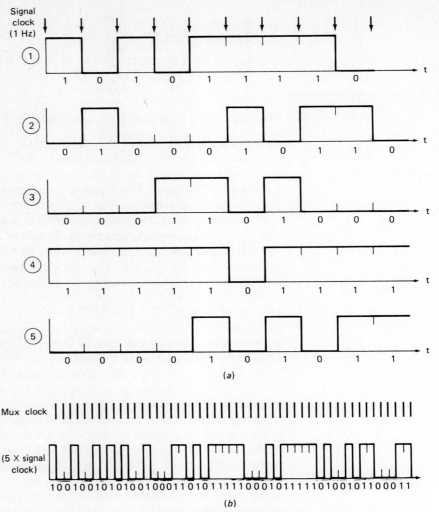

Fig. 5-19 Time division multiplexing of digital signals: (a) five digital signals, with a clock at five times the bit rate of each user; (b) the TDM result.

clock period, the sampled binary value is taken and passed to the first user, then the second clock period occurs and the next bit is passed to the second user (this next bit is the first bit of the second user's signal), and so on. After the first 5 bits are sampled, the demultiplexer takes the next bit, representing the second bit of the first user signal, and begins again. In this way, the signals of the original five separate users are split up and restored to their original form (Fig. 5-20).

There are two key points to this process. First, to multiplex N number of users, the multiplexer must use a clock rate of N times the rate of any single user. For example, if 10 user signals are to be multiplexed together, and each user signal is the result of sampling at 10 kHz (of a 3-kHz-bandwidth voice signal) then the multiplexer and the rest of the communications system must operate at a 100-kHz rate. The increased rate corresponds to the increased bandwidth needed, which in

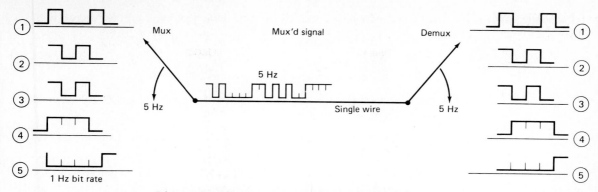

Fig. 5-20 The multiplexing and demultiplexing process for TDM can be represented by a switch which scans through the digital inputs, in sequence (multiplexing) and another switch which sorts the received, multiplexed signal into one of several outputs, in synchronization with the first switch (demultiplexing).

turn corresponds to the increased rate of information flow that the system is supporting (10 users versus a single user).

The second point is that the multiplexer can sample any particular digital input at any time within the bit period. This is because the value of the bit is unchanged during the bit period. It does not matter whether the multiplexer samples user number 3, for example, just after a bit time slot begins, in the middle of a time slot, or near its end (Fig. 5-21). This means that a multiplexer can sample the fifth bit of each user by going from one user signal to the next user, and so on, using a fast clock to drive the sampling operation.

This is in contrast to any sampling operation involving analog signals. For analog signals, the sampling time must occur at precisely known intervals so that Nyquist's theory can be used. If several analog signals are being combined by TDM, the timing between successive channels becomes crucial. As more and more analog channels are combined, it becomes technically more difficult to meet the required specifications. Digital signals allow a great deal of design flexibility in the way the sampling is done for the purposes of multiplexing.

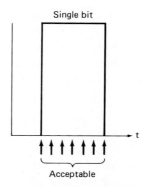

Fig. 5-21 The TDM process can sample the digital signal at any time within the bit period, since the bit value is constant (logic 1 or 0) during that time.

Digital Communications 149

Synchronization and Framing

At the receiving end of the system, the demultiplexer user must do two things:

Provide a clock at the same rate as that used at the multiplexer.

Provide some synchronization indication to the user, to show where the first bit of user number 1 is located. This is often called *framing*.

The problem of providing a clock at the receiving end may be solved in one of several ways:

The multiplexer and demultiplexer may be in a common electronic chassis, so the same clock signal may be available to both.

It is possible to use sophisticated circuitry such as a PLL to derive a clock from the incoming stream of data bits electronically. For this to work properly, there should be no long groups of only 0s or only 1s, because a constant signal level provides no timing information.

Use encoding, such as Manchester, so that the data bits effectively carry their own clock signal with them.

The synchronization to the bits of the first user is provided by some type of framing signal, which is a special bit pattern which the multiplexer inserts into the overall data stream just before the first user begins.

This bit pattern is distinctive, for example, a 010101 grouping. The demultiplexer looks for this pattern and knows that the first data bit occurs after it (Fig. 5-22). Of course, it is entirely possible that the user clock will also have this pattern. The system designer can do two things:

Make the pattern longer, for example, 16 or 32 bits, so it is less likely to occur naturally.

Use an encoder at the multiplexer which checks whether any of the data patterns are the same as the framing group pattern. If one is, then use a predefined scheme to modify the data bits, so as to prevent confusion. At the receiving end, the decoder restores the original pattern.

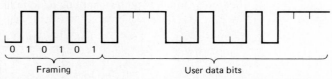

Fig. 5-22 A TDM system uses a specific binary pattern as framing bits, to provide synchronization to the overall frame of multiplexed bits.

Some systems use a framing group for a relatively few user data bits, such as 16 or 32 user data bits. This makes framing synchronization simpler and more reliable but means that a large percentage of channel time is wasted (not carrying actual user data). By contrast, it is possible to use a single framing group for a long string of user data bits, perhaps 128 or 256, and this is more efficient in using channel time. However, the synchronization becomes more difficult technically and more susceptible to problems due to noise, jitter, and other channel

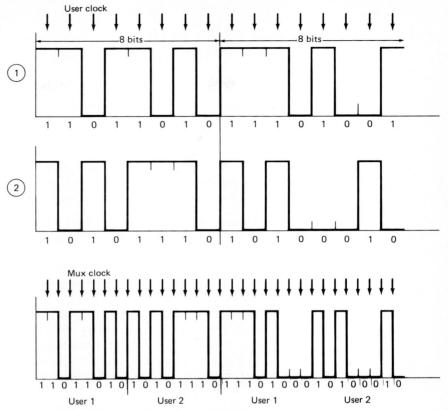

Fig. 5-23 An alternative TDM approach is to take a group of bits from each user at a time, rather than scan through the users on a bit-by-bit basis. Here, two users have their signals combined in groups of 8 bits each.

problems. If these problems cause framing to be lost, many more bits will be demultiplexed in error as compared to using framing groups more frequently.

Some systems use a variation of the bit-by-bit multiplexing of digital signals. Instead, these systems take all of user number 1's bits, then go on to user number 2 and take all of those bits, and so on. The result for this type of multiplexing is shown in Fig. 5-23. The multiplexer still requires a clock rate that is equal to the number of users times the number of bits, but the multiplexing is done on a user-by-user basis instead of a bit-by-bit basis. The result is still the same: a single high-speed channel can handle many users. This method may be chosen because it has some technical advantages in a specific situation, or because it is more compatible with existing equipment that is already installed.

The Bell Telephone System installed digital multiplexers called *T-1 digital carrier communications systems* beginning in the 1960s. This system was used to connect one local office to another local office and could handle 24 voice signals with a bit rate of 1.544 megabit/second (Mbits/s). The system is still used extensively. It digitizes each voice signal into $2^7 = 128$ levels, using 7 bits, with

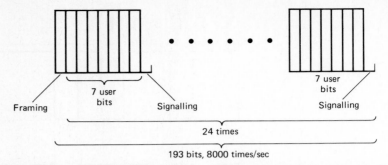

Fig. 5-24 The T-1 TDM system used by the Bell System (among others) handles 24 users with seven digitized voice bits and one signaling bit per user, and a framing bit for the entire group of 24 users, for a total of 193 bits. This is repeated 8000 times per second.

a sampling rate of 8 kHz. An eighth bit is added to the end of each user group of 7, to serve as a marker for interoffice signaling. Each group of 8 bits times 24 users is called a *frame* and contains 192 bits. Every frame is preceded by a special framing bit, which allows proper synchronization at the receiving end (Fig. 5-24). Thus, 193 bits must be sent 8000 times/s, for a total rate of 1,544,000 bits/s.

It is also practical to take groups of users who already have been multiplexed and remultiplex them into supergroups. This method is used when groups of users are gathered together in local areas, and then grouped again for transmission over a high-performance link. The telephone system predominantly uses this method, since there are many individual callers going in the same general direction. Individual callers from a single central office are digitized and combined via TDM and sent to a regional office as a group. Several central offices are then combined at this regional office in a supergroup, via TDM. The advantage of digital signals and TDM in this type of application is that the regional office can do this effectively by using a very high speed multiplexer and a high-bandwidth link to the larger office on the receiving end (Fig. 5-25). Suppose each central office digitizes 10 user signals at an 8-kHz rate. An 80-kHz rate is then used to perform the TDM operation and combine these 10 users into a single signal, to go to the regional office. The regional office can combine 5 similar local offices, each with 10 users, by multiplexing at a rate of 5 x 80 kHz, or 400 kHz. There is a clear building block approach in which users are combined and groups of users are combined again, that is modular and flexible. This is essential to efficient operation of a system such as the phone system.

Digital Modulation

Digital modulation is more specialized and much easier to understand than the more general analog modulation. The modulator has to impose only a select group of variations in amplitude, frequency, or phase to accommodate the data which it is getting from whatever source.

A modulator for binary signals has to shift only the amplitude, frequency, or phase between two values. The effects of AM, FM, or PM are shown in Fig. 5-26. For AM, the amplitude is shifted between 70 percent and 100 percent of full scale, for the example in the figure. For FM, the frequencies are 1000 Hz and

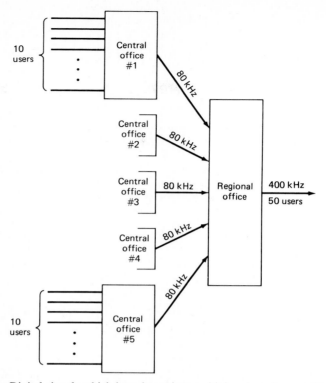

Fig. 5-25 Digital signals which have been time multiplexed can be remultiplexed at a higher rate with similar TDM signals, to travel over a higher-bandwidth, higher-rate link. The figure shows the way that five central offices, which each have multiplexed 10 users with an 80-kHz clock rate, are combined into a single bit stream at 400 kHz representing 50 users.

2000 Hz. For PM, the phase shift is 0° or +180°.

Mixed forms of modulation can be used. A system may be designed to use FM for the even-numbered bits and AM for the odd-numbered bits. This may seem like a complicated way of achieving a goal. However, this combined modulation may be needed in applications in which it is difficult to build a digital modulator that can operate at the needed full bit rate. If every second bit is used to modulate, the effective modulation rate for either the AM or the FM is reduced to half the original value. The total bandwidth required for combined modulation is approximately the same, since the required bandwidth is determined by the total amount of information sent, in theory. In practice, however, there are some techniques which can reduce the bandwidth by carefully designed modulation methods, and so combined modulation may allow a system to send data at closer to the theoretical maximum for the bandwidth available; or it may allow a lower-bandwidth system to handle more of the information. These designs can be complicated, but are sometimes justified in expensive systems, such as space vehicles and satellites, which require the best use of channel and bandwidth resources under difficult conditions.

Digital Communications

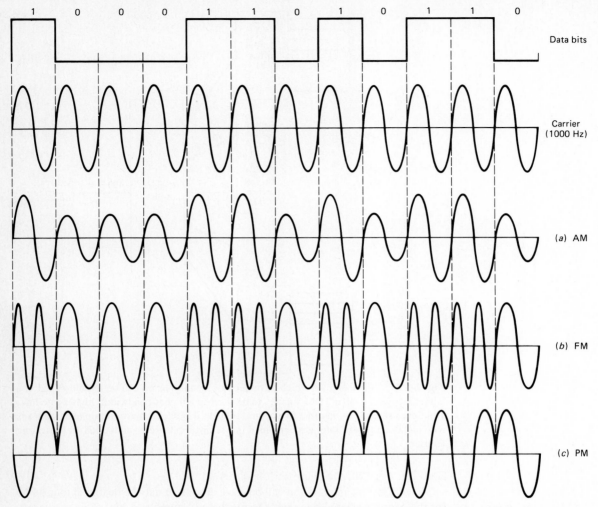

Fig. 5-26 A modulator for digital signals causes the carrier to shift between specific values only. For binary, only two outputs can result from the modulation process. For the binary pattern and carrier shown (a) is the result of AM from 70 to 100 percent of full scale, (b) is FM between 1000 and 2000 Hz, and (c) is PM with values of 0° and 180°.

The Data Communications System to This Point

The aspects of data communications examined in this chapter and the preceding chapters define a great deal of the complete data communications system. There are a source of data (either data that is in computer format or the digital equivalent of analog signals), an encoder, and a modulator. If required, there is also a scheme for multiplexing.

There are, however, several more pieces to be provided to create a complete system that meets the needs of today's data communications system user. What has been discussed so far are some of the lower levels of a system. Nothing has been said yet about the way that the data is put into the right organization or structure so that the system can handle it properly. There are also issues that relate to the characters which should be used to represent the data, the way the system prepares for and handles errors, and the response of the receiver to any message from the sender to let the sender know the message has or has not been received properly. These are issues of format and protocol and are discussed in subsequent chapters.

What has been discussed is analogous to using mail service and letters to send data. Even though the sender and receiver have established that communications will be written, and letters will be sent to a certain address, there is more to be agreed on. What alphabet should be used: English, Chinese, Russian, or another choice? Even if the alphabet is defined, what language is chosen—English, French, Italian? What punctuation should be used to indicate where sentences end and whether they are statements or questions? For successful communications, both parties must agree, in advance, on these and many other related issues.

QUESTIONS FOR SECTION 5-6

1. Why are digital signals attractive for multiplexing and modulation?
2. Does SDM or FDM differ significantly for analog versus digital signals? Explain.
3. How does TDM work for digital signals?
4. What clock rate is needed for a TDM system for digital systems? How is it determined?
5. How is demultiplexing performed for digital signals in a TDM system?
6. What are two features needed for successful TDM of digital signals?
7. When does the multiplexer have to sample the digital signals it is multiplexing? Explain how this differs from analog sampling.
8. What is framing? Why is it needed?
9. How is framing done?
10. What is the tradeoff in choosing the number of bits per frame and the number of framing bits used? What performance features are affected?
11. What happens if framing is lost?
12. What is an alternative way of interleaving users in a TDM process? Why is it sometimes used?
13. How can a signal which is the result of a TDM operation be remultiplexed again with another TDM signal?
14. Why might mixed modulation be used?

PROBLEMS FOR SECTION 5-6

1. Four identical signals, each at 10 kbits/s, are to be multiplexed. What clock rate is needed?
2. Two identical user signals are to be multiplexed. The signals have a 5-kHz bandwidth and are being sampled at exactly the Nyquist rate. The signals are then digitized with a resolution of 8 bits.

   a. What bit rate does each signal require?

b. What clock rate does the multiplexer require?
   c. What is the clock rate needed to multiplex 10 such signals?
3. A simple system is being designed to multiplex 10 users. Each user has a 10-kHz bandwidth. Sampling is at twice the Nyquist rate, and the digitization is to 7 bits.
   a. What bit rate does each user signal require?
   b. What clock rate does the multiplexer require?
   c. The resolution is increased to 8 bits. What multiplexer clock rate is needed?
4. A system can support a data rate of 100 kbits/s. How many users can it multiplex if each user is a 3-kHz-bandwidth signal, sampled at the Nyquist rate, and with 7-bit digitization?
5. A system uses one group of 10 framing bits. The multiplexing takes five signals, each of which has been digitized to 8 bits, and combines them. What percentage of the overall frame is spent with framing bits?
6. A system is designed for high efficiency in using channel time. The goal is to have framing take up no more than 5 percent of the frame time. The system supports 10 users, each of whom has been digitized to 12 bits. How many frame bits can be used?

SUMMARY

Digital communications systems are designed to handle digital signals, which can take on only a limited set of specific values. This is in contrast to analog signals, which can take on any value within the overall range. Digital systems offer advantages in many areas. They are much more noise-resistant, allow signals to be restored to original quality by repeaters, and enable error detection and correction schemes to be employed and to be computer-compatible. Digital signals are also easier to analyze, multiplex, and modulate. They can be multiplexed into groups, and each of these in turn can be multiplexed into larger groups.

Data which is in digital format can be used directly with digital systems. Analog signals, however, must be converted to digital format. First, the signal must be sampled, at the Nyquist rate of at least twice the signal bandwidth. Then, the signal is digitized by an analog to digital converter. At this point, the signal is in digital format. Some communications systems require that the digital signal be encoded to prevent the data stream from having more 1s than 0s, which would cause a change in the average or dc value. This would cause distortions as the signal passed through the system. In addition, encoding is sometimes needed to help the system synchronize with the clock signal that was used for the digital signal at the sending end, so it can be received properly and the original digital bits recovered without error.

Multiplexing of digital signals requires a switch which has the effect of scanning, in order, from one user data value to another, through all the users.

The nature of digital signals allows the digital signal value to be sampled at any time by the multiplexer. The multiplexer provides a digital data stream to the modulator. This stream represents the interleaving of all the individual data streams. It can also be combined with other multiplexed signals for an even higher speed supergroup of user signals.

END-OF-CHAPTER QUESTIONS

1. How is a digital system similar to an analog one? How is it different?
2. What is a binary system? How is it related to a digital one?
3. What is data? Why is the term used with digital systems?
4. What is the function that a digital signal receiver must perform? How does this compare to the function of an analog system receiver?
5. What is noise? How does it affect the signal in the system? What are some of its sources?
6. Explain why a digital system is superior to an analog one with respect to noise, measurement, attenuation, error detection, error correction, TDM, and computer compatibility.
7. Explain why digital systems have some immunity to noise. Why can a digital system actually restore the signal to its original value exactly? Where is the boundary between 0 and 1 usually set in a binary system?
8. What is the difference between noise and signal attenuation? How does each affect the way the original signal is received?
9. How can the error rate of a digital system be specified? What are typical values?
10. What is the concept of sampling of an analog signal?
11. How frequently must the sampling be performed, according to Nyquist theory? What is the key specification in this theory?
12. What is the significance of the Nyquist rate in determining how many users a TDM system can support?
13. What is jitter? What is its implication in sampling?
14. Why is sampling performed at a rate higher than Nyquist theory requires?
15. What is the name of the problem that occurs if sampling is performed at a rate below the Nyquist rate? What happens to the reconstructed signals?
16. Why do most systems have a low-pass filter at the input?
17. How are analog signals converted to digital format?
18. Give two applications in which the digitized signal must be converted back to analog format, and two in which it does not.
19. What is the resolution of a 3-bit converter? An 8-bit converter?
20. Why don't systems simply use more bits? What is the tradeoff in bits of resolution versus system communications capacity?
21. What is the effect on the apparent accuracy of only a few bits of resolution?

22. What is quantizing error? What causes it? Explain why it is not a defect in the ADC circuit operation itself.
23. Why are buffers for bits sometimes required for faster system operation? For slower operation?
24. Explain why a combination of sampling and digitization forms a PCM system.
25. What is dc drift? How can it occur in digital systems which send out a number of 0s and 1s?
26. What part of an incoming signal contains information for synchronization?
27. What is the function of encoding? What is the encoding name for regular binary 1s and 0s?
28. Compare NRZ and RZ. What is the difference? Why is RZ sometimes adequate, and sometimes not, in overcoming the weakness of NRZ?
29. What is Manchester encoding? Compare the bandwidth it uses to that of NRZ. What are the advantages of Manchester encoding?
30. How are digital signals and PCM signals multiplexed in a TDM system?
31. How is the clock rate for a digital TDM system determined?
32. How is the demultiplexing operation performed?
33. Explain the difference between sampling an analog signal and a digital one.
34. How is a system receiving a digital TDM signal synchronized to the overall group?
35. Why are framing bits kept to a minimum? What is the risk in doing this?
36. Explain how a multiplexed TDM signal can be combined with another similar signal. What clock rate does the combined signal require?

END-OF-CHAPTER PROBLEMS

1. A digital system is designed to operate with noise levels of up to 1 V. Binary 1 is represented by +10 V. What is the closest value for binary 0 that will give error-free performance?
2. A digital system is designed with nominal voltage levels of 0, 2, 4, and 6 V. Where would the digital decision voltage levels be set?
3. A system is designed with digital levels of 8, 9, 10, and 11 V. Digital signals arrive with these amounts of noise: -.55, +0.3, -1.1, -.45, and +0.75. Which of these noise values will cause an error at the receiver?
4. A binary system uses signals of 0 and 5 V, to represent binary 0 and 1, respectively. The binary pattern is 10100. For each of the noise values shown, sketch the 10100 pattern before and after regeneration: −2.0, +3.0, -0.4 V. Which signals are received in error after regeneration?
5. What is the theoretical Nyquist sampling rate for signals with these bandwidths: 10 kHz, 90 kHz, 6 MHz?
6. At what bandwidth will aliasing occur, in theory, with these sampling rates: 10 kHz, 1 MHz, 250 kHz?

7. A system is designed to handle analog signals with 10-kHz bandwidth. What sampling rate is needed to sample at 50 percent above the theoretical rate? Five times the theoretical rate?

8. A system is designed to sample five video signals, each of 6-MHz bandwidth. The sampling is performed at twice the Nyquist rate. What number of samples per second must the system support?

9. A system is capable of supporting 150,000 samples/second. How many users of 10-kHz bandwidth can be supported if sampling is performed at a rate 50 percent greater than the Nyquist rate? How many at a rate 100 percent greater than the Nyquist rate?

10. A 3-b ADC is used over the range of 0 to 2 V. What analog range does binary 011 represent? How about 100?

11. By what factor is the resolution of a 10-bit converter better than that of an 8-bit converter?

12. What is the resolution, in volts, of a 4-bit converter designed for the 0- to 10-V range? Into how many subranges is the 10-V range divided?

13. A 4-bit converter is used for signals that span from -1 to $+1$ V. What binary value will occur for these analog values: $-1, +1, -0.3, +0.4$ V?

14. A communications system uses an 8-bit converter at exactly the Nyquist rate for signals with a 3-kHz bandwidth. What bit rate is needed to sample and digitize the analog input? What is the bit rate if the converter is a 10-bit unit?

15. For the preceding problem, what is the bit rate if the system has to support 10 users with 8-bit conversions?

16. A binary system sends out this pattern: 1011 0010 1111 0000. It uses 0 V to represent binary 0, and 5 V to represent binary 1. What is the average voltage after each successive bit?

17. Repeat Prob. 16 for -1 V representing binary 0, and $+1$ V representing binary 1.

18. Sketch the signal represented by the values in Prob. 16.

19. Repeat Prob. 16, using RZ encoding.

20. Repeat Prob. 16, using Manchester encoding.

21. For a binary pattern of 1011 0100, sketch the NRZ, RZ, and Manchester encoding patterns. Use signal levels of 0 and 1.

22. A system is designed to multiplex, via TDM, 24 identical users. Each user is presenting 8 kbits/s. What clock rate is needed at the multiplexer?

23. A system is converting voice signals to digital format and then multiplexing many such identical signals. Each signal is a 3-kHz-bandwidth voice signal. It is sampled at twice the Nyquist rate. The ADC is a 10-bit unit.

    a. What is the sampling rate for each signal?
    b. How many bits per second does each signal cause after it is digitized?
    c. What clock rate is needed for the multiplexing if there are eight identical users?

24. A system can transmit 500 kbits/s. The users who are multiplexed into this system have 5-kHz bandwidth. They are each sampled at twice the Nyquist rate and digitized to 10 bits.

   a. How many users can be handled?
   b. How many can be handled if the sampling is performed at three times the Nyquist rate?
   c. How many can be handled at twice the Nyquist rate, but for 10-kHz bandwidth?
   d. How many can be handled at 10-kHz bandwidth, twice the Nyquist sampling rate, and 12 bits digitization?

25. A system uses 10 framing bits for every group of 25 user digital signals. Each user is digitized to 8 bits.

   a. What percentage of time is devoted to framing bits?
   b. The number of framing bits is increased to 20. What percentage of time does this represent?
   c. The system goal is to have efficiency of 99 percent. What can be changed to achieve this? How many framing bits should be used? How many bits of resolution per user, or users?

6

Communications, Media, and Problems

The physical path for the data communications signal energy is the medium, *which can be wire, air or vacuum, or optical fiber. Each of these offers technical benefits and drawbacks, as well as cost differences. The choice of the medium is sometimes made by the system designer, and sometimes by the application alone. There are several key performance criteria that each medium must be judged against, to determine how well it can send the signal the required distance, in the application environment.*

The relative strength of various signals is measured by using the decibel (dB) scale, which is a convenient way of expressing the ratio between any two signals. The decibel scale is also useful for comparing signal strength to noise strength. The medium bandwidth determines the bit rate that the medium can support, and digital signals need large amounts of bandwidth so that each bit is received without excessive smearing or distortion, which would confuse the receiver. An eye pattern *helps quickly judge how well the medium is performing.*

Some applications may suffer from problems with undesired voltages on both signal wires. These can be minimized by using differential or isolation techniques, which considerably reduce this voltage, as well as the danger of equipment or user damage due to high voltage.

6-1 The Role of the Medium

The medium is the actual physical path for the electromagnetic energy of the link or channel of the communications system. Through the medium, the energy representing the data of the sender can reach the receiver. This path can take many forms: an electrical conductor such as wire, air and/or vacuum, or optical fiber. Each of the three main categories can be further divided into classes which specify in more detail the nature of the medium, such as the many ways that wire can be made in cables for communication.

In some applications the designers and users of the system have very little choice about which type of medium is employed. For sending data from remote locations, or from users on the move, *radio links* using air and vacuum as the medium are the only practical choice, since a physical wire would be very restrictive. In other applications, the determination of which medium to use is based on cost, performance, and ease of installation issues. The link which connects the phone system regional offices on both the east and west coasts of the United States uses all three media. Each has specific advantages and disadvantages which the telephone system takes into account.

Each of the choices for the medium can provide acceptable performance in the right situation. Each type of medium, however, is also capable of providing poor performance if not chosen, installed, or used correctly. Some of these problems are due to laws of physics and electronics that cannot be changed. Others are due to the way the medium is installed, handled, or used. Sometimes the problem comes not from the medium itself but from the connectors that are usually needed to attach it physically to the rest of the system.

The key medium characteristics that affect performance are

Range: The distance over which the medium can be used

Bandwidth: The bandwidth that the medium can support

Loss: The amount of attenuation of the data signal in the medium

Noise resistance: The effectiveness of the medium in preventing electrical noise from corrupting the signal that is being carried

Security: If the medium is resistant to an unauthorized user "tapping in" and picking up the data

Cost: The cost to install the medium

Connection: The way the medium is connected to the system physically and the way it can be spliced together

It is important to understand these characteristics in more detail, so they can be related properly to the specific medium used in the communications system.

The range that is achievable with any medium is determined by what the medium does to the signal, or how well the signal fares while traveling through the medium. The range can be as short as a few feet (for signals that are traveling through ordinary copper wire) or as long as millions of miles (for signals traveling through a vacuum in space). Some of the reasons why range may be limited are electrical noise that the desired signal picks up as it passes through the medium, attenuation of the signal in the medium, and practical reasons such as the difficulty in building circuits which can recover a weak, noise-corrupted signal.

Different media have different bandwidths; that means that they can support a wider or narrower range of signal frequencies. Any signal component beyond the bandwidth of the medium is cut down very sharply in amplitude and so barely reaches the other end of the link. The effect is that the medium acts as a low-pass filter. The cutoff frequency for the filter depends on many factors, including the way the medium is fabricated. A good example is copper wire used as signal cable. The copper can be made into simple pairs of wires twisted around each other and having a bandwidth of about 100 kHz. The same copper wire can be made into coaxial cable, which has one wire running inside another one, and have a bandwidth of hundreds of megahertz. Although in general wider bandwidth is desirable (why?) the simpler twisted pair may be used for an application, if it has sufficient bandwidth, since it is cheaper and easier to install.

Even within the bandwidth that the medium can carry, the signal is attenuated as it travels by wire resistance or energy absorption by the atmospheric molecules. This loss of signal strength means that at some distance the value of the signal voltage (or energy) will be too small and the receiving circuitry will be unable to recover the original data. This loss distance can be anywhere from a few feet to many thousands of feet, depending on the signal type and medium construction.

All communications must be concerned about electrical noise. Even if the signal is perfect, the environment that the signal travels through may have electrical noise from other wires, motors, power supplies, and other sources. The electrical noise adds to the desired signal and corrupts it. Some types of media are better able to prevent this outside noise from reaching the actual signal-carrying wires; others offer very little resistance to noise pickup. In general, producing noise-resistant cables is expensive, and the shielding which keeps the noise out also causes some performance problems in range, bandwidth, and loss. Thus, the choice of which type of wire or cable to use is a tradeoff, juggling the required performance in each of these areas. Of course, the use of air or vacuum for radio transmission means that noise in the atmosphere and space cannot be avoided.

Many systems require some sort of security against unwanted listeners picking up the data communications signal. There are many ways to achieve this, and some systems use a combination of methods for better security. The first level of protection is to make it difficult for someone to tap into the wire or cable, at least without the sender or receiver knowing it. For some types of media, such as radio waves, this is impossible, since anyone can put up an antenna and pick up the signal. Some wire and cable types, in contrast, are very hard to splice into without detection, since adding a new splice upsets the signal electrical balance as it travels down the wire. The newest communications medium, using glass fibers to carry the signal, has a great advantage in this area. It is extremely difficult to tap into the fiber without ruining the whole system.

Cost, of course, is an important consideration. Some media, such as air and vacuum of space for radio transmission, are low in cost, although the sending and receiving equipment is expensive. Other media can be expensive to build and install, such as thick, heavy cables used for carrying signals long distances, including across the ocean floor from the United States to Europe. These cables have many wires in them for many users and also have thick protective jackets to keep the copper wire of the cable safe from rocks, animals, and mechanical equipment. Ordinary copper wire itself is relatively inexpensive, and a few feet of wire for connecting a computer to a printer costs only a few dollars, including the protective jacket.

Any type of medium has to be connectable to the rest of the system circuitry, in order to be practical in real installations. For each medium, there are different types of connections and connectors that are used. Some of these methods, such as using solder to connect copper wires to circuitry and to connect two pieces of wire together, are simple and straightforward. Other media have more complicated connectors and require more involved mechanical schemes to take the signal energy from the medium to the rest of the system.

The next sections discuss, in more detail, some specific types of media used for data communication and the performance characteristics they have in the areas highlighted. These media include twisted pair wire, flat multiconductor cable, coaxial cable, microwave waveguides, fiber optics, and air/vacuum.

QUESTIONS FOR SECTION 6-1

1. What is the role of the medium in a communications system?
2. What are the three classes of media that are of interest?
3. What is the meaning of these characteristics: range, bandwidth, loss, noise resistance, security, cost, and connectors?
4. What span of ranges is available? What limits range?
5. What is the effect of bandwidth limitation? What determines the bandwidth of a medium?
6. What are the causes of attenuation?
7. What are the security issues for various media?
8. What has an effect on cost, besides the medium itself?
9. How do mechanical connectors affect the choice of medium?

6-2 Wire and Cable

Copper is an inexpensive metal that is easily made into wire. Fortunately, it is also an excellent conductor of electricity. When separate wires are collected into bundles, or given special protective jacketing, or more than plain wire is needed, the term "cable" is often used. For our purposes, the words "wire" and "cable" are essentially interchangeable.

There are many ways that wire can be used for communications. Each type of cable that is made from wire has certain characteristics that make it a better or worse choice for the application. The main types of wire and cable used in data communications are

Twisted pair

Multiconductor flat cable

Coaxial cable

Twisted pair is the simplest and lowest-cost cable. It consists of two insulated wires, usually no. 22 or 24 gauge, twisted around each other in a continuous spiral (Fig. 6-1). The reason for the twist is to minimize the susceptibility of the cable to external electrical noise. Many circuits are designed to cancel out any noise or interfering signal that appears on both wires identically. Through spiral

Fig. 6-1 Twisted wire pairs are an inexpensive and effective communications medium in many applications.

wrapping of the two wires of the twisted pair, the noise appears more nearly identical on the two wires, and the communications circuitry can suppress most of it.

Twisted pair can be run for distances of many miles. It is used, for example, between telephones and the central office, and between central offices themselves. It has a bandwidth of several hundred kilohertz depending on wire size. If a separate metal shield is used around the twisted pair to shield it further from noise, the noise resistance is increased, but the bandwidth is reduced by a factor of 2 or 3. The loss factor for twisted pair is moderate—not as good as for parallel, spaced wire but better than for some types such as coaxial cable. Since twisted pair is simple wire, it is easy to strip, connect, and solder (or use with mechanical connectors), but the security is only fair, since unauthorized users can do the same (although sometimes a tougher outer insulating jacket makes tampering harder).

One problem with twisted pair is that it is difficult to use when many signals must be brought from one place to another, especially within two subsections of a system. It may be difficult to route all the pairs, and connectors must be used for every pair, or else such wire must be stripped and soldered, processes which are expensive and time-consuming, and make taking the pairs apart for repair or test difficult. Many times, *multiconductor flat cable* is used. This kind of cable consists of many parallel wires in a common plastic jacket (Fig. 6-2). A cable of this type can have any number from 10 to about 50 wires. The wires all are a single group mechanically, and they can be used with a single connector at each end. This connector crimps onto the cable, pierces the insulation, and provides contact to all the wires in the cable in one quick operation. Since the wires are all in one cable, with a single connector, this type of cable is very convenient for

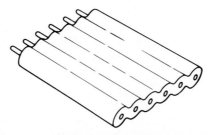

Fig. 6-2 Multiconductor flat cable is more costly than twisted pair, but it has wider bandwidth and can be easily terminated in a connector.

assembling and disassembling parts of a system, procedures which are needed in installation and repair activities.

Flat cable is more expensive than twisted pair, and that is one reason why it cannot always be used. The bandwidth of the flat cable is much larger than that of twisted pair, typically about 10 MHz. The loss of flat cable is high, and signals cannot travel more than about 10 to 100 feet (ft) before being severely attenuated. As a result, flat cable is used mainly within a system, as a data interconnection. It is not technically practical, or inexpensive enough, to use over longer distances. It is also much more susceptible to electrical noise, because the long parallel wires in the cable act as antennas and pick up undesired signals. Since it is used within a system, the integrity and security of the flat cable are usually not issues as they would be between widely separated users.

For some applications, *coaxial cable* (''coax'') must be used. Coaxial cable consists of a single wire centered in a tube of copper braid, which acts as the second wire of the pair and electrically shields the center wire from electrical noise (Fig. 6-3). This shield is almost always connected to signal ground for this purpose. The coaxial cable diameter can be as narrow as 0.15 inches (in) or as wide as 0.75 in, depending on the type used. Coaxial cables are usually identified by an RG number. The electronics industry has assigned these RG numbers to specific types of coax, and each number identifies a certain inner wire gauge, overall diameter, and what kind of insulation, if any, is in place within the tube. The most common types of coax are RG-58, RG-8, and RG-174. Each of these coaxial cables offers different performance.

Coax is fairly resistant to electrical noise because the entire cable carries its own shield. It is fairly expensive to buy (about $1/foot) and can be difficult to install because of its diameter and mechanical stiffness. It does offer very high bandwidth, up to several hundred megahertz, and this feature makes it very attractive for many data communications systems. It is hard to tap into along its path, but it requires special connectors at each end. It is most often used for a link from one point to another where the link will be permanently installed and high data rates are needed. Typically, coaxial cable has moderate levels of attenuation. (A signal in a coaxial cable loses about one-half of its original value after 1000 ft of travel, depending on signal frequency.)

A system designer can choose among many tradeoffs with coaxial cable. The larger-diameter cable has wider bandwidth and less attenuation, but is more

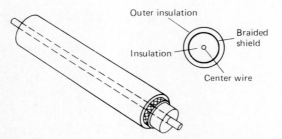

Fig. 6-3 Coaxial cable consists of a center conductor wire, insulation, outer braided shield, and outside insulating jacket.

costly and harder to install. It also is bulkier, and therefore there is room for fewer cables in a trench or duct. The cable is mechanically rugged (and special ruggedized jackets are available), and the physical construction of the cable with an inner wire, insulation, and outer tube means that the cable can be run in curves and bends and still have the same electrical characteristics. Different types of internal insulation can provide various attenuation and bandwidth factors.

It should be clear now that the choice of the type of wire and cable to use is influenced by many factors. The system designer weighs all the choices against the needs of the system and in some instances must modify the system's desired goals to meet a cost or connector requirement. However, there is usually an acceptable choice to be found among twisted pair, flat cable, and coaxial cable.

QUESTIONS FOR SECTION 6-2

1. What is twisted pair? What are its benefits? Drawbacks? Where is it used? How is it connected? What is its sensitivity to electrical noise?
2. Repeat Ques. 1 for flat cable.
3. Repeat Ques. 1 for coaxial cable.
4. How is coax designated? What are some of the differences from one type to another?

6-3 Air and Vacuum

The earth's atmosphere and the vacuum of space are very useful, low-cost, and effective ways of transmitting the energy of the user data. There are some similarities between the two media, and there are some key differences. The entire electromagnetic spectrum from extremely low frequencies to extremely high ones can be used in both air and vacuum because the bandwidth of both is very wide. (See Chapter 2.) In a vacuum, such as in space, the electromagnetic energy can travel without any impediments. The range of signal travel is many millions of miles, as deep-space satellites and probes have proved in actual use. There is very little signal loss in a vacuum, but only a small fraction of the original signal can reach the receiver antenna. This small amount is not due to loss, or waste, in the medium. It is a result of the fact that the signal from the sender spreads out over a wide area, and the receiver antenna can only intercept a small percentage of this signal (Fig. 6-4a). Even a dish antenna, which is designed to focus all the sender electromagnetics energy at the receiver, cannot completely solve this problem, since over great distances even a small energy beam spread can be significant (Fig. 6-4b). For example, if all the transmitter energy is focused by the antenna into a beam only 1° wide, after several thousand miles only about 0.00001 percent of the energy is intercepted by the antenna at the receiver. The rest of the energy is in other directions, still as electromagnetic energy, but unreachable.

Electromagnetic energy can travel great distances through air, but not so easily as through vacuum. The various molecules in the air (carbon dioxide, oxygen, nitrogen, and others) absorb some of the electromagnetic energy, and there is

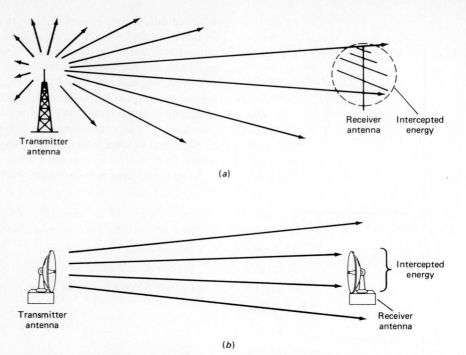

Fig. 6-4 (a) A simple wire antenna radiates electromagnetic energy in all directions; (b) a dish antenna focuses the energy pattern at both the transmitter and the receiver.

less left to reach the receiver. The amount absorbed depends on the frequencies of the electromagnetic energy. The molecules have differing effects, depending on the specific frequencies being used. The situation is most severe for the higher frequencies of about 300 MHz and above, usually lumped into the term "microwaves." These waves are very severely attenuated by the absorption of the energy by water vapor in the air. If there is a microwave link between two users, and a rainstorm moves in between these users, the link may become useless as most of the energy is absorbed by the water droplets and never reaches the receiver.

Through some subtleties in the laws of physics, this effect of absorption is not uniform through the microwave bands. Some specific wavelengths are just the right size to move past the water vapor without being absorbed. Longer wavelengths and shorter wavelengths, however, are absorbed.

Energy in both vacuum and air travels in straight lines. This is in direct contrast to energy in wires and cables, which can be directed to flow in the desired path by the copper conductor. This means that air and vacuum are most usable when there is a direct line of sight between the sender and receiver. Examples of this are signals from a spacecraft on the moon, or signals between two high towers with microwave antennas. (Why are antennas installed on high structures?) In general, the higher frequencies are not reflected except for solid metal or tight mesh reflectors. If a 300-MHz signal must travel from the United States to Europe, a single line-of-sight path is impossible. A communications satellite is used, to act as a signal reflector. It picks up the line-of-sight signal from the transmitter, amplifies it, and retransmits it down to the receiver on earth.

Lower frequencies (up to about 30 to 50 MHz) can be reflected by the various layers of the earth's atmosphere. These lower frequencies can be used for worldwide communications by carefully bouncing the signals from the sender, to the atmosphere, to the receiver. This type of communication is not extremely reliable, though, and is subject to severe day-to-day and hourly variations, depending on conditions in the layers of the atmosphere. This phenomenon is called *fading*, and the received signal strength goes up and down, sometimes going quite low.

Air and vacuum media do not offer much security. Anyone can put an antenna in the path of the signal and pick up some of the data. This is one drawback of using air and vacuum for transmitting energy. Although the medium itself is free, it doesn't offer protection against unauthorized listeners. These listeners can even make it impossible for the original sender and receiver to communicate. Since there is no shielding in the medium, someone can interfere with (''jam'') the channel by broadcasting at the same frequency. Then, the energy of the jamming signal and the original signal mix together, and the receiver may find it difficult or impossible to sort out the desired signal.

There are solutions to both the unauthorized listener and the jamming problems. If the data is encoded with a secret scheme, someone can pick up the actual electromagnetic energy but not be able to determine what it means. This is commonly done. It is the same as sending messages in a language which no one else knows. The sender and receiver can communicate, but an outsider receives only a jumble of meaningless letters. Jamming can sometimes be overcome by special signal transmission techniques, which give the receiver some idea, in advance, of what the received signal looks like. Then circuitry at the receiver can try to use this as a pattern for locating the real data, even though the jamming noise is severe. It is not a guaranteed way to overcome jamming, but it does sometimes help, especially if the jamming energy is not great.

Of course, the medium is free. But, the connections to it are not. When electromagnetic energy is transmitted and received, antennas are needed to couple the energy from the transmitter into the air or vacuum. These antennas can range from a simple stiff wire or tube a few feet long, to complex arrays that are tens or hundreds of feet long. For some frequencies, dish antennas anywhere from 3 to 100 ft in diameter, are needed. These larger dish antennas are used for long-distance communication when the signal energy must be precisely aimed and focused by the sender transmitter, and the receiver must intercept as much of it as possible to have usable signal strength.

Waveguides

Sometimes air or vacuum is used as the medium, not by choice but through consequences of laws of physics. If the communications system uses the microwave or higher frequencies, the circuitry at each end must handle signals at these frequencies. It would be easiest just to use wire, as any other circuit does. However, at the short wavelengths of microwaves (about 1 to 10 centimeters

[cm]) the wires in the circuit act as small antennas. Instead of confining and directing the electromagnetic energy, the signal radiates off the wires into the surrounding area and system. This causes two problems: the energy that is needed is lost, and the rest of the circuit sees undesired signals and malfunctions.

The solution to this problem is called a *waveguide*. A waveguide is a hollow metal tube, either a cylinder or a rectangle in cross section, that acts as a pipe for the electromagnetic energy (Fig. 6-5). The diameter of the waveguide is about the same as the wavelength. Since the waveguide is solid metal, any electromagnetic energy put into it cannot escape and travels through the guide to the other

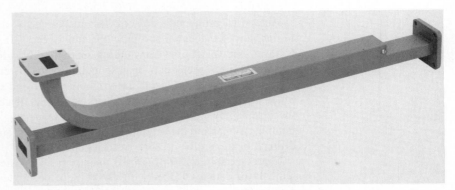

Fig. 6-5 A waveguide is a hollow metal tube which directs and contains the electromagnetic energy at microwave frequencies. (Courtesy Hewlett-Packard Co.)

end. A system using waveguides looks more like a plumbing installation than an electronics one (people who work with waveguides in electronics often call each other plumbers) with pipes, pipe fitting, pipe taps, and so on. The waveguide is a very effective way to confine the energy to the microwave signals and ensure that the signals travel only where they should.

Certainly, mechanical structures such as waveguides are much more expensive than ordinary wire and cable. Why are the microwave frequencies used, if they require such elaborate plumbing? The answer is that the wide bandwidths needed for high-rate data communications are available only at the microwave frequencies, where frequency bands are tens and hundreds of megahertz wide. The lower frequencies, from 1 to 30 MHz, can offer a maximum bandwidth of 30 MHz—but only if the system can take the entire spectrum for itself, and that is unlikely. High data rates require bandwidth, which means using up hundreds of megahertz, and that requires going up to the microwave region.

Like wire and cable, air and vacuum are not the ideal media. For many applications, they offer the best cost and performance choice, and for some applications they are the only choice, since it is impossible to run a cable to a ship, airplane, or spacecraft. Even for land-based data communications systems, the low cost and flexibility of air and vacuum often outweigh the problems with antennas, security, and reliability.

QUESTIONS FOR SECTION 6-3

1. What is the relative bandwidth available with air and vacuum?
2. Why does only a small part of the transmitted signal reach the receiver? What is sometimes done to improve this?
3. What causes attenuation in air? Which frequencies are most affected?
4. How can the transmitted energy be made to pass between the two points that are not in a straight-line path for very high frequency (VHF) and higher frequencies?
5. What is fading? Why is it a problem?
6. What is the security of air or vacuum? What techniques are used to increase it?
7. What is jamming? How can it sometimes be overcome?
8. Why are dish antennas sometimes needed?
9. What are waveguides? Where are they used? Why?

6-4 Fiber Optics

The concept behind using fiber optics for communications has been known since the 1960s, but many technical developments had to occur before the method could become practical. Now, the use of optical fibers for data communications is considered among the best alternatives that a system can use. It offers wide bandwidth and many other advantages.

The idea of *fiber optics* is to use light, instead of current or voltage, as the energy which carries the data, with the light as a carrier that is turned on and off, with binary amplitude modulation. The problem is to direct the light from the source to the receiver. The solution is to use a hair-thin strand, or fiber, of glass or plastic as a light pipe. If a light source is put at one end, any light that enters the fiber stays in that fiber and travels through the fiber to the other end. The light does not pass out of the walls of fiber as it travels. This is because of a property called *total internal reflection*: if a light wave is traveling through a material with a high index of refraction compared to an adjacent material, and it hits the interface between them at certain low angles, the light does not cross the boundary but completely reflects back (Fig. 6-6). The same effect occurs when

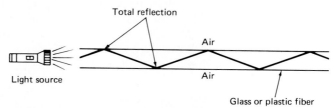

Fig. 6-6 Total internal reflection is the key to optical fibers. The light energy bounces off the fiber/air boundary and reflects internally only; it does not escape.

Communications, Media, and Problems

one is looking at the sky while under water—from some angles it simply can't be seen.

At the receiving end of the fiber, a light detector senses the light. Thus, the communications medium is the fiber, and the energy used is light energy.

The technical problems in implementing this method have been largely overcome. These problems involved various issues: the light source had to be small, reliable, and easy to turn on and off (for data) at high rates. The glass or plastic fiber had to be smooth and pure. Any impurities would be the equivalent of wire resistance and would cause some of the light put into the fiber to be lost before it reached the other end. Modern fibers have low loss, and about 50 percent of the original light passes through a 1000-ft fiber. This is enough for a practical communications system. Another problem involves the receiver. Some sort of reliable and light-sensitive component is needed to sense the light. A general problem at the sending and receiving ends is devising a way to physically couple the light source (or receiver) to the fiber since the light source and light receiver must be precisely lined up with the ends of the fiber, so that the maximum amount of light is transferred into the fiber and out of the fiber. This problem has been overcome with new, specially designed connectors and mechanical couplings.

Now that many of the initial problems with fiber optics have been overcome, optical fiber is becoming a preferred medium for data communications. There are many reasons for this:

- The bandwidth of the fiber and light beam is extremely wide. It is possible to handle signals which turn on and off at gigabit per second rates (1 gigabit [Gbit] = 1000 Mbits).
- The fiber itself is very thin and not expensive (about $0.10 to 0.50 per foot). The thinness means that it is easy to handle, and many fibers can be put in thin trenches or narrow conduits.
- The light signal is absolutely immune to electrical noise from any source. Even if there are sources of electrical noise directly touching the cable, the electric fields of the noise source cannot affect the light beam in the fiber.
- The signal in the cable is secure from unauthorized listeners. It is relatively hard to tap into the cable without being noticed, and the entire light signal is confined within the fiber. No light escapes to the outside where someone else could see it.
- Since there is no electricity or electrical energy in the fiber, it can be run in hazardous atmospheres where the danger of explosion from sparks may exist. Also, the fiber itself is immune to many types of poisonous gases, chemicals, and water (which would short-circuit regular cables unless they had waterproof jackets).

Fiber Optic Weaknesses

Of course, fiber optics does have some drawbacks. The major problem still is the method used to terminate the fiber cable. Unlike wire, which can be stripped of insulation easily, then mechanically connected and soldered if necessary, fibers have to be cut very carefully so that the cut end is smooth and straight. Special

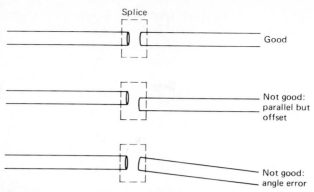

Fig. 6-7 Alignment is the key factor in splicing optical fiber cables. Misalignments cause severe loss of signal energy.

tools are needed for this. If the fiber has to be spliced to another fiber for longer length, this involves carefully cutting and polishing the cut ends and joining them in a special fitting, often with special optically clear glue. The ends of the two fibers must be in perfectly straight-line alignment (Fig. 6-7), unlike copper wires, which can be simply twisted together, then soldered or crimped. It is ironic that this difficulty in splicing and connecting fibers also makes a fiber optic system very secure against eavesdropping.

Another drawback to fiber optics is that it is hard to have multiple users on the line. For mechanical and optical reasons, at present there is no satisfactory way to have a party line, whose users can be attached or removed easily. This is in contrast to wire and cable, where these multiple-user situations are common and very easy to implement. Because of this limitation, fiber optics is used mainly for *point-to-point communications*, in which data is going from one point to another, with no new users picked up or dropped off along the way. These point-to-point applications include links between telephone central offices, cross-country communications links, and computers connected to their printers nearby. Fiber optics would not be a good choice for a system that carries messages around a building from a single computer to many terminals at people's desks. But for applications in which large amounts of data go from point *A* to point *B*, the fiber is an excellent choice. Multiplexing is presently achieved electronically, before the data signal is converted to light.

A majority of the applications for fiber optics at this time are for the high-data-rate users; in these applications the difficulties of splicing the cable and connecting it are overridden by the bandwidth and channel density that fiber optics can provide. In the last few years, several manufacturers have developed lower-cost connectors that look like connectors for ordinary cable and are easy to use. These do not provide the full-bandwidth capability of fiber optics, but they do allow the other benefits of this new technology to be realized, and many applications do not need full bandwidth (a computer talking to another computer nearby, for example). Some manufacturers have even developed special kits, which plug into the existing wire connector and convert the electrical signal to a light signal, so that it can be used with fiber optics. The kit also contains a small unit which

Communications, Media, and Problems

performs the reverse process by taking the light signal and reconverting it to an electrical signal compatible with the existing connectors. In this way, a user can convert quickly and cheaply to fiber without designing a whole new system. This is very useful for applications in which the system needs the noise immunity or security of fiber optics as part of a system performance upgrade.

QUESTIONS FOR SECTION 6-4

1. What form does the energy of the data communications signal take in a fiber optic medium?
2. What is used as a guide for the light? Why does the light stay in the medium?
3. What are the components of a fiber optics system?
4. What four technical problems had to be overcome for successful fiber optics systems?
5. What is the bandwidth achievable in a fiber optics system?
6. What is the advantage of fiber optics in size and cost?
7. Why are fiber optics systems immune to electrical noise?
8. Why are optical fibers secure, compared to other media?
9. What is the greatest problem of fiber optics systems, as compared to copper-wire-based systems?
10. What are two other problems with fiber optics systems?
11. Where is fiber optics a good choice, in general? Why?

6-5 Noise

Electrical noise is a general problem that any data communications system must accommodate. The effect of noise is to corrupt the desired signal and make it more difficult to determine correctly what the original data value was. Noise itself is a complicated signal. It can have many shapes, forms, causes, and implications for system performance.

Many systems are designed to perform reasonably well in the presence of certain types of noise. Noise falls into three groups:

Wideband noise, which contains many frequency components and amplitude values. When wideband noise is viewed in the time domain, it looks like a very random, chaotic picture (Fig. 6-8a). In the frequency domain, the noise looks like Fig. 6-8b and has energy components extending over some wide range of frequencies. If the amplitude of the noise values across all the frequencies within the bandwidth is equal, it is called *white noise*. This wideband noise is very unpredictable at any instant. Knowing its value at any one time gives no information about its value at the next instant. The effect of this noise is that occasionally a data bit is corrupted and lost, at those times when the amplitude of the noise is large enough to cause the circuitry to make the wrong decision on what digital character was received.

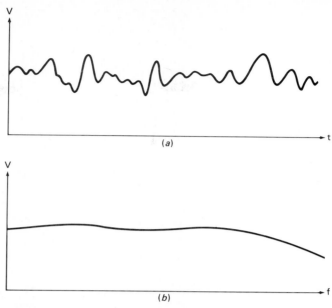

Fig. 6-8 Wideband noise (*a*) in the time domain is very random in amplitude and changes between different amplitude levels; (*b*) in the frequency domain, its spectrum is relatively uniform over the bandwidth of interest.

Impulse noise, which is a burst of noise in the time domain (Fig. 6-9). (Its frequency domain picture varies considerably with the exact nature of the burst but is usually a wide band of frequencies.)

It is caused by a sudden change in electrical activity, such as an electric motor starting up. This noise is a problem because it may have a large value for a relatively long amount of time, perhaps up to 10 to 20 ms. During this time period, it completely overwhelms the data signal and makes many data bits in a row useless. The effect of this type of noise is that synchronization may be lost, or framing may be affected. In addition, it is hard to figure out what the data was, since so many bits in succession are affected. It is like trying to figure out what someone wrote in a sentence. If a single character in the sentence is missing, it is usually possible to determine what that character should have been. If an entire word is missing, it is much harder to figure out what the word should have been on the basis of the context.

Frequency-specific noise, which has a constant frequency but may vary in amplitude, depending on the distance of the communications system from the noise source, the amplitude of the noise source, and the shielding that is in place. Figure 6-10*a* and *b* shows this noise on both time and frequency domains. The most common cause of this noise is 60-Hz power line signals radiating energy, which is inadvertently picked up by the communications system. Any power lines, circuits to lamps and test equipment, and alternating

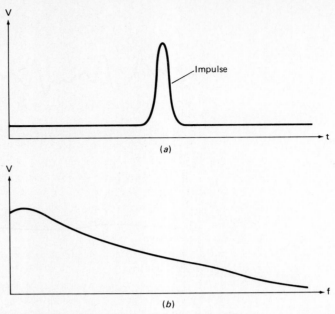

Fig. 6-9 Impulse noise consists of (*a*) a sharp spike or pulse in the time domain and (*b*) a wide bandwidth with decreasing amplitude versus frequency in the frequency domain (typical).

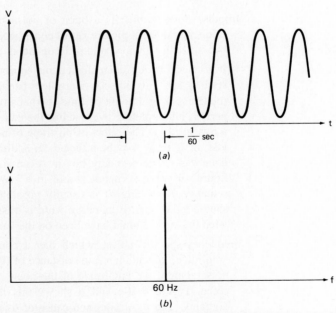

Fig. 6-10 Some noise is frequency-specific. Noise from power lines (60 Hz in the United States) has (*a*) a sine wave shape in the time domain and (*b*) a narrow frequency spectrum in the frequency domain.

176   Chapter 6

current (AC) power wiring can act as antennas to transmit this noise. This noise can either affect the received data voltages or cause parts of the communications system itself to malfunction. In many cases, this noise can be minimized by moving away from power lines and other sources. In addition, since the noise has a known frequency, it can often be reduced to a small value by filters which do not allow 60-Hz signal components to pass, but allow only desired frequencies through the system.

The important point about noise is that its value is unpredictable at any instant in time. Since the noise is random, designing the communications systems to be resistant to it is very difficult. After all, if the system could somehow know what the noise value would be, it could subtract this value from the received signal and compensate for it. The more that is known about the characteristics of the noise, the easier it is to design noise resistance into the system. That is why noise from the power lines can be easiest to deal with: as the frequency is known, defensive action can be taken against it.

Depending on the general kind of noise that the system will see, the designer may put in circuitry and methods to deal with that kind of noise. For wideband noise, in which usually only a single bit is corrupted, there are relatively simple methods available to encode the data so that the receiver can determine that an error has occurred, and even correct it. Impulse noise, which affects several bits in a row, is less frequent, but harder to handle.

Another type of noise in a system is caused by *crosstalk*. When many wires are running parallel to each other, each wire acts as both a sending antenna and a receiving antenna. Some of the signal energy on the wire does not stay in the wire, but instead radiates from it. An adjacent wire may pick up some of this energy and add it to the signal energy that it is carrying. The crosstalk noise is usually wideband, but it can be impulse. It really depends on the type of signal that the adjacent wires are carrying. Telephone cables often have to be concerned about crosstalk, since hundreds or thousands of wires are laid in thick cables, going from one office to another. Sometimes, when you are talking to someone on the phone, you can hear another conversation. This is often a result of this crosstalk. The solution is to shield the individual wires or space them apart. Shielding is expensive and makes the wires thicker, as does spacing. Both would result in fewer wires and users in a cable of fixed thickness. Crosstalk can also occur in parallel tracks on a printed circuit board.

The sources and types of noise are varied. Someone using a system that is subject to large amounts of noise, or is poorly designed for coping with noise, may think the system is failing. This is not the case. Nothing in the circuitry is broken, but the circuitry is being called upon to handle a situation that is beyond its capabilities. The user sees intermittent problems, in which data is occasionally bad or synchronization sometimes fails. These noise-caused problems can be very hard to localize and fix. Is the noise coming into the system at the transmitting end, within the media, or at the receiving end? The service person working with the system can spend many frustrating hours dealing with the problem. An important tool in understanding the effect of noise is understanding noise measurements.

QUESTIONS FOR SECTION 6-5

1. What does wideband noise look like in the time and frequency domains?
2. What is impulse noise? Describe its time domain appearance. What is its effect on data communications?
3. What is frequency-specific noise? What affects its amplitude? What effects does it have on a system? How are these sometimes reduced?
4. Why is frequency-specific noise easiest to deal with?
5. Why is burst noise hardest to handle?
6. What is crosstalk? Where does it occur? How can it be minimized?

6-6 Noise Measurements

Noise is difficult to measure and characterize. Because it is a random process, there is no simple, single equation which can represent it properly. The way that noise is specified is by some of its general characteristics. These are then usually related to the characteristics of the data communications signal, so that system performance can be determined or calculated.

The obvious question to ask is, how much noise is there? This question has many equally good answers. Recall that the value of the noise signal at any instant of time can be anywhere within the overall range of signals that the system can handle. Therefore, the greatest noise value possible is often fairly large, as compared to the data signal value, but this large value occurs rarely. The average or mean value of the noise may also be of interest. For various electrical reasons, it turns out that the mean value is 0, and noise itself takes on many values equally on the plus and minus sides (Fig. 6-11a). Even if the noise does not have a zero mean, a simple circuit can easily be built to remove the steady, nonzero mean value (often called the *dc component of noise*) (Fig. 6-11b).

The most common measure of the noise is the *root-mean-square* (rms) value. This is calculated by taking noise values (plus and minus) and squaring them. The squared values are then averaged, and the square root is taken. The reason for the squaring is to make all the values positive. Otherwise, any average value would be zero since the noise has equal plus and minus values and is zero mean.

A simple example shows the meaning of this. Suppose a noise signal has 10 values of $+1$, -0.5, -0.2, -0.3, $+0.7$, -0.5, -0.1, $+0.1$, -0.2, and -0.4 V. The average value is 0 V. The rms value is calculated as the square root of the average of the square of each value or the square root of the average value of ($1 + 0.25 + 0.04 + 0.09 + 0.49 + 0.25 + 0.01 + 0.01 + 0.04 + 0.16$). This average value is then 2.34/10, or 0.234, and the square root is 0.484. For these noise values, then, the mean value is 0 V, but the rms value is 0.484 V.

Of course, it is not practical to measure many noise points and go through these calculations for many noise signals. Fortunately, there are instruments available which can electronically determine the rms value of any signal quickly and easily.

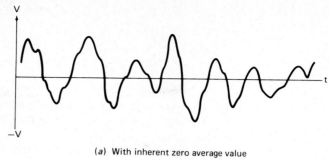

(a) With inherent zero average value

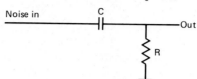

(b) Simple resistor-capacitor circuit can make any noise have zero average

Fig. 6-11 (a) Most noise inherently has an average value of 0 V. (b) Where it does not, the nonzero component can be easily filtered out.

Once the noise values have been characterized, it is necessary to compare them to the desired signal value. This is because a receiver can recover the data properly only if two conditions exist: the magnitude of the received signal is large enough for the receiver circuits to operate; and the ratio of the signal to noise is high enough that the receiver can process the received signal and not be misled by the noise. In theory, no matter how small the received signal is, many amplifier stages can be used to boost it to a larger value so that the actual circuitry which decides which data value is sent can work properly. However, if the received signal has noise, then the noise will also be amplified. Also, the amplifying circuitry of the receiver adds some noise to the signal, so that very minute signals will not be improved if they are amplified. Therefore, the received signal must be large enough so that additional amplification does not add significant noise, and the ratio of the desired signal to the undesired noise must be high.

The concept of signal strength to noise strength is called *signal-to-noise ratio* (SNR). It appears in most data communications systems as a key performance parameter. It is measured in decibels (dB). The decibel can also be used for specifying the strength or value of any signal, and understanding how decibels are calculated and how they can be used is critical in communications systems.

Calculation of Decibels

The decibel is used to measure the ratio between any two signals quickly. These signals can be the data signal and noise, or the data signal before and after amplification, for example. Very often, these ratios have very large ranges. It is common, in some systems, to have a ratio of signal to noise of 150,000 to 1. These numbers can become hard to handle, and difficult to use in additional calculations. The decibel uses a logarithmic scale to measure the ratio. For signal voltages, the formula is

$$dB = 20 \log \frac{\text{signal } A}{\text{signal } B}$$

[**Note:** For signal power, replace the 20 with a 10. The reason for this is that signal power is related to the square of the signal voltage. In many systems there is a difference: at some points the signal power is the critical factor, and at other places it is the voltage.]

The practical meaning of measuring in decibels is seen with a few examples: Suppose a data signal is at 1 V, and the rms value of the noise is 0.01 V. Then the SNR is 1/0.01, or 100. In decibels, this is 20 log 100 or 40 dB. This signal has an SNR of 40 dB. In another case, the noise has an rms value of 0.001 V, and the data signal has a value of 10 V. The SNR is 80 dB, representing a 10,000 to 1 ratio.

A communications system and its performance can then be specified by saying that the *bit error rate* (BER) is a certain value at some SNRs. A typical performance specification may be as follows:

SNR, dB	BER, Errors per Million Bits
40	5
60	3
80	1
100	0.1

As the SNR increases, the BER decreases. (Why?) The user must ensure that the SNR value of the system is at least as good as the value needed to make sure the BER performance is what the application requires.

Of course, it is possible to perform the reverse mathematical operation, and determine the actual ratio and signal values, if the decibel value is known:

$$\text{Signal } A = B \times 10^{dB/20}$$

$$\text{Signal } B = \frac{A}{10^{dB/20}}$$

[**Note:** for power, replace the 20 with a 10.]

If the specification says that the SNR is 50 dB for signal power to noise power, and the measured noise rms value is 0.05 watts (W), then the value of the signal power is

$$\text{Signal power} = 0.05 \times 10^{50/10} = 5000 \text{ W}$$

The units of measurements for most signals are voltage, watts, or current. However, it is sometimes more convenient to express the actual signal value itself in decibels, because it allows for easier system calculations. This can be done as long as the value of a reference, called the 0-dB point, is clearly defined. In audio work, for example, the industry has defined the 0-dB point as a

1-milliwatt (mW) signal into a 600-ohm (Ω) load, which then means a 0.775-V signal (since power = V^2/R). All other audio signals can be measured in decibels, referenced to this value. An audio signal of 10 dB is really a 2.5-V signal.

The decibel can take on negative values. If signal A is smaller than B, the ratio is less than 1, and the decibel value is negative. This shows that signal A is smaller than signal B, or that the signal has somehow been attenuated. In some systems, the incoming signal power is too great to be used directly, and it has to be reduced by one-half its value. Then the ratio of signal A to signal B is 0.5, and the reduction is -3 dB. In fact, 3 dB is a very common value. A ratio of 3 dB is close to a 2:1 power factor, and -3 is close to a 1:2 power factor. The corresponding 2:1 factor for voltage ratios is 6 dB or -6 dB, shown by the formula.

Decibel Applications

The decibel scale is often used for comparing any two signals, not just data signals and noise, or seeing what amplification a signal gets as it passes through various system amplifiers. This highlights another convenience of using decibels. If a system has several stages (blocks) of amplifiers (as most real ones do), then the overall gain can be calculated by simply adding up the decibel gain of each stage. As Fig. 6-12 shows, there are three stages. In stage A, the low-level signal voltage from the antenna is boosted by 20 dB. In stage B, the signal is amplified further to be more practical for this system, by 30 dB. Stage C takes the signal and provides the final boost of 10 dB, needed to interface to the computer. The total gain is 20 + 30 + 10 = 60 dB. Using ratios only, the gains would have to be multiplied. Stage A would be 10:1, stage B would be 32:1, and stage C would be 3:1. The overall gain is the product of these, 960:1. If for any reason the gain of stage B is reduced to half its previous value, then the entire multiplication is repeated. But using decibels, the gain is changed from 10 to 7 (recall that 3 dB is a factor of 2), and it is simple to see that the total gain is now 57 dB. Similarly, decibels are used to specify attenuation in media. A typical specification would be in decibel loss per 100 ft, or per kilometer, to indicate the ratio of signal to original signal at that distance.

Decibels are very powerful for making quick estimates of performance. Since decibel values add (and subtract) it is easy to see that every 3 dB corresponds to a doubling (or halving) of the previous power value. A system with a power gain factor of 16 dB has four times the gain of one with 10 dB. A power attenuation of -9 dB means that the signal has been cut in half three times, and is down to one-eighth of its original value. Many times two signals have equal values. This means that the ratio of the two signals is 1:1, or 0 dB.

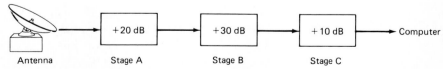

Fig. 6-12 A multiple-stage system has a specific gain figure in dB for each stage. The total overall gain is the sum of the dB values.

Modern data communications systems can provide almost error-free performance if the SNR is greater than about 20 dB. Space satellite systems which are receiving data from millions of miles away have to perform with SNRs as poor as −100 dB, which means that the desired data signal power is about 10 billion times weaker than the received noise. It may seem impossible to recover a data signal that is weaker than the noise. It is difficult, requiring special techniques, but not impossible. The data signal uses synchronous techniques which allow the receiver to find an inkling of a coherent pattern among all the random noise. The data rate is very low to make receiving easier, as low as 10 b/s. The data may be encoded in special ways so that any errors can be recognized and corrected. Designing a system to perform with SNRs below 20 dB is difficult and costly. Fortunately, most systems, such as computer links, telephone systems, and computer subsystem links, have SNRs of 50 to 60 dB or higher, which means that high data rate and low-error performance can be achieved at reasonable cost.

QUESTIONS FOR SECTION 6-6

1. Why is noise hard to measure? What is the average value of a noise signal, in most cases?
2. What is rms? Why is it used?
3. How is noise measured electronically? Is a calculator required?
4. What are the two conditions for proper reception of a signal?
5. Why is it not possible to use as much amplification as desired?
6. What is SNR? What is the unit of measurement for SNR?
7. Why is the decibel scale used? What is the formula for decibels? Power?
8. How can a decibel value be converted to a ratio?
9. What is bit error rate (BER)? What happens to BER as the SNR increases?
10. What is the use of the decibel scale for absolute measurements? What does the 0 dB point mean?
11. What does a 3 dB ratio represent? A −3-dB ratio?
12. How is the decibel scale used conveniently for multistage systems?
13. How can an SNR of less than 0 dB be used?
14. What is the typical SNR of a computer and its associated data communications system?

PROBLEMS FOR SECTION 6-6

1. Find the average and rms value for these noise voltages: 0.01, 0.03, −0.02, 0.0, −0.05, 0.04, 0.08, −0.09, 0.06, and −0.06.
2. Find the average and rms value for these noise voltages: 0.3, 0.7, −0.6, −0.4, 0.9, −0.3, −0.7, −0.4, 0.0, and 0.5.
3. What is the decibel value for these voltage ratios:
   10:1, 15:1, 125:1, 0.001:1, and 0.7:1?
4. Find the decibel value that corresponds to these voltage values of signals A and B:

A	B
20	5
25	5
20	4
32	15

5. What is the decibel value for these power ratios: 12:1, 20:1, 125:1, 3:1?

6. Find the decibel value that corresponds to these power values for signals *A* and *B*:

A	B
25	5
20	4
10	20
10	15
100	2

7. For the decibel value shown, and the value of voltage *A* given, find voltage *B*:

Decibel	A
10	2
6.8	3
11	6
−3.9	1

8. For the decibel value shown, and the value of voltage *B*, find voltage *A*:

Decibel	B
6	2
10	2
−4	7

9. For the decibel values shown, fill in the missing power value in column *A* or *B*:

Decibel	A	B
2		1
6		7
10		3.5
15		3.5
−4		6.4
−7		9

10. If 0 dB is defined as 1 V, what are the voltage values of 3-dB, 4-dB, −1-dB, and 10-dB signals?
11. Repeat Prob. for 0 dB = 5 V.
12. With 0 dB defined as 0.1 W, what are the power levels of 4 dB, 7 dB, 20 dB, −0.3 dB, and 0.4 dB?
13. Repeat Prob. 12 for 0 dB = 0.5 W.
14. A multistage system has four stages, with voltage gains of 10:1, 6:1, 20:1, and 15:1.
    a. What is the gain, in decibels, of each stage?
    b. What is the total gain ratio?
    c. What is the total gain, in decibels?
15. The three stages of a system have power gains of 20, 30, and 40 dB.
    a. What is the total gain, in decibels?
    b. What is the gain ratio of each stage?
    c. What is the gain ratio of the total?
    d. The last stage has a new transistor installed, which increases its gain by 2 dB. Repeat parts **a, b,** and **c** for this situation.
16. A system has the following voltage gain values, in decibels. Estimate (no calculators or equations) the gain ratio: 6 dB, 13 dB, 17 dB, −11 dB.
17. A system has power gains shown, in decibels. Estimate the gain ratios: 3.5 dB, 6.7 dB, −2.8 dB, −9.8 dB.

6-7 The Effects of Bandwidth Limitation (and Related Problems)

Every communications system has some maximum bandwidth that it is capable of supporting. The limitation may be in the circuitry at either end of the system or in the medium itself. Just like a chain, the bandwidth of the total system is determined primarily by its weakest link, or lowest-bandwidth portion.

In theory, a system would be capable of handling frequencies up to a value f_c (for cutoff frequency) and pass no frequency components beyond that value (Fig. 6-13a). In reality, the frequency domain representation of the system, or subsystem, shows that the ability to pass frequencies decreases as the frequency value extends higher and higher (Fig. 6-13b). This degradation is not a sudden drop, but gradual. How then is the effective *bandwidth* defined? The common definition of the bandwidth used in industry is that the bandwidth cutoff point f_c is the region where the power of the frequency components has dropped to one-half its maximum value. This is often called the "3 dB" point, since 3 dB represents a power factor of 2, so when the power amplitude is one-half, it is −3 dB from its peak value. Some examples are shown in Fig. 6-14. For voltage, the corresponding 3-dB factor is a decrease to 0.707 (one-half the square root of 2) of the peak voltage value.

Different types of circuitry and different media have varying rates of signal

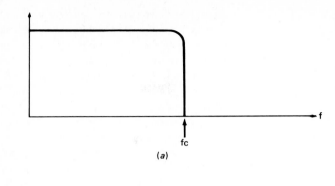
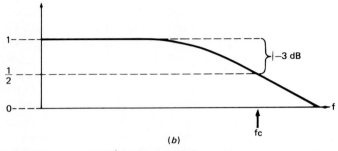

Fig. 6-13 (*a*) An ideal system would pass all frequencies up to some cutoff f_c equally; (*b*) a real system has a gradual rolloff to the 3-dB frequency.

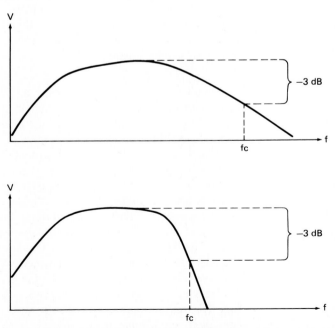

Fig. 6-14 The 3-dB bandwidth point shown for two examples of frequency response and bandwidth.

degradation. Sometimes, the drop is relatively sudden, although in other cases it is more gradual. The bandwidth is still defined by the 3 dB point. The rate of drop is determined by physical characteristics of the medium. For circuitry, it depends on the electronic components used in the system, and the way the circuit is designed. The design engineer may in fact choose, for technical reasons, to design in one rate of drop-off or another, to meet some other system needs.

Most data communications systems need much more bandwidth than may seem necessary at first intuitive glance. Consider a simple system carrying a voice signal with a 3 kHz bandwidth. If this were carried by a purely analog system, then a 3 kHz channel would be sufficient. Now take the same signal, and digitize it at an 8 kHz rate, with 8 bits of conversion. The bit rate that results from this is 8 kilosamples/second times 8 bits/sample, or 64 kbits/s. This means that the digital system must send bits at this rate. The bandwidth needed to support a bit rate with reasonable accuracy and freedom from distortion is usually 5 to 10 times the bit rate, so a channel with a bandwidth of 320 to 640 kHz is required. This is far greater than the bandwidth needed for the original analog signal.

It is useful to understand the effect of bandwidth limitations on the frequency and time domain graphically. A bandwidth limitation is similar in many ways to a low-pass filter. It allows frequencies up to the cutoff f_c to pass with little attenuation, but the reduction in amplitude is greater than half beyond the cutoff (3-dB) point. When a digital signal is sent through such a filter, its corners will be rounded, since the higher frequency components contain the sharp corner information. As the bandwidth cutoff frequency gets lower, this rounding is more severe. Eventually, the sharp-cornered digital signal is nearly reduced to a sine wave, at the frequency of the digital bit rate. Figure 6-15 shows the frequency and time domain graphs for a pair of digital bits, as the cutoff frequency is moved lower and lower.

Intersymbol Interference and Eye Patterns

Communications, however, involves a string of bits. The effect of limited bandwidth on a series of bits is that in the time domain, the bits begin to "smear" together, with long tails on either side of each bit (Fig. 6-16). These tails can add to the observed signal value of the adjacent bits and in fact confuse the receiver, which is trying to determine the value of a single bit. The bits, or signaling symbols, begin to interfere with each other when viewed in the time domain. This phenomenon is called *intersymbol interference* (ISI).

There are some other causes of ISI. The frequency components of the signal travel through the channel after starting out at the same time, but the speed of travel may differ slightly for the different frequency components. This is due to some subtle characteristics of the medium, and there is nothing the system designer can do about this. In the frequency domain, all the components are seen to arrive with the intended amplitude, but in the time domain the arrival times are not the same. The effect on the received pulse, as viewed in the time domain, is that it is smeared and distorted in a way similar to the effect of a bandwidth limitation, and ISI may result.

Whatever the cause or causes, the effect on the receiving circuitry is that it is harder to make the correct determination about what bit value was sent. In a

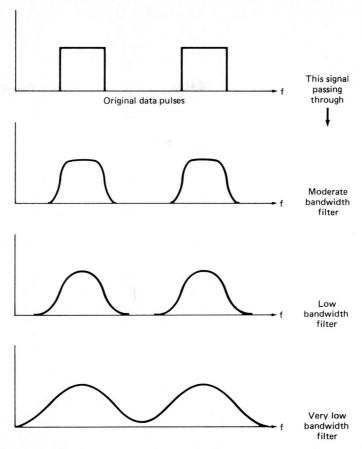

Fig. 6-15 The effect of low pass filtering on data pulses. As the filter cutoff goes lower, the pulses lose their sharpness and look more like sine waves.

severe case, the bits blur together so badly that it is nearly impossible to determine the bit value correctly. Fortunately, there is a simple way to measure how severe the ISI is, and in some cases compensate for some of it at the two ends of the system. The technique used to measure the ISI is called an *eye*

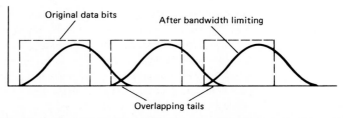

Fig. 6-16 The effect of low bandwidth on a series of data bits is a smearing or overlapping of the start of a new bit and the tail of a preceding bit.

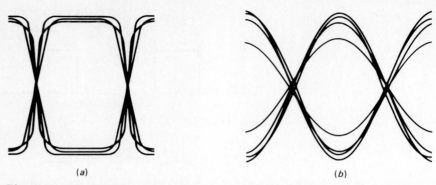

Fig. 6-17 Eye patterns are a convenient way to see intersymbol interference and other bit problems: (*a*) the eye pattern for nearly ideal bits; (*b*) the pattern for bandwidth-limited bits with noise.

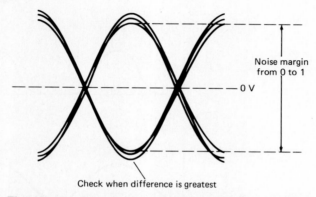

Fig. 6-18 The eye pattern shows where the best point in time is to decide whether a binary 1 or 0 was received. The noise margin is greatest where the eye is most open.

pattern. This eye pattern can be set up easily on an oscilloscope, and then ISI characteristics can be measured on the screen of the scope.

The eye pattern is created by overlaying each bit of the data on top of another. This is done by setting the scope sweep so that a single bit period occupies the entire screen. The scope is then set to trigger on each bit (this is done by using either an external trigger derived from the bit stream clock or the trigger controls of the scope). The effect of this overlaying is to superimpose all bits on each other, so that an eye-shaped picture is seen on the scope screen.

For data bits with the ideal, near-rectangular shape, the eye is as shown in Fig. 6-17*a*. As the bits become rounded, because of bandwidth limitations or for any other reasons, the eye pattern has the appearance of Fig. 6-17*b*.

As the bits are reduced in amplitude, the eye begins to close. By examination of the eye pattern, a great deal can be learned about the characteristics of the communications channel. The amount of eye closing shows the difference between the binary 1 and 0 levels. As the eye is more closed, the difference becomes less, and therefore there is less noise margin between 0 and 1; a wide open eye means that the binary 1 and 0 voltages are separated by a larger amount

(Fig. 6-18). The roundness of the eye shows how much the original signal has been distorted. The receiving circuitry should make its determination on what was sent, by looking at the received bits at the time when the eye is wide open. Depending on the distortions of the channel, this may not be in the middle of the bit.

How does the eye pattern help? It quickly summarizes what is happening to the data bits as they are sent, and how they have been changed from the ideal. This helps the receiving circuitry, which can electronically study the eye pattern (without using an oscilloscope screen, of course) to calculate the best time to determine what was sent, the quality of the received signal, and the likelihood of error. It also sets the stage for circuitry which compensates for problems, called *equalizers*.

Equalization

Equalizers make use of some knowledge about what is happening in the system to try to compensate at the receiving (and sometimes transmitting) end of the system. If the received signal has a certain kind of distortion, from whatever cause, it may be possible to put it through some special filters that reverse some of the distortion, and thus make the received signal look better. This reduces the chance of an error in the receiver. The circuitry therefore compensates, or equalizes, the signal, to improve it. An example is a communications channel which has different signal propagation speeds for different frequency components. The compensation circuitry provides an equalizing delay, so that all the signals effectively arrive at the same time. If the three frequencies used take 100, 120, and 110 nanoseconds (ns) to reach the receiver, the receiver adds 20-, 0-, and 10-ns delay, respectively, to each component, so that all would arrive at the decision circuitry in 120 ns.

There is another approach, which is technically more difficult but can provide very dramatic improvement. If the transmitter and its circuitry knew what kind of distortions were occurring, the transmitted signal could be deliberately distorted so that it arrived in good shape. This *precompensation* is a very powerful tool in data communications, because it allows the transmitter, which knows what the correct values of 1s and 0s should be, to make its best effort to ensure that the signals that arrive at the other end of the channel are in the best form. For example, if the channel bandwidth means that some of the higher frequencies are attenuated, then the transmitter boosts the amplitude of these frequency components so they arrive at the receiving end at the same amplitude as the lower frequencies (Fig. 6-19). If some of the frequency components are delayed by the characteristics of the medium, then the transmitter deliberately sends out the slower ones before the faster ones, so that they all arrive at the same time.

One effect of the equalization at the transmitting end is that the bits as sent look very different from the ideal. Instead of resembling a rectangle, the transmitted bit may appear badly distorted (Fig. 6-20*a*). The communications channel, however, will introduce its own mirror image distortions, and the signal at the receiver is closer to ideal, Fig. 6-20*b*. The circuitry to perform this type of equalization and compensation is very sophisticated. Modern integrated circuits can do it, fortunately, at reasonable cost.

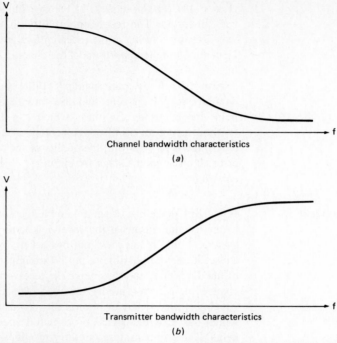

Fig. 6-19 Equalization compensates for channel characteristics. If the channel has a spectrum as shown in (a), the transmitter circuit can boost the frequency values so the resulting sum is even across the entire bandwidth of the channel.

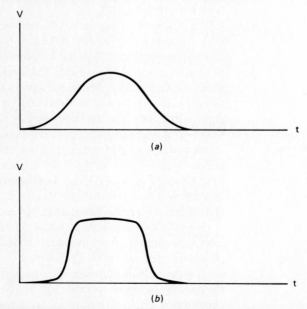

Fig. 6-20 The effect of equalization: (a) the bit at the transmitter; (b) the bit at the receiver. Note that the received bit looks more like an ideal data bit.

One of the key points to making the precompensation and equalization work is that the transmitter must have some idea of what the problems of the channel are. This is done in two ways. First, the system designer usually knows, to a large extent, what the channel will do to the signals and can make a good estimate of what should be done at the transmitter. A more advanced method involves having the transmitting side send some special test signals down the line and seeing what happens to them. This allows the transmitter to tune itself and also gives the equalizing circuitry at the receiver an opportunity to make its best adjustment on a known pattern. A simple approach is to send several thousand alternating 1s and 0s. Since the receiver knows what the data is, it can internally adjust and fine-tune on the basis of this knowledge of the correct results. It is much harder to adjust when the incoming data is random, representing signals from someone talking or sending numbers, although it can be done. Many communications systems with equalizers and compensation circuitry have this initial training period, in which no real data is sent, only test patterns.

One limitation to this initial pattern method is that the characteristics of the channel may not hold steady during the entire period of communications. If the channel, or any part of it, involves radio or satellite links, for example, there may be distortions and noise that vary with time or signal fading. In this case, the equalization may be correct for one instant, but not the best choice for others. Some very advanced communications systems use a sophisticated method called *adaptive equalization*, in which they constantly measure and monitor what the channel is doing. This constant measurement leads to continuous changes in the equalization and compensation, to try to always have the optimum signal. Adaptive equalization is a very powerful technique, but it involves complicated circuitry. It can be difficult to work with and troubleshoot, since the data signal as seen on an oscilloscope keeps changing in shape and amplitude, as the equalization factors continuously change to match channel conditions. Nevertheless, it can provide outstanding performance under many difficult conditions and may be worth the cost and difficulties.

QUESTIONS FOR SECTION 6-7

1. What does the bandwidth term "f_c" mean?
2. How is f_c defined usually?
3. Why are bandwidth needs very large? What bandwidth factor is needed for a given bit rate?
4. What is the effect of limited bandwidth on a string of bits?
5. Describe ISI. What are its causes?
6. What does the eye pattern show? How is it set up?
7. What does an equalizer do?
8. What is precompensation? How does it work?
9. How is the compensation system set up?
10. What is an adaptive equalizer? Why is it needed? What are its drawbacks?

PROBLEMS FOR SECTION 6-7

1. A signal is distorted as it passes through the medium. Its various frequency components take different amounts of time to reach the receiver, as follows:

1000 Hz	15 ms
2000 Hz	16 ms
4000 Hz	18 ms

   What compensation should be used at the receiver to improve and restore the shape of the original signal?

2. In another medium, signal components take the following amounts of time to reach the receiver:

1 kHz	120 ms
10 kHz	130 ms
20 kHz	125 ms

   What compensation should be used at the receiver to improve and restore the shape of the original signal?

3. A signal amplitude is distorted by these gain values at various frequencies:

1 kHz	+10 dB
10 kHz	+ 8 dB
20 kHz	+ 7 dB

   What sort of amplitude equalization should be used? Sketch the effective filter shape of the medium and the amplitude versus frequency shape of the equalizer.

4. Repeat Prob. 3, with the following values:

1 kHz	+10 dB
2 kHz	+ 5 dB
4 kHz	+ 0 dB

   Sketch the resultant amplitude versus frequency after equalization, for a data signal which has a constant amplitude versus frequency graph from 0 to 4 kHz.

6-8 Common Mode Voltage

A typical, common data communications situation is the case of a computer that needs to send data to a nearby printer, perhaps only 10 ft away. The simplest, most straightforward way to do this is with a channel that uses wire to link the computer and printer. If done properly, this can be a low-cost, reliable method.

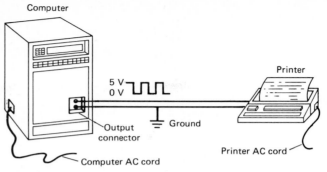

Fig. 6-21 A typical computer-to-printer communications link uses a signal wire for the data bits and a common ground wire.

Even in this simple case, however, a wire with the necessary circuitry at each end may not work properly. A problem called *common mode voltage* (CMV) often arises in practical installations; it can cause errors or even damage the computer or printer. Common mode voltage and its associated problems can sometimes make a system that was working suddenly work unreliably and cause many hard-to-identify problems. It is important to understand what it is, and the simple ways that it can be overcome in many data communications installations.

Consider the situation of Fig. 6-21, in which the computer is sending characters to the printer, located nearby. The computer is using binary patterns to represent the various letters, numbers, and punctuation symbols, with binary 0 at 0 V while binary 1 is at 5 V. The 0 or 5 V level is impressed on the wire, and felt at the printer, which interprets it and prints the correct character. This all seems fairly simple, low-cost, and clear. Unfortunately, it may not work.

The electrical reality is that the ground voltages of the computer and the printer may not be the same. They may differ by a few volts or even dozens of volts. Yet, each system—the computer and the printer—defines voltage with respect to its own ground point. (Remember that *voltage* is defined as the potential difference between two points, so a voltage value exists only with respect to another point.)

At the computer end, the computer data circuitry sends out 0 or 5 V signals with respect to the computer ground point. The printer ideally sees the same 0 or 5 V signals. But since its ground may not be of the same voltage as the computer ground, instead of receiving 0 and 5 V, it sees 0 or 5 V plus the *ground potential difference*. This difference in the ground voltages is common to both the higher voltage and ground point of the printer, and thus is called the *common mode voltage* (CMV). The electrical circuit that really exists is shown in Fig. 6-22. Note that V_{cm} causes the voltage as seen by the printer to be either 0 plus V_{cm}, or 5 plus V_{cm}. Instead of seeing the 0 or 5 V intended, the printer circuitry sees two completely different voltage values, values that it cannot deal with and recognize properly. What CMV does is change the voltage values that were sent into values that are incorrect.

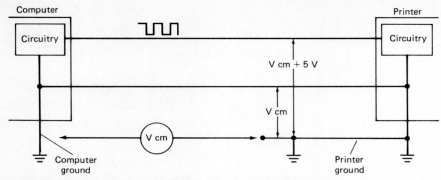

Fig. 6-22 The actual electrical equivalent of the computer-to-printer circuit, showing the common mode voltages that can exist and affect operation and equipment.

There is another consequence of CMV that has to be dealt with. Since the printer and computer are connected, any voltage that appears at the input to the printer also appears at the output of the computer. The CMV at the printer, which can be as high as 50 to 100 V, is put onto the output circuitry of the computer and may severely damage it (or even endanger anyone connecting the two units). The reverse situation also applies: if the CMV exists at the computer, it may pass this high voltage over to the printer input and damage the printer. Even if the communications circuitry can be made to work despite the CMV corrupting the data signals, it may be damaged by the higher values of CMV.

Differential Signals

Typically, two units plugged into the same outlet have only small values of CMV, less than 5 V, yet even this small value is enough to corrupt the 0 and 5 V data signals. If the outlets of the two systems are separated or on separate branch circuits within the building, then the higher, damaging CMV values are more likely. Fortunately, there are some low-cost, technically simple solutions to the problems of CMV. It is not necessary to use a radio or more complicated link to work around the problem. The solutions are differential signals and isolation.

In a *differential system,* the ground of the two units is not used as a reference for the data voltage. Instead, two wires, distinct from the ground, are used. One is called the *high lead*, and the other the *low lead* (Fig. 6-23). The voltages used for the data are impressed across this pair of wires. If a 0 V value is needed, then the high lead is made to be 0 V with respect to the low lead; when 5 V is needed, the high wire is made 5 V with respect to low. At the receiving end, the circuitry looks for the difference between the two wires—hence the term ''differential.'' The CMV that may exist between the systems is not a problem, since the ground values and their difference are not used in sending signals or receiving them properly. Even if the two leads themselves have some CMV (and they do, since their respective circuitries use their own chassis grounds) it does not matter. The received signals on the two leads are (high value plus CMV) and (low value plus CMV). But the circuitry used in the receiver takes the difference of these two, so that the actual voltage seen by the receiver is simply (high value + CMV) − (low value + CMV), which equals (high value − low value). Note how the CMV has dropped out of the equation, because of the difference factor.

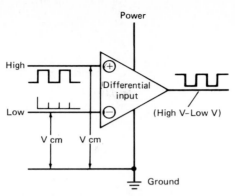

Fig. 6-23 A differential input uses two signals wires, separate from ground. The data signal voltage is the difference between these two wires.

The ability of the differential circuitry to suppress the CMV is called the *common mode rejection ration* (CMRR) and is measured in decibels. The CMRR compares the CMV before and after the differential circuitry. Ideally, the CMV would be completely eliminated, but practical circuits cannot do this. They can reduce it a great deal. A typical differential circuit may take a 10 V CMV and reduce it to 0.1 V, which is almost insignificant compared to the 0 and 5 V levels used for the data bits. The CMRR would be the decibel ratio of 10 V and 0.1 V, or 40 dB. A circuit that reduced the CMV by a factor of 10,000 (for example, reducing 15 V to 0.0015 V) would have a CMRR of 80 dB.

A system which uses differential signals does this by using special ICs which are specifically designed for the purpose. These ICs are called *differential line drivers* (for the transmitting end). The drawback to using differential ICs and circuitry is not the cost of the IC (which is about the same as the cost of nondifferential ICs) but the necessity of having a pair of wires for each signal. In the simple, nondifferential method, only one wire is needed per signal, since the second wire was the ground of the power line. For this reason, the nondifferential scheme is called *single-ended*. If more signals have to be handled at the same time, a single-ended system needs only one wire for each one. A differential scheme needs two wires for each one. Although wire itself is relatively inexpensive, the additional wires can be a problem when the cable has to be run in tight, crowded areas. Connectors for cables with more wires also cost much more than smaller connectors.

In general, the single-ended method is used for very short distances or applications in which the two units being connected share a common power supply so that there is no possibility of CMV. Most often, this is between two subsections of a single system, such as a computer communicating with its video display screen located on top.

Neither the single-ended nor the differential method can overcome the problem that can occur when the CMV is so large that actual damage is a problem. The ICs used for differential circuits are capable of withstanding moderate amounts of CMV, typically 5 to 20 V. Beyond that, the differential circuitry within the IC malfunctions, and soon after that the IC itself is damaged. If the

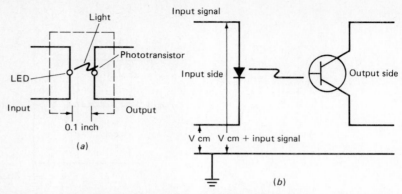

Fig. 6-24 An optoisolator is a nearly ideal differential element: (a) the physical concept; (b) the electrical schematic. Note that there is no electrical signal connection between input and output, and electrical ground is not used in any way.

CMV is higher than about 50 to 100 V, then a more sophisticated technique, called *isolation*, is used.

Isolation is the technically superior solution to both the problem of communicating reliably when CMV is present and that of preventing damage from excessive CMV. Up to about 10 years ago, it was difficult and expensive to do. However, low-cost ICs called *optoisolators* which can do the job at low cost and with excellent performance are now available. The optoisolator is related to the fiber optics components discussed earlier.

An *optoisolator consists* of a light-emitting diode (LED) and a light-sensitive transistor, housed in the same package (Fig. 6-24a). The two are separated by a distance of about 0.1 in. The LED gives off light when there is a voltage across its two terminals. This light travels the short distance to the phototransistor, which converts it back into a voltage. The result is that a voltage can be sent from one side of the optoisolator to the other, without an actual electrical path between the two sides. The light beam forms part of the path.

How is this useful? The optoisolator acts as the nearly perfect differential circuit element, with a CMRR of about 150 dB. It has no connection or reference to ground. The LED inside the optoisolator sees only the differential voltage of its two leads, and in effect wipes out any CMV (Fig. 6-24b). It does not matter whether the voltages are 0 and 5 V, or 1000 and 1005 V. The optoisolator converts only the difference to light and then to an output voltage. There is no ground or electrical path between input and output, so a CMV on one side is not seen by the other side. The only limiting factor is the amount of voltage the optoisolator can withstand before its insulation fails. Typically, this is hundreds or even thousands of volts. Optoisolators are used as part of the communications circuitry when unknown or large amounts of CMV are present. They eliminate the effects of CMV on signal quality while protecting equipment (and people) from danger. Even if the circuit uses only low voltages and has no CMV problems, optoisolators are often used for protection. If a failure or misapplication causes line voltage of 120 V to be put on the communications wire, someone touching the wire may be electrocuted. But the optoisolator does not allow 120 V to pass. It allows only the voltage difference on the input to reach the other side, but without an electrical path. Even if the LED is destroyed, no electricity crosses over. The light beam within the optoisolator acts as a barrier to unwanted voltages.

A typical connection from a computer to printer uses optoisolators in each signal line between the two. This means that two wires are needed for each signal, as with a differential circuit, but the performance is better. The cost of the optoisolator is only slightly more than the differential line driver/receiver pair, so they are used in most cases in which CMV problems may exist. From a conceptual standpoint, a fiber optic link is very similar to an optoisolator. The difference is that in an optoisolator, the optical path is short (0.1 in.) and within the component. In a fiber optic link, the optical path is longer, made of a light-carrying fiber. Both have a voltage-to-light source at one end, and a light-to-voltage converter at the other.

Using a wire and simple single-ended circuitry for sending data signals even a short distance may work in some limited cases, especially those in which the installation and unwanted effects are easily controllable. In other cases, differential circuits and optoisolators provide a more reliable and safe interconnection and path for the data signals. However, when testing systems using differential and optoisolator circuitry, special care must be used not to defeat the purpose of the circuits by connecting test equipment grounds to the signal lines in such a way that the differential circuit is defeated (by short-circuiting) or the isolation of the optoisolator is bypassed. Misleading results and even physical danger to equipment and users can result. An oscilloscope probe that is grounded at the computer end, for example, with the high side of the probe connected to the printer end, effectively puts an electrical path around the isolating, optical path (Fig. 6-25).

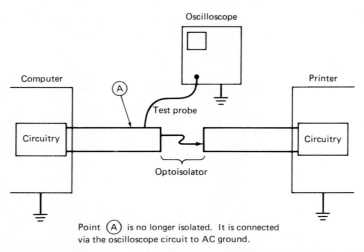

Fig. 6-25 The probe of a test instrument can inadvertently defeat the isolation, by connecting two isolated units through the test instrument circuitry. Note that point A is no longer isolated. It is connected via the oscilloscope circuit to AC ground.

Communications, Media, and Problems   **197**

QUESTIONS FOR SECTION 6-8

1. What is common mode voltage? What are two problems it can cause? What causes it?
2. What are typical CMV values for two units in the same outlet? In different outlets?
3. How does a differential scheme work? What are its advantages? Disadvantages? When does it no longer provide a good solution?
4. How is the ability to reject CMV specified? What does the specification mean?
5. What is isolation? What are its advantages and disadvantages?
6. What is an optoisolator? How does it work? Why does it result in a large rejection of CMV?
7. How does an optoisolator protect against electrocution?
8. How is an optoisolator similar to a fiber optics system?

PROBLEMS FOR SECTION 6-8

1. A differential input circuit provides a CMRR of 110 dB. To what voltage is a common mode voltage of 120 V reduced?
2. The preceding circuit is improved with some design changes, and the CMRR increases to 116 dB. To what value is the common mode voltage now reduced? (Note: Think for a moment before doing calculations again.)
3. A system is using 5 V signals for data. The circuitry works as long as the CMV is below 1 V. The actual CMV is 25 V. What value of CMRR is needed?
4. A system is using 10-V signals for data and can withstand CMV of 10 V. The CMRR of the input circuitry is 15 dB. How much CMV can be applied before the allowable limit is exceeded?
5. A system can handle a CMV of 10 V, but the CMV is coming in at 20 V. By what factor must the CMV be reduced, additionally, by the input circuitry? By how many decibels?

SUMMARY

The performance of any data communications system is determined by the performance of each piece of the system. The medium through which the signal energy passes is a critical element. Wire, air/vacuum, or optical fiber can be used. Each has some technical strong and weak points. In some cases, the system designer can choose which one is desired; in others there is little choice for some practical reasons.

Each medium is judged on its range, bandwidth, loss, noise resistance, security, cost, and ease of physical connection. Copper wire offers a wide span or range, moderate bandwidth, moderate loss and noise resistance, poor security, low cost for short distances and high cost for large ones, and easy physical connection. A wide diversity of cables can be made from copper wire. Twisted pair is low in cost, flat cable somewhat higher, and coaxial cable the most expensive. Fortunately, each provides a mix of operating characteristics, so the best performance choice is not always the most expensive.

Although air or vacuum is a free medium, there are costs associated with the transmitting and receiving stations. The security is poor, but often there is no alternative. The signals are also subject to losses due to moisture and impurities in the air and fading resulting from varying atmospheric conditions.

The use of fiber optics for point-to-point communications has exploded in recent years, now that some technical problems have been overcome. Optical fibers offer a unique combination of high bandwidth, long range, low cost, excellent noise resistance, and security, but only fair interconnectability. A light wave is sent down a hair-thin glass or plastic fiber, and the digital modulation of this light beam carries the information.

Any communications system is concerned with noise. Noise is difficult to characterize and is generally divided into three groups: wideband random noise, impulse (burst) noise, and frequency-specific noise. Each can cause different problems in a system, and there are different techniques for dealing with each type. Systems can also create their own noise problems, via crosstalk from adjacent wires. This can be solved by increasing wire spacing or shielding, both of which are costly.

The relationship between signal and noise is called the *signal-to-noise ratio*, usually expressed in decibels. The decibel is a simple way of measuring ratios of any two signals and uses a logarithmic scale to compress these ratio values. It is also a very convenient way to define system performance when there are many stages or blocks in the system, since the decibel changes groups of multiplications to simple additions or subtractions. Comparing signals measured in decibels, is valid only if they have a common reference point, called the 0 dB value. This 0 dB point can be different in different applications.

Any system can provide usable bandwidth only up to a certain cutoff frequency. Wide bandwidths mean that high data bit rates can be supported. If the bandwidth is too low for the rate, the data bits become severely distorted and blur into the adjacent bits. The receiver then cannot properly determine what bit value was sent. An eye pattern overlays all the data bits and graphically shows the effects of channel bandwidth and other problems. Equalizers and compensation are used to compensate for medium-induced problems, by correcting the received signal shape or predistorting it so that it arrives looking closer to ideal.

Whenever wire is used as the medium, unwanted voltages common to both the high and low sides of the wire pair can occur. These are called *common mode voltages*; they can range from a few volts to hundreds of volts. The CMV upsets the circuitry of the system and can cause data errors, malfunctions, or damage and danger to users. Differential techniques allow most of the CMV to be eliminated, but they only work for small CMV values. Isolation is technically superior and is used in most cases in which the wire and cable go between systems for any substantial distance. Both the differential and isolation methods require two wires per user signal, as compared to single ended, which needs only one wire per user plus a ground common to all.

END-OF-CHAPTER QUESTIONS

1. What are some of the factors that affect the range that can be attained in a medium?
2. What factors affect the bandwidth? Why is wide bandwidth needed for digital signals?
3. What is attenuation? Why is it an important parameter in determining whether a medium is suitable?
4. Give examples of media in which security is poor, fair, or good. Give examples of communications installations where security is an issue, and where it is not.
5. Compare the cost of various media. What are the issues involved?
6. What are the differences between twisted pair, flat cable, and coax? How do their performance and application compare?
7. Compare the amount of physical space taken up by a coaxial cable, versus flat cable, versus twisted pair? How many users can each cable support? How does the physical space comparison change if you consider space occupied per user?
8. What are the advantages and disadvantages of using air or vacuum as the medium?
9. What path do radio signals follow? How can this path be modified?
10. What are the effects of atmospheric dirt, moisture, and so on, on radio signals?
11. What is the purpose of jamming? Explain how the techniques used to overcome it may also be used for recovering extremely weak signals, such as those from space vehicles.
12. How are microwave signals confined to the desired path in a circuit and system? Why does the method work?
13. What is fiber optics? What keeps the data signal energy going along the desired path?
14. Why are fiber optic signals immune to outside interference?
15. What are three major advantages of a fiber optics system? Three disadvantages?
16. Give two examples of applications for which fiber optics would be a good choice, and two for which it would be a bad choice.
17. Compare the installation of fiber optics to the installation of a copper wire system. Which is easier? Which is more difficult?
18. Compare wideband noise and impulse noise. Which has higher amplitudes, generally? Which affects more bits in succession?
19. Explain why knowing the value of noise at one instant does not help determine it at the next instant, to a large extent. Why is it important to try to know a lot about the characteristics of the noise?
20. What causes frequency-specific noise? How can this noise be overcome in many cases?

21. Why is crosstalk a problem in cables with many wires going to a telephone central office?
22. What is the value of the dc component of most noise signals? What is the average value of the noise?
23. How is noise usually measured? Why?
24. How is the relationship between the signal level and the noise level usually specified?
25. What is a decibel scale? What numbers are compared by decibels?
26. Why are decibels often used for analyzing the performance of a system with multiple stages?
27. What is the ratio of two signals with a 0-dB SNR? With a 3-dB SNR?
28. What is the usual relationship between BER and SNR?
29. Explain why SNRs of less than 0 dB can still be used in communications. What are the steps that have to be taken to make this signal-to-noise ratio work properly?
30. How is bandwidth usually defined? What is the cutoff frequency?
31. What happens to the frequency components beyond f_c?
32. What happens to data bits if the channel bandwidth is too low?
33. What are two causes of intersymbol interference? How does looking at the eye pattern help?
34. What is compensation? Equalization? Precompensation?
35. What are the benefits and drawbacks of an adaptive equalizer?
36. Give the two reasons that CMV is undesirable. What can cause it?
37. What is the typical value of CMV in two chassis plugged into the same outlet? Different AC branch circuits?
38. What are the comparative benefits and tradeoffs among single ended, differential, and isolated media?
39. What does the specification for CMRR tell about the circuitry?
40. How does isolation with optoisolators work? What provides the electrical isolation?

END-OF-CHAPTER PROBLEMS

1. Find the average and rms value for this group of noise voltages: -2, 1, -8, 0, 2, 5, -6, -4, 3 and 9 V.
2. What are the average value and the rms value for these voltage values: -3, -1, 4, 1, 3, -5, 2, 5, and -2 V?
3. Convert these voltage ratios to decibels: 25:1, 50:1, 75:1, 500:1, 1:1, 0.375:1, and 0.1:1.
4. Convert these voltage ratios to decibels: 17:25, 20:10, 10:2, and 100:25.
5. Find the decibel values for voltages A and B:

A	B
75	25
30	10
10	30
5	2

6. Convert these power ratios to decibels: 100:1, 20:1, 25:1, 30:1, 0.8:1, and 0.5:1.

7. Find the decibel values for power values *A* and *B*:

A	B
100	30
30	100
25	50
25	100

8. For voltages *A* and *B*, calculate the third missing value in the chart:

	Decibel	A	B
a.	20	20	
b.	30		10
c.	10	15	
d.	−5	6	
e.	−9		10
f.		4	9
g.	0		2
h.	0	9	

9. For power values *A* and *B*, calculate the third missing value in the chart:

	Decibel	A	B
a.	0	1	
b.	10		7
c.		9	15
d.	5	7	
e.	8		3
f.		9	12
g.	0		14

10. A multistage system has voltage gains of 3, 7, 12, and −2 dB in each stage.

    a. What is the total gain, in decibels?

    b. What is the gain ratio of each stage?

    c. What is the total gain ratio?

11. A multistage system has power gains of 3, 2, and 10 dB with a fourth stage that has adjustable gain of -3 to $+3$ dB.
    a. What is the total gain, in decibels, including the adjustable stage?
    b. What is the gain ratio of each stage?
    c. What is the total gain ratio, including adjustment?

12. A medium and associated circuitry have these delay factors versus frequency:

1 kHz	25 ms
2 kHz	20 ms
3 kHz	22 ms
4 kHz	26 ms

    What compensating delay factors should be used to restore the original waveform shape?

13. A system receiver can provide compensation for time delay up to 10 ms. By how much will it fall short of the ideal, and for which components, with these delays in the medium?

10 kHz	50 ms
20 kHz	45 ms
40 kHz	57 ms
50 kHz	58 ms
100 kHz	63 ms

14. A medium passes the various frequency components with varying gains, although the goal is to have 0 dB gain, as shown:

1 kHz	3 dB
2 kHz	1 dB
4 kHz	-2 dB
8 kHz	-5 dB

    What compensation values for amplitude versus frequency should be used?

15. A system is needed to produce a 10-dB gain over its entire range of 0 to 25 kHz, but the actual gain values at the various frequencies are the following:

0 kHz	9 dB
5 kHz	7 dB
10 kHz	6 dB
15 kHz	12 dB
20 kHz	13 dB
25 kHz	15 dB

What gain compensation should be used at each frequency?

16. A system uses -25 and 25 V to represent 0 and 1. The values received are 0 and 50 V. What is the CMV?

17. A system has a CMV of 190 V. The system can tolerate a maximum voltage of 10 V. The data signals are 0 and 5 V. What CMRR is needed? Why?

18. A designer must choose between differential input with CMRR of 120 dB and an optoisolator with CMRR of 150 dB. The system requires that the CMV of 100 V be reduced to less than 0.00001V. Which input meets the design needs? Why?

7 Communications System Requirements

In this chapter, the communications concepts of the previous chapters are applied to real data communications situations, subsystems, and systems. The issues that were discussed are now put into practical form to allow data to be communicated reliably and efficiently from one user to another. In these actual systems, performance specifications and limitations are based on electronic circuitry capabilities and on outside influences such as noise and bandwidth. Therefore, many technical choices and tradeoffs are made in the design and operation of a data communications system to provide the performance needed in the application.

Equally important, both users must agree to a common set of rules for encoding and communicating the message. These rules cover various levels of the overall communications structure. The finest system designed is useless if there are differences in voltage, data rate, or method used to represent the data.

7-1 Data Communications System Issues

The path from data communications concepts to practical, reliable, useful data communications systems is not short. Along this path there are many decisions that have to be made to produce a properly working system. This applies to a data communications system whether it is used in a local situation, such as sending data from one subsection of a single chassis to another, or to a system used for long-distance communications.

There are two reasons that many decisions have to be made. First, there are technical issues and capabilities that must be taken into account. A single wire pair could not effectively be used for communications across the country, since the bandwidth, attenuation, and noise pickup would be too restrictive. Second, any communications, whether it is between computers or between people, requires that there be mutual agreement on many of the aspects of the communications scheme, to ensure that both parties can actually send and receive messages to and from the other. These choices must be made and agreed to, even when there are several communications schemes that are technically capable of providing the performance needed. If the two parties do not accept a mutually satisfactory set of rules, then communications may be impossible, very difficult, or inefficient, or the application may require additional circuitry at one or both ends to overcome the differences.

The essential issues in a complete data communications system are as follows:

The physical connector that holds the wires and mates with the other wires or chassis must be compatible.

The voltages (or currents) used must be the same, so that the circuitry can handle them without damage and knows what values represent binary 0 and binary 1.

The format or code used to represent each character must be agreed to, since the system has to have a way of representing letters, numbers, and symbols using just binary (or digital) values.

The rules for starting and stopping conversation, and for determining whose turn it is to speak, must be defined. This ensures that each party is ready to listen when it is receiving a message, and that messages are not sent faster than the receiver can accept them. This set of rules is called *protocol*. There may also be some special characters or symbols used to establish this protocol.

The contents of the message must be in known, predefined style or form. Each party should know what part of the message actually contains the data, and which part has some characters that are needed by the communications system for proper operation but do not represent actual data for the user.

These elements form a series of levels, or *hierarchy*, in a communications system. The user, of course, is concerned only with the level which has the data of the message itself and doesn't care about (or is unaware of) what voltages, connectors, formats, and so on, are used. There is an analogy, or parallel, to most of these items when someone writes a letter to another person. For the message in the letter to reach the reader, and be understood by the reader, many elements have to be established. Assume the sender wants to tell the receiver that he will be arriving at 2 o'clock. All the sender cares about is that this message be received and understood by the person at the other end. The sender puts pen to paper and writes out, I WILL ARRIVE AT 2 O'CLOCK, puts this message into an envelope, addresses and stamps the envelope, and mails it away. How is this process similar to a data communications hierarchy?

The paper size must fit into the envelope, and the envelope must be the acceptable size for the postal system. If it is too large, it may not fit into the mail slot or mailbox of the receiver. This step is similar to the connectors used for the communications cables.

The ink used on the paper must be the kind that will not smudge or disappear, especially as the postal mail handlers and others pick up, handle, and sort the mail. This procedure is like using the proper voltage values.

The symbols used must agree. This book uses the English version of the Latin alphabet, with letters A through Z, numbers 0 through 9, and some special symbols along with other ones, such as accent marks for Spanish or French.

Actually, an entirely different set of symbols could be used, such as the Greek alphabet or Chinese characters. The postal clerks do not have to understand the symbols of the message (only of the address), but the reader of the message must. This is called the *message code* or *format*.

The protocol may require that the sender first check whether the receiver is home to actually read the message once it arrives, that the receiver's mailbox not be overflowing, and that the receiver send a message confirming that the message has been received. There is no point in sending a message if the receiver is not ready for it. In many protocols there is a need for confirmation that the message has actually arrived, such as a receipt. Messages that are misunderstood or have errors must be identified.

The receiver must be able to separate the message address and greeting and closing (DEAR JOHN, YOURS TRULY), from the actual data part (I WILL ARRIVE AT 2 O'CLOCK) and distinguish the message from the surrounding characters which are required for successful communications but carry no actual user information. This means the reader must know what is appended to the message for purposes of communication. It also includes special symbols, such as punctuation marks, that are used to separate and segment parts of longer messages. The reader must understand what these mean. Early telegraph systems did not have the ability to encode punctuation, so the word "STOP" was substituted for the symbol period (.). This often caused confusion in the message since the recipient of the telegraph message could confuse the "STOP" representing end of sentence with the word "STOP" in a sentence. The protocol may instead require that each sentence have exactly ten words, so there will be no confusion.

The relationship between the elements of a data communications system and sending a letter is shown in Fig. 7-1.

There are three other issues which are even further removed from the user. These issues are at the level where the actual signal transmission and reception occur:

The rate at which the bits are sent must be known to both parties, so that the receiving circuitry can be set to look for bits coming at that rate. If the sender is sending 1000 bits/s but the receiver is set for 300 bits/s, the receiver either gets only every third letter (approximately) or misses the message entirely, because the protocol is not established properly.

The circuitry at both ends must be designed for either parallel or serial communications. *Serial systems* require only a single wire for the data itself but are slower in general since each bit of a character must be sent one at a time. A *parallel system* uses more wires or links, but allows many bits of a symbol to be sent simultaneously.

The system is designed for either synchronous or asynchronous communications. The decision to use one or the other is based on many technical issues, but the choice also affects the protocols that can be used effectively.

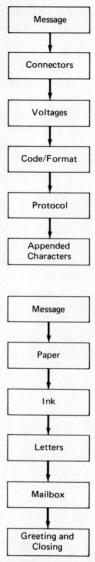

Fig. 7-1 Any communications system, whether transmitting data electronically or sending handwritten letters, involves many levels of message translation.

The issues briefly discussed in this section are explored in much more detail, as practical items, in the subsequent sections of this chapter. The overall hierarchy is shown in Fig. 7-2.

Even with all the hierarchy levels in agreement, there is no guarantee that the message will be understood. If the user at the receiving end is a piece of equipment that understands only messages in which a certain word order is specified and it receives a message with a different word order, the characters may be received properly in every respect but the sense of the message will not

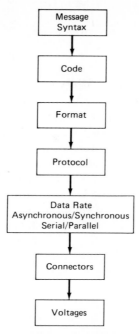

Fig. 7-2 The overall communications hierarchy shows the many levels through which a message must be translated and passed before it is sent.

be comprehended. An example is a voltmeter whose range is selected by a computer controller. The voltmeter may expect to see the range value first: 10V RANGE. However, the computer may assume that the order is reversed and send: RANGE 10V. The message will get through but make no sense. It would be like a census listing with the columns in the wrong order. This correct order is called *syntax*. It is the responsibility of the message originator to provide correct syntax, since the communications system passes the message along as it is and does not change or correct the syntax.

QUESTIONS FOR SECTION 7-1

1. What are two reasons that a data communications system requires many decisions in its design and use?
2. What is format?
3. What is protocol?
4. What is the hierarchy of levels in a data communications system?
5. How is sending a message by mail similar to sending one by using formats and protocols in data communications?
6. What other technical issues must be agreed on in a data communications system, with respect to data rate, serial or parallel transmission, and asynchronous versus synchronous signals?
7. What is syntax? Why is the communications system "blind" to incorrect syntax?

7-2 Codes and Formats

The goal of any code is to put the characters of a message into a digital, usually binary, format. For each character of the message, there must be a unique code that can be sent over the data communications channel. The message I WILL ARRIVE AT 2 O'CLOCK contains 26 characters (spaces within the message count as characters, too), and there must be a format so that a receiver can associate a received binary pattern with the character it represents.

One very common format used by the data communications industry for representing alphanumeric characters is the American Standard Code for Information Interchange (ASCII) (pronounced ASK-KEY). In the ASCII code, a field of 7 or 8 bits is used. Every letter, number, and special symbol is assigned a bit pattern. For example, the letter A is 1000001, letter B is 1000010, the number 0 is 0110000, the number 1 is 0110001, and so on (Fig. 7-3). Appendix B has a complete ASCII chart for all the characters. The characters which the ASCII code can represent are divided into logical groups: control characters, numbers, uppercase letters, and lowercase letters. The unused spots in the letter and number groups are used for punctuation.

The control characters are sometimes called the *nonprintable characters*. These characters were introduced in the early stages of the development of data

Character	ASCII
A	1000001 (LSB)
B	1000010
C	1000011
D	1000100
a	1100001
b	1100010
0	0110000
1	0110001
2	0110010
3	0110011
+	0101011
!	0100001
?	0111111
/	0101111
—	0101101
#	0100011
$	0100100
CAN (Cancel)	0011000
BEL (Bell)	0000111
CR (Carriage Return)	0001101
LF (Line Feed)	0001010
DEL (Delete)	1111111
BS (Backspace)	0001000

Fig. 7-3 The ASCII representation of some letters, numbers, symbols, and functions. A complete listing is given in Appendix B.

	S	E	Y	
Last bit	1010011	11000101	1011001	First bit

Fig. 7-4 The ASCII sequence for YES. The least significant bit is sent first, so the bit sequence as sent (and received) is from right to left.

communications for applications in which a computer was sending a message to a printer. Special codes were needed to tell the printer to perform a line feed (LF) or a carriage return (CR). These codes are not part of the message itself but are essential for proper overall operation. They are nonprintable because they do not cause anything to be produced on the printer (or screen of a video terminal), but instead cause an action to be taken. There is even a code to ring the bell (BEL).

When ASCII codes are used to send characters, the least significant bit (LSB) of the code is sent first, followed by all the other bits, in sequence. A message with characters YES would use the following bits:

Y	E	S
1011001	1000101	1010011

(using the ASCII representations from Appendix B) and would be sent as shown in Fig. 7-4.

An alarm message to print the word "ALARM" on the screen and ring a bell repeatedly would look like the following:

BEL	0000111
A	1000001
L	1001100
A	1000001
R	1010010
M	1001101
CR	0001101
LF	0001010

with only the letters ALARM printed. The terminal would ring the bell, print ALARM on the screen, then perform a carriage return and line feed to prepare for the next message.

How many different characters can the ASCII code set represent? Since there are 7 bits in the ASCII field, there are $2^7 = 128$ unique possibilities. Some variations of the ASCII set allow 8 bits in the field. This means that 256 characters can be represented, and 8 bits is more compatible with the standard computer byte of 8 bits. However, both sides of the communications link must agree to use either the 7- or the 8-bit format. The symbols often represented by the additional 128 possibilities in the 8-bit format are used in languages other than English, and include accent marks, umlaut (¨), and similar symbols. Most European data communications systems support the 7- and 8-bit versions for this reason; many U.S. systems support only the 7-bit version.

When working with a system that uses a format that supports both nonprintable and printable characters, it is important to use special equipment if there is a need to diagnose a problem. Consider the case of a system that is sending messages from a computer to a nearby terminal. The user complains that it seems as if the messages are getting garbled since the characters on the screen often do not make sense. Looking at the screen of the terminal alone reveals only the printable characters that were received—the terminal screen cannot show the nonprintable control characters, such as carriage returns. Perhaps this nonprintable character is being misinterpreted by the terminal and causing the screen to become garbled. Special test equipment and monitors must be used to show the technician or service engineer what nonprintable characters are being sent; this kind of equipment is available from many manufacturers.

The ASCII code is not the only one in use, although it is the most popular for low- to medium-performance communications systems. It is used between computers and their printers or plotters, between computers that are sending small to moderate amounts of data to each other, and between computers and special external devices such as temperature controllers. Another code that is sometimes used is the *Extended Binary Code for Data Interchange* (EBCDIC), developed by IBM. It is commonly used between IBM computers and their terminals, and between other IBM-supplied devices and equipment. (IBM also uses ASCII in some systems.) The EBCDIC format uses 8 bits only and assigns symbols to the bit patterns as shown in Fig. 7-5 for some letters and symbols. See Appendix C for a complete EBCDIC code chart.

Since a wider field of bits can represent more symbols or characters, it may appear that the best thing to do is make the field as large as needed. After all, an 8-bit field can represent 256 characters, and a 10-bit field can represent up to 1024 characters. Why, then, are the field sizes kept to a low value? The answer is that sending extra bits in a field takes time, and time is a precious resource on a communications link. If the user is only sending standard American English and

Character	ASCII	EBCDIC
A	1000001	11000001
B	1000010	11000010
C	1000011	11000011
D	1000100	11000100
0	0110000	11110000
1	0110001	11110001
2	0110010	11110010
3	0110011	11110011
.	0101110	01001011
,	0101100	01101011
/	0101111	01100001
−	0101101	01100000

Fig. 7-5 The EBCDIC representation of some letters, numbers, and symbols, along with the ASCII representation.

can use 7-bit ASCII, then a 10-bit field would add an unneeded 3 bits and increase the time to send each character by 3/7, or about 43 percent. For many applications, even the eighth bit of the 8-bit ASCII code reduces the number of characters that can be sent in a given amount of time by too much.

The situation is basically this: a larger bit field lets the system handle more characters, since a field N bits wide can represent 2^N unique characters. However, every additional bit imposes a penalty on the number of characters that can be sent in a given amount of time, assuming a certain number of bits/second is transmitted.

In some applications, the message consists of numbers only. It may be more efficient in these cases to use straight binary coding instead of a formal code such as ASCII or EBCDIC. In *straight binary coding*, the binary field directly represents the number. To send the number 56, for example, the code is 111000, which is the binary equivalent of 56. The advantage of using straight binary coding is that it is much more efficient in use of bits. A 16-bit field for straight binary coding can represent any number up to 65,535. By contrast, ASCII requires five bit fields, one each for 6, 5, 5, 3, and 5, for a total of 40 bits.

Straight binary coding is much more efficient, but it does have some drawbacks. First, it can be used only for numbers, and many practical systems require both numbers and letters, punctuation, and similar symbols. A more serious problem is that there are no nonprintable control symbols, which are often needed in a system to ensure proper, effective operation. How does the sender tell the receiver to start a new line, or a new column, for example? This is the function served by the special *control codes*, which are the nonprintable characters, of the ASCII set.

Code and Format Conversion

Systems which use different formats are not compatible with each other, even if the connectors, voltages, and transmission rates are the same. It is analogous to receiving a properly addressed letter in a language that you don't know. The solution to this problem is called *code conversion*. Special circuitry which converts one type of representation to the other type is used. This can be a single integrated circuit (IC) which stores the relationship in its memory. When pattern 1110010 (which is r in ASCII) is received, it returns 10011001, which is r in EBCDIC. Other systems use a software program within the computer to perform the conversion. Both methods are practical but have different cost and performance factors. The specially designed IC can be used for the job but costs money and requires circuit board space. The software solution requires that the microprocessor of the system devote some of its time to perform the conversion. It therefore cannot be doing its regular, primary job while it is performing the code conversion. The IC solution is called *hardware conversion*, and the program solution is called *software conversion*.

The system designer decides which is better on the basis of cost and performance needs, although having to perform code conversion is always a burden. It is much better if the systems inherently use the same format and are directly compatible. In fact, there are cases where the systems must be compatible, because some system circuit designs simply do not accept codes that they do not understand. If the code is not accepted, it cannot be put into the hardware or software converter. Many ASCII terminals have circuitry specifically designed for ASCII codes and no other. If a non-ASCII code is received, the terminal simply ignores it or indicates that some error has occurred.

Code conversion can also be done manually. If a character is in ASCII format, for example, and must be converted to EBCDIC, then Appendixes B and C must be used together. The ASCII character bits are first converted to the character itself by using Appendix B, and then converted over to EBCDIC by looking up the EBCDIC representation for the character in Appendix C. This is a two-step process. Special tables are available which list the characters and the ASCII and EBCDIC codes (or whatever codes are being used) alongside each other to make this lookup process easier.

QUESTIONS FOR SECTION 7-2

1. Why is a code needed? What function does it perform?
2. What is ASCII code?
3. What are nonprintable characters? Give three examples of printable characters and nonprintable ones. Why are these nonprintable ones needed?
4. How many unique characters can be represented by 7 bit ASCII? By 8 bit ASCII?
5. Why is special test equipment needed to observe the nonprintable characters?
6. Why is the character field kept as short as possible? What does a longer character field allow?
7. Where is straight binary coding used? What are its advantages? Drawbacks?
8. What is code conversion? What are the merits and drawbacks to hardware code conversion versus software code conversion?

PROBLEMS FOR SECTION 7-2

(Use Appendixes B and C for ASCII and EBCDIC codes.)

1. What are the ASCII codes for these printable characters? (Commas separate characters.)

    A,   a,   Q,   t,   3,   9,

2. What are the ASCII codes for the following nonprintable characters?

    bell,   line feed,   carriage return,   null

3. What is the EBCDIC code for the following characters?

    P,   p,   7,   1,

4. Convert these bit patterns, which represent ASCII codes, to their EBCDIC equivalents:

    0111111,   1010100,   0001000

5. Express the decimal number 245 in ASCII. Also express it in straight binary code. By what percentage is binary code more compact than ASCII code?

6. Repeat Prob. 5 for decimal 100.
7. How many characters could a 9-bit field represent? A 10-bit field?
8. A set of characters is needed to represent the 26 alphabet letters only, without numbers, punctuation, or nonprintable symbols. How many bits should the field have, for maximum efficiency? If 7-bit ASCII is used instead, what percentage of the field is wasted?

7-3 Protocol

For successful communications to occur, it is not enough for the sender of the message simply to transmit and hope that the receiver gets the message and there are no errors along the way. This is analogous to picking up the phone, dialing, and then talking, without bothering to check whether the other phone has rung, has been picked up; and the party can listen without distraction. There are rules that must be followed for communications. These rules form the protocol of the data communications system. Each party to the communication knows the rules to be followed and knows that the other will also follow them.

Many types of protocols are in use in modern data communications systems. Different applications have different performance needs and cost requirements. There is no single protocol used that is superior to all others in most applications. Therefore, it is more important to understand what a protocol is, and what it does, than the details of many specific protocols in use today.

The basic concept behind a protocol is handshaking. By *handshaking*, each end of the communications link shows the other end that it has something to send, it is ready to accept messages, a message has been received, and the reception has been successful. If any of these steps is not completed properly, the handshaking shows this and each party follows a predefined set of rules for this special situation. Figure 7-6 shows the handshaking plan that a simple protocol might use to send a message from A to B (one direction only). First, A signals B to say "I've got a message for you"; this is called a *request to send* (RTS). Then B responds, if it can accept a message, by sending a *clear to send* (CTS), which tells A to go ahead. After the message has been sent, B may check it for transmission errors. While this checking is going on, B does not signal CTS, but

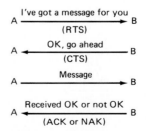

Fig. 7-6 The concept of handshaking in a simple protocol allows the message sender to make sure that the receiver is ready and able to accept a message and that the receiver can indicate that the message is received properly.

instead shows it is busy. Then, B sends some type of *acknowledgment* (ACK) showing that the message was received without error, or a *negative acknowledgment* (NAK) indicating a problem. Depending on the particular protocol, A may repeat the cycle if an NAK is returned, or it may just go on to the next message.

There are several important points in this process. First, even though the user message is going only in the A to B direction, there must be some method of having B send back its protocol messages, such as CTS, ACK, or NAK. Second, the protocol does not look at the contents of the user message itself. The protocol is designed to handle any message that the user may wish to send. The checking for correct reception is done by various techniques that do not require the protocol (and therefore the data communications circuitry) to know which messages are allowed and which ones make sense. The protocol is said to be *transparent* to the user, since the user simply enters the desired message into the system. How the message is handled by the system, and what the system does to it, is not seen by the user at the A end.

Timing Diagrams

A *timing diagram* is often used to show the interaction of the various data and signal lines in communications. Arrows are used to show which signal on a certain line causes another line to change. These diagrams are very useful for showing cause and effect in protocol interaction. They are especially helpful when a data communications system is not working properly. For example, if a system is not sending out data, it may be that its transmitter is malfunctioning. But the timing diagram also shows that this transmitter will not send out any data until it receives a CTS signal from the receiver, which in turn is prompted by an RTS. Perhaps the RTS signal is not going out, or the CTS response is not present. The timing diagram shows the logic flow of requests, responses, and data flow.

How is handshaking accomplished? There are two techniques. One technique, called *hardware handshaking*, uses extra signal wires to represent the handshake messages such as CTS or RTS. This is easy to implement and relatively inexpensive when the two users are close together, for example, a computer talking to its printer. Even though using more wires requires additional interface circuitry and a thicker cable between the units, it is technically acceptable and practical. A typical installation of this type would look like Fig. 7-7a, with the time sequence in operation shown in Fig. 7-7b. Some installations actually need more than the relatively few handshake lines shown. This is the case in full-duplex operation, in which A sends messages to B and B simultaneously sends messages to A. Also, some types of data communications circuitry have several levels of handshaking in each direction. They may be used to indicate which specific parts of the other side are ready or not ready to communicate. In the printer example, one handshake line may show the computer that the printer interface circuitry is ready, and another may show the status of the physical mechanism of the printer. It is possible to have a system that can accept messages when the interface circuitry is ready even though the printer itself is still printing, through the use of buffers (discussed later in this section).

It is not practical to use multiple lines for longer distance communications, however. This would require that additional channels of the system be dedicated to relatively infrequent messages indicating the operating status of each side of the link. Certainly, this process is very costly and inefficient. The solution is to use special *handshake messages* which can be sent over the actual user message channel. These messages, which are different from the user messages, can perform the handshaking function. Each side checks the incoming messages very quickly and continuously to see whether any of the messages is one of these handshaking ones instead of an actual user message. The timing diagram for this operation is shown in Fig. 7-7c.

Handshaking not only ensures that the message is sent when the users are ready, but that the system functions properly if channel problems occur during transmission of the message. For example, suppose a printer is printing out text coming from a computer. It may seem that once communications are established and the printer is going ahead, there is no need for handshaking until the job is complete. This is not so. The printer may run out of paper or ribbon in the middle and be unable to print what it is receiving. If the computer kept sending, part of the text would be lost. To overcome this problem, handshaking functions even during the message. As soon as there is a printer problem, the printer uses its CTS line to signal the computer that there is a problem and that transmission

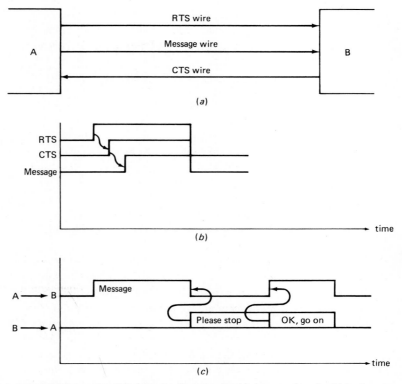

Fig. 7-7 Hardware handshaking implemented with extra wires besides the message signal wire: (*a*) the installation; (*b*) the relative timing of the RTS, CTS, and message lines, when the CTS line responds to the RTS request; (*c*) the way messages themselves can be used for handshaking, eliminating the need for additional wires.

should stop. When the problem is corrected, the printer again uses its CTS line to indicate to the computer that the system is okay.

The ability to use handshaking to indicate problems even during the message is very important. The problem may not be a mechanical one. It may be that the receiver is getting the message faster than it can handle it, and so the sender must stop transmitting while the receiving end has a chance to "digest" what it has received so far. The printer, for example, may be working fine but falling a little behind in printing the message. Or a computer at the receiving end may have to perform some operation on the received data and be unable to pick up the next character of the message immediately as it comes in. Rather than let the sending end keep sending, and therefore allow characters to become lost, the sender stops until the printer or computer catches up.

Buffers

The problem of the receiving end not being able to keep up with the transmitting end is common. To avoid this problem, the sender could transmit one character at a time and wait for some confirmation that the receiver has properly received it and is ready for the next character. Handshake lines or messages would be used to confirm this and indicate readiness. However, this scheme is very inefficient in use of channel time. Every time the flow of characters stops and starts, some valuable time is lost. Lost time can never be recovered. It is much more effective to send a continuous, nonstop string of characters.

To accomplish this continuous transmission, a buffer is used at the receiving end. A *buffer* is special circuitry dedicated to receiving characters at high rates and storing them as they are received. The receiving device (printer, computer, or other) can then take the received characters from the buffer at a slower rate than they are being sent, as its time allows (Fig. 7-8a). From the viewpoint of the sender, the channel is available to send at the higher rate, without any stopping. Of course, if the receiving system does not take the received characters, eventually the buffer fills up. When that happens, the receiver handshakes to the sender to stop. The effect of the buffer is to allow higher-speed, continuous transmission with better use of the capabilities of the circuitry at both ends. The buffer can be filling with new characters while its other side is simultaneously emptying (presenting characters to the receiving system).

Typical buffer sizes range from a few characters to several thousand. A majority of buffers are in the 128- to 512- character range. In the case of a 128-character buffer, for example, up to 128 characters can be received before the buffer is full and transmission must stop. As the receiving system takes a character from the buffer for processing, all the received characters shift over so that a new empty buffer space is created (Fig. 7-8b). It is like a waiting room with chairs, waiting for service at a doctor's office. As each person is served, everyone else moves down one seat and a new person can sit down. The chairs are like the character storage locations of the buffer, the door to the waiting room is like the connection from the data link, and the door to the doctor's office is the same as the actual system (the printer or computer) that is using the data. This kind of buffer is sometimes called a first in, first out (FIFO), since received characters are passed along in the order received, with room made for any new characters as each old one is taken and handled.

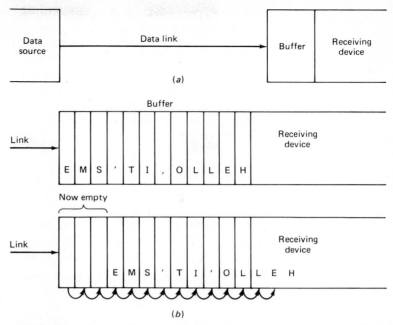

Fig. 7-8 Buffers allow messages to be sent faster than the receiving circuitry can process them: (*a*) the physical location of the buffer; (*b*) the operation of the buffer as received characters enter the buffer at one end and are passed to the receiver circuitry at the other end. As each character is shifted out, the entire buffer contents shift over to make room for another incoming character.

The size of a buffer is determined by the system designer on the basis of several factors. Larger buffers cost more in circuitry, circuit board space, and power consumption. However, larger buffers allow longer continuous strings of characters to be sent and result in a much more efficient system and communications link. The system designer chooses a buffer size to provide storage for as many characters as may come in during a "dry spell," when no characters are taken out of the buffer by the receiving system. Of course, there is some small possibility that the dry spell will be longer than planned, so the designer can either increase the buffer size or accept this possibility (in some cases it is possible to calculate the possibilities in advance).

For example, a computer may be sending characters to a nearby printer at 1000 characters/s. The printer is relatively slow, and can only print at 10 characters/s. The designer may want to make sure that the transmitter at the computer can go on for at least 1 s before the buffer fills up, so the buffer would be 1000 characters deep. The printer then takes about 100 s to print the contents of the buffer, but while it is doing this the computer could be performing other tasks. As the buffer empties, the computer begins to refill it. Typically, a system is designed to signal the transmitter that it can accept more characters only when the buffer is emptying and is past the half-full point. This way, the 1000-character buffer has 500 empty locations, and the computer can send 500 characters at the high rate to refill the buffer.

Communications System Requirements

Message Protocols

Whenever a protocol takes the user message and adds special characters before and after the message to change it into the style required by the date communications system, the protocol becomes more complex and powerful. This more powerful protocol allows systems to be built that have greater efficiency, can pass more characters per second, and can give more specific indications to the sender of the condition of the received message. This protocol is transparent to the user who actually originates the message.

The type of data that the protocol attaches to the message consists of a preamble and a postamble (Fig. 7-9a). The preamble is sometimes called the *header* and contains information such as the following:

A message number, so that the receiver can identify messages uniquely when there is a problem in reception to report.

A receiver number, so that many receivers may share a channel. Each receiver examines the first part of each received stream of data and then ignores any messages that are not addressed to it. If the message is addressed to it, the receiver allows the message to enter the receiver buffer.

The postamble may contain information on the following:

The number of characters in the message, so the receiver can check that the entire message has been captured by the buffer. (That information may be in preamble, instead.)

Some special checking information which helps the receiver determine whether any of the received characters have any errors, even if the character count is correct.

Other special information about the end of the message which the system may require.

Each of the preamble and postamble data groups is assigned a field of bits. The data (message number, character count, receiver number, and so on) is then put into the assigned field. A typical field may be 8 bits. This allows for character counts from 0 through 255, and receiver numbers over the same range. The

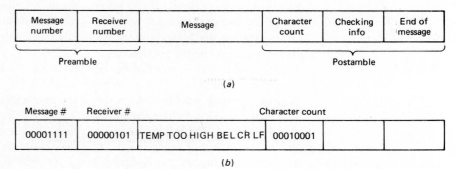

Fig. 7-9 The protocol associates additional information along with the message, in the preamble and postamble: (*a*) the general location and information that preambles and postamble contain; (*b*) a specific example with a message number, receiver number, and character count. (Note: BEL, CR, and LF are single characters.)

overall message seen on the link then looks like Fig. 7-9b, where the user-supplied message is surrounded by the protocol messages. The user message in this case is simply TEMP TOO HIGH, followed by BEL, CR, and LF. This is what the person at the computer terminal typed. The protocol for this communications link requires that the user message be preceded by a receiver number (5 in this case), a message number (15 here), and a character count (BL, CR, and LF count as characters). The 8-bit fields have the binary equivalent of the preamble and postamble data.

At the receiving end, the receiver checks the received message for errors and then can acknowledge. It can send back the message number with the acknowledgment, so that the sender knows which message was received properly or improperly. The sender then follows its protocol rules and resends the entire message if there is an error, along with the same message number.

The handshaking of requests to send and acknowledgments take time. Some of the link time is devoted to the preamble, postamble, and retransmissions. The sender also must devote some of its resources to looking for the response from the receiver. Both of these activities take away some of the potential performance of the system running at full speed, sending only the user message itself. However, such a system would not be very useful. Without the information contained in the protocol around the message, and the return acknowledgments, the actual performance of the system would not provide guaranteed, reliable, useful data. The sender needs to know that the data is properly received, the receiver must know that it has correctly received the user message, and the sender has to verify that the message is properly received. The loss in efficiency is more than made up for by the increased usefulness of the overall system.

Not all data communications systems require complex protocols. The computer and its nearby printer usually do not, since the probability of an error is low. Simple handshake lines tell the computer to stop sending when the printer buffer is full or the printer has a mechanical problem. However, systems which send large amounts of critical data from one computer to another, often over long distances and at high rates, need the additional capabilities that protocols with preambles, postambles, and acknowledgments can provide.

QUESTIONS FOR SECTION 7-3

1. Why is protocol needed?
2. What does protocol accomplish?
3. What is handshaking? What does it do? How does it work?
4. What is a timing diagram? What does it show? Why is it useful?
5. What are two methods of handshaking? How do they differ?
6. Why is a system with many handshaking lines often needed?
7. How is message handshaking accomplished?
8. Why is handshaking needed even when communications has been successfully established?
9. What is a buffer? Why is it needed? How does it work?
10. How does a buffer improve system performance and efficiency?

11. How is buffer size determined?
12. What roles do the preamble and postamble play? What kinds of information do they contain?
13. What are the drawbacks of messages with preambles and postambles? The benefits?

PROBLEMS FOR SECTION 7-3

1. A communications system is used to send data from one computer to another. The data consists of 1000 character blocks. How large a buffer should the receiving system use to ensure that the computer at the transmitting end can send 2 blocks in a row, without any waiting?
2. A system can send data at 200 characters/s. The receiver is not able to handle characters at this rate, but it can handle 50 characters/s. After 2 s, how many characters will the transmitting end have waiting to send? How large a buffer will eliminate this?
3. A 1000-character buffer is full. The receiving system is taking out characters at 500 characters/s. The buffer is designed to allow more new characters to enter, as soon as it is half-full. How long will this take?
4. For Prob. 3, how long, approximately, will the buffer take to refill if the transmitter is sending 10,000 characters/s?
5. A protocol has eight bit fields each for the message number and character count as part of its preamble and postamble. For the following message, called *message 4*, what is the binary pattern in each field: I WILL BE LATE!?
6. The next message is number 5. It is: CAN YOU MEET ME LATER? What are the two fields, in binary code?
7. A protocol uses four bit fields for the message number, receiver number, and character count. What do the fields look like for message 10, receiver 7, when the message is: COME PLEASE.?

7-4 Synchronous and Asynchronous Systems

Modern data communications systems use both synchronous and asynchronous forms for data transmission. Each has advantages and drawbacks. Often large systems use a mix of the two methods, with the best scheme used in each part of the system. The two methods are not compatible, so special circuitry is used to bridge the gap between them where needed. In addition, the protocols used for asynchronous and synchronous systems differ in most cases, so some sort of protocol conversion is also used.

In an asynchronous system, the data message is sent one character at a time. Each character is preceded by a *start bit*, which is a signal transition from the *idle value* to the start value. In the figure shown (Fig. 7-10), the idle value is logic 0, in this case 0 V. A start bit is represented by this signal's going to the logic 1 level of 5 V. After the start bit has been sent, the transmitter sends the bits of the character itself. If this is ASCII code, then 7 or 8 bits follow the start bit. Finally,

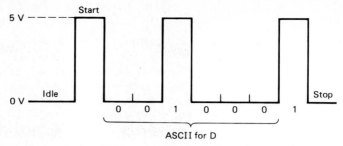

Fig. 7-10 In an asynchronous system, the character bits are preceded by a start bit and followed by a stop bit. The letter D is shown in ASCII bits.

the signal generates a *stop bit*, which is a return to the idle value. The figure shows an asynchronous system sending the letter D by using ASCII code, since D is 1000100. (The LSB is sent first.)

There are some enhancements and variations used with asynchronous systems, especially in the area of the number of stop bits and any error-checking bits. These are specific to certain systems and are discussed in Chapter 8.

The time between characters in an asynchronous system is not fixed at any specific amount. It can be very short if there is another character to be sent. It can also be very long, if there is no new data. Asynchronous is essentially a start/stop type of communications and is used when the source of data may not be providing a steady stream of new characters. A good example is someone typing at a terminal connected to a computer. Each key of the terminal keyboard is a character to be sent to the computer. The typing rate is not constant, and there can be long gaps when there are no new characters being typed. The data thus comes to the computer at unevenly spaced intervals, without reference to some master clock, hence the name "asynchronous."

In an asynchronous system, the receiver must be prepared at all times for the start bit to appear. As soon as it sees the start bit, it must begin to accept the character bits as they come. It knows that the character bits have finished when the stop bit arrives. The receiver is designed in advance, or set up by the system user, for the exact number of data bits per character. The circuitry of the asynchronous receiver cannot handle one character that is 7 bits of data, followed by one that is 10 bits. The stop bit serves mainly as a confirmation that transmission of the character bits is finished.

How does the receiver know at which rate the bits of the character will arrive? Both the transmitter and receiver are set, in advance, to the same rate. For example, the transmitter may be set to send the character bits at 1000 bits/s, so a new bit is sent every 1 ms. The receiver has also been set to expect bits at this rate. The receiver therefore knows to look for the first character bit 1 ms after the start bit, the second character bit 2 ms after the start, and so on. The receiver system uses its own clock circuitry to generate the timing signals and gets its starting time synchronization only from the appearance of the start bit.

Synchronous systems use a different approach than asynchronous ones. The synchronous system is designed to transmit long blocks of characters, without constant starting or stopping. This does not mean that it never starts or stops, but

Communications System Requirements

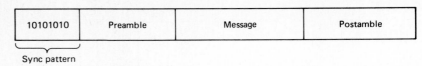

Fig. 7-11 A synchronous system uses a synchronization pattern and is able to transmit a long group of characters, instead of a start and stop bit per character.

that it is specifically intended for situations in which there are many characters to be sent. An example is a computer sending the equivalent of a one- or two-page report to another computer. This may have 500 to 1000 characters. Synchronous systems are designed to handle this efficiently and quickly. The entire 500- or 1000-character group is not necessarily sent as one huge group. It can be divided by the synchronous system into several smaller blocks, such as groups of 100 characters.

In the synchronous system the signal line can be idle, just as in an asynchronous one. The transmission begins with a synchronization pattern of bits (Fig. 7-11), similar to the start bit of the asynchronous system in its intended function. This pattern alerts the receiver that bits are about to come down the signal line and allows the receiving circuitry to establish timing synchronization with the transmitter clock. Here is another major difference between the two systems. The synchronous receiver system knows the clock frequency of the transmitter (as does the asynchronous system), but the receiver wants to lock onto the frequency precisely. Therefore, the receiver must minutely adjust its own clock to synchronize with the transmitter clock, so that it can properly receive each bit as sent. This adjustment process is performed by having the receiver actually extract the clock of the transmitter from the received signal. (In some applications, a separate *clock only* signal is sent on another link in parallel with the data, but this is less common.)

The purpose of the synchronization pattern at the beginning of the transmission is to give the receiving circuitry time to become synchronized to the transmitter clock, so that both clocks are "in tune" exactly when the first data bit arrives. Once the data bits start, many bits, representing many characters, can be sent. The protocol of the system usually requires that the sync pattern be followed by some preamble, then the actual characters of the message, and then the postamble. The receiver captures all of these and follows its protocol rules to make correct use of the information contained in the preamble and postamble, as well as the message. The goal of sending many bits (representing protocol information as well as the user message) has been achieved. After the last bit of the postamble, the communications line returns to the idle state. If the transmitter has more data, and the receiver is ready, they both follow the protocol rules and more blocks of characters can be sent.

Comparison of Asynchronous and Synchronous Features

Both methods of data communications are necessary to the overall system. Each has strong and weak points in a particular application.

Asynchronous systems are very inexpensive to implement, and low-cost ICs are available that easily check for the start bit, store the character bits, and identify the stop bit. These ICs then signal the remainder of the system that a new

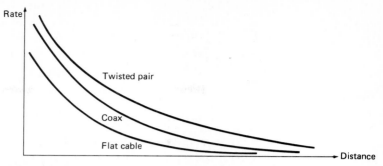

Fig. 7-12 The data bit rate that can be effectively achieved decreases with cable length. It is also a function of the cable type itself.

character has been received. In applications in which the rate of data is low (less than a few thousand characters/s) an asynchronous system can provide the required performance at low cost. It is also attractive in cases where the data is very sporadic, such as from a keyboard. For sending a single character, the total *overhead* (nondata bits needed for start and stop, but not part of the actual character) is low, and the character is sent to the receiver as soon as it occurs (as soon as the user types it, for example). There is no need to wait for an entire block of characters to be generated by the user. A computer system that does not display the characters as soon as they are typed, but only after every 50 to 100 characters, would be very difficult to use.

However, asynchronous systems have weaknesses. First, from an electrical standpoint, as the signaling rate exceeds about 20 to 40 kbits/s, it is impractical to have the receiver function properly and capture the bits of the character based solely on the basis of the timing synchronization provided by the start pulse. This is because any slight mismatch in the clocks at either end becomes more significant at higher rates. Any tiny jitter also becomes more of a problem. A chart (Fig. 7-12) is often used to show the achievable data rates versus cable length, since the length distorts the signal and makes timing jitter more severe.

Second, the character at a time structure of an asynchronous system does not fit well into the more advanced protocols. These protocols allow the transmitter to identify messages by number, indicate their length, and add some error-checking postambles to the message. The asynchronous system is oriented toward character-by-character situations, and not toward stepping back and looking at messages in their entirety. It is possible to use an asynchronous system with these protocols, but performance is poor. A considerable amount of communications time is lost between characters, waiting for the protocol response. There are some low-speed systems in noncritical applications that use these protocols successfully with asynchronous systems, but the application has to be carefully chosen and the limitations must be fully understood by the system designer and user.

Asynchronous systems, despite these limitations, are very widely used. The typical applications for asynchronous systems include the following:

Those from a computer terminal to a nearby computer.

Those from a computer to a printer..

Those between computers that are located in the same area.

Those between a computer and a specialized peripheral device, such as a temperature controller or pump with electronic control.

Those between computer users who are using the standard telephone line for communications with each other. (Some additional interface circuitry is used, but the communication is asynchronous.)

Synchronous systems are capable of very high performance. *High performance* means that they can send characters at a high rate, with few undetected errors. They make good use of the time available on the channel. The rates that can be reached on a synchronous system range from 100,000 bits/s to many Mbits/s, depending on the specifics of the circuitry used, the channel, and the medium. The efficiency of the synchronous system is high, because many characters can be transmitted in a long block, with a single preamble and postamble. Error-checking and acknowledgment schemes work well.

For example, consider a synchronous system sending 100 characters as 8-bit bytes. Suppose one additional byte is needed for each of the synchronization characters, message number, character count, and postamble. Then, only 4 bytes of overhead are needed for every 100 characters sent, or 4 percent. Contrast this to an asynchronous system, which needs a start and stop bit for each character, or 2 bits per 7 or 8 bits (about 20 to 25 percent), not including any protocol information. The key point is that an asynchronous system needs the overhead for each character, whereas the synchronous system puts some overhead up front and at the end and does not require any overhead for each individual character.

In addition, an asynchronous system has a significant amount of idle line time between characters, and this is wasted time. A synchronous system has idle time only between blocks. The result is that a synchronous system has much higher efficiency in use of time, and a higher throughput of characters (throughput is the number of actual message characters that can be sent in some time period).

However, synchronous systems require more complex circuitry at both the transmitting and receiving ends. The circuitry to generate data bits at a high rate is more complex, as is the circuitry needed to extract the clock and synchronize to it at the receiver. Both sides need buffering. The transmitter must buffer up enough characters so it can send a block (synchronous systems usually cannot send only one or two characters), and the receiver must have a buffer to accept all the characters at the rate at which they are sent, even if the rest of the system cannot handle them immediately. For some applications, such as the terminal, it is not desirable to wait until a full block of characters is typed, and thus a character-by-character scheme is more friendly and convenient. Finally, the speed advantages of synchronous communications may not be attainable if the medium is low-bandwidth. The standard telephone line is an example—it cannot support more than about 2000 to 20,000 bits/s, depending on the specific line type. The medium must specifically be a wide bandwidth type for the higher bit rates to be usable.

Some synchronous systems are designed for handling very large amounts of data, on an almost continuous basis. These systems often use a constant number of characters/block. Both the transmitter and receiver know what this number is, by design. The advantage is that there is no need to send a character count, and some other patterns in the receiver system can be established in advance since the exact amount of incoming data, per block, is known. Therefore, some additional efficiency and throughput are possible, and there is simplification of the circuitry on both ends of the system. There is some loss in flexibility, of course, but some applications do not need the flexibility since they have so much data to send all the time. For cases in which less than a full block must be sent, or there is a small "leftover" block of characters from a longer message, the system fills the unused character locations with blanks, called *nulls*. These are simply placeholders and carry no information, but they occupy the space as required.

Typical applications for synchronous communications include the following:

Transmission of large amounts of data from one computer to another

Telephone system central offices, which take the digitized signals from many callers and group them to be sent together to another telephone office

Situations in which there have to be absolute assurance that every character was received correctly and protocols to define what should be done if an error has occurred

Applications which require more than about 1000 characters/s over a distance of more than 50 ft

QUESTIONS FOR SECTION 7-4

1. What is the role of the start and stop bits in asynchronous systems?
2. What determines the time between characters in a synchronous system? How short can this time be? How long?
3. For what kinds of applications is an asynchronous system a good fit?
4. Why is a system called asynchronous?
5. How does the receiver know the rate of incoming bits in an asynchronous system? From where does it define its synchronization and timing?
6. Where is synchronous communications used?
7. What is the equivalent of the asynchronous start bit in synchronous systems?
8. How does a synchronous system become synchronized to the clock of the transmitter? How is this different from an asynchronous system?
9. Compare the pros and cons of asynchronous versus synchronous systems in three areas. Explain each pro or con.
10. What is the maximum practical speed for an asynchronous system, in bits per second? For synchronous systems?
11. Compare the overhead in an asynchronous system versus a synchronous system. How is the overhead partitioned in each? Which is more efficient, in terms of the overhead, to send several characters? Why?
12. What are the advantages of a constant block length? The disadvantages?

PROBLEMS FOR SECTION 7-4

1. A system is used to send 100 characters using 7-bit ASCII representation for each character. How many bits are overhead (start and stop) in the asynchronous approach? What percentage of the total is this? Compare this percentage to the percentage when sending 200 characters the same way.

2. A system is using synchronous communications to send the same 100 characters by using 7-bit ASCII. The characters are being sent in one large block. The preamble has 8-bit fields for message number and receiver number, and 8-bit fields for the postamble character count and error character. There is also an 8-bit synchronization pattern at the beginning. How many bits are used for overhead? What percentage of the total is this?

3. Repeat Prob. 2 for the case in which there are only 50 characters to be sent as a single block, and for the case in which the 100 characters are split into two groups of 50 characters each.

4. A system has to transmit 20 characters, using 7-bit ASCII. Which is more efficient: using asynchronous format or synchronous format with two 8-bit fields in the preamble, and two in the postamble? Why?

5. A designer needs to choose between asynchronous and synchronous formats. The data is sent as 7-bit ASCII characters and is coming from the source at the rate of 10 characters/s. How long would the receiver have to wait for the first character, if the synchronous format uses a fixed-length data block of 30 characters (not counting the preamble and postamble)? Compare this to asynchronous format at 1000 bits/s.

7-5 Data Rates and Serial and Parallel Communications

Both the sender and receiver of data bits must use the same data rate for the data to be received properly. The *rate* is the number of bits/s that are sent. The communications industry has standardized on several different rates, for various applications. Normally, a system uses the highest rate practical. Some of the factors that limit the rate that can be used include the cost of circuitry, the bandwidth of the medium, and the number of errors that can be accepted.

The measurement unit used for the data rate is usually *baud*, named after Jean Maurice Emile Baudot, who in 1874 invented a 5-bit on-off code to be used with electromechanical teletypes. His code is still in use today since it requires only five bits and can be very efficient in use of channel time for some applications. His scheme was designed to overcome the problems of standard Morse code, which used variable-length on-off signals (which we call *dots and dashes*). It turns out that people are very good at distinguishing by length of time, but electromechanical equipment has trouble doing that. Baudot's scheme was an early attempt to make a modern digital system, because it used five evenly spaced time intervals for the five code bits. Electromechanical equipment could handle this much better than dots and dashes. A rate of 1 baud corresponds to 1 bit/s, if each bit of the transmission represents one signaling unit. This is the case for most communications systems, so "baud" and "bits per second" are

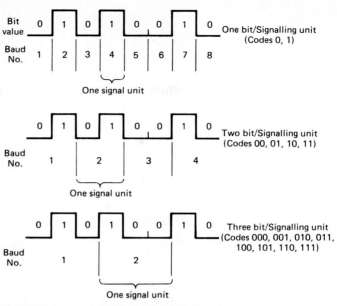

Fig. 7-13 The relationship of bit rate, baud, and the number of bits per signaling unit. Shown are 1 bit, 2 bits, 3 bits per signaling unit.

equivalent in these cases. Some communications systems use clusters of 2 or 3 bits as the basic signaling unit, in which case 1 baud is one signaling unit per second. Figure 7-13 shows the situation for regular binary coding in which the signaling unit is one unit per bit, and for 2- and 3-bit cases. In the 2-bit case, the baud is one-half the bits/s value; in the 3-bit case, the baud is one-third the bits/s value. Usually, the signaling unit is 1 bit, and the term ''baud'' in this book means bits/s unless otherwise specified.

The communicating industry has settled on some standard baud values. These are the following:

50	2,000
75	2,400
110	3,600
150	4,800
300	7,200
600	9,600
1,200	19,200
1,800	38,400

The large number of values available reflects the fact that different data communications systems traditionally used various subgroups of these rates. The electromechanical teletype was a 110- or 300-baud device; telephone lines and associated equipment used 150 or 300 baud in the past. Fortunately, most modern applications concentrate on 300, 600, 1200, 2400, 4800, 9600, and 19,200 baud for asynchronous and lower-speed-range synchronous use.

Communications System Requirements 229

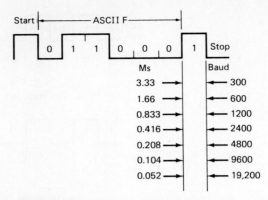

Fig. 7-14 An ASCII character, sent asynchronously, with start and stop bits. The time period corresponding to common data bit rates (baud) is shown.

The letter F represented by ASCII code is shown with start and stop bits in Fig. 7-14. The data rate defines the speed at which the individual bits are sent, in succession. The figure also shows the time period corresponding to some of the common baud values.

Synchronous systems are capable of higher data rates. Typical synchronous systems begin at about 2400 baud, and can go up through millions of bits/s (megabaud [Mbaud]). There are some standard synchronous rates in use, such as 1.544 megabits per second (Mbits/s). However, synchronous high-speed systems are usually dedicated to use between two specific users, such as phone company regional offices. Since the system is not general-purpose and open to use by outsiders, the designers have the freedom to choose the highest rate that they can, without worrying about compatibility with outsiders. For this reason, there are fewer standard rates in use for synchronous systems.

Data Rates and Throughput

In order to send as many characters as possible in a given amount of time, such as 1 s, higher bit rates are used. However, bit rate alone does not indicate the actual number of characters that will be transferred. The term used for actual number of characters is "throughput." The throughput compares the efficiency of the channel to what the maximum number of characters/s would be, in theory, for the data rate.

Consider an asynchronous system that is using the 9600-baud rate. This corresponds to a data bit every 0.1 ms (to a close approximation). Therefore, if each character is represented by a start bit, 7 ASCII bits, and a stop bit, it takes 9 bits or 0.9 ms, to send each character. If the system could send characters without any interruptions (from handshaking, protocol, or any other source) then the character rate would be 1/0.9 s or about 1100 characters/s. Many asynchronous systems use an eighth bit for the ASCII or use an extra bit for an error check (called *parity*, discussed in a subsequent chapter), and so it takes 10 bits for each character. Thus, the character rate is one-tenth the bit rate. A very good initial estimate of the maximum number of characters/s throughput is 10 percent of the baud value.

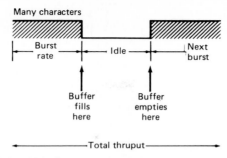

Fig. 7-15 The relationship of burst rate and average throughput. Idle time between bursts of data results in an average throughput of data that is less than the burst value.

In real systems, unfortunately, this maximum rate cannot be achieved. Many factors reduce it. Real systems need time for the proper handshaking to occur, for the receiver to accept the next character and indicate it can take another one, for any protocol characters in the preamble and postamble, and for other factors.

One such additional factor is the *receiving buffer*. The transmitter may send a group of characters at high speed, with almost no idle time between characters. Once the receiving buffer fills up, the transmitter must stop for a relatively long time while the system at the receiving end processes the received characters and empties the buffer (Fig. 7-15). The high-speed rate is called the *burst rate*. The overall rate which averages both the burst and the idle time is the *actual throughput*.

An example shows the difference. A system may send 100 characters at a burst rate near the maximum for 9600 baud, about 1000 characters/s. After 100 characters the receiving buffer is full, and it takes 0.4 s for the buffer to empty enough that transmission can resume. The burst rate is 1000 characters/s, but the average rate is 100 characters in the 0.1 s it takes to send them, plus the 0.4 s of idle time, or 100 characters in 0.5 s. The throughput is then 100/0.5, or 200 characters/s, much slower than the burst rate.

Some applications require high burst rates for short messages and can accept much lower throughputs for longer ones. One example would be an alarm system, which must get the basic alarm data to the fire station immediately, although additional details can take a little longer. The system designer can adjust buffer sizes and other system factors to trade one capability for another in many cases: higher burst rate but much lower throughput versus lower burst rate closer to the overall throughput.

The efficiency of the system indicates the closeness of the system throughput to the theoretical value. A system which spends 20 percent of the time idle between characters is 80 percent efficient. The throughput of an asynchronous system is often determined by the amount of data being sent. If the person at the keyboard is typing slowly, the 9600-baud link may send only two or three characters per second, compared to the maximum of just below 1000 characters/s.

Synchronous systems tend to lose efficiency in a different way than asynchronous ones. In asynchronous systems, a large portion of the time is lost when the

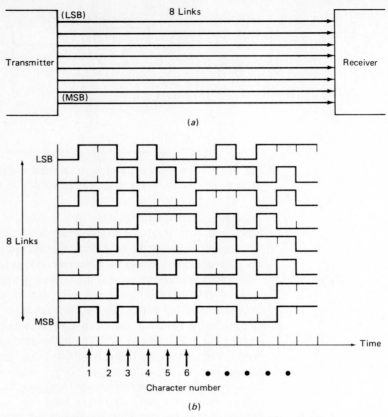

Fig. 7-16 Increased throughput results from using a parallel signal path shown physically in (a). The timing of sending a group of characters is shown in (b).

sender (for example, the typist) has no new data or during handshaking. Synchronous systems, in contrast, lose time in the preambles and postambles, since a synchronous system is designed for applications in which there is already a large block of data waiting to be sent. For this reason, synchronous systems try to keep the preambles and postambles short, and the data blocks long. A synchronous system that uses 20 characters for the preamble, 30 for the postamble, and a block of 50 message characters in the middle is at best 50 percent efficient.

Serial and Parallel Communications

Another way to obtain high throughputs at a fixed baud value is to send the data on several parallel channels at the same time. In a *serial system*, a single channel is used. Each bit of the data is sent in succession, one at a time. In a *parallel system*, there is a circuit and link for each bit of the character (Fig. 7-16a). A typical parallel system uses eight such circuits and links and sends entire characters at the bit rate. In the first time period, the first character is sent, and in the next time period, the second character is sent, and so on (Fig. 7-16b). The effective increase in throughput is determined by the number of parallel lines—an eight-channel-wide system can send an entire character in one bit period, as

Communications Scheme	Characteristics	Typical Application
Parallel synchronous	Parallel data paths Highest performance Highest cost	Buses within computer system Short distance only
Parallel asynchronous	Parallel data paths Moderate performance Moderate cost	Buses within computer system Computer to moderate peripheral device
Serial synchronous	Single data path Moderate to high performance Moderate to high cost	Computer to computer Telephone central office intercommunication
Serial asynchronous	Single data path Low performance Low cost	Computer to printer Computer to telephone line

Fig. 7-17 Characteristics of various serial, parallel, asynchronous, and synchronous communications schemes, in order from highest to lowest performance.

compared to eight bit periods for a serial system. The throughput is increased by a factor of eight as a result.

Parallel systems provide major improvements in system throughput, without increasing the bit rate or link bandwidth. There is a price for this. Much more circuitry (cost, circuit board space, reliability) is needed, and most important, multiple links are needed. This is not a problem for a system which is sending data from one subsection to another, for example, between circuit boards of a computer on a backplane or bus (which interconnects all boards). It is impractical for applications in which there is any distance over a few feet, and multiple conductor cables or multiple radio channels are required. Parallel communications is only practical over short distances, as a result. There are costs for the interface circuitry and costs for running cables with many conductors for hundreds of feet or even many miles.

The concepts of synchronous and asynchronous systems can be combined with those of serial and parallel systems to give different levels of technical performance, at various costs. Highest performance is achieved with a parallel, synchronous system. Lower performance results from serial synchronous, parallel asynchronous, and serial asynchronous designs, in descending order. Each type of system has applications in which it is the best choice. Figure 7-17 shows in brief the characteristics of each and typical applications. As with most systems, higher performance costs more, of course. If the cost were the same, designers would naturally choose the higher performance system for the same money.

As shown before, the exact nature of the system is transparent to the user in many cases. The person or equipment actually generating the data or message does not know what communications method or data rate is used. Similarly, the person or equipment at the receiving end is unaware of what the interposed system is. The user forms the message, and it is sent by any of the four basic approaches. The user only cares whether the system performance is poor (throughput too low) or many errors result.

QUESTIONS FOR SECTION 7-5

1. What is baud? What is the relation between the baud value and the bits per second rate for regular binary communications?
2. What are three of the common baud values in use?
3. What is the baud range that can be achieved with asynchronous systems?
4. Why are the rates for synchronous systems less standardized than those for asynchronous ones?
5. What is the meaning of the term "throughput"? How does it compare to the maximum number of characters per second that can be transmitted successfully?
6. Give three reasons why the actual throughput is less than the maximum value.
7. What is the burst rate? What is the effect of buffer size on throughput and burst rate?
8. How does the efficiency of asynchronous systems compare to that of synchronous ones? Why?
9. What is the physical difference between serial and parallel systems? What is the difference in operation?
10. By what factor does a parallel system improve the throughput over a serial system, all other factors remaining the same?
11. Give the relative performance rankings of these communications modes: serial synchronous, serial asynchronous, parallel synchronous, parallel asynchronous.
12. Give a typical application of each of the preceding modes.

PROBLEMS FOR SECTION 7-5

1. By what factor is each of the first baud values greater than the second one in the pair?

    1200/300;   4800/1200;   9600/600;   19,200/600

2. A system is sending 7-bit ASCII characters. What is the maximum number of characters that can be sent per second asynchronously at 300 baud? At 2400 baud? At 9600 baud?
3. A computer is sending 8-bit ASCII characters to a very slow printer. The printer is capable of printing 30 characters/s. What is the effective throughput at 300 baud? At 1200 baud?
4. What amount of time does the computer spend to send 100 characters at 300 baud? At 1200 baud? At 9600 baud?

5. A system can send characters at 4800 baud. The receiver can accept up to 10 characters in a row before its buffer fills. When the receiver buffer fills, it takes 1 s to empty before more characters can be accepted. What is the burst rate? What is the throughput?

6. A change is made to the transmitting circuitry of the system in Prob. 6, so that it can send at 9600 baud. What are the new burst rate and throughput? What would be the advantage of using this higher baud value?

7. A system can send 8-bit ASCII characters at 1200 baud. The receiver can handle only one character at a time (no buffer), and this takes 0.05 s. What is the throughput in characters per second that can be achieved? What is the increase in throughput if the baud value is increased to 4800? What is the decrease if the value is reduced to 300 baud?

7-6 Protocol Examples

Many protocols are in use in modern data communications systems. Some follow industry or government standards; others are used exclusively by one manufacturer of equipment. This section will examine several of these in detail.

Computer to Nearby Printer

One of the most common applications of data communications is to send messages to a nearby printer. This is a low- to medium-performance application. The format and protocol used are usually asynchronous serial ASCII, with an absolute minimum of protocol handshaking. To send a message such as HELLO (followed by a return to a new line) the ASCII representation for each character is sent, followed by the ASCII for CR and for LF (Fig. 7-18). Each character, of course, has a start bit and a stop bit surrounding it. The transmission rate is most likely to be 300, 600, 1200, 2400, or 9600 baud. The printer usually has a built-in buffer that lets it store incoming characters at a high rate, since the print mechanism is usually slower than the data communications.

Handshaking can be performed with the hardware handshake lines (request to send, clear to send, and similar lines), which allow the computer to indicate to the printer that it has a message to send, and the printer to respond that it can or cannot take the message (depending on whether or not the buffer is full and the

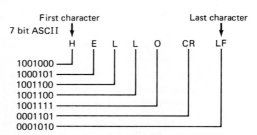

Fig. 7-18 The relatively simple protocol from a computer to a nearby printer involves sending the characters to be printed, followed by a CR and LF.

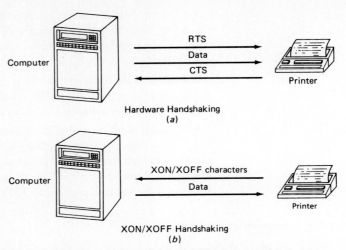

Fig. 7-19 Handshaking can be performed by either (*a*) hardware with signal lines or (*b*) software with special ASCII characters XON/XOFF.

printer is working properly) (Fig. 7-19*a*). Since the data flow is to the printer only (handshake lines are not considered data) the communications channel is simplex.

Some computer-to-printer installations use special ASCII characters for the handshaking, instead of control lines. The ASCII characters *XON* and *XOFF* (also called *DC1* and *DC3*) are used in most of these cases. These characters are sent back to the computer on a reverse direction data communications channel, to indicate whether the printer can or cannot accept new characters (Fig. 7-19*b*). The communications between the computer and printer must be full duplex to support XON/XOFF, since the indication to stop sending characters can come at any time while the computer is sending characters.

Computer and Graphical Plotter

Many computers can send their output data to a *plotting device*, which draws graphs, designs, and figures based on the computer instructions. The plotter also allows the user to indicate and signal important points of the figure to the computer, for highlighting. The user does this by moving an indicator, or *cursor*, over the plotter drawing and then pushing a button to say "I want this point." The plotter then sends the *X* and *Y* coordinates of the designated point to the computer.

This operation requires a full-duplex channel. It is usually implemented with the same type of asynchronous communications used for the simpler printer situation, with ASCII characters representing the data and messages. The handshaking can be either through signal lines (for each direction) or through XON/XOFF. What is different is that a more complicated protocol is required.

The reason for this is that the plotter draws what it is told to draw, but it has to know what it is being told. For example, to draw a box on the plotter paper, the user may type a command indicating the desired figure shape, size, and location (using the coordinates of the lower-left corner):

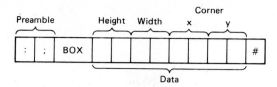

Fig. 7-20 The protocol between a computer and a plotter may have a preamble, figure name, and then the numeric information about the figure, all followed by an ending character such as #.

DRAW BOX, HEIGHT 3, WIDTH 7, CORNER 2 AND 4

The program in the computer converts this to the protocol required by the plotter. The plotter may require that all messages begin with two ASCII characters :; and end with an ASCII #. It may also require that the data be put into a format with two-character data fields for each piece of plotting information, as shown in Fig. 7-20. The final message looks like the following:

: ; BOX 03 07 02 04 # [**Note:** Spaces are shown here for easier reading. There are no spaces in the actual message.]

Of course, the ASCII equivalent of each character is sent to the plotter.

The reverse direction of the duplex channel is used for sending the coordinates of the point indicated by the user, from the plotter to the computer. The protocol may require that this message begin with a #, followed by the X and Y coordinate, and end with a :; pair of characters. Thus, if the user pushes the button when the cursor is at $X = 5$, $Y = 2$, the message is:

05 02 ; :

Synchronous Data-Link Control (SDLC)

The *SDLC protocol* was established by IBM Corporation to allow very high performance data communications of the type that occurs when two computers have large amounts of data to transfer between them. It is supported by many data communications and computer companies, for both IBM and non-IBM equipment. It allows for many types of message "give-and-take" situations and ensures that the communications system can handle any occurrence.

The basic unit in the SDLC format is the *frame*, shown in Fig. 7-21. The frame consists of beginning bit group, 01111110, used for synchronization, followed by an 8-bit address field and a special control field, also 8 bits. The user data comes next, in an information field. This field can have as many bits as

Synchroni-zation	Address field	Control field	User data	Error checking field	End field
01111110	8 bits	8 bits	Any number of bits	16 bits	01111110

Fig. 7-21 SDLC allows large amounts of data to be transferred efficiently. The basic frame format has a synchronization field, an address field, a control field, and the actual user data bits, followed by an error-checking field and an ending field.

desired—the SDLC protocol does not restrict it to any specific value. The user data is sent as a string of bits. The bits can represent characters in ASCII form, numbers in binary code, or anything else. This is because SDLC is a bit-oriented protocol. It is entirely the system user's option how long a string of bits should be used to represent the message. This is in contrast to the protocol used for the simple printer, in which all the data is on a character-by-character basis, and it is impossible to send a single bit.

The SDLC format ends with a 16-bit field used for checking and error-detection characters, and an ending bit group, 01111110, which is the same as the beginning group. The end of the link which has responsibility for managing the link is called the *primary station*, and the other end is the *secondary station*. Data messages can be initiated by either station, not only from primary to secondary. SDLC can be used with both full-duplex and half-duplex channels.

The control field is a key to SDLC operation. SDLC has three modes: information transfer mode, supervisory mode, and nonsequenced mode. The control field indicates which mode is being used. The control field also has as 1 of its 8 bits the *poll/final bit* (P/F). A frame with a P/F bit set to 1 is sent from the primary station to a secondary station to authorize transmission; when the bit is 0, the secondary station sees that it is not allowed to send any data to the primary station. A frame with the P/F bit set to 1 is sent by the secondary station to the primary station as a response to a *poll*, or request, from the primary station. In a typical exchange, the primary station sends a number of frames of data to the secondary station, each with the P/F bit set to 0. This tells the secondary station to listen, not send. When the primary station is finished, it changes its P/F bit to 0. The secondary station sees this and can begin to send data back to the primary station.

The three modes indicated by the control word have specific functions. The *information mode* is used for regular transmission of data. It uses 3 bits of the other bits of the control field as a message number, 0 through 7 (binary 000 through 111). Each frame that is sent is numbered so the receiving station can determine whether any are missing; this number is called Ns. The receiver also counts the frames as they are successfully received, with a value called Nr. By using these numbers and keeping track of the transmission, the receiver can make sure it properly receives all frames, and also report to the sender which ones are received properly and which are missed.

The *supervisory mode* is used to send messages which control the operation of the link and the transfer that is about to take place. Between the actual groups of information frames, the link may have to be turned around (if it is only half-duplex) or some other management activities may have to be performed. This is the role of supervisory messages. The third mode, *nonsequenced*, is used for the special messages that are needed only at start-up time, when the equipment on both ends of the link, as well as the computers connected to the equipment, has to be initialized, reset, or set to a specific operating mode. This is before the normal flow of information and supervision begins. Supervisory and nonsequenced modes do not have message numbers. Only the information mode uses message numbers.

A Half-Duplex SDLC Example

Some of the operation of SDLC can be illustrated by a half-duplex example. SDLC also supports full-duplex, but the operation protocol sequences are more complicated. In the example, shown in Fig. 7-22, the sequence begins when the primary station sends a supervisory frame to the secondary station. This frame has a command field which indicates that the supervisory station is ready to receive, has Nr of 1, and has a poll bit set to 1. There is no Ns since this is a supervisory frame, not an information frame. The primary station is saying, "I am ready to receive, I have accepted your frames through frame count 0, and I am waiting for frame 1. This is a poll command so that you can initiate your transmission."

After receiving this message, the link must be *turned around*, since it is half-duplex. After the turnaround, the secondary station begins its transmission. It sends information frame 1 (I1) and an Nr of 4, to indicate that it will count the next transmission from the primary as frame number 4. Since the primary station has no information it uses a supervisory frame to initiate transmission from the secondary. On the third frame, the secondary includes the P/F bit set to 1 (for *final*) to indicate the last frame of the transmission.

Suppose this communications system has other secondary stations. The primary turns the half-duplex line around again. It then polls the next station by saying, in this case, that it is ready to receive the next frame, number 5, from that secondary station. The station has no information to transmit and responds with a supervisory frame that says, "I am ready to receive, I am looking for your information frame 0, and since I have no information to send you this is a final frame."

If the SDLC protocol appears complicated, that is because it is. The preceding example represents only a small part of what SDLC does. Why this complica-

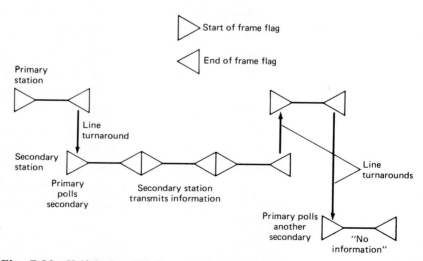

Fig. 7-22 Half-duplex SDLC operation requires passing of control and signaling information from the primary station to the secondary, and frequent reversing of the channel direction.

tion? The reason is that a high-speed, high-performance, low-error system requires many special protocol messages, to handle the many planned and unplanned events that can occur. There can be errors in the data, communications can suddenly stop in the middle of a message (equipment or channel problems), the system may have some higher-priority information to send which must override the planned message flow, and similar surprises may occur. The complexities of the protocol cover every conceivable occurrence and even make provision for problems whose cause is not clear. The goal of the protocol is to provide reliable communications at high speed, in which all error conditions are handled, errors corrected wherever possible by retransmission, and, most important, no "hang-ups" or crashes occur. A *crash* is a situation in which one side or the other is waiting for a message and the message never arrives or never completes. The receiver may wait forever, and meanwhile the system at the receiving end is stuck in an idle mode. The transmitter system eventually also becomes stuck, since at some point it waits for a response from the receiver. The danger is that a problem that is not anticipated and handled via a defined protocol will cause one and then the other end of the link to come to a halt, and then the communications system will be frozen. In some cases, it may not be possible to break the freeze without the equipment operators having to restart everything manually—sometimes by shutting power off and then turning it back on! Certainly, this is a very undesirable situation, but it can happen if the protocol is too simplistic.

QUESTIONS FOR SECTION 7-6

1. How is computer-to-printer data communications often achieved?
2. Why does the printer usually have a character buffer? Why is this desirable?
3. How is handshaking performed with special characters?
4. Why does special character handshaking require a full-duplex channel?
5. Why does a plotter with cursor need a full-duplex channel?
6. How does the plotter protocol differ from the simpler printer protocol?
7. What is the basic frame of SDLC?
8. What does it mean to say that SDLC is "bit-oriented"? What was the orientation of the plotter protocol?
9. What is the role of the poll/final bit in SDLC?
10. What is the SDLC primary station? The secondary station?
11. What is the SDLC supervisory mode? Information mode? Nonsequence mode?
12. What are two of the SDLC fields? What are they used for? What purpose do they have?
13. Why is SDLC relatively complicated?
14. What is a crash? How does it happen? Why must it be prevented?

PROBLEMS FOR SECTION 7-6

1. Show the string of characters that are sent to a printer to print the message HERE IS OUR REPORT, followed by the bell, a carriage return, and three line feeds.

2. The printer being used with the simple protocol has a very small buffer. It can hold only two ASCII characters. Show the timing diagram for the case in which the computer is sending a total of 4 characters at the rate of 10 characters/s, but the printer can only print 5 characters/s. Use hardware handshake lines RTS and CTS.

3. Repeat Prob. 2, but use XON and XOFF instead of hardware handshake.

4. Show the string of characters that would be sent to the plotter to indicate a box that is 2 units high and 6 units wide, with corners at 10 and 12.

5. What string of characters is sent to the plotter for the cursor at $X = 15$, $Y = 9$?

6. Show the timing for RTS and CTS handshaking if the computer is sending two strings of instructions to the plotter (for two boxes) but the plotter buffer can only hold 14 characters. It takes 1 s for the plotter to actually do the drawing and be ready to accept the next string.

7. Repeat the preceding case, using XON/OFF handshaking.

7-7 Hardware Versus Software: Protocol Conversion

Protocol involves following a set of rules that have been defined in advance. This makes protocol a good candidate to be implemented by a microprocessor or other computer as part of the *software*, the program instructions that the computer processor executes in a specific order and sequence. The program can instruct the computer how to prepare the message, put the various bits of the preamble and postamble in place, and send the message. It can also instruct the computer to receive incoming messages, take off the preambles and postambles, use them to check the message, and prepare any required response. Different protocols can be handled by different software programs.

Although all these functions are practical in theory, performing them is often a poor choice in practice. The reason is that a large portion of the processor resources (time and memory) has to be allocated to performing these tasks, and there is very little processor resource left over for the main functions of the computer, such as processing numbers.

In fact, handling communications and protocols is very complex. Computer processors cannot keep up with handling the characters at any except the slowest rates, typically below 300 baud. A few sample calculations will show why. Suppose a system is using simple ASCII format, with no preambles or postambles. A computer processor has to determine the correct bit value to send for the start bit, ASCII bits, and stop bits (assume no error-checking bits). Each determination may take five computer instructions, and each instruction takes the computer about 20 μs. Therefore, the processor needs 100 μs per bit. At best, then, the processor can transmit characters at 1 bit every 100 μs, or 10,000 bits/s. This seems fairly fast. However, this is an ideal case, and processors really cannot devote all their time to this activity. In practical applications, only a small percentage of total time can be devoted to communications. The actual baud value that the computer can support is only a small fraction of the ideal value. Also, this is a simple case with very little protocol to perform. Finally, synchronous schemes require that the bits be handled without delay.

In very few real systems does the processor perform all the protocol manipulation directly. Instead, specialized circuitry and ICs are used. This is called the *hardware approach*. These circuits and ICs are designed solely for the task of handling the protocol and interfacing efficiently with a computer processor. The circuitry of this hardware relieves the processor of many of the details of the protocol itself. In a typical case, the processor only has to put the characters of the message into this communications interface circuitry. The interface circuitry takes care of putting on the preamble and postamble information and coordinating the protocol handshaking with the receiver. It also takes care of the timing for the data bits. It then signals the processor that the entire message has been successfully (or unsuccessfully) sent and asks for the next message. At the receiving end, the circuitry provides the receiving clock, stores the incoming message, checks it for correctness, and signals the processor that a new message has been received. The processor can then ask, at its next convenient point in time, for the message itself (stripped of protocol characters). This reduces the overhead, nonproductive task of the processor to the absolute minimum required.

Different protocols require different amounts of this specialized support circuitry. The simple asynchronous format used between a computer and printer employs a relatively simple IC that takes care of the handshake lines, converts the character information from the processor into the ASCII bits, adds the start and stop bits, and sends out the whole group of bits at the proper rate (if the handshaking lines indicate that the receiver is ready). The more complex protocols, such as SDLC, use sophisticated ICs that actually implement the many rules of the protocol directly. In these cases, the processor usually passes the entire message to the IC, which then has a complicated job of developing the final bit pattern.

The software method does have some advantages. First, the protocol can be changed simply by using a new computer program. This gives the final product flexibility for use in different situations, since the same hardware can serve many needs. It is low in cost, since little or no special communications protocol circuitry is needed. If the application can tolerate the low baud values and the use of computer processor resource for communication, then the software method can be useful. The hardware method is capable of greater performance, but at a higher cost.

Some systems use a compromise between the two. Communications ICs are used to take care of the lower-level, highly detailed aspects, and program software is used for some of the general message formulation. The plotter example of the previous section is a good case (Fig. 7-23). The plotter protocol requires using the special preamble :; and postamble #, as well as putting the actual message into the required format (shape, size, and location numbers in a

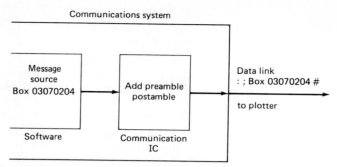

Fig. 7-23 Many communications systems use a computer program and processor to generate the message and have a specialized communications IC to add the necessary preamble and postamble bits and fields.

certain arrangement). The computer program is used to convert the message that the user types into the specific style needed. This message is passed, completely formulated, to an IC, which then manages the asynchronous communications. This IC adds the start and stop bits and sends the complete character at the specified bit rate. It also manages the handshake lines and instantly stops sending characters when the plotter indicates that it cannot accept more characters (and resumes when the indication is to go on). This is a very effective tradeoff between a fully software and a fully hardware focused approach. If the plotter is replaced by a different model, which requires another preamble, message structure, or postamble, the program can be changed. The time-consuming details of directly forming the ASCII characters and sending them at the correct rate do not have to change.

Protocol Conversion It is often necessary to connect two systems which use different protocols. (The term "protocol" here means both protocol and format.) Special devices, called *protocol converters*, act as interfaces which can do this. There are many different protocol converters available to handle the large number of protocol combinations. A typical converter application is to convert asynchronous ASCII format to synchronous format (Fig. 7-24).

Protocol conversion is a very complicated operation. There are many operational subleties that must be considered when one protocol is converted to another. For example, a slow asynchronous protocol is at odds with the intent of the higher-speed synchronous protocol, and many problems which the protocol converter must accommodate can arise. Sometimes, the conversion is so difficult that a practical protocol converter does not exist. In this case, a user wishing to change from one protocol to another must either change one of the protocols at the sending or receiving system or give up! It is never an easy choice.

Because protocol converters are dedicated to performing one task, and performing it well, they are usually implemented with circuitry to give them the maximum performance possible. Some aspects of the protocol conversion are handled by software within the converter. The entire conversion process should be transparent to the user. The message goes into the *black box* of the protocol converter in one format and protocol and comes out the other side converted. The

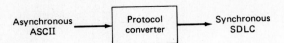

Fig. 7-24 A protocol converter converts one format and protocol to another one, so that the originator of the message does not have to be concerned about the differences in protocols. Asynchronous ASCII conversion to synchronous SDLC is shown.

message sender has no idea of what the message looks like on the other side, only that it has been sent. Some converters, as a practical matter, must place some restrictions on the nature of messages that can be sent. In this case, the converter is no longer transparent to the sender. The sender must then make sure that the messages fall into the allowed category, and communications problems (errors, system hang-ups, and so on) can occur if invalid messages are sent.

QUESTIONS FOR SECTION 7-7

1. How is protocol determined in software?
2. What are the drawbacks to performing protocol in software? The advantages?
3. Why is hardware-based protocol a good alternative to software-based protocol?
4. On what basis are the choice and tradeoff between hardware-based and software-based protocol made? Relate this to the plotter protocol.
5. What is protocol conversion?
6. Can protocol conversion always be accomplished? Explain.

PROBLEMS FOR SECTION 7-7

1. A computer system microprocessor takes 10 instructions to determine whether it should send a 1 or 0 bit under a specific set of circumstances. Each instruction takes 35 μs to execute.
   a. How many bits per second can this system generate?
   b. How many 7-bit ASCII characters/s can this system send?
2. A receiving system uses a microprocessor to handle incoming bits. Each bit interrupts the processor, which then goes to a special routine to accept and process the input bit. The interrupt takes 20 μs to activate the routine. The routine itself takes 80 μs to process the bit.
   a. How many bits per second can this system accept?
   b. How many 8-bit ASCII characters/s can be processed?
3. A system is using asynchronous ASCII to send a two-character message, the single word "NO."
   a. Sketch the binary pattern for sending this with 7-bit ASCII. (Use 1s and 0s.)
   b. Sketch the binary pattern for a protocol which uses 4-bit fields for message number in binary, character count in binary, and postamble character # in ASCII code. The message is number 6. (Use 1s and 0s.)
4. A system sends messages with 6-bit fields for character count (binary),

message number (binary), and error-check field. The message, number 4, is sent as ASCII characters. It is, I'M LATE!

  a. What is the 1,0 pattern sent?

  b. What is the 1,0 pattern if this is sent using asynchronous ASCII for the message alone, without any preamble or postamble?

5. A system is designed to send numbers using ASCII format for each digit. It is sending to a system that can accept only binary format, in 16-bit fields. For the following numbers, what is the 1,0 pattern for each format that the protocol converter must handle: 75, 125, 257?

6. A protocol converter converts regular binary numbers to ASCII format, one character per digit. The received numbers are in 8-bit fields. What is the output of the protocol converter for the following received patterns: 11110000, 01010101, 10000001?

SUMMARY

A complete communications system has many layers between the originator of the message—a person or computer—and the point at which the message is actually sent onto the link. The message characters must be put into a format. This format must be used with a protocol, which establishes the rules of the conversation and also puts necessary characters before and after the message. The protocols in use range from very simple to quite complex, to serve a wide variety of cost and performance needs.

Figure 7-25 shows the layers of the communications hierarchy clearly. The user of the system is unaware of the kinds of transformations and manipulations, such as handshaking and conversion to ASCII, that the message undergoes. Not all messages go through all of these steps. The result of the protocol activity is

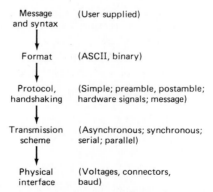

Fig. 7-25 The hierarchy levels of data communications show how a message must be generated and put into proper syntax, format, and protocol. Handshaking is added, and the data bits are then put through the transmission circuitry (often modulated) and then to the actual physical interface to the link.

Communications System Requirements

that a digital pattern is available that will be used to drive the modulation circuitry, which puts the signal data on the link. This completes the activity at the transmitting side.

The receiver circuitry must accept the data bits, check the protocol, strip off the nonmessage bits, and pass the message to the actual user at the receiving end. Both synchronous and asynchronous transmission can be used. The synchronous method is preferred for long blocks of characters that must be sent at high rate, but it costs more to implement than the slower asynchronous method. Asynchronous transmission is better suited to cases in which the message has time gaps between characters, such as a typed string of characters.

The rate at which bits are transmitted is one important factor that determines how many characters per second are actually sent. Other factors that affect the actual throughput are the complexities of the protocol preambles and postambles, the handshaking sequence, the ability of the receiver to handle incoming bits promptly, and the number of bits per character. Buffers are usually used at the receiving end to make sure that the transmitter can send characters without having to wait a relatively long period of time between characters. The receiver circuitry processes each character as received.

The preambles and postambles of some protocols serve several purposes. They can include message numbers, character count, and other checking information so the receiver can determine whether the message is correctly received. The receiver, in turn, can respond and tell the sender which messages are received properly and which are not. The sender may, in some protocols, resend those that are received with errors.

For situations in which the format and protocols of the two systems at the ends of the communications systems are different, protocol converters can act as an interface. These are dedicated circuits, designed for performing format conversion and protocol conversion and accommodating the necessary handshaking. There are many protocols in use, and therefore many converters available. Not all protocols can be interfaced successfully. Often there are restrictions placed on one end to ensure proper conversion.

To this point, the text has shown the way a user message becomes a group of data bits to be sent onto the link. One more piece remains: a discussion of the specific signal voltages, currents, or frequencies that are used. This is covered by using a specific example, in Chapter 8.

END-OF-CHAPTER QUESTIONS

1. What is the meaning of *hierarchy* in data communications? In any form of communications?
2. Explain the difference between format and protocol. Give a parallel to both in daily life, similar to the text example of mailing a letter to someone.
3. Besides format and protocol, how do data rate, parallel and serial methods, and synchronous and asynchronous schemes affect the ability of a message to be received by the other user?
4. Why does having all the format, protocol, and related issues correct still not ensure that the receiver of the message actually understands it?

5. What are the roles of codes in formatting data?
6. What can data consist of? What can binary number code data represent?
7. How many characters can a 5-bit code represent? An 8-bit code?
8. Explain the difference between printable and nonprintable characters. What role do the nonprintable characters have? Why are they essential?
9. What is the tradeoff in making the character bit field longer versus shorter? Explain the benefits and drawbacks.
10. What is ASCII code? EBCDIC code?
11. Why do data communications systems need protocol? Give two reasons.
12. What is handshaking? Why is it needed? What happens if there is none?
13. Explain hardware handshaking. Compare this to handshaking with messages in terms of the way each operates.
14. What are the relative benefits and drawbacks of hardware versus message handshaking?
15. Explain why handshaking is needed even during the data communications transmission, as well as at the beginning and end.
16. Why do buffers improve communications performance, as seen by the transmitter of data?
17. What is the role of a larger buffer? When is it desirable?
18. Explain the way a buffer works at the receiving end on a system.
19. What is a preamble? A postamble? What kinds of information do they carry?
20. Explain the ways that the preamble and postamble are transparent to the originator of the message.
21. Are there any drawbacks to preambles and postambles? Explain.
22. What is the key difference between a synchronous system and characters? How does the receiver determine the bit timing?
23. What are the relative benefits and drawbacks of each method?
24. How does an asynchronous system send characters? How does the receiver determine the bit timing?
25. Repeat Ques. 24 for a synchronous system.
26. What range of bits per second is practical with a synchronous system? With an asynchronous system?
27. Explain why synchronous transmission may be inefficient for short, sporadic characters.
28. When is 1 baud the same as 1 bit/s?
29. What are some common baud values for asynchronous systems?
30. What is the maximum throughput that a system can have, in relation to its baud value?
31. What factors reduce the throughput from the theoretical maximum value?
32. Explain how a system can have high burst data rates, but lower throughput.

33. Explain how a synchronous system can have higher throughput for long strings of characters.
34. What are the differences in operation between a serial and a parallel system?
35. What are the differences in performance between the two? In cost?
36. Where is a synchronous parallel system most practical? Least practical?
37. Explain where more complex types of protocols are needed. Why?
38. Where are simple protocols sufficient?
39. Where is SDLC used? What are some of its key characteristics?
40. Why is a system communications crash a possibility if the protocol is not suited for the application?
41. Why does XON/XOFF protocol require a full-duplex channel?
42. What are the benefits of performing protocol and format in software? The drawbacks?
43. What compromise is often used in the hardware/software tradeoff?
44. What is the role of protocol conversion? Give two examples of conversion.
45. Can protocol conversion accommodate any protocols? Explain.

END-OF-CHAPTER PROBLEMS

1. What are the 7-bit ASCII codes for the following: q, #, LF, BEL, DC2?
2. What ASCII characters are represented by the following patterns?

   1110000, 0010110, 1100100, 0000000, 1111111

3. How would the following decimal numbers be expressed in the 1,0 pattern of ASCII: 298, 100, 35?
4. How would the following three messages be expressed with ASCII bits?

   25 CENTS, NO!, MAYBE?

5. How many unique characters could a 4-bit field represent? A 12-bit field?
6. A buffer is used at a receiver. The characters are arriving at 1500 per second. The largest group of characters is a message of 25 characters. The receiver takes characters from the buffer at 100 characters/s. How long a buffer would guarantee that three messages could be transmitted without waiting?
7. A transmitter is sending characters at 100 characters/s, to a 10-character buffer. The characters are taken out of the buffer at the rate of 50 characters/s. Sketch the number of characters in the buffer, versus time, for 2 s.
8. What is the binary pattern of 1s and 0s that will be sent for a protocol that has a character count in binary, a receiver number in binary, and a message number in ASCII code, for receiver number 4, message #7 OK?, with 4-bit fields for the message, character count, and receiver number?
9. A system is using an 8-bit asynchronous format to send a 25-character message. How many bits have to be sent, including stop and start bits?

10. The system in Prob. 8 is changed to synchronous transmission, with an 8-bit synchronization frame, a 4-bit message number, and an 8-bit character count. How many bits have to be sent for the entire message?

11. Repeat Probs. 8 and 9 for a 50-character message.

12. What is the bit time for each of the following baud values: 300, 600, 9600, and 19,200?

13. A system is sending 7-bit ASCII characters asynchronously. How many characters/s can be sent, in theory, at 1200 baud? At 19,200 baud?

14. A system has to send a 100-character message by using 8-bit ASCII asynchronously. Can this be done in 1 s at 600 baud? At 1200 baud?

15. A system can send data at 9600 baud, using 7-bit ASCII, asynchronous. The receiver can accept 20 characters and then takes 2 s to process these before accepting more. What is the burst character rate? The throughput? What is the efficiency?

16. The system in Prob. 15 is changed to a 300-baud system. What are the burst rates and throughput?

17. Show the ASCII 1,0 pattern for this message, sent with simplest protocol: bell, line feed, line feed, OK. Repeat for a protocol that uses ## for preamble and !! for postamble.

18. A system is sending characters by using hardware handshake lines RTS and CTS, at 10 characters/s. After 5 characters, the receiver signals that the sender must stop, for the 1-s time that it takes to process these. Sketch the timing diagram for 15 characters.

19. The system in Prob. 17 uses XON/XOFF. Sketch the timing diagram.

20. A system uses software only to process incoming bits. It takes 12 instructions of 15 μs each to process these bits.

   a. How many bits per second can be processed?

   b. How many 7-bit asynchronous ASCII characters/s can be processed?

   c. A change is made to the receiver design, so that the start and stop bits are processed by hardware, in 5 μs. Repeat part b.

21. A system is using binary fields of 16 bits to send numbers. The receiver uses 7-bit ASCII. The formats must be converted. What is the 1,0 pattern for the following numbers, before and after conversion: 100, 254, 128?

22. For the preceding case, how many bits does it take to represent each number in binary versus ASCII? How many bits are saved using binary? What percentage of saving is this?

8 The RS-232 Interface Standard

This chapter examines in detail one particular data communications interface: the RS-232 standard interface, which is among the most common. Understanding the signal voltage levels, physical implementation, connectors and pinouts, and operation of this type of interface is important. Some other critical issues include the many ways that the interface can be implemented, the problems that can occur, and the method used to work through the problems to result in a working communications interface. The integrated circuits used to meet the electrical rules of the standard are also discussed. Finally, the drawbacks of this standard are studied, as an introduction to the need for other standards in applications with different requirements for performance and reliability in various situations.

Using a standard such as RS-232 does not ensure that the data communications between two systems will operate the first time they are connected. It provides a framework that has the potential for proper operation. In this sense, it is a tool. Using the right tools does not promise high-quality cabinetry; a good job will result from good tools applied and used properly.

8-1 Introduction to RS-232

A successful and useful interface standard must define the lowest-level details of the interface, such as the voltage (or current) values, the shape of the signals, and the performance of the interface under many normal and abnormal conditions. One of the most common interface standards is the *RS-232 standard*, which is published by the Electronics Industry Association (EIA), an industry group based in Washington, D.C., which helps form and publicize standards related to all aspects of electronic components, equipment, systems, packaging, and communications. The formal name for the standard is "Interface between Data Terminal Equipment and Data Communications Equipment Employing Serial Binary Data Interchange," but the industry shorthand for it is simply "RS-232." (The latest revision is the third, so the exact designation is RS-232C.)

The prefix "RS-" stands for *recommended standard*, since the EIA publications do not have any legal force. They are simply voluntary standards that electronic equipment manufacturers may decide to follow, for the general convenience of others.

There are many ways to implement an interface for binary serial data. RS-232 defines one of these ways, one that an enormous number of companies have elected to use.

The RS-232 interface is used in many types of applications for data communications, where relatively low performance is *acceptable to the user or all that can realistically be provided because of cost and technical constraints*. Applications for RS-232 include the following:

A connection between a computer terminal (screen and keyboard) and computer.

A computer sending data to a nearby printer.

A specialized device, such as a bar code reader, sending data to a computer.

The link between a computer and a nearby communications box, which in turn communicates with a distant communications box using some other methods. This second box then uses RS-232 to connect to a nearby terminal (Fig. 8-1).

The RS-232 interface is a good choice for many applications, but there are situations for which it is completely inadequate or just marginal. These applications involve high speed (greater than 20 kbaud), or long distances. If used properly, however, RS-232 can provide very good performance at low cost. There are other, better standards for the areas in which RS-232 is not the proper choice.

Since RS-232 is not the universal standard, why study it in detail? There are many good reasons:

RS-232 interface is among the most common interfaces in use.

Studying the issues related to a working RS-232 interface illustrates many aspects of an actual link. These issues have parallels and similar choices in other interfaces.

Studying RS-232 shows the benefits and drawbacks of using standards set by industry.

Like many sets of rules, the RS-232 standard does not define everything and every conceivable situation. Communications issues related to use of RS-232 fall into three categories: those for which the standard is absolutely explicit and

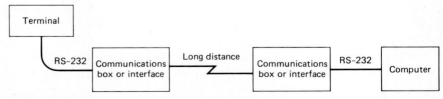

Fig. 8-1 The use of the RS-232 interface between a computer and a nearby communications interface.

precise, those for which the standard lays out some guidelines but allows flexibility, and those for which the standard sets no rules. RS-232 carefully and fully characterizes the voltage levels, the signaling baud values, the shape of the binary signals, the performance specifications of the interface circuitry at the receiving end and transmitting end, and the functions of signal wires in the cable. In other areas, such as the many ways the signal wires can be used in the areas of handshaking, the RS-232 specification allows many variations. Finally, there are areas in which the RS-232 standard says nothing at all! These include the protocol, the encoding of the user message into characters, and the connector to be used.

The later sections in this chapter show how it may range from easy to difficult to get two pieces of equipment, both of which meet the standard, to communicate with each other. Often, some careful detective work by the technician or engineer can reveal where the problem lies, and some adjustments or modifications can be made so that noncommunicating devices can be made to communicate. Techniques for making these modifications are discussed. The emphasis is on the more common implementation of RS-232 interfaces. RS-232 is designed for both asynchronous and synchronous communications, but the asynchronous method is more commonly employed. Unless otherwise indicated, all references to RS-232 in this chapter assume asynchronous communications.

QUESTIONS FOR SECTION 8-1

1. Who publishes the RS-232 standard? What legal force does it have?
2. What are three typical RS-232 applications?
3. What are two applications not mentioned in the text?
4. Why study the RS-232 standard in detail? (Give three reasons.)
5. What are the three levels of RS-232 explicitness? In what areas?

8-2 The RS-232 Voltages

Two voltage values are needed to represent binary data. The RS-232 standard specifies these values. Binary 1 is called a *mark* and can be any voltage in the range from -3 to -25 V. Binary 0 is called a *space* and is within the range of $+3$ to $+25$ V. The voltage range between the allowable mark and space values is -3 to $+3$ V and is not a valid voltage value. There should be no mark or space signal voltages within that 6-V-wide zone at any time (Fig. 8-2).

The transmitting circuit can send any value within the mark or space range. The receiver must recognize any value within the range and correctly identify it. For example, a mark may be sent as -7 V. It may arrive at the receiver, as a result of noise, attenuation, or other causes, as -3, -8, or even -25 V, but the receiver can correctly characterize it as a mark. The parallel situation applies to sending a space.

Of course, it is in the best interest of error-free performance that the largest value possible be used for mark and space. In this way, a larger value of electrical noise is needed to cause a mark to look like a space, or vice versa. If

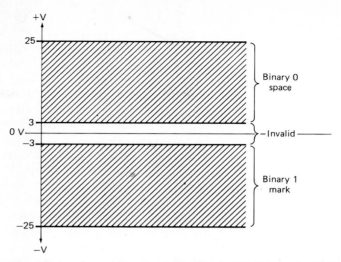

Fig. 8-2 The voltage values and ranges defined by the RS-232 standard for mark and space.

the space signal is sent as +3 V, it is only 6 V away from looking like a mark. If the space signal is sent at 10 V, it is 13 V away; at 25 V, it is 28 V away. The use of the minimal values is very error-prone, and they are rarely used. The key is that the signal must arrive in the proper range in order to be correctly interpreted as a mark or space, regardless of the value that is used at the transmitter. Using values closer to +25 and −25 V at the transmitter provides "insurance" against noise.

The voltages sent by using the RS-232 standard are sent with reference to a common ground, or *0-V point*, between the systems. This is called *single-ended*. As shown in Fig. 8-3, the transmitter sends its mark or space voltage along a signal wire to the receiver. The cable between the two systems has several other wires, used by the RS-232 standard for handshaking and special functions. All of

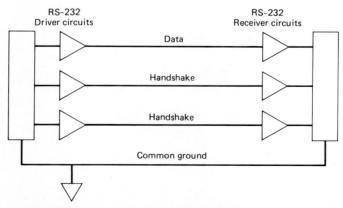

Fig. 8-3 All RS-232 signals—data and handshake—share a common ground with the systems at both ends of the link.

these wires share the common ground wire. The advantage of this is that it saves wires in the cable. The disadvantage is that there can be severe common mode problems if the ground points of the two systems are not at the same voltage with respect to earth ground. It is entirely possible for one chassis to be tens of volts different from the other with respect to earth ground. In this case, what looks like one voltage at one end appears very different at the other. This causes errors. However, in many applications in which the two systems are close, plugged into the same power socket, or in a properly wired building, the common mode voltage is low and the single-ended scheme is adequate.

Finally, the RS-232 standard specifies the rate at which the mark and space signals can make the transition from one value to the other. This is called the *slew rate* (Fig. 8-4a). A fast slew rate corresponds to a faster change, and, as shown in Chapter 2, this means that increased bandwidth is needed. The RS-232 standard assumes that the bandwidth of the system is limited, and attempting to use greater bandwidth to send fast-changing signals can cause signal distortion. Therefore, the standard specifies that the actual electronic circuitry used at the transmitter should limit the slew rate to certain specific values, even if the digital signal from the transmitting end sends binary signals with fast transitions from 0 to 1 and 1 to 0. As shown in Fig. 8-4b, the data bits of the message are coming

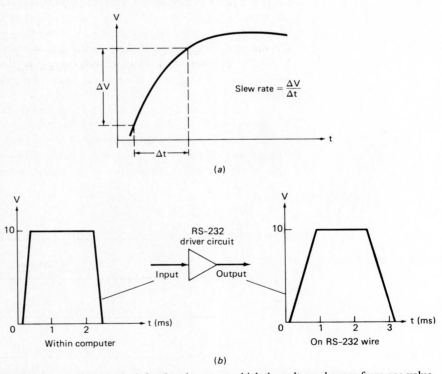

Fig. 8-4 (a) *Slew rate* is defined as the rate at which the voltage changes from one value to another; (b) the RS-232 specification allows only a limited slew rate signal. The high-slew-rate signal within the computer is deliberately reduced to a lower slew rate by the RS-232 driver circuit.

from the electronic circuitry with very fast rise and fall times. The RS-232 interface circuitry deliberately decreases this slew rate (measured in volts per second [V/s]) to a maximum allowed value. The figure shows a slew rate limited to 10 V/ms, as an example, by the RS-232 circuitry, although the original slew rate was greater.

Note that slew rate is not the same as baud value, although there is an indirect relationship between the two. The *slew rate* is the transition rate from one digital value to another. The *baud value* indicates the number of bits/second (bits/s) that can be sent. Of course, the baud value is limited by the slew rate, since the bit per second rate cannot be faster than the rate at which the voltage can swing from the binary 1 to binary 0 value. For example, consider a signal which uses $+10$ V and -10 V. If the slew rate is 20 V/ms, then the transition alone takes 1 ms (Fig. 8-5a). Therefore, the bits per second rate cannot be greater than 1 bit/ms, or 1000 bits/s, since higher rates would mean that the signal never has time to reach the $+10$- or -10-V level, before it is forced to change to the other state (Fig. 8-5b). By contrast, a signaling rate of 100 bits/s means that the bit remains at its high or low value for 10 ms (Fig. 8-5c). This is longer than the time it takes to reach either value due to slew rate, so the bit is able to reach and hold the voltage value.

The RS-232 specification states that the specification is intended for connections of 50 ft or less. Beyond 50 ft, the signals may be degraded by noise, attenuated, or "rounded," and reliable transmission is not assured. In fact, many installations can go longer distances, up to several hundred feet. This is possible because the specification tries to cover performance in worst-case, difficult situations. If a lower baud value is used, the communications system can tolerate

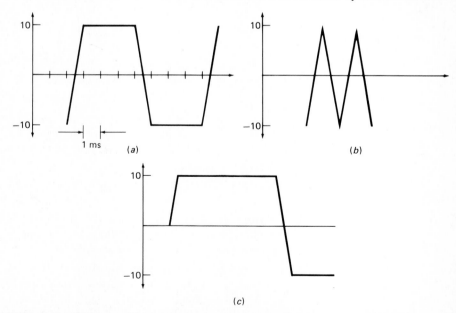

Fig. 8-5 Three examples of slew rate and its effect, on the achievable bit rate *(a)* 20 V/ms, *(b)* 1000 bits/s with 20 V/ms slew rate; *(c)* 100 bits/s with 20 V/ms slew rate.

and perform effectively even with noise and signal degradation. It is difficult to use RS-232 at 20 kbaud for more than 50 ft, but relatively easy to use it for 150 ft at 2400 baud. For longer distances than RS-232 can support, special communications boxes which convert the RS-232 signals are used. These boxes convert the RS-232 signals into types of modulation that can be sent long distances (miles or around the world) at the desired baud value.

QUESTIONS FOR SECTION 8-2

1. What does "mark" represent in binary coding? What about "space"?
2. What is the voltage range allowed for mark and space? What range is not allowed? Mark (−3 to −25V) Space (+3 to +25V)
3. What must the receiver of RS-232 signals recognize?
4. Why are the large voltage values preferable?
5. What does the term "single-ended" mean? How is the ground used? *Interface voltage sent with reference to a common ground.*
6. What are the advantages and disadvantages of the single-ended method?
7. Why is common mode voltage unlikely in many installations?
8. What is slew rate? What is the effect of high slew rate? What does RS-232 specify about slew rate? *The rate at which the mark and space change values from one to another (0 to 1, 1 to 0)*
9. What is the relationship between slew rate and maximum baud value?
10. What is the distance limitation for RS-232? How does it vary with baud value? 50 ft.

PROBLEMS FOR SECTION 8-2

1. An RS-232 system is using −10 and +10 V. What is the noise margin? (*Noise margin* is measured against the lowest value allowed that is still valid.)
2. Another RS-232 system is using −5 and +20 V. What is the noise margin in each direction?
3. If the maximum RS-232 voltages are used, what is the noise margin? What is the margin if the minimum voltages are used?
4. Sketch a −10- to +10-V signal with a slew rate of 10 V/s. Label the time and voltage axes of the graph.
5. Sketch a −5- to +5-V signal with a slew rate of 20 V/s. Sketch a −10- to +10-V signal with the same slew rate alongside it. Which takes longer to reach its full value?
6. A signal has a slew rate of 100 V/ms. How long does the digital signal take to go from 0 to 10 V? From −10 to +10 V?
7. A system is sending 1000 bits/s, using −5- to +5-V signals. The time to slew from negative to positive should be less than 10 percent of the bit time. What is the minimum slew rate acceptable?
8. A system has a signal that goes from −10 to +10 V. The slew rate is 10 V/ms. If the slew time must be less than 5 percent of the bit time period, what is the maximum bits per second rate that can be supported?

8-3 Data Bits

The asynchronous operation of RS-232 is the most common way that the standard is applied. It is very useful to study the way a group of bits, which may represent a number or a character, is actually sent. The RS-232 standard does not define the number of bits sent between the start and stop bits, but 5, 6, 7, or 8 bits are most commonly used.

When no character is being sent, the signal line is at an idle state, indicated by the mark voltage. The transmission begins when the line goes to the space value, as the beginning of the start bit (Fig. 8-6). The start bit lasts for one bit time period. Next come the user data bits. These are the user 1s and 0s, with the negative voltage of mark for binary 1, and the positive voltage of space for binary 0. This use of positive voltages for 0 and negative for 1 is defined by the RS-232 standard and may seem in conflict with the common use of a 0 V for binary 0 and another, higher voltage for 1, such as the 5 V used in many circuits. However, it is not a problem, since the RS-232 receiver circuit converts the received voltages into the voltages and binary levels compatible with the receiving user system.

After the data bits, there may be a parity bit, which is used for determining whether there has been an error in any of the previous bits. Both sides must agree in advance whether or not a parity bit will be used, so that the receiver knows the meaning of the bit after the data bits. Finally, after the last data bit (or parity bit), the signal line goes to the mark value. This is the beginning of the stop bit. There may be one or two stop bits, depending on the system (in rare cases, systems use one and one-half stop bits). Until the end of the last stop bit, there can be no new characters. Once the stop bit or bits are completed, the transmitter is ready to begin the entire process again with a start bit, using the mark-to-space value transition.

Since this is asynchronous transmission, there is no limit on the amount of time between the stop bit of one character and the start bit of the next. If there are additional characters to send, the transmitter normally sends them out immediately after the end of the stop bit(s), if the receiver using handshaking indicates it is ready.

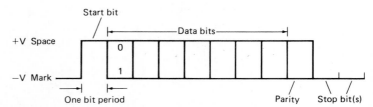

Fig. 8-6 RS-232 signaling begins with *start bit*, which is defined as a voltage transition to the space value. It ends with the *stop bit(s)*, as the voltage goes to the mark value. The data bits are sent between the start and stop bits.

The RS-232 Interface Standard 257

Sending the letter D, for example, by using ASCII (D is 1000100) with 7 bits, no parity, and one stop bit, would look like Fig. 8-7a. Note that the least significant bit (LSB) is sent first. The same letter, using parity bit, is shown in Fig. 8-7b. The parity bit may be 1 or 0, depending on the parity selected and the data bits. Sending D with two stop bits, no parity, is shown in Fig. 8-7c.

The data bits can be used for any purpose, as long as both sides agree on their meaning. Most applications use ASCII format to represent characters. Some situations use natural binary numbers, with the 5, 6, 7, or 8 bits representing from 0 up to 31, 63, 127, or 255 (Fig. 8-8a). In other cases, the RS-232 interface is used with some specialized type of instrumentation and the bits may represent the position of switches. An example of this is an alarm system which monitors a few doors and windows, which are wired to a small alarm box (Fig. 8-8b). When an alarm is triggered, the box reports this to the central alarm office by using RS-232, with a bit set to 1 for each door or window that is open (and triggering the alarm). This allows the central office to identify the exact source of the alarm in the building.

The RS-232 specification allows for baud values from 0 to 20 kbaud. In practice, most applications use the values discussed in Chapter 7, such as 300, 600, 1200, and 2400. Some new installations get higher throughput by using higher rates, such as 38.4 kbaud, but with the RS-232 voltages and start/stop format. These installations are not RS-232-compatible, strictly speaking. They are often talked about, however, as RS-232 interfaces, although incorrectly. They are RS-232 with respect to voltage levels, use of start/stop bits, and characteristics of the signal line driving circuitry, but they are not fully in compliance with the rules.

Break and Break Detect

A special signal indication, called *break*, is often used with RS-232. Between characters, the RS-232 line goes to the idle state (mark voltage). During the character, it makes transitions from mark to space several times, depending on the exact data bits, parity, and stop bits. Therefore, the line should never have the space voltage for more than a few bit periods in succession. This provides a

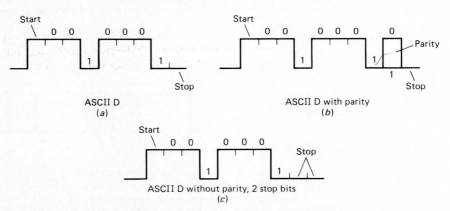

Fig. 8-7 The letter D sent by using RS-232, in various ways: (a) seven data bits, no parity, one stop bit; (b) as in (a) but with parity; (c) with two stop bits and no parity.

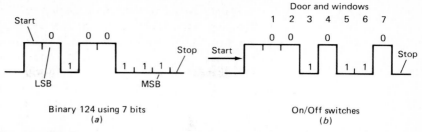

Fig. 8-8 RS-232 use is not only for ASCII characters: (*a*) the natural binary representation of 124; (*b*) the RS-232 data bits representing the position of doors and windows in an alarm system.

scheme for interrupting the receiver and getting its attention, by holding the signal line in the space condition for longer than a full character time. This is a *break*, and receiver systems which are equipped to use it have *break detect*. The break detect circuitry is usually independent of the circuitry which actually handles the start, data, parity, and stop bits. In this way, there is a separate path that can be used to get the attention of the receiver, in case of some system problem or communications crash. The transmitter can send a break signal and know that it will be seen by the receiver, regardless of almost any other problem that may be interfering with proper receiver operation.

RS-232 Error Conditions

Any communications system can have various types of errors. The most obvious one occurs when a data bit is changed from a 1 to a 0 or a 0 to a 1. There are some other types of important error conditions that can occur. These are framing errors, overrun errors, and parity errors. These errors are characteristic not only of RS-232 systems; they can occur in almost any system. They are easiest to describe in an asynchronous system.

A framing error exists when the receiver expects a stop bit but does not get one. (The receiver is counting the data bits as they arrive and knows when there should be a stop bit.) This may be caused by one or two factors. Either some noise or signal degradation has caused the stop bit to look like the opposite signal state, or the receiver is set for one number of data bits and the transmitter is set differently to send another number of data bits. The result is a misunderstanding, in which the receiver counting bits does not see the stop bit as it expects. The receiver character is therefore not framed properly by the start and stop bits.

An overrun error occurs when a new character appears at the receiver circuitry before the previous character has been handled and disposed of completely. If the computer is sending characters to a printer buffer, the handshaking or baud values may be incorrect so that the computer sends characters while the buffer is still full and cannot accept new ones. The character in the buffer is overrun by the new one coming in. Both will be in error.

A parity error occurs when the parity bit received with the character bit indicates that one or more of the data bits has an error. Parity is discussed in detail in Chapter 12.

The RS-232 Interface Standard

It is important to understand that a framing error or overrun error is fundamentally different in nature from a parity error. A framing or overrun error means that the receiver is unable to receive or handle the incoming character bits properly, either because there is no stop bit or because the character is overwritten. Therefore, there is no point in looking at the received bits. In contrast, the parity error indicates that the entire group is successfully received, but one of the received bits may be wrong. The reception of the character itself is implemented and performed properly. If a data bit is corrupted by noise along the link, the character comes in without framing or overrun errors but may cause a parity error. The framing or overrun error invalidates the receiving sequence itself, and the parity error indicates that the receiving sequence has been executed successfully but there appears to be an error within the data. It is analogous to receiving a letter which is torn and illegible compared to receiving a letter which arrives in good condition, but with one part smudged or wet so that it cannot be read with certainty.

QUESTIONS FOR SECTION 8-3

1. What is the signal state of the data line when there is no data?
2. What is a start bit?
3. What is a parity bit?
4. What is the stop bit or bits? How many are used?
5. Describe the difference in meaning when the bits of a character represent ASCII versus binary versus independent data bits.
6. What are common RS-232 baud values?
7. To what baud values is the RS-232 standard sometimes extended? How is this possible?
8. What are break and break detect? Why are they needed? How are they used?
9. What is a framing error? What causes it?
10. What is an overrun error? What is its cause?
11. What is a parity error used for?
12. What is the difference between what a parity error represents and what a framing or overrun error represents?

PROBLEMS FOR SECTION 8-3

1. Sketch an asynchronous signal representing the letter Q in ASCII, with start bit, seven character bits, two stop bits.
2. Repeat Prob. 1 for one parity bit and one stop bit.
3. Sketch the percentage symbol (%) which is being sent with 7-bit ASCII, one stop bit, no parity.
4. Sketch the number 123 sent in binary format, with eight character bits, one stop bit, no parity.
5. Repeat Prob. 4 for seven character bits.
6. A group of bits represents the position of doors and windows in an alarm system. A binary 0 means open; a binary 1 means closed. Sketch the bit

pattern sent with no parity, and one stop bit, if the first four doors and windows are open, and the next four are closed.

7. A break detect circuit assumes that a break has occurred if the line is held in space for two character periods. How long a time is this for a system which is sending 8 bits/character with a parity bit and two stop bits, at 1200 baud?

8. Repeat Prob. 7 for 5 bits/character, no parity, one stop bit, at 2400 baud.

8-4 RS-232 Signals

Basic communications requires only a single wire and ground wire. The RS-232 standard, however, makes provision for various handshaking arrangements and interface functions that go beyond just sending bits on a wire. There are actually 19 signals that the interface standard discusses, plus two grounds. Most applications use only a few of these, typically from 4 to 8. Different applications may use different subgroups, depending on the specific device that is connected via RS-232.

It is important to understand the words and phrases that are used in the RS-232 specification for signals. Some of these terms came about from the original use of the RS-232 standard and may seem unconnected to present-day use. Nevertheless, these are the terms used in industry today, even though they may not be the most convenient way to describe the signals in state-of-the-art systems. Historically, the RS-232 specification was used as a method of interface between a user's terminal or computer, called the *data terminal equipment* (DTE), and a device which allowed the terminal to communicate with the computer, as shown in Fig. 8-9. This communications device, which is identical on both sides of the link, is called the *data communications equipment* (DCE). The acronyms ''DTE'' and ''DCE'' are used extensively when discussing the functions of signal lines under RS-232.

In many of today's installations, the computer and terminal are connected directly to each other, without any dedicated communications devices in the middle. Alternatively, a computer may be connected to its printer directly. However, a system which is a DTE can only talk to a DCE. A DTE cannot talk to a DTE, and a DCE cannot talk to a DCE. It is like trying to connect two power

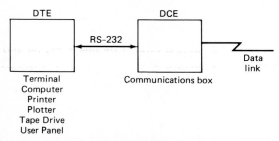

Fig. 8-9 The meaning of DTE and DCE. The *DTE* is the data terminal equipment, or source and user of the data. The *DCE* is the communications box or interface to the actual data link.

line plugs together—they must be of the opposite types (plug and socket) to fit.

This does not mean that a computer, acting as a DCE, cannot ever communicate with a terminal, also acting as a DTE. It does mean that some modifications are needed. These are discussed in a later section of this chapter. The important point is that the terms "DTE" and "DCE" are used to designate one side or the other of an interconnection, and signal lines are spoken of with respect to these terms.

The RS-232 standard calls for a maximum of 25 wires in the cable and connector. Within the connector, the 25 wires are assigned specific pin numbers. The list of 25 wires and their names is shown in Figure 8-10. The standard actually divides the many signal wires into three functional categories: a data group, a control group, and a timing group. Figure 8-11 shows the RS-232 signals again, this time divided by groups. It also shows the direction of a signal on any wire within the group.

Here is a clear case of a situation in which the terms "DCE" and "DTE" are needed. Look at wires number 2 and 3. One is called *transmitted data*, and the

RS-232C LINE DESIGNATION

Line Pin No.	Circuit Name	Description
1	AA	Protective ground
2	BA	Transmitted data
3	BB	Received data
4	CA	Request to send
5	CB	Clear to send
6	CC	Data set ready
7	AB	Signal ground (common return)
8	CF	Received line signal detector (data carrier detect)
9	—	(Reserved for data set testing)
10	—	(Reserved for data set testing)
11		(Unassigned)
12	SCF	Sec. rec'd. line sig. detector
13	SCB	Sec. clear to send
14	SBA	Secondary transmitted data
15	DB	Transmission signal element timing (DCE source)
16	SBB	Secondary received data
17	DD	Receiver signal element timing (DOE source)
18		Unassigned
19	SCA	Secondary request to send
20	CD	Data terminal ready
21	CG	Signal quality detector
22	CE	Ring indicator
23	CH/CI	Data signal rate selector (DTE/DCE source)
24	DA	Transmit signal element timing (DTE source)
25		Unassigned

Fig. 8-10 The signal line designations of the 25 wires within an RS-232 standard, arranged by line number.

RS-232C LINES: TYPES AND SOURCE

	Line Interchange Circuit	Pin No.	Description	Gnd	Data		Control		Timing	
					From DCE	To DCE	From DCE	To DCE	From DCE	To DCE
Data	AA	1	Protective ground	X						
	AB	7	Signal ground/common return	X						
Control	BA	2	Transmitted data			X				
	BB	3	Received data		X					
	CA	4	Request to send					X		
	CB	5	Clear to send				X			
	CC	6	Data set ready				X			
	CD	20	Data terminal ready					X		
	CE	22	Ring detector				X			
	CF	8	Received line signal detector or data carrier detect (DCD)				X			
	CG	21	Signal quality detector				X			
	CH	23	Data signal rate selector (DTE)					X		
	CI	23	Data signal rate selector (DCE)				X			
Timing	DA	24	Transmitter signal element timing (DTE)							X
	DB	15	Transmitter signal element timing (DCE)						X	
	DD	17	Receiver signal element timing (DCE)						X	
(Secondary Group)	SBA	14	Secondary transmitted data			X				
	SBB	16	Secondary received data		X					
	SCA	19	Secondary request to send					X		
	SCB	13	Secondary clear to send				X			
	SCF	12	Secondary rec'd line signal detector				X			

Fig. 8-11 The RS-232 signal lines, arranged by function: data, control, and timing. Also shown are the directions of the signal wire on each wire.

other is *received data*. Does this mean that both the DCE and the DTE transmit on wire 2 and receive data on wire 3? No, it does not. The signal that is transmitted from one end must go to the corresponding receiver point on the other. If both DTE and DCE send out a signal on wire 2, two signals collide on the same wire. Certainly, this is not a workable case. Instead, the signal wire which is called *transmit*, number 2, is defined by the RS-232 standard to carry a signal from the DCE end to the DTE end. The signal wire which is called *receive*, number 3, is used for the DCE to receive, and therefore the DTE transmits on that wire. Even though the wire is called the *transmit signal*, the direction of signal flow depends on whether the end is DCE or DTE. To avoid this confusion, the RS-232 standard defines the direction of signal flow for each of the signal names wires in the cable. It clearly shows that the received data line is used for data from the DCE end to the DTE end. The wire called *transmitted data* is used for the DTE to transmit, and for the DCE to receive.

The RS-232 Lines and Their Functions

The figures of the previous section showed how the signal lines of the RS-232 standard are divided into groupings. Each of these groups represents a specific function of the interface.

The ground wires are the simplest. One ground is the *protective ground*, which is connected to the electrical chassis of the equipment. Its function is to provide protection to the user against shock in case of some electrical fault. This is similar to the third wire on a standard alternating current (ac) line plug and socket. The second ground is the signal ground. It is this wire which establishes the 0-V point for the single-ended signal transmission of the RS-232 standard. The voltages that are transmitted and received on all the other wires are measured with reference to this point. The signal ground may be connected to the protective ground, but it also may be disconnected. The choice is made by seeing which provides better performance and minimizes electrical noise, which is a function of the ac line noise, the signal noise, and the particular type of wiring configuration within the DCE and DTE systems.

The transmitted data and received data lines are the wires which carry the actual message bits between the DTE and DCE ends. The direction of signal is as shown in Fig. 8-11. There is also a pair of lines for secondary received and secondary transmitted data. These secondary lines are provided for applications in which there are two channels in each direction. One channel is the primary, higher-speed, and higher-performance channel, and a second channel is needed for some less critical, separate message. The secondary channels may, for example, transmit information to the other end about the condition of the data link, such as the number of errors detected. This is separate and independent from the system user's message. Most installations do not use the secondary channels, but some DTE and DCE equipment can make use of them.

The timing lines are used when the RS-232 specification is used in designs in which a timing signal must be sent along with the data. This timing signal allows the receiver to synchronize and know when to look at the data bits. Most RS-232 installations are asynchronous and do not use these lines.

The largest group of lines are the control lines. There are nine control lines for the primary channel and three for the secondary channel. Most pieces of equipment use only a small group, or *subset*, of these lines. Effective handshaking and control can be performed with only a few of these lines. The additional lines are for some functions that are needed in some special applications of the RS-232 interface. Since the control lines are important and varied, it is useful to study some of them in more detail.

Of the nine primary control lines, six are for the DCE-to-DTE direction and three are for the DTE-to-DCE direction. There are technical and historical reasons for this. The original and still common purpose of the RS-232 standard is to allow the interconnection of a computer or terminal (DTE) to a communications box, the DCE. Many of the communications boxes then use telephone lines or radio links to connect to the corresponding communications box at the other end of the link. Each box serves not only to put the user's data signals onto the link, but also to keep the DTE informed of link conditions and operation.

There is another interesting point about the control lines. When a control line is indicating that the condition it represents is true, it takes on the space value, the +3- to +25-V range. When the control line wants to show that the condition it represents is not occurring, it uses the mark value, from −3 to −25 V. This is different from the sense for the data, in which space is used to show binary 0 and mark to show binary 1. Therefore, the signal sense for the data lines is opposite to that for the control lines. This poses no problem since both ends of the link have agreed to this. However, when a technician is probing with a voltmeter or other test equipment, this must be kept in mind to avoid confusion.

The control lines which are used for handshaking of the type discussed in previous chapters are request to send (RTS), clear to send (CTS), data set ready (DSR), and data terminal ready (DTR). The *data set* is the communications box which acts as the DCE; the *data terminal* is the computer or terminal which is the DTE. The RTS, CTS, DSR, and DTR lines form the basic handshake group. A handshaking protocol may use RTS and CTS for the DTE device to request clearance to send data to the DCE device, and for the DCE device to grant clearance. The DSR and DTR lines indicate operational status (ready or not ready) of the DCE to the DTE and the DTE to the DCE paths, respectively.

The ring indicator, received line signal detector, and signal quality detector are used by the DCE device to indicate some additional items to the DTE. When any data is coming into the DCE from the telephone lines or whatever link is in use, the *received line signal detector* shows this. It also shows whether the quality of the received signal is adequate for low-error performance and indicates this via the *signal quality detector*. If a telephone line is in use, the DCE can show the DTE that someone or something is calling in when it detects the bell-ringing signal on the line, and it uses the *ring indicator* to show this.

Many installations do not use the ring indicator, received line signal detector, or signal quality detector. If the DCE is not being used on telephone circuits, the ring indicator may be meaningless. The signal quality and received line signal detector may not be necessary in cases in which the communications link is a direct piece of wire only a few feet long.

QUESTIONS FOR SECTION 8-4

1. How many signals are defined by the RS-232 standard? How many are typically used in an application?
2. What does a DTE represent? What about a DCE?
3. Explain the problem of a direct connect, such as DTE to DTE.
4. How many wires, maximum, in an RS-232 connector?
5. What are the three groupings of RS-232 signals? What is the directional meaning? Why is this possibly confusing at first?
6. What is the role of the protective ground? The signal ground? Can they be connected together? Why would it be done?
7. What are secondary channels? Why are they sometimes used?
8. When are the timing lines used?
9. What group represents the handshake lines? Why are there more DCE-to-DTE lines than DTE-to-DCE lines?

10. What is the signal sense on the control lines? Compare this to that on the data lines.
11. What is a data set?
12. What is the function of the ring detector? Of the received line signal detector and the signal quality detector?

8-5 Some RS-232 Examples

Different application needs, performance requirements, and cost considerations have resulted in many different ways to use the RS-232 standard legitimately for interconnection. This section will show examples and explain how these operate, within the standard's guidelines.

Temperature Meter

In a *temperature meter* application an electronic meter is used to monitor the temperature at a remote, dangerous location. This temperature value is converted to digital format and then sent once per second as ASCII characters by wires to a computer which records the values as received (Fig. 8-12). Researchers can study the temperature values and analyze them at any time, by using the computer.

The interconnection can be very simple, using just a single wire pair for the data and a ground wire. There are no handshake lines needed in this application. Why is this so? What if the computer is not ready for the data? The answer is that it does not matter. Since the meter is updating its temperature reading every second, it must pass the old reading on or it will be lost. It makes no difference whether the meter sends its reading to a computer that is not ready (or whether the wire between the two units is broken) or waits 1 s and then has a new reading that overwrites the old. The only function for the meter is to obtain a new temperature reading every second, pass it on, and get the next one.

The advantage of this type of interconnection is that only two wires are needed. The disadvantage is that data may be lost, if the computer is not ready. But this data would be lost anyway, by the design of the temperature meter. If this meter had some built-in data storage, then it would be meaningful for the meter to hold off sending data until it was assured, via handshake, that the computer was ready.

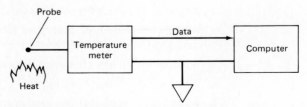

Fig. 8-12 An RS-232 connection from a temperature meter to a computer. Only the data line and ground wire are used.

Computer to Printer *Computer-to-printer* transmission is a very common application of RS-232. The computer is directly connected to the printer. The printer is used exclusively by this computer (not shared by any other computers). The interconnection requires that the computer know whether it is okay to send additional characters to be printed to the printer. There is no flow of data in the reverse direction, from the printer to the computer.

In many of these interconnections, a single handshake line is used (Fig. 8-13). This is the CTS line from the printer, which shows the computer the status of the printer. The RTS line from the computer to the printer really isn't needed, since the only function the printer has is to accept data from the computer. Therefore, proper management of the situation only requires that the computer know whether the printer is ready or not.

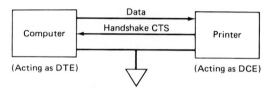

Fig. 8-13 The RS-232 connection between a computer and printer. There are a data line from the computer to the printer and a single handshake line indicating the printer status to the computer.

The timing diagrams (Fig. 8-14) show the conditions of the data line and the CTS line for two different printers. One is a slow printer, without any internal buffer. At the end of any line, the printer must return to the beginning of the next line, and this takes about 0.3 s. The printer is receiving characters at 300 baud, and can print 30 characters/s. Therefore, the CTS line drops after each character

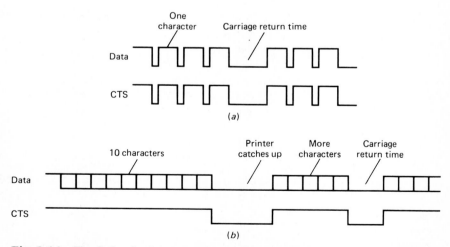

Fig. 8-14 The timing for the computer-to-printer interface: (*a*) a printer with no buffer; (*b*) a printer with a 10-character buffer. Note that the CTS line indicates that it is ready to accept or not accept characters very differently in each case.

is received, until it is printed, since no new character can be accepted until the previous one is actually on paper. At the end of a line, the CTS line drops for the time it takes for the printing mechanism to return to the beginning of the next line. Note that even if the baud value were higher, the timing diagram would look about the same, since the most important factors in the timing are the printing speed and carriage return delay.

The second case is a printer with a 10-character buffer. Here, the printer is printing at the same rate, but it can accept characters from the computer in a burst. This is especially noticeable at the end of a line, since the computer does not have to wait until the carriage returns. In fact, the computer does not even see a long time period drop in CTS from the printer, since the buffer acts to accept characters while the carriage is returning.

Computer to Terminal

The communications between a computer and its terminal is two-way: data from the keyboard to the computer and from the computer to the screen. The interconnection can look like Fig. 8-15a. The handshake line from the terminal to the computer tells the computer that the terminal is available (power on and ready). The handshake line in the reverse direction tells the terminal that the computer can accept the next typed character. The amount of data flowing from the terminal to the computer is low, and the rate is slow, since typing on a keyboard

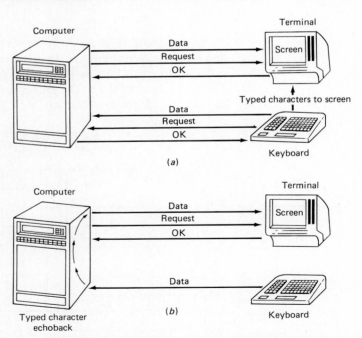

Fig. 8-15 A computer-to-terminal interface. In the simple method (a) there is a handshake line in each direction, from the computer to the terminal screen and from the keyboard back to the computer. The alternative is (b), in which the typed data is sent to the computer and then echoed back to the screen. This requires fewer signal lines and is a better indication that the typed character has been accepted by the computer.

is a low-rate activity. The reverse direction is usually much busier: the computer may be putting many characters up on the screen in response to a few keystrokes. Handshaking is needed in both directions, however, since it is important that the user know that the keyboard data can be accepted by the computer. There is no point in typing if the typed characters cannot be accepted, and it would be frustrating for a user to find out that keystrokes have been ignored because the computer is busy or unavailable.

To overcome this problem, many computer-to-terminal interconnections do not use a handshake line from the computer to keyboard, but use a more effective method of letting the user know that the typed characters have been received properly. This is called an *echoback scheme* (Fig. 8-15b). It works this way: instead of having the terminal itself put the character on the screen as it is typed, and send it to the computer, the computer is responsible for accepting the character and then echoing it back to the terminal screen, which then displays it. Therefore, if a character is typed and appears on the screen of the terminal, it means that the character is received and accepted by the computer. This is much better than just sending it to the computer on the basis of the handshake line status. This echoback scheme is used more commonly than the simpler alternative. When it is used, there is no need for a handshake line from the computer to the terminal, since as the user types, he or she immediately sees whether or not the characters are being accepted.

Computer to Computer

The requirements for a computer-to-computer data communications system are usually the most demanding. The two computers often have large amounts of data to send, and the protocol must ensure that the link does not cause problems or crash on either side. The way to ensure this is through complete handshaking in each direction, using hardware lines or XON/XOFF control characters.

The *XON/XOFF method* requires no handshake lines, but it does require a full-duplex channel (Fig. 8-16). It also requires that the receiving circuitry at each computer be especially designed for XON/XOFF. This means that the receiving circuitry must be designed to look immediately at each incoming character of a message for the presence of the XON or XOFF character. If an XOFF is received, the other end is saying that it cannot accept any more characters, and transmission of new ones must stop immediately. The circuitry must therefore provide an operating channel link between the receiver section, which gets the XOFF, and the transmitter, to shut it off as soon as the XOFF is received.

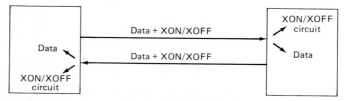

Fig. 8-16 The hardware handshake can be replaced by special characters XON and XOFF. The receiving circuitry must be designed to see the XON or XOFF immediately and take the appropriate actions.

In some system designs, the XOFF is sent before the receiving buffer is full, to allow for some delay. This enables a few characters to be sent out by one end of the link, after the other side indicates that it cannot accept any more. The reason for this is that there are some propagation and link delays that are completely outside the control of the computers at each end. It may take several microseconds or even milliseconds for the XOFF character to reach the intended computer. Even if the character is processed and identified immediately, several more characters may have been sent out in the interim. The system designers set the point at which the XOFF indication is sent out depending on the expected propagation and link delays. A system used within a building has a very short delay; two computers at opposite sides of the United States have a much larger one, as long as 200 ms for a cable interconnection.

The original intent of the RS-232 specification was to satisfy needs of systems with a DTE-to-DCE link, a communications path from the DCE at one end to the DCE at the other end, and then a DCE-to-DTE link at the receiving end. The control lines indicate that there is data to be transferred and whether the transfer can proceed. They also indicate the condition of the communications line and its situation. In a typical situation, a terminal is communicating with a computer, through intervening DCE data sets.

Suppose a user at a terminal wishes to "dial up" a computer located in another building. Figure 8-17 shows the sequence of events. The user first picks up a phone and dials the phone number of the other computer (actually, of its data set). The data set electrically sees the ringing signal and indicates it to its DTE, the computer. If the computer is willing to take the call, it indicates this with its DTR line. The data set then sends an audible tone to the other side, so the user knows that it has been picked up. At this point, the electrical path has been established, but no data has been sent. The user's data set indicates to the terminal that this is so, by turning its DSR line on.

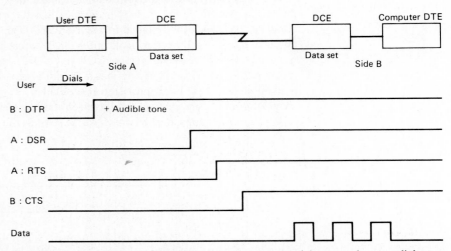

Fig. 8-17 The sequence of signal events when a user picks up a phone to dial up a remote computer. There is considerable handshaking before data is sent, to make sure that both sides are ready.

The terminal then wants to send the typed keyboard characters. The terminal issues an RTS to its data set. If the phone line is not in use, the data set responds with CTS. The data is then passed to the data set from the terminal, from the data set along the phone lines to the other data set, and onto the computer. The computer's data set, meanwhile, uses its received line signal detector and signal quality detector to show the computer that the link is active and to indicate the signal quality.

In many applications, this dialing sequence is performed without any user action. The terminals and computers are intelligent devices, with microprocessors and built-in programs that can actually dial, answer, and respond by themselves. The same sequence of RS-232 lines is used, though, as with manual operation.

A large number of RS-232 applications do not go over phone lines, but use a dedicated link from one end to the other. A common example is the computer and printer at opposite ends of a building. These applications do not have the dial-up and answer sequence, and the signal wires for these are not even built into the DTE system (computer or printer) in these cases. The installation is much simpler, and only a few of the RS-232 lines are needed at each DCE.

Many other types of legitimate RS-232 interconnections are possible. All can provide proper operation if implemented correctly. The challenge for system engineers and technicians is to connect two devices, both of which individually meet the RS-232 standard, and make them work. This can take anywhere from a few minutes to many hours. Neither side is at fault, yet the interconnection process is difficult. The next section details some of the concerns and considerations in making two such pieces of electronic equipment talk to each other reliably. It also shows what approach is used in getting the system to work.

QUESTIONS FOR SECTION 8-5

1. Why is there no need for handshaking in some applications?
2. Why does a computer-to-printer system often need only one handshake line?
3. Where might a printer have to indicate to a computer to stop sending data?
4. How can terminal keystrokes be displayed on a screen?
5. How does echoback provide better assurance that the keystrokes are received? How does it eliminate the need for one handshake line?
6. What are the system requirements when using XON/XOFF?
7. Why is XOFF sometimes sent before the buffer is full?
8. How does the RS-232 standard function with phone lines? What is the sequence of events?
9. What advantages does a dedicated line have over a telephone line?

PROBLEMS FOR SECTION 8-5

1. Show the timing diagram for the temperature meter, as it sends one reading of two characters every second, at 150 baud. Sketch the signal activity on the data line.
2. Sketch the data line and handshake line activity for a computer-to-printer

hookup, when the printer has a single handshake line showing it is ready. The printer has no buffer, prints 50 characters/s, and is receiving characters at 300 baud.

3. In Prob. 2, the printer takes 1 s to return from the end of the line. Sketch the activity on the data line and the handshake line.
4. The printer is improved with the addition of a 50-character buffer. Repeat Prob. 3.
5. Repeat Prob. 3 for the case in which the printer uses XON/XOFF on a reverse data channel instead of a handshake line.
6. Repeat Prob. 4 for the application using XON/XOFF instead of a handshake line.

8-6 Making RS-232 Interconnections Work

The RS-232 is a standard which is very specific and rigorous in some areas, and less strict in others. It defines with absolute clarity the signal voltages, signal waveform shapes, and functions of all the lines that the standard allows. It does not, however, require the user of the standard to implement all of these signal lines in all situations. The meaning of what the signal voltages represent (ASCII, binary, number of bits) is also left to the user.

The result is that two devices can meet the RS-232 standard and still not communicate properly with each other. Making these two pieces of equipment successfully communicate can be frustrating. Fortunately, there is a logical approach that can be used to make the task easier and usually successful. There are three areas which must be checked by a technician or engineer trying to make the interconnection work: bit configuration and baud; connector and pinouts, and handshaking. They should be checked in the order listed, since it is better to go from the easier areas to examine to the harder, and some of these easier areas are prerequisites for the next level of checking.

Bits and Baud

The first item to be checked is the bit and baud setting. Both systems must be set for the same baud value and bit configuration. The settings for these items are provided in either of two ways: through hardware (using switch settings or jumper wires on one of the system circuit boards) or software (issuing special setup commands through a keyboard or operator panel of the system). Neither method has any inherent advantage over the other, although the hardware setting method can be checked with the power off, the unit disassembled, or even when the system is having some other problems. For the software method to work, the equipment has to be functioning properly. Many service people prefer the hardware method for this reason, but many systems use the software method because it is less costly and gives the user of the equipment more flexibility to make changes without having to pull out circuit boards.

The baud values are usually in the 300-baud and higher range, using one of the standard values that have been discussed previously. Note that in a system that is sending data in both directions, either half duplex or full duplex, the baud value does not have to be the same in both directions. The keyboard-to-computer channel may use a low baud value since characters are provided relatively slowly; the computer-to-screen channel uses a much higher rate, so the computer can transmit characters to the terminal screen much faster, as it needs to do. The situation regarding the bit configuration is more complex. The choices are as follows:

Number of bits per character: 5, 6, 7, or 8.

Number of stop bits: 1 or 1. (Although 1½ is occasionally seen, it is rare.)

Parity bit: even, odd, or none. (The meaning of these bits is discussed in Chapter 12.)

A system set for 7 bits/character, one stop bit, no parity, does not work or works erratically when connected to one which is set for 8 bits/character, two stop bits, even parity.

Some systems allow the parity bit to be set to no parity, but instead of not sending any parity bit, they use a fixed bit value (1 or 0) in its place. This is because 7-bit ASCII is commonly used for the character representation, and computers often like to see a full byte of 8 bits. The 7-bit ASCII is therefore ''padded out'' with an eighth bit where the parity bit would be, but it is not used for an actual parity bit (Fig. 8-18). Other systems, in contrast, convert the 7-bit ASCII as received into a byte, and there is no need to use a filling bit as a placeholder. Recall that each bit needed in the transmission decreases the number of characters that can be sent per second.

If the systems do not have a common setting of bit configuration and baud values, they are not able to communicate reliably, ever.

Connectors and Pinouts

The RS-232 specification does not state that any special connector should be used. It does recommend a 25-pin connector. The industry has generally settled on a D-shaped, 25-pin connector (Fig. 8-19). This connector has male and female forms. The male end has 25 pins, and the female end has 25 receptacles. These are also known as the *plug end* and the *socket end*. Just as with ac line cords, a plug must go into a socket. Usually, the RS-232 side that is acting as a DTE uses the plug connector, and the DCE end uses the socket connector, but this is not always the case. It is possible to have both DCE and DTE with the same connector type, and therefore they cannot be plugged directly together without changing one connector or making an adapter cable. To make matters

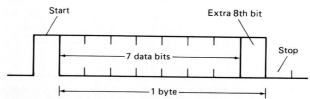

Fig. 8-18 The 7-bit ASCII code may have to be ''padded out'' with an extra eighth bit, when the receiver expects to see eight data bits.

Fig. 8-19 The D-shaped 25-pin connector is the most common one used for RS-232 connections (Courtesy AMP Incorporated).

even more difficult, changing the connector itself may not solve the problem. Consider two units, both acting as DTEs. Both may or may not have the same connector, but by the definition of DTE both have their data signal flowing out of the equipment on the same pin, number 2 (Fig. 8-20a). Therefore, even if the connectors are a *mating pair* (plug and socket) the electrical signals will clash. Both units transmit to each other on the same wire and expect to see signals coming in on another pin, number 3.

The solution here is not simply a connector change, but a special adapter called a *null modem*. This adapter is a cable which transposes some of the wires as they appear at the connector, to make a DTE look like a DCE, and vice versa. The simplest null modem swaps wires 2 and 3, the transmit and receive data lines (Fig. 8-20b). This allows data from one unit to flow into the receive data pin of the other, and vice versa. The electrical signals then are properly matched (transmitter to receiver), and no collision occurs.

How does a technician or engineer determine when a simple connector change is needed, or when a null modem adapter is needed? The only way is to study the equipment's documentation manual, which usually states whether the RS-232 port on the equipment is designed as a DTE or DCE. The documentation may also show the direction of signal flow in a *pinout diagram*. This is a very useful type of diagram, much more helpful than a circuit schematic for making the RS-232 interface work properly (Fig. 8-21). The schematic shows the wires and associated electronic components, but the direction of signal flow from these wires is not always clear. The signal flowchart shows the more useful informa-

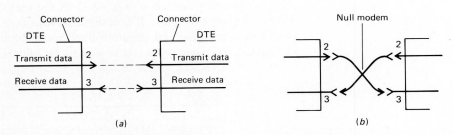

Fig. 8-20 An RS-232 conflict in connecting two DTE devices together: (*a*) shows the way signal sources would clash, and signal receivers would get no signal; (*b*) shows the way a null modem interchanges the connector wires so that signal sources and receivers match.

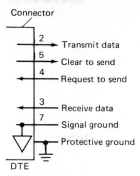

Fig. 8-21 A physical pinout and signal flow diagram is very useful in making RS-232 interfaces function, since it shows the signals that are present on the connector and direction of signal flow.

tion: which signals are present on the RS-232 connector and which direction they have.

In the ideal world, one unit would be a DTE with a plug (or socket) connector, and the other would be a DCE with a socket (or plug) connector. In reality, there are several other possibilities that the user may have to deal with. Both units may have mating connectors but be configured as both DTE or both DCE. Then, a null modem is required. Or both units may be a matching DTE–DCE pair, but the connectors may not mate. Then a simple connector change is needed. (In fact, special connector assemblies are sold which have a socket connector at each end, with all 25 wires brought straight across, or a plug connector at each end and all 25 wires carried across. These can be used to plug into a plug or socket connector and change its sense from one to the other quickly.) Finally, the connectors may not mate, and so the DTE-DCE pair does not exist. For these cases, a null modem that also changes connector sense is needed. Before making any connections or rewiring, the technician or engineer should look at the documentation that is supplied with each unit. Some test aids, such as breakout boxes, are also available to make the job easier. These are discussed in detail in Chapter 13.

The 25-pin connector is most common, but not the only choice. Since many systems use only a few of the RS-232 wires, it is wasteful of circuit board space and costly to use a 25-pin connector. Some smaller personal computers do not have room at the edge of their circuit cards for a 25-pin connector. These systems often use a smaller connector, such as a 9-pin D-shaped connector. When the technician is connecting to this, the system documentation must be consulted to determine which signals appear on which pins. Connecting a 9-pin D to a 25-pin D unit requires an adapter cable that takes the signals on one connector and brings them across to the correct pin on the other. The DTE-DCE problem may still occur, as well. Figure 8-22 shows the schematic of a system which uses a 9-pin connector, connecting to a printer that has a 25-pin connector. The technician or engineer may have to make up this adapter cable specifically for the application.

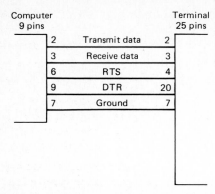

Fig. 8-22 An adapter cable is needed to connect a 9-pin interface to a 25-pin interface.

Handshaking

Setting the bit configuration and baud value and making sure the connectors and DTE-DCE pair are compatible are the first two steps to a successful RS-232 interconnection. The third step is more difficult and requires an understanding of the operation of the RS-232 interface on each side of the connection.

As was shown in the previous section, the amount of handshaking that is done varies with the particular equipment and application. The temperature meter uses no handshaking, although the computer-to-computer connection uses handshaking in both directions. Suppose the temperature meter is to be connected to a computer that expects to see RTS from the meter and then to issue a CTS indication. The communications port on this computer does not even accept characters until it goes through the RTS-CTS cycle, yet the meter does not even have an RTS line to use. If the meter is connected to this computer, the computer will never accept any RS-232 characters. What can be done?

The solution is to make the RTS line, as seen by the computer, always indicate that the remote device is issuing an RTS request. This is done at the computer, by wiring the RTS line to a space (positive) voltage point. From the perspective of the computer, the temperature meter is issuing the RTS, and it responds accordingly. The handshake needs of both sides are satisfied.

Many installations are not so straightforward. One very popular printer used with personal computers employs pin 11 to show that it is ready. The personal computer, however, is looking at its CTS, pin 5, to see whether the printer is ready. A direct connection between the two units cannot work. A special cable which connects pin 11 at the printer to pin 5 at the computer must be made (Fig. 8-23a).

More complicated cases occur when multiple handshake lines are used, as in half- or full-duplex systems. Here, the two sides of the connection are each looking for proper handshake lines. It is possible to get into a "which comes first—the chicken or the egg?" situation, in which one side won't start until it receives the handshake from the other, and the other won't provide the handshake until it sees the handshake from the first. If the lines are not connected in corresponding pairs, then a lockout of the communications is possible. The solution is to study exactly which handshake signal is required by each side. If the other side does not provide the handshake, the side looking for it must be "fooled" into seeing that handshake. This can be done by either hardwiring the handshake line, as was done in the meter/computer example, or self-handshaking. In *self-handshaking*, the outgoing handshake line from the interface is looped back and looks like an incoming handshake line (Fig. 8-23b).

In some cases, the equipment at one end has several handshake lines that must be satisfied, but the other end provides only one of these. For these cases, a special cable which takes the single line and branches it out to both points at the other side is required (Fig. 8-23c).

These various methods of correcting any handshake mismatch may have some possible weaknesses. The temperature-meter-to-computer approach certainly ensures that data from the meter gets to the computer interface. Some of the self-handshaking schemes, however, effectively defeat the purpose of handshaking. The handshake, instead of being a true indication of the situation and status at the other end, is just a "self-induced" guess at the status. There are situations in which this can cause characters to be lost or received improperly. Suppose a computer is capable of sending characters at higher baud values to a low-baud-value printer. The printer does not have handshaking, but the computer issues an RTS and expects to see a CTS. To get around the handshaking mismatch, the computer user wires the RTS line to the CTS line. Therefore, whenever the computer issues an RTS, it receives a CTS and begins sending data. However, it never finds out that the printer is unable to keep up with the data rate or is not available. It keeps sending characters and assumes they are being printed, since its own handshake lines apparently tell it everything is okay. The solution may be to get another printer or at least slow down the baud value and transmission rate to the printer so that it is more likely that the printer can keep up with characters from the computer. It is better to send slowly and reliably than quickly with missed characters.

This section has shown that the existence of the RS-232 standard is a step toward making two electronic systems talk, but it does not mean that they can simply be plugged together. Problems can arise at various levels. Some other industry standards have the same set of potential problems as the RS-232 standard, and for this reason it is useful to study the operation of RS-232. Other

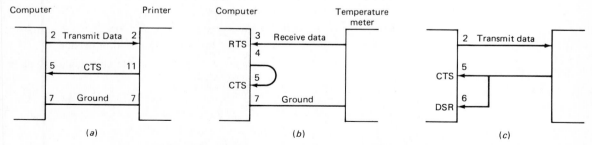

Fig. 8-23 Special RS-232 interface cables are often needed: (*a*) an interface with the correct signals, but on differing pins; (*b*) a computer which needs handshaking connected to a device that provides none, in which case the computer must self-handshake; (*c*) a single handshake line branched out to handshake to two signal lines at the other end.

industry standards, however, are much more specific in the nature of the interconnection and virtually guarantee that two devices that are connected are able to pass characters, even if the characters make no sense to the user. These stricter standards provide higher performance with fewer problems, but usually cost more to implement and have less flexibility. The simple temperature-meter example which only uses two wires would not be possible. A more expensive interface, with more circuitry and wires, is needed.

Not all RS-232 devices can be successfully connected. A computer which has no handshaking lines and uses XON/XOFF exclusively is very difficult to connect reliably to a printer which uses only handshake lines and does not support XON/XOFF characters.

QUESTIONS FOR SECTION 8-6

1. In what areas is the RS-232 standard strict? In what areas is it not specific?
2. What are the three areas to check to get an RS-232 interconnection working? Why is the order important?
3. What is the difference between setting the baud value in software versus hardware? What are the pros and cons of each?
4. Are baud values in a duplex system the same in both directions? Explain.
5. What are the ways to have and not have a parity bit? What is "padding out" the parity bit? What are the pros and cons of this?
6. What is the potential for conflict between RS-232 connections?
7. What are the four connector–DCE-DTE combinations that are possible? Which is preferred for easiest interconnection?
8. What is the general handshake problem that may occur? What is the approach to overcome it?
9. How can lockout occur? How is it taken care of?
10. What are the drawbacks to self-handshaking and similar schemes?
11. Explain why two RS-232 devices may never work together properly.

PROBLEMS FOR SECTION 8-6

1. Two RS-232 systems are to be connected together. One is a DTE; the other is a DCE. Both use pin connectors. Sketch the interconnection needed.
2. A DTE and a DTE are to be connected. Both use socket connectors. Sketch the interconnection that must be built.
3. A system is configured as a DTE. It needs to talk with another DTE. Each system uses a single handshake line, the DTR line, to let the other side know it is ready. Sketch the cable that must be made up for this.
4. Two DCE devices are supposed to send and receive data from each other. They each have the RTS and CTS handshake lines. What cable schematic is used to make an interconnection?

8-7 Integrated Circuits for RS-232

The circuitry to perform the operations of an asynchronous RS-232 interface can be built of discrete components such as transistors, resistors, and capacitors, along with some smaller-scale integrated circuits (ICs) which provide simple boolean operations. However, this type of interface is so common and popular that it was among the first electronic functions that were built into larger-scale ICs. The larger IC eliminates nearly all the individual components and thus saves cost, board space, and power and is also more reliable.

Integrated circuits for communications fall into two groups: those which provide the formatting and protocol and those which are the physical interface to the communications link or outside world (Fig. 8-24). There are good reasons for this. The part of the communications circuitry which is connected to anything besides the circuitry itself must be electrically rugged, since it may be accidentally misconnected or abused. The *physical interface* IC is designed to withstand, without damage, voltages and currents that may damage an ordinary IC. In addition, the physical interface IC may have to provide voltages, such as the 25 V of RS-232, that are not handled by regular ICs, which usually can deal with only 5 V. The interface IC is specifically designed to provide the necessary voltages (or currents) and withstand excessive voltages (or currents). Even when the interface IC fails, through misapplication, it is designed to fail in a way that does not cause the rest of the electronics system to be damaged. In effect, the interface IC acts as a safety fuse. It blows open and prevents the damaging signals from reaching the rest of the circuitry.

The more complex IC is the one which provides the *formatting and protocol*. Many ICs are available, each providing different amounts and types of format,

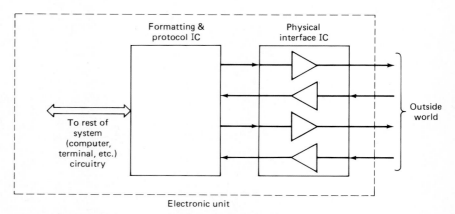

Fig. 8-24 A typical communications interface uses two ICs. One provides the necessary formatting and protocol for the data bits from the main part of the system, and the second is the electrical interface to the outside world data link.

The RS-232 Interface Standard 279

protocol, and general flexibility. The system designer chooses one which provides the features and type of operation needed for the application. A simple RS-232 interface using ASCII characters in asynchronous operation uses a fairly simple IC; one that implements SDLC needs a more complex IC.

This points out another reason why a data communications interface uses two ICs—one for the format and protocol and one for the physical interface. It allows the system to be designed with the best combination of the two. For example, a system may need to use asynchronous ASCII *internally* (between two parts of a single circuit) but not need the voltage levels of RS-232. An IC which provides the asynchronous ASCII is used, but without the higher-voltage RS-232 interface IC. If the system needs to be connected to another system via RS-232, then only the interface IC is replaced by one specially designed for RS-232. Similarly, if the system requires a complex protocol, but with the RS-232 voltage levels, the advanced format/protocol IC is paired with the RS-232 interface IC. This gives maximum flexibility at lowest cost.

It is useful to study, in a little more detail, a typical IC pair used for ASCII asynchronous RS-232 communications. A pair that is commonly used is the Intel 8251 Universal Synchronous/Asynchronous Receiver/Transmitter (USART) and the industry standard 1488 and 1489 RS-232 line driver and receiver combination. A full-duplex channel which uses these ICs between a computer and terminal is shown in Fig. 8-25. There is an 8251 at each end which provides a transmitting and a receiving point for asynchronous ASCII characters. Each 8251 is connected to a source of characters (the keyboard or computer) and a device which receives characters (the terminal screen or computer). The 1488 line driver IC (with two independent drivers) takes the characters from the 8251 and puts these on the interconnection cable by using RS-232 voltages and signal shapes. The 1489 is a line receiver IC (with two independent receivers) which takes the RS-232 voltages as received and passes them on to the receiver portion of the 8251. Note that a duplex channel requires only one 8251 at each end, but a pair of 1488/1489s.

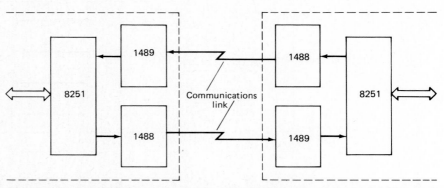

Fig. 8-25 A full-duplex system using 8251 UARTs for character handling and 1488/1489 ICs as the RS-232 line drivers/receivers.

The Intel 8251

The Intel 8251 is a 28-pin IC which provides many of the functions needed to take characters from any source and prepare them for data transmission (Fig. 8-26). It also can accept characters as received to prepare them for the computer processor or other equipment which makes use of the characters. The functions of the 8251 are selected by the system computer or microprocessor software. This means that the 8251 can be addressed through the computer bus and given directions as to which modes of operation it should use. No mechanical switches are required. The computer does this setup itself or may do it in response to user actions at the keyboard.

The features and choices of the 8251 include the following:

Operation up to 19.2 kbaud. There are three rates that can be selected.

Choice of 5-, 6-, 7-, or 8-bit characters.

Parity or no parity.

Provision of one, one and one half, or two stop bits.

Support of various handshake lines.

The overall function of the 8251 is twofold. In the transmission direction, the 8251 is given a character by the source (keyboard, computer, or other) represented by the ASCII of the character. It takes the character bits; adds the start bit,

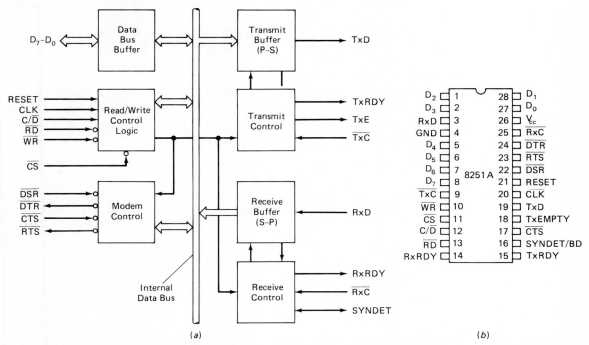

Fig. 8-26 (*a*) Block diagram and (*b*) pinout of the Intel 8251 UART, showing the internal functional parts of the IC and the pins of the IC as seen by the user and technician.

parity bit (if any), and stop bit(s), and then checks the handshake lines to determine whether the receiver can accept a new character. If yes, the 8251 sends the character bits out serially, at a bit rate determined by the user-selected value. The 8251 gets a master clock for this purpose, from which it can electronically derive the baud values it needs. It next tells the data source that the character is sent, and it is ready to handle another one.

In the receiving direction, the 8251 manages the handshake lines to the external world and shows the other transmitter that it can receive. When the bits of the character start arriving, the 8251 captures them in sequence by using its clock. When the entire character has been received, the IC strips off the start and stop bits, checks for errors (if any) with the parity bit, and then signals to the rest of the system that a new character has arrived. (It also indicates whether there is a framing or overrun error.) While this new character is in the 8251, the IC sets the handshake lines to show that it cannot accept any additional characters. When the rest of the circuitry takes the character in the 8251 holding buffer, the 8251 then shows, via handshake lines, that it can accept another character.

By doing this, the 8251 relieves the system of many of the small but essential details of communications: adding or checking start, stop, and parity bits; managing the handshake lines; and providing precise timing at various baud values. As far as the system processor is concerned, sending a character is a matter of simply passing the character to the 8251, and receiving a character involves only seeing the "new character received" signal and getting that character from the 8251. The time-consuming details do not burden the main part of the system. At the same time, the 8251 imposes no constraints on what can be sent or received. It can be used with any protocol or characters. The use of the 8251 is totally transparent to the rest of the system and source of the messages. It is like having a good assistant or secretary.

How does the 8251 communicate with the rest of the system of which it is a part? It uses a very common technique of *registers*, which are one-character buffers. The 8251 has several buffers, each of which is located at a unique address in the memory of the system processor (Fig. 8-27). The processor can

Address Line	Processor Read or Write	Result
0	Read	Read buffer data passed to system processor
0	Write	System processor passes data to write buffer
1	Read	Status buffer data bits passed to system processor
1	Write	System processor passes command bits to command buffer

Fig. 8-27 The internal address locations and registers of the 8251 allow the system processor to read or write data, command bits, and status information to and from the UART.

then put characters or data into the desired buffer by writing data to the appropriate address. There are registers for each direction of characters: one into which a system can place the next character to be sent and another to hold the character received from the external world. These registers are cleared automatically by the 8251 when the character is actually transmitted or taken from the register into the remainder of the system circuitry. Once cleared, they are ready to be used again.

The other critical register pair is the command and status pair. Through the *command register*, the system using the 8251 can set up the various operating specifics: baud value, stop bits, parity. Individual bits of the 8-bit register represent various desired conditions during this setup phase. The bits of the command byte are assigned as shown in Fig. 8-28a. Two bits are used to select the baud value, which is derived from the master clock of the system. The baud value is either equal to the clock or $\frac{1}{16}$ or $\frac{1}{64}$ the clock value. These are referred to in the 8251 data sheet as 1x, 16x, or 64x values. If the master clock is at 76.8 kHz, selecting 16x provides a value of 4800 baud.

The parity desired is set by two more bits. The choices are whether parity is on or off (when off there is no parity provided) and whether parity should be odd or even. Finally, 2 bits of the command register are used to set the number of stop bits: 1, 1½, or 2.

To set a baud value of 4800 (16x), seven character bits, even parity, two stop bits, the following pattern is sent by the processor of the system to the 8251: 00111010.

The other key register is the *status register* (Fig. 8-28b), which is also 1 byte. This register lets the system processor find out the status of the 8251 and some key RS-232 lines connected to the 8251. The bits of the status register represent the following:

What the state of the DSR line is.

Whether a break has been detected by the 8251.

Whether a framing error has occurred.

Whether an overrun error has occurred.

Whether a parity error has been determined by the 8251.

Whether the received data buffer is empty or full. (If full, a new character has arrived and should be taken by the processor from the 8251.)

Whether the transmit buffer is empty or full. (If empty, the last character was sent out and the 8251 can take a new character from the processor to send on the channel.)

A status byte that indicates the DSR line is ready, with no errors and both buffers empty, would look like this: 0000011X.

[**Note:** X means ''don't care if 0 or 1, is not relevant in this case.'']

If, however, a framing error and overrun error have occurred, it would be the following: 0011011X.

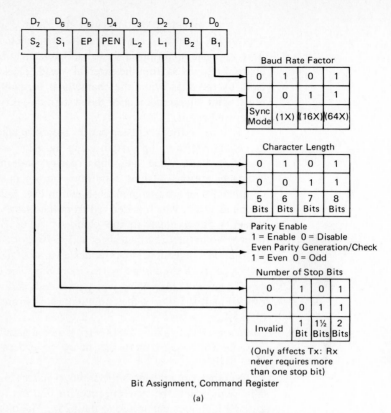

Bit Assignment, Command Register
(a)

When a new character is properly received into the 8251, the status word would be the following: 1000010X.

[**Note:** The DSR says "not ready," since the receive buffer is full and the external device should know that no new characters can be received.]

Once the system has read the newly received character from the buffer, the status changes to this: 0000011X.

Finally, when the system wishes to send a new character, it checks first to see that the transmit buffer is empty. The status word looks like the following: 000001XX.

The character is sent to the 8251 by the system processor, and the status byte becomes 000000XX.

After the character is actually transmitted, the transmitter buffer is empty and the next character can be accepted from the system processor. The status byte becomes 000001XX.

This shows how the processor of the computer (or keyboard, or screen) can use the 8251 to manage the transmission and reception of characters, through the use of the command byte for setup, and the status byte for actual information about what is occurring. The processor of the system reads the actual data register for incoming characters and deposits characters to be sent into the transmit register.

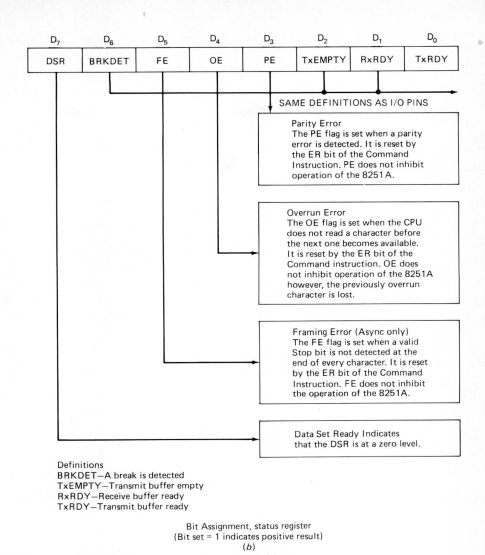

Fig. 8-28 The functions of the individual bits of the command registers are shown in (*a*); (*b*) shows the status register bit assignments.

Line Drivers and Receivers

In contrast to the sophisticated and complex operation of the 8251, the 1488/1489 pair is functionally very simple. The *1488* takes digital signals of 0- and 5-V levels and translates them to the -25- and $+25$-V maximum levels of RS-232. For this reason, the IC is called a *line driver*. If the system designer instead chooses to use -10 and $+10$ V, the designer simply uses a smaller-value power supply on the 1488. The 1488 takes the 0- or 5-V digital signal and puts it on the link at the higher, RS-232 voltage. It also can withstand some typical conditions that may exist on a communications link:

Connection to a powered system (via the cable) while its own power is off. This occurs when a printer which is off is connected to a computer which is on and sending characters. There should be no damage.

Connection to voltages higher than the 25 V of the RS-232 interface.

Accidental short circuit of any wire to ground (this happens quite often either by a misconnection or by a failure in the circuitry at the other end of the interconnection cable).

The design of the 1488 also makes sure that the signal being put onto the cable does not have slew rates which exceed the RS-232 specifications (and thus have excessive bandwidth). Even if the 0- and 5-V signal has very fast rise and fall times, the 1488 slows them down before transmission.

The *1489* is a corresponding complementary IC used to receive RS-232 signals and is called a *line receiver*. It can withstand the same fault and problem conditions that the 1488 can handle. The function of the 1489 is to take the RS-232 voltages, as received, and translate them down to 0- and 5-V signals, so they can be passed to a communications IC such as the 8251. The 1488/1489 ICs are manufactured by approximately half a dozen companies, and the ICs from each one of these companies are essentially interchangeable. This means that a user of the 1488/1489 type of line driver/receiver can easily find a source for the parts.

A communications interface using sophisticated ICs such as the 8251, along with the simple line driver/receiver such as the 1488/1489, is the general design used for this type of interface. The 8251 provides the "brain," and the 1488/1489 provides the muscle. If the system needs a different type of driver or receiver, such as one designed for fiber optics, then the 1488/1489 is replaced by a fiber-optic driver/receiver. If the system uses the RS-232 voltages but needs a more advanced protocol implemented at the interface, the 8251 is replaced by a more advanced IC.

QUESTIONS FOR SECTION 8-7

1. What are the two types of ICs needed for data communications? Why?
2. Why must the outside connected IC be rugged and accept voltages or signals different from the internal ones?
3. What flexibility does the two-IC scheme offer?
4. What is a line driver? A line receiver?
5. How is the 8251 set up? What things must it be initialized with?
6. How does the 8251 get a character which is to be sent out? What does it do when sending to both the outside world and its own processor?
7. How does the 8251 receive? What does it do when it receives a complete character? When it passes on that character?
8. How does an IC like the 8251 relieve the processor of many chores? What restrictions does it impose on the character set? On protocol?
9. How does the 8251 communicate with its processor? How are characters passed in or out of the 8251?
10. How is the 8251 set up? What is set?
11. How is the 8251 status checked? What can be determined from the status?

12. What action should be taken by the processor when the status indicates that the receive buffer is full? When the transmit buffer is empty?
13. What does the 1488 do? What is its role in the system?
14. What conditions does the 1488 have to withstand?
15. Why is the 1488/1489 pair an industry standard?

PROBLEMS FOR SECTION 8-7

1. Set up the 8251 for 6 bits, one stop bit, even parity, 16x clock. What bit pattern should be written to the IC?
2. The 8251 is to be used in 7-bit format, odd parity, 64x clock, two stop bits. What does the command byte look like?
3. The 8251 status byte is showing that all registers are full, that a framing and overrun error has occurred, but that the DSR line is ready. What does the bit pattern to the processor using the 8251 look like for this condition?
4. The 8251 receives a new character, so its receive buffer is full. The IC determines that this character has a parity error. What does the status byte look like?

SUMMARY

The RS-232 standard allows many applications to have successful, low-cost, reliable communications. To understand fully the working of this popular interface, it is necessary to study the following:

The voltage levels for the data

The data rates

The handshake lines and the many ways they are implemented

The physical pinouts

The terminology used

The role of the many lines of the interface

The many legitimate ways to use the interface standard

The ICs which are used to build it in a real system

The applications for short and long distances

The areas in which it is limited or weak in performance

Although the RS-232 standard is among the most common in use, it does have limitations. Some of these limitations are not drawbacks in the standard itself, but come about because users try to apply it to areas beyond its original design. These limitations include the following:

Distance: The specifications state that the RS-232 standard is for distances less than 50 ft. It is often used for larger distances, especially at the lower rates.

Noise pickup: The single-ended signals of the standard are very susceptible to electrical noise pickup. Higher noise resistance means using mark and space voltages closer to the +25- and −25-V levels. These are difficult and costly to obtain on modern digital systems, which often have only 5-V supplies.

Baud values: The 20,000-baud limit of RS-232 is too slow for many applications. Once again, some applications use higher rates, such as 38,400 baud, for shorter distances. This may work reliably, but it is outside the standard.

To overcome these limitations, some additional standards are used by industry. These standards do not change the basic start/stop nature of the RS-232 specification. They do change the details of the electrical interface (voltage, single-ended) to provide higher-speed, longer-distance communication with lower voltage values.

END-OF-CHAPTER QUESTIONS

1. What is defined by the RS-232 standard? What is suggested? What is left entirely to the user?
2. What is the source of the RS-232 standard? Is it mandatory?
3. Where is the RS-232 interconnection commonly used?
4. Why is studying this standard useful in understanding data communications?
5. What is the name of the signal state that represents binary 0? Binary 1?
6. What voltage ranges are used for binary 0 and binary 1?
7. What is the advantage of using the largest voltage value with the range for a binary 1 or 0?
8. Explain the way a voltage is actually sent down the wire in a single-ended scheme such as RS-232 uses.
9. What is the advantage of single-ended? The drawback? Where is it most likely to be satisfactory?
10. What is slew rate? How is it measured?
11. What is the effect of a fast slew rate? What is the limitation on system performance that a slow slew rate causes?
12. Explain the relationship between the slew rate and the maximum bits/s value that can be achieved.
13. What is the distance that RS-232 is intended for? How can users often go for longer distance with the same RS-232 voltages?
14. What is the idle state in an asynchronous RS-232 system? What voltage range is used to represent it?
15. What is done to indicate that a new character is about to begin?
16. How is the end of a character indicated? How many such bits may be used?
17. Where does the parity bit go in the bit sequence, if there is one used?
18. What can the bits within the RS-232 character represent? Explain how they can be ASCII characters, binary numbers, or independent bits.
19. What is the range of allowable baud values for RS-232? To what range do users sometimes extend it, outside the specifications?

20. Why is the break function sometimes needed? What is a break? How is break detected at the receiver?
21. Compare the meaning and cause of a framing error versus an overrun error.
22. Compare a parity error to either a framing or an overrun error.
23. What are the three groups of signals that RS-232 defines?
24. How many signals are in each group?
25. How is the direction of these signals determined?
26. Explain the potential confusion about terms such as "transmitted data" and the way the RS-232 specification overcomes this.
27. What are secondary channels? Are they used often?
28. What do the acronyms "DCE" and "DTE" mean? What is their importance when discussing RS-232 communications?
29. What voltages are used to indicate when a control line is active, or true—such as "The data set is ready"? How about when the line condition is false, such as "The data set is not ready"? Compare this to the voltages for binary 1 and 0 data bits.
30. Explain why most RS-232 installations do not use the timing lines.
31. Where are the ring detector, received line signal detector, and signal quality detector indications needed?
32. Explain why different amounts of handshaking are needed in different applications. Use some examples.
33. What is the echoback from a terminal to a computer? Why is it very powerful?
34. What is the use of XON/XOFF as compared to handshake lines? How is it implemented?
35. Why can there be problems in making two RS-232 devices communicate successfully?
36. What are the three areas that must be checked for successful data transmission? In what order? Why?
37. Why do engineers and users of systems often prefer setting baud values in software, although technicians prefer hardware?
38. Why is even a simple item such as a connector often a source of RS-232 confusion and problems?
39. Explain why a full RS-232 interface uses two IC types working as a pair? What flexibility and advantages does this offer?
40. How does an IC such as the 8251 relieve the system processor of many details of signal transmission and reception? How does the 8251 pass data to and from the processor?
41. How does the processor indicate to the 8251 what operating mode it wants? What types of things can be selected?
42. How does the 8251 indicate to the processor what is occurring at the IC and

the RS-232 interface? What types of information can be provided?

43. What is a line driver? Line receiver? Compare the complexity of these types of ICs with those like the 8251.

44. What line conditions must the line driver and receiver tolerate? Why?

END-OF-CHAPTER PROBLEMS

1. What are the noise margins for mark and space in a system that uses -9 and $+12$ V? (Recall that noise margin is measured against the boundary of the "not valid" voltage.)

2. By how many volts is the noise margin less when using (plus and minus) 5 V as compared to (plus and minus) 12 V?

3. Compare the time to reach full voltage of a signal with a slew rate of 10 V/ms versus 5 V/ms.

4. How long does a signal going from -8 to $+8$ V take, at a slew rate of 25 V/μs? From -16 to $+16$ V?

5. Sketch a -5- and $+5$-V binary signal with a slew rate of 10 V/μs, alongside one with a slew rate of 20 V/μs.

6. Sketch a signal with a slew rate of 50 V/μs for a -10- to $+10$-V bit, versus a -20- to $+20$-V bit. How long does the smaller signal take to reach its full value, compared to the larger one?

7. The slew rate of a bit should be less than 5 percent of the bit period for best performance and maximum acceptable rounding. Using this standard, what is the maximum bits/second rate that can be supported by a signal that must go from -5 to $+5$ V at a 25 V/μs slew rate?

8. Sketch a group of 4 bits, alternating from -10 to $+10$ V, with a slew rate of 10 V/ms, and a bit rate of 2000 bits/s.

9. Repeat Prob. 8 for a bit rate of 500 bits/s.

10. Show the bit pattern, in 1s and 0s, for the letter X in ASCII with seven character bits, one start bit, two stop bits, and no parity.

11. Show the 1, 0 pattern for the symbol &, with 7 character bits, 1 start and 1 stop bit, and parity bit.

12. Repeat the previous problems for the number 25, represented by a 6-bit pattern.

13. How long must a two-character-long break detect be at 9600 baud, for a 7-bit character, 1 stop bit, no parity? For the same character with 2 stop bits and parity?

14. Show the data and handshake line activity for a printer which has a single handshake line to the computer. Characters are arriving at 1200 baud, can be printed at 200 characters/s, and the printer has no buffer.

15. The same printer is now receiving characters at 2400 baud. Do the same sketches.

16. Sketch the four basic types of interconnection possible, for DTE and DCE devices, with plug and socket connectors.

17. Sketch the schematic of a null modem cable that accommodates a DCE-to-DTE device. The DCE has RTS and CTS handshaking, the DTE expects to see handshaking on its DSR line.
18. An 8251 USART is used for communications at 9600 baud, from a clock running at 614.4 kHz. Set up the IC for communicating with 2 stop bits, no parity, 6 bits per character.
19. Set up the command byte for the 8251 in the case of 7 bits per character, 1 stop bit, odd parity, and 4800 baud (from a 76.8-kHz clock).
20. The status byte from the 8251 should show that all the registers are empty and no errors exist. What does this byte look like?
21. The status byte should now show that a character has been properly received, but with a parity error. What is the status byte in this case?
22. The character, with parity error, is taken from the 8251 by the system processor. What does the status byte now look like?

9 Other Communications Interfaces

The weaknesses of the RS-232 standard in the areas of speed, noise resistance, distance, and other specifics means that many other communications interfaces are used. These other interfaces are designed to provide superior performance to RS-232 in one or more areas. Each one has a clearly defined position where it is the best choice for the application, but each also has some drawbacks. In data communications, "one size fits all" does not apply.

This chapter will study some of these interfaces and examine their technical and application features. It also shows ways these interfaces are some of the key building blocks for a network, which is a larger interconnection of many users.

9-1 Additional Interface Needs

The RS-232 interface studied in Chapter 8 provides acceptable performance in many applications. However, because of its various limitations other interfaces have been developed by the communications industry and users. These interface standards overcome one or more of the RS-232 limitations in the following areas:

Distance: The 50-ft distance of RS-232 often is not enough for many small industrial plants or large offices.

Data rate: The 20-kbaud rate (sometimes increased beyond the standard definition to 38.4 kbaud) does not allow large amounts of data to be quickly transferred.

Multiple users: There is a need for allowing many users to share the same wire, since the cost of wire and media is often the largest part of the overall system expense. The RS-232 specification is designed for two users in a configuration in which one user is connected directly to the other.

Overall performance and flexibility: Many systems need an interface standard which has better performance (distance, speed) but also greater flexibility.

This means that many interconnection arrangements are allowed, and the one best suited to the application's need can be selected.

Noise resistance: The single-ended wiring of RS-232 is susceptible to noise pickup, even at short distances, if the environment has electrical noise. This means that communications have many errors. Detecting these errors is possible, but as a result of the errors the overall communication is not efficient. An interface which minimizes the number of errors is a better choice.

Voltages: The +3- to +25- and −3- to −25-V range of RS-232 is not directly compatible with the voltage levels in many digital electronic circuits. Special power supplies must be used to provide these voltages, and they add cost, physical space, and unreliability to the electronics of the system.

No single interface standard exists which overcomes all these shortcomings at the same time. The system designer examines what the needs of the system are, what the various choices are, and what users are already accustomed to before choosing the "best" interface for the application. Some of these interface standards are combined with specific protocols and formats to form communications networks, which are examined in detail in later chapters. The *interface standard* is the lowest level of a complete system. It defines the currents, voltages, signal timing, and capabilities of the overall system at the foundation level.

This chapter examines interfaces for multiuser operation, current-driven interfaces (as opposed to the voltage-driven RS-232), and higher-performance interfaces. For each interface, there is a discussion of the technical details, along with the advantages or drawbacks of the interface. Some of these interfaces have industry standard numbers assigned to them, just as RS-232 has. Others are popular and exist only through informal agreement among many users.

QUESTIONS FOR SECTION 9-1

1. What are some of the limitations of RS-232 in terms of the following:
   a. Distance?
   b. Data rates?
   c. Number of users?
   d. Noise resistance?
   e. Voltage levels?
2. Why is there no single "best" alternative? What factors enter into the choice of the interface used?
3. Are all interface standards formal? Explain.

9-2 Multidrop Communications

The term "multidrop" is used to describe a communications system in which many users share the same medium, usually wire (Fig. 9-1). The multidrop

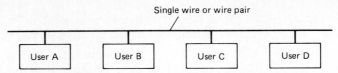

Fig. 9-1 In a multidrop system, many users share the same medium to reduce wiring cost and simplify cabling.

system is a sort of party line, on which any of the users may send or receive messages. Usually, a multidrop configuration has one user as the *senior*, or *master*, *station*. All other users take their cue from this master station.

Multidrop systems are needed in applications in which the cost and inconvenience of providing many paths for the messages, in parallel, are too high. One typical example is a building temperature and ventilation control system. This application may have some temperature-measuring and -control electronics for each group of rooms or area of the building. The central computer of the building needs to be informed about room conditions on a regular basis. The computer can run a set of wires to each electronics package, but in a large building this is expensive and the wire cable is very bulky. Another solution is to use a single set of wires and have all the individual room electronics connected to this. This is very economical, if it is technically feasible.

It is reasonable to ask why a standard such as RS-232 cannot be used for multidrop communications. The answer is that the RS-232 standard has each transmitter circuit putting out a voltage for a 0 or 1 at all times. If two users on the line try to communicate, the data line of the third is an interfering signal (Fig. 9-2*a*). There is a *contention* problem, in which the voltage output of the users who are not communicating fights against the voltage outputs of the users who are trying to communicate.

In the reverse direction the problem does not exist. It is electrically feasible to have many users who are only listening to a message from a single user who is the message source (Fig. 9-2*b*). This is called a *broadcast mode*, in which there is one talker and many listeners. Even here, however, RS-232 is not a good choice. There is a problem with the handshake lines, by which the listeners inform the sender of their status. There are contention problems on the status lines even if the transmitted data line is clear and not seeing contention.

The solution to the problem is to have an interface circuit that allows the transmitter voltage output to be turned off, so that it can be physically on the line but not electrically interfere with the two users who are trying to talk. This takes care of the electrical problem. There are several ways to achieve this goal. Among them are three-state outputs and current loop communications.

A *three-state output* (sometimes called *tristate output*) is a binary output that really has three values, as opposed to the normal two. There is a voltage level corresponding to binary 0, and one corresponding to binary 1. There is also a third state, called the *high-impedance mode*. In this state, the output circuitry puts no voltage on the line and looks instead like a very high impedance (Fig. 9-3). From the viewpoint of other users on the line, the high-impedance output is simply not present. It is effectively disconnected from the line.

A standard for multidrop operation must define what the 1 and 0 voltages are,

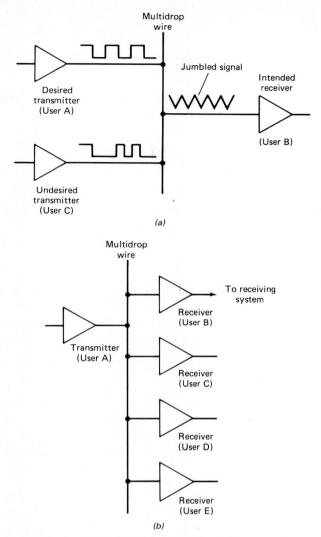

Fig. 9-2 A contention problem exists when two transmitters put signals on the multidrop wire at the same time, as shown in (*a*). There is no contention when multiple receivers listen at the same time, as in (*b*).

and also this third, high-impedance state. Several standards exist for this, such as Electronics Industry Association (EIA) RS-422 and RS-423. There are also some informal standards which use current, instead of voltage, to achieve the same goals. Both the EIA and current loop operations are discussed in subsequent sections of this chapter. But there is more to proper multidrop operation than an electrical interface that allows a user to be "out of the picture" electrically. There must also be a set of rules and protocol for defining when the user should turn on, putting a 1 or 0 voltage on the line, and when the user should turn off and go to the three-state mode.

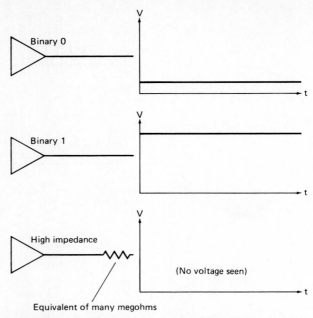

Fig. 9-3 A three-state output has the regular binary 0 and 1 output levels, plus a third mode in which the output looks like a very high impedance and is effectively out of the circuit.

These necessary protocols can be implemented in two ways. The simpler way is to use a command/response type of protocol. Here, no user speaks, unless the message has spoken to it. In implementation, it works as follows:

All users listen to all messages from the master unit. The master unit puts a header on each message with a *user number*, corresponding to the number of the intended receiver. All users first check the incoming message for user number. If this number agrees with the physical number of the user, the user listens to the rest of the message and responds. If the number disagrees, the user ignores the rest of the message. Users who are listening and ignoring keep their output circuitry in the high-impedance mode. The user who is being addressed and is responding turns on its output circuitry and puts the response onto the wire.

In the case of the building temperature-control system, the units may be numbered with a first digit corresponding to each floor, and a second digit for the zone of the floor. Unit 17 is floor 1, zone 7; the unit at zone 6 of the fifth floor is numbered 56. The protocol sequence would then be as follows:

1. All units are listening, with transmitters in high-impedance mode.
2. Central computer, as master station, issues a message asking for information from a specific unit, such as "17—temperature please?"
3. All units see the number 17. They all ignore the message and remain in the high-impedance mode, except unit 17.

4. Unit 17 responds with the temperature information, after turning on its circuitry for output. After the response, it goes back to high-impedance mode.

This is called *command/response* because no user ever speaks without first being commanded to give a response. It is a very effective protocol for some applications.

Unfortunately, it is not a good protocol for many applications in which unexpected events may occur. For example, if a dangerous condition occurs in one zone of the building, the central computer does not find out unless it asks. Meanwhile, the dangerous condition (temperature too high or low, water leak, and so on) can continue unnoticed by the computer that is supposed to be in charge of the entire building. The protocol that is preferred here is one that allows any user to talk at any time, but without contention problems and the errors they can cause.

The method used to implement this type of protocol is to use circuitry which allows each user to check the data lines to see whether anyone else is talking. If the line is clear, then the user gives the message. If it is busy with another user, this new user must wait. In some installations the user checks a handshake line instead of the actual data line to see whether the line is free, but the effect is the same. The protocol rule is, "You can speak if no one else is. If someone else is speaking, keep checking until the line is free; then you may go." This method offers improved performance over the simpler command/response. Unfortunately, there are many subtle problems that can occur, such as when a user has a very long message, or has an internal failure that causes it to keep using the line and sending useless data. These complications lead to very formal protocols which can handle all these situations, which form the basis of networks, covered in detail in Chapter 11.

Multidrop communications is really a variation on multiplexing. In multiplexing, the goal is to have many users, at both ends of a link, share the medium. This reduces the cost of equipment at both sides and the cost of the medium itself. In multidrop communications, some of the same sharing and savings occurs. The difference is that the users of a multidrop system are related to each other through the application. In a multiplexed system, the users are usually independent of each other, for example, the many users of a telephone system. The various callers know nothing about, and have no interest in, the other phone users. In a multidrop system, the various users share a common application or need as the glue that binds them together. The circuitry used to implement multidrop communications has many electronic similarities to multiplexing circuits, however. Multidrop systems can be very low in cost for the capabilities they provide.

Multidrop communications does have some drawbacks. The main weakness is that the line is shared, so the amount of data that each user can send is limited. Also, the efficiency of the system is low, since a lot of time is spent in the protocol itself (such as the commands to each user to report and respond). There are many situations in which this is not a limitation. In the building temperature and ventilation system, the changes generally occur at a slow rate compared to the rate of the digital signals. It is enough for the application if each floor and zone is checked every 10 s, since room temperatures don't change very quickly. On the other hand, a science laboratory that is gathering data from 100 points on some experiment, at high rates, does not use a multidrop system, since each data collection point is not able to report frequently enough, and data may be lost.

QUESTIONS FOR SECTION 9-2

1. What does the term "multidrop communications" mean?
2. Where is multidrop communications needed? Why?
3. What is the role of the master unit?
4. Give three instances in which multidrop communications may be needed.
5. Why can't RS-232 be used for multidrop communications?
6. What is the broadcast mode? Where is it technically practical? Where does it have technical limitations?
7. What are the two types of solution to the contention problem?
8. What is the effect of the high-impedance (three-state) mode?
9. What is command/response protocol? How does it work?
10. Where is command/response protocol not useful? What is its weakness?
11. What is an alternative scheme to command/response protocol? What are some of the complications?
12. Compare multiplexing and multidrop communications. How are they similar? How do they differ?
13. What are the weaknesses of multidrop communications?

9-3 Other Key EIA Standards

The RS-232 standard is common and effective, but the increasing needs of users have caused system designers to develop new circuitry for interfaces that can provide improved performance. The improvements are in the areas of speed, distance, and multiple users. Recall that the RS-232 standard uses single-ended signals (one wire is ground) to achieve distances of up to 50 ft, at rates of up to 20 kbaud. Resistance to noise is only fair, and the receiver must see a voltage of at least $+3$, or more negative than -3, for a valid binary 0 or 1. Only two users are allowed: a single transmitter and a single receiver.

The EIA standards which have been developed are RS-422, RS-423, and RS-485. Each can provide improved performance in one or more areas. Special line driver and receiver ICs available from many companies provide the very carefully detailed electrical characteristics for each of these interfaces, as well as the desired modes of fail-safe and power-off operation.

The RS-423 Standard

The *RS-423 standard* expands on RS-232 in allowing multiple receivers, longer distance, and faster data rates. Like RS-232, it is single-ended and requires only one signal wire and a ground for data. The ground is the same ground that the rest of the circuitry uses. The multiple receivers do not make RS-423 into a full multidrop system specification, since there cannot be multiple transmitters. In other words, it is a sort of broadcast system.

A typical application is that for a computer which updates display screens at an airport. The computer may have to drive many screens, located at various spots in the check-in and waiting areas. The data direction is entirely from the computer to the many receivers (TV screens). The receivers have no need to send data back to the computer. The schematic of an RS-423 line driver, connected to RS-423 line receivers, is shown in Fig. 9-4.

The electrical specifications that RS-423 defines allow for the following performance specifications:

Distances to 4000 ft

Data rates to 100 kbits/s

Up to 10 receivers

The voltage levels used in RS-423 are much lower than those used in RS-232. This partially accounts for the higher data rates that can be achieved. Recall that one of the limiting factors on data rate is the *slew rate*, the rate at which the voltage can make the transition from one value to the other. Because of the large signal voltages of RS-232 high slew rates are needed for high data rates. Yet, many systems cannot provide these slew rates, since the driver circuitry is too complex, and the capacitance of the medium itself limits the slew rate. Lower voltage levels need proportionally lower slew rates, for the same number of bits per second. The same slew rate, but with lower voltages, can achieve a much higher bit rate.

RS-423 uses voltage levels of $+3.6$ to $+6$ V for binary 0, and -3.6 to -6 V for binary 1. The same slew rate that the medium and drivers can support with RS-232 voltage levels provides about four times the data rate with RS-423 voltages. In addition, the binary 1 and 0 decision points at the receiver for RS-423 are set much lower. A $+200$-mV signal is interpreted by the receiver as a

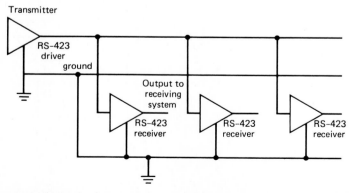

Fig. 9-4 When using RS-423 drivers and receivers, a single driver can provide the signal to many receivers, using one active wire and a common ground.

Other Communications Interfaces

binary 0, and a -200-mV signal as a binary 1. This means that more voltage loss in the wire can be tolerated by a communications system using RS-423.

A few calculations show the effect of transmitted signal voltage levels and decision points at the receiver on data rate versus slew rate capability. Suppose a driver is capable of developing a signal with a slew rate of 1000 V/s. To make an RS-232 transition of -25 to $+25$ V takes 50 V/(1000 V/s) = 0.05 s. The same signal can go from -6 to $+6$ V in 12/1000 = 0.012 s. Therefore, the same driver slew rate allows for much higher data rates because of the lower signal voltages.

At the receiving end of the link, similar calculations can be used. The line driver may be capable of very high slew rates, but the capacitance and inductance of the cable may act to limit the slew rate as seen at the receiver. For a fixed slew rate, it takes only 6.7 percent of the time for a signal to traverse the -200-mV to $+200$-mV decision region of the RS-423 as compared to the -3- to $+3$-V range of RS-232. (Why?) The medium becomes much less of a limiting factor.

If this seems to be a situation of higher slew rates, and thus higher data rates, at no cost, why is RS-232 used at all? The answer is partially historical—there are millions of RS-232 systems in use, with well established uses. Many older systems had the larger RS-232 voltages, not the lower RS-423 levels. Many newer interfaces offer both RS-232 and RS-423 signal levels, so either one can be selected, depending on the application and installed equipment.

The RS-422 Standard

The *RS-422 standard* is similar in many ways to the RS-423 standard, but with some major differences. Like RS-423, it can have one transmitter and up to 10 receivers and can be used for distances of up to 4000 ft. The voltages used are also the same in RS-423. The transmitted voltage has to be only $+2$ or -2 V.

The difference is that RS-422 uses differential signals rather than single-ended signals. A typical RS-423 configuration is shown in Fig. 9-5. (Note that although the data lines do not use a common ground, the transmitters and receivers are not isolated. Isolation and differential signals are different.) The benefit of using differential signals is that the signal quality on the line is much better controlled.

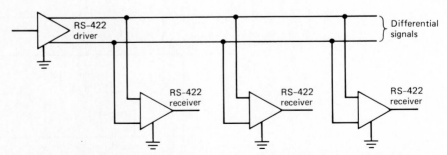

Fig. 9-5 RS-422 is similar to RS-423 but uses differential signals rather than sharing a common ground. This results in higher-quality signals and the capability for longer distances at higher rates.

Noise is much less of a problem, and the effects of ground problems on the signal shape are fewer, since the ground wire is not used to carry any of the signal energy.

The improvement in signal quality, combined with the lower voltage levels, means that the RS-422 signal can be used at very high data rates. This standard can have data rates of up to 10 Mbits/s.

A typical RS-422 application is a case in which a computer has to send large amounts of data to several terminals simultaneously. The use of differential signals (which require one extra wire) allows the 10 Mbits/s rate to be used. This is needed when time is a critical factor in the application, as in recording signals from a data collection system and informing many users of the results.

The RS-485 Standard

The *RS-485 standard* has the highest performance and flexibility of this group of four general-purpose EIA-sponsored standards. (There are highly specific standards that are designed for even higher performance in specific situations and applications.) It uses differential signals and allows the same distance and data rates as RS-422. The key enhancement is that it supports true multidrop operation. Up to 32 transmitters and 32 receivers can be connected to a single pair of wires, as long as these transmitters and receivers meet the RS-485 specification. The specification defines what signals should be used, and what a high-impedance mode of an RS-485 transmitter should look like electrically.

A typical RS-485-based configuration is shown in Fig. 9-6. The voltages used by this standard differ slightly from those of RS-422. A +1.5-V signal represents binary 0, and a −1.5-V level is used for binary 1. The receiver uses the same +200-mV and −200-mV levels to decide what signal has been received. The differential wires help ensure that the noise picked up by the wiring is small enough in value that this small decision level is satisfactory for most applications. Ten megabits per second is the maximum disk rate.

Each of the standards in the RS series provides good performance at reason-

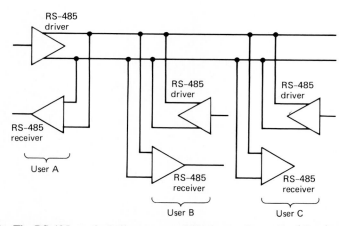

Fig. 9-6 The RS-485 standard allows true multidrop operation, with differential circuitry and three-state outputs.

Feature	RS-232	RS-423	RS-422	RS-485
Mode	Single-ended	Single-ended	Differential	Differential
No. of drivers and receivers	1 Driver 1 Receiver	1 Driver 10 Receivers	1 Driver 10 Receivers	32 Drivers 32 Receivers
Max. distance (ft)	50	4000	4000	4000
Max. data rate (bits/s)	20K	100K	10M	10M
Signal levels (V)	3 to 25 -3 to -25	3.6 to 6 -3.6 to -6	2 to 6 -2 to -6	1.5 to 6 -1.5 to -6
Receiver decision point (V)	$+3$ -3	0.2 -0.2	0.2 -0.2	0.2 -0.2

Fig. 9-7 The key characteristics of RS-232, RS-423, RS-422, and RS-485 summarized and compared.

able cost, in the right application. Figure 9-7 summarizes and compares RS-232, RS-423, RS-422, and RS-485. Note that none of the standards allows the maximum distance and bit rate to be used at the same time. There is a graph published with each standard that shows the maximum bit rate versus distance. This graph has the shape shown in Fig. 9-8 and indicates the way in which the bit rate drops off with distance.

For applications in which a voltage representing binary 1 or 0 is technically a poor fit, the data communications system designer has an alternative, called a *current loop*. The current loop provides some advantages over voltages but also has some limitations.

An RS-422 and RS-485 IC

Many integrated circuits are available from different manufacturers to provide the required voltage levels and characteristics for interface standards. One such

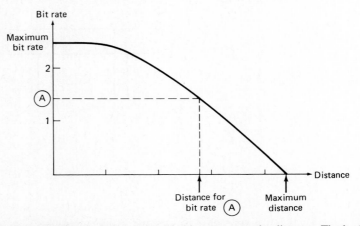

Fig. 9-8 A typical graph of the achievable bit rate versus the distance. The horizontal and vertical axis numbers depend on the particular standard used.

IC is the SN75176 from Texas Instruments. This provides one differential receiver and one differential three-state driver in an 8-pin package. The functional schematic is shown in Fig. 9-9a. The IC is primarily intended for RS-422, but can be used with RS-485.

To understand the operation of the SN75176, look at the operation for transmitting. The data bits from the communications IC are connected to pin 4, called D (for driver). The differential signal wires are connected to pins 6 and 7, called simply A and B. The data is converted to the standard specification voltages by the driver function within the IC. When the data bit to be sent is 1, output A is at a high voltage and output B is at a low voltage (A is greater than B). When the data bit is a 0, then the relative voltage between A and B is reversed, with B greater than A. This provides the differential output required.

The driver can also be put into high-impedance three-state mode, to meet the requirements of multidrop operation such as RS-485. The *driver enable* (DE) line is used for this purpose. When DE is a binary 0, the outputs A and B go to the three-state mode, regardless of the state of data bit at D. The operation of the driver is summarized by the function table (Fig. 9-9b).

The operation as a line receiver is shown in the function table in Fig. 9-9c. The incoming data arrives differentially on A and B. When the difference voltage $A-B$ is greater than 0.2 V, the receiver output R to the communications circuitry is a binary 1. Conversely, when the difference $A-B$ is greater than -0.2 V (meaning that B is greater than A), then the output R is binary 0. If the difference between A and B is less than 0.2 V, then the incoming differential signal falls into the undefined zone of the specification, and the output at R can be either a 0 or a 1.

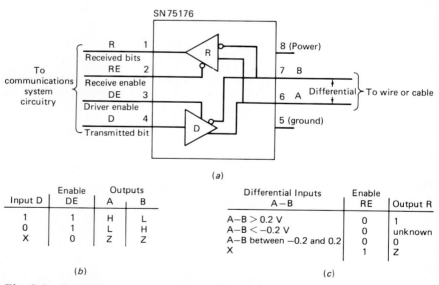

Fig. 9-9 The SN75176 IC provides an RS-422 line driver and line receiver in a single 8-pin package: (*a*) the function block diagram and pinouts of the IC; (*b*) and (*c*) the driver and receiver function tables, respectively, for operation of the IC. (Note: X means irrelevant; Z means high impedance mode.)

Like the driver, the receiver has the ability to be put into the three-state high-impedance mode. This is useful in some circuitry configurations. Pin 2 (called *receive enable* [RE]) of the IC controls this mode. When RE is at binary 1 then the receiver output R is high-impedance, regardless of the inputs at A and B. Only when RE is at 0 does the line receiver present the correct 0 or 1 output, depending on the state of A and B.

QUESTIONS FOR SECTION 9-3

1. Compare RS-423 to RS-232 in terms of (a) bit rate, (b) distance, (c) single-ended versus differential signals, (d) number of users, (e) voltages.
2. What are three potential applications for RS-423?
3. How does RS-423 support higher data rates than RS-232? How do the larger voltage values of RS-232 limit the data rate, for a given slew rate?
4. What voltages are used for RS-423?
5. What are the decision points for binary 1 and 0, at the receiver?
6. Compare RS-423 and RS-422 in terms of (a) bit rate, (b) distance, (c) single-ended versus differential signals, (d) number of users, (e) voltages.
7. What voltage levels are used for RS-422?
8. What are some typical RS-422 applications?
9. Compare RS-485 to RS-232 in terms of the items of Question 6.
10. How many users does RS-485 support? What types of users?
11. Explain the operation of the SN75176 line driver/receiver between the communications system circuitry and the actual medium. How does it convert the differential signals to signals compatible with the communications system?
12. How does the driver in the SN75176 put a binary 1 data bit on the differential output? What is the voltage at A with respect to B for this?
13. What is the relationship between differential inputs A and B when the SN75176 indicates that a 0 was received? How is the receiver output put into three-state mode?

PROBLEMS FOR SECTION 9-3

1. A driver circuit and medium can support a slew rate of 100 V/ms. If the signal levels are $+25$ and -25 V, how long does it take for the binary 1 to binary 0 transition?
2. Repeat Prob. 1 for RS-423 levels of $+6$ and -6 V; RS-422 levels of $+2$ and -2 V; RS-485 levels of $+1.5$ and -1.5 V.
3. What is the noise margin, in volts, for RS-232, RS-423, RS-422, and RS-485?
4. A system is using RS-422. A noise voltage of $+1.5$ V appears on a signal representing binary 1. Show why this will or will not be misunderstood by the receiver.
5. Repeat Prob. 4 for an RS-485 system sending a binary 1.

6. Prove the statement in the section that the transition time between RS-423 levels is only 6.7 percent the time for RS-232 levels, for a decision fixed slew rate value.

9-4 The Current Loop

Most interface standards involve the use of various voltage levels to represent the binary 1 and 0 signals. Voltage signals are easy to generate and easy to observe with a voltmeter or oscilloscope, and they can give the required performance in many applications. There is an alternative to using voltage for binary data: Current can instead be used effectively in some situations.

The electronics industry has an informal standard using the presence or absence of current to represent a binary 1 or 0. The most common value is 20 mA, so the interface is called a *20-mA loop*, or simply a *20-mA interface*. The connection for a single-user pair is shown in Fig. 9-10. Note that there are two loops, one for data in each direction. Somewhere in the loop there must be a power supply that provides not voltage, but a fixed amount of current. This is called a *current source*. The transmitter of the message simply opens and closes an electronic switch which allows current to flow or prevents it from flowing. The receiver circuit must detect the flow of this current, to find out whether a 1 or 0 is being sent. The presence of current is a binary 1; the absence is a binary 0.

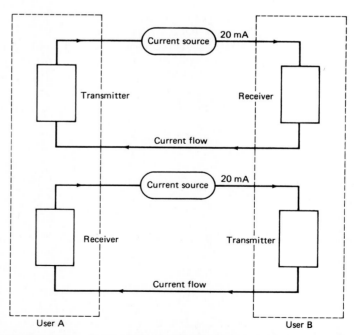

Fig. 9-10 The basic current loop for sending data in both directions. The presence or absence of 20 mA represents a 1 or 0.

Other Communications Interfaces

What are the advantages of using current, as compared to voltage signals? The primary advantage of using current is that signal lines which carry current are much less sensitive to electrical noise. The electrical noise has to generate substantial current in the line to affect the desired signal; this means that the noise must have considerable electrical energy—which it usually does not. In contrast, it does not take much energy to generate a large voltage signal and impose it on the wire carrying the voltage which represents the data.

Current signals can travel for several thousand feet in regular wires, at rates to about 20 kbaud. The electrical characteristics of the wire (resistance, capacitance, inductance) are less of a factor for a current loop than for voltage signals, since a properly designed current source forces the 20 mA down the loop regardless of these characteristics. A distance-versus-baud figure similar to Fig. 9-8 is used to show the current loop capabilities.

There is another major advantage of using current. The electronic circuitry commonly used for current sources employs an *optoisolator*. This is a small, inexpensive, rugged device that interfaces the system circuitry with the signal wires and at the same time provides isolation, which is important in many applications. Any harmful effects of common mode voltage (CMV) are eliminated by this device, and any damage to electronic equipment is prevented. An inexpensive optoisolator can withstand typically 1000 V or more of CMV and allow communications despite it.

Implementation of a Current Loop

A typical current loop uses the circuitry shown in Fig. 9-11. One optoisolator acts as the transmitter. The *light-emitting diode* (LED) is driven by the circuitry of the system which is sending data. When the LED is on, the phototransistor of the optoisolator conducts current, so the 20 mA flows in the loop. This 20 mA is received at the other end, where it provides the current to turn on the LED of the receiver optoisolator. The current in turn causes the phototransistor to turn on, letting current flow in the receiver circuitry itself. Note that there is no direct electrical path from receiver to transmitter. The optoisolators break this path and use photons instead, thus providing electrical isolation.

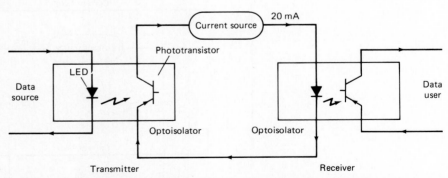

9-11 Actual circuitry of a typical current loop, for one direction of data. Optoisolators are used to isolate the transmitting system and the receiving system from the loop and any loop problems, as well as differences in ground voltages.

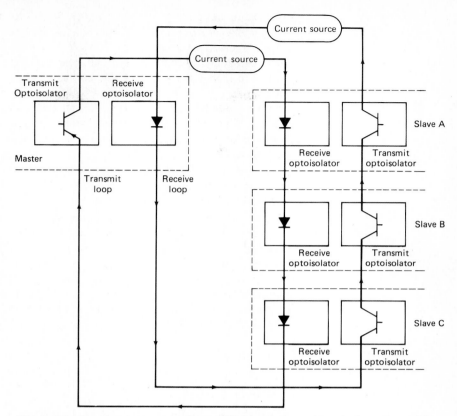

Fig. 9-12 A current loop system for multidrop operation, with a single master and three slave units. Note that the details of each optoisolator are not shown—they are the same as in Fig. 9-10.

Multidrop Operation on a Current Loop

Current loop operation lends itself to multidrop operation very easily. The schematic of a single-master/three-slave user configuration is shown in Fig. 9-12. The current that is controlled by the master when transmitting flows through the LEDs of all the slaves, which are part of the overall loop. All the slaves can therefore listen to any message from the master.

In the slave-to-master direction, a slave that needs to send data to the master must control the flow of current in the slave-to-master current loop. All the current switches of the other slaves, which are implemented by the phototransistors of these users, are in the on position so that they allow the 20-mA current to flow. The master unit LED sees the current and so sees the digital data.

Current loop operation makes multidrop operation fairly straightforward, but there are several areas of concern. First, if any unit should fail or lose power, and its optoisolator becomes stuck in the off (or open) position, then the flow of current through the loop is broken and all units are unable to communicate. Therefore, it is not a good choice in very critical applications, unless the circuitry is specifically designed so that any problem with the units leaves the optoisolator in the on (closed) position. The optoisolators themselves are very rugged and reliable, and rarely fail.

Second, adding or removing any unit from the loop requires that the circuitry of the loop momentarily be opened, and a jumper must be put in place of a removed unit (since the loop must always be a continuous circuit). This means that while the unit is removed or replaced, communications temporarily cease. Simply pulling out a unit breaks the loop, and installing a new unit involves cutting the loop to make an opening.

Another issue with current loops is compliance voltage. By the laws of physics and the principles of electricity, current does not exist independently of voltage. Where there is one, there must also be some amount of the other. They are, of course, tied together by Ohm's law: voltage = current × resistance. When the 20-mA current source of the loop is forcing current through the loop, it encounters various sorts of resistance from the loop wire, and voltage drops across the LEDs of the optoisolators. Therefore, the current source must be able to supply the current, but at this amount of voltage, called the *compliance voltage*. It is the parallel situation to a voltage power supply which must supply not only a specified amount of voltage (such as +5, +15, or +24 V) but also the amount of current that the system it is supplying draws from the supply. The current source must provide both the current and enough corresponding voltage at the same time.

A typical situation for a current loop in multidrop operation is a loop resistance of several ohms, and a voltage drop of 1½ V across each LED. Thus, supplying current to a three-user system requires (20 mA × 2 Ω) or 0.04 V, plus 3 × 1½ V = 4½ V. The minimum compliance voltage required is therefore equal to 4.54 V. A current source capable of supplying at least 20 mA at this voltage is sufficient, just as a voltage power supply can be used in a system if it can provide the required voltage with at least the required amount of current. If the supply of the current loop does not have enough compliance, then it is unable to supply the 20 mA through the loop, because its overall capabilities (voltage and current) are insufficient.

Some current loop applications do not have optoisolators. Instead, the receiver has a small resistor in place of the LED (Fig. 9-13). As the 20 mA of current is forced through the resistor, a voltage is developed across the resistor. The receiving circuit senses this voltage and sees that a binary 1 bit has been sent. In a

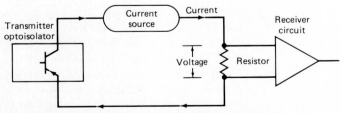

Fig. 9-13 In some applications, a resistor is used instead of an optoisolator. The receiver input sees the voltage which is developed across the resistor when current flows.

typical case, the receiver wants to see 5 V for a binary 1, so the resistor has a value of 250 Ω. The 20-mA current flowing through the resistor develops this 5 V. This also means that the compliance required for each receiver is 5 V, and multidrop installations require 4 V per receiver. A small system with four or five receivers needs 20 to 25 V of compliance; this amount is often inconvenient or expensive to build into the supply. The resistor is much cheaper than the optoisolator but has a higher voltage drop and thus requires greater compliance. It also provides no isolation. In some specific cases, though, in which providing higher compliance is not a problem, isolation is not needed, or in which cost must be absolutely minimum, the resistor is used.

Handshaking on Current Loops

One of the advantages of current loops is that they can be run long distances without significant noise pickup. Most applications which have long distance needs do not want to run many wires, since it is costly and inconvenient to do so. As a consequence of this, most current loop installations use only four wires: two for the transmit loop and two for the receive loop. There is no handshaking. The data rate must be slow enough to ensure that the receiver can process each character as received, or the receiver must have a buffer. The most common format for data on current loops is asynchronous serial communications, with start and stop bits and optional parity. The data character itself can be ASCII or binary. In effect, the RS-232 voltages are replaced by the presence or absence of 20-mA current.

Current Loop Installation Issues

The current loop method of communications does not have an industry standard specification equivalent to that of the RS-232 or similar standards. As a result, although the current loop is common for some types of applications, there can be many problems getting the installation functioning due to "mismatches" at both sides of the loop.

First, there is no standard or most common connector for the current loop. Sometimes the D connector of RS-232 is used. Virtually any connector can be found in a current loop installation, and there is a high probability that two devices using current loop use different connectors, necessitating a special adapter cable.

The most prevalent source of current loop problems at installation is related to the actual location of the current source. The voltage specification for RS-232 requires that the source of the voltage for the digital data be the transmitter circuitry, so there is no ambiguity. For a current loop situation, the current source can be anywhere in the loop. Some devices put it in the transmitting device, some in the receiving device, and some expect it to be separate from either device (Fig. 9-14). This causes all sorts of confusion, since there can be one and only one loop current supply in a loop. If there is none, there is no current flow; if there is more than one, the two current supplies fight each other.

A transmitting or receiving port which is capable of supplying current is called active; one which has no built-in current source is called passive. Anyone installing a current loop system must make sure that only one end is active. If neither end is active, then a separate current source must be placed in the loop.

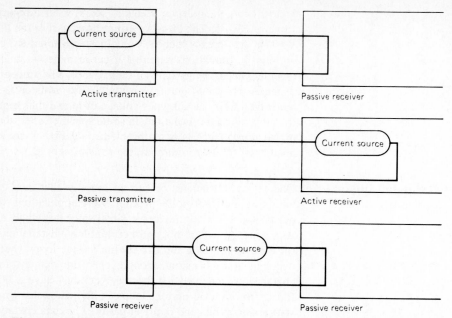

Fig. 9-14 The current source location can be a source of confusion and operational problems. There should be only one current source in the loop, and it can be located at the transmitter, in the loop, or at the receiver.

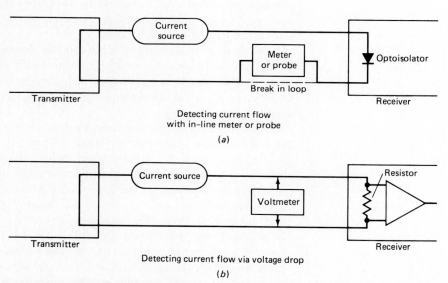

Fig. 9-15 Troubleshooting current loops requires sensing the flow of current. (*a*) The loop may have to be broken to insert a current meter, or (*b*) a voltmeter may be used when the loop develops a voltage across a resistor.

Many electronic units that support the current loop operation can be either active or passive, selected by a switch on the unit. This has to be checked by the technician or engineer performing the initial installation. Also the compliance of the source, whether it is in the transmitter, receiver, or external, must be checked to make sure it can handle the number of units in a multidrop installation.

Finally, troubleshooting a current loop installation requires a different technique than that for a voltage-driven system. A voltmeter cannot be used directly, since current flow, not voltage, must be monitored. The current can be checked with a current probe or a multimeter which is inserted into the loop to show current (Fig. 9-15a). If the current loop has an accessible component across which a voltage drop is developed (such as a resistor input on a receiver), then the voltage mode of the meter can be used (Fig. 9-15b). If the current loop uses optoisolators, then special care must be taken so that the voltmeter (or any other test instruments used) does not violate the isolation. This can cause malfunctions and equipment damage or even endanger the test operator.

QUESTIONS FOR SECTION 9-4

1. What is a 20-mA current loop?
2. What is a current source? Where is it located in the loop?
3. What are the advantages and disadvantages of using current instead of voltage to represent data signals?
4. Why are current signals more resistant to having noise imposed on them than voltage signals?
5. What is an optoisolator? How does it provide isolation?
6. How does a system use an optoisolator to control the flow of current?
7. How does an optoisolator sense the flow of current?
8. How does a current loop system implement multidrop operation?
9. What is a concern about the reliability of multidrop current loop?
10. What is the problem with replacing or repairing a system which is interconnected via current loop?
11. What is compliance voltage? Why is it needed if current, rather than voltage, is being used?
12. Compare a current source and its compliance to a voltage power supply and its current capability.
13. How is the compliance need determined?
14. Where and when is a resistor used for current loop?
15. What are the pros and cons of using a resistor instead of an optoisolator?
16. What is the number of wires in a typical current loop interface?
17. How is the lack of handshaking overcome?
18. What are the areas of difficulty in installing a current loop interface, besides the different connectors?
19. Where is the current source located in a current loop installation? What is an active user? A passive user?

20. What are two of the troubleshooting differences between voltage-based interfaces and current loop interfaces?

PROBLEMS FOR SECTION 9-4

1. Sketch a current loop system with one master and two other users using optoisolators. The system is a broadcast system.
2. Repeat Prob. 1 with resistors at the receivers instead of optoisolators.
3. The system of Prob. 1 may have a compliance problem. The wire has a resistance of 20 Ω. There is a 1.5-V drop across each optoisolator input. What compliance voltage is needed?
4. What compliance voltage is needed for the system of Prob. 2, with 250-Ω resistors?
5. A 20-mA multidrop system uses a combination of optoisolator inputs and 500-Ω resistor inputs. The loop has a resistance of 10 Ω. Each optoisolator has a 2-V drop across it. The system has three optoisolator inputs and 3 resistor inputs. What compliance voltage must the 20-mA supply have?
6. A 20-mA power supply has a compliance of 5 V. Can it be effective in a system with only one source and one receiver, if the receiver is an optoisolator and the loop itself has a resistance of 10 Ω? If the receiver is a 250-Ω resistor? Each optoisolator has 1.5-V drop.

SUMMARY

The many types of applications for data communications require different levels of performance at different costs. The single-ended RS-232 specification is effective for low data rates over limited distances. Additional standards, such as RS-422, RS-423, and RS-485 support faster rates, longer distances, and even multidrop operation. These improvements occur as the single-ended line is upgraded to differential, lower voltage levels are used, and three-state driver circuitry is employed at the transmitter.

Multidrop systems allow many users to share the same set of wires. This sharing reduces wiring cost and space but decreases the amount of time during which any single user has access to the data link. Many applications, however, do not need to use the channel 100 percent of the time. A broadcast system has one transmitter and many receivers. A full multidrop system has many transmitters as well as receivers. Multidrop operation is similar to multiplexing, although it is done on a different scale.

For some types of environments, it is difficult to send voltage signals without signal loss or error over the required distance. The electrical chassis of the two systems may be at different ground potentials, and signal isolation is required. An alternative to voltage signals is the current loop, which uses a fixed amount of current to represent a binary 1 or 0. The current loop allows simple multidrop operation, simple isolation (if desired), and good noise immunity. It does require understanding of the many possible configurations for the location of the current source and a different installation troubleshooting approach than voltage signals.

The use of voltage values or current values to represent binary 1s and 0s is effective for many installations in data communications. It does not allow data to be sent over telephone lines, or over very long distances, because of signal degradation and some electronic constraints in the phone system (or any very long set of wires with intermediate amplifiers). The solution is to abandon steady current and voltage levels and to use amplitude, frequency, or phase modulation of various waveforms instead. This is discussed in detail in the next chapter.

END-OF-CHAPTER QUESTIONS

1. What are the shortcomings, in some applications, of RS-232?
2. Which EIA standard overcomes the speed and distance limitations of RS-232, but not the number of users? How are slew rate limitations overcome?
3. Which standard allows the greatest distance, data rate, and number of users? With what specific values?
4. What is multidrop operation? Where is it needed?
5. What is the difference between a broadcast type of multidrop operation, and a half-duplex or full-duplex type?
6. What are some of the technical difficulties that occur in multidrop broadcast systems?
7. What is contention? How can it be overcome?
8. How can a binary circuit have three output states? What is the third state? Why is it needed?
9. What are some multidrop protocols? How do they work? What are their limitations?
10. Is multidrop the same as multiplex? Explain the similarities and differences.
11. What is a performance limitation inherent in multidrop operation? What benefit offsets this tradeoff?
12. What are the key changes in bit rate, distance, single-ended versus differential signals, voltage signal levels, and decision points for RS-232, RS-423, RS-422, and RS-485?
13. Give an example of a typical application for each of the four EIA standards discussed.
14. How does a current loop of interface differ fundamentally from a voltage interface?
15. Explain a key advantage and disadvantage of a current loop interface, compared to a voltage interface.
16. Where is the current source located? Why does this issue never arise in a voltage interface specification?
17. What is an active user? A passive user? Are all transmitters active? Explain.
18. Why is isolation sometimes needed? How does a current loop easily provide this?
19. If isolation is unnecessary, how can the optoisolator be eliminated at the receiver?

20. How does a current loop installation provide multidrop operation?
21. What is compliance voltage? Why is it a concern in a current loop system?
22. How are compliance voltage requirements determined?
23. What are the installation problems common in current loop systems? What different troubleshooting techniques are needed?
24. How is handshaking avoided in current loop systems? Why is it done this way?

END-OF-CHAPTER PROBLEMS

1. Sketch a multidrop system (broadcast) with four users using RS-423.
2. Sketch a half-duplex multidrop system, with four users, using RS-485.
3. Sketch a signal with a slew rate of 100 V/s, going from -25 to $+25$ V. How long does it take for the transition?
4. Alongside the sketch of Prob. 3, sketch a signal with the same slew rate, going from -6 to $+6$ V. How long for this transition?
5. Draw a binary 1 and 0 signal using minimum RS-422 signal levels and RS-485 signal levels. Also show where the decision points that the receiver uses are located.
6. A current loop system is designed to broadcast to 10 receivers. Each receiver has an optoisolator input with a 1.5 V drop. What compliance voltage is needed?
7. To save cost, the optoisolators are replaced in a new design with 250-Ω resistors. What compliance is now needed?
8. The compliance voltage needed in Prob. 7 is too high for the current source, so 100-Ω resistors are used. What compliance voltage is now needed? What is the voltage that the receiving circuitry will see when 20 mA is flowing?
9. A receiver circuit in a current loop system has circuitry such that it must see at least 10 V for a binary 1. The voltage drop across the input resistor must therefore provide this 10 V when 20 mA is present. What is the minimum resistor value?
10. Several receivers of Prob. 9 are linked in a multidrop system. What compliance voltage is needed when the system has five such receivers? What value is needed if the receivers need only to see 2 V each, instead of 10 V? What value of input resistor can provide this?

10 Telephone Systems and Modems

The telephone system is an important part of many data communications systems. Although it is in many ways less than ideal for data communications, telephone connections are everywhere. The telephone system is often used for the same reason Mount Everest is climbed: "because it's there."

Understanding how the phone system is used for data communications requires knowledge of the technical aspects of the phone system. The areas of most interest are the nature of signals on the phone lines, functions of the phone company local office, characteristics of phone wires, and dialing. The connection between a data communications device and the phone system requires a device called a modem, *which puts data signals into an electrical form compatible with the analog phone system. Modems are available in several forms, both physically and electrically. Although they have performance limitations based on the limitations and operation of the phone system itself, properly chosen modems can provide very good performance for the user application. Modern integrated circuits (ICs) allow modems to offer greater performance in smaller space, with lower power use and cost, than earlier-generation modems.*

10-1 Basic Telephone Service

The telephone and its system are things that most people take for granted. The telephone system covers the United States, Canada, most of the Western Hemisphere, Europe, and other parts of the globe. There are only a few places where there is not some level of phone service.

As a system, the telephone system is extremely complicated. There are many technical subtleties and advanced types of technology that make this system work smoothly, efficiently, and reliably on a countrywide and worldwide basis. Fortunately, making use of the telephone system for data communications does not require understanding every detail of the way the system works, any more than using a television requires a detailed knowledge of the way the TV set operates. However, there are some interface issues that affect every user, technician, or engineer who is attempting to use the telephone system for the communication of data.

The use of the telephone system and lines for communication of data is often referred to as *telecommunications*. Why use the telephone system at all for data, since it was really designed primarily for voice? There are several reasons. The major reason is that the system is available nearly everywhere. The connections that equipment should make to the system are standardized throughout the country, and therefore the same interface connectors, signals, and formats can be used easily in any location. The system is relatively low in cost, compared to the value it provides and the cost of trying to build a duplicate system. It may not be perfect, but it is acceptable enough, with low effort, for many types of applications, especially when there is some reasonable distance to cover, such as between adjacent buildings or neighboring towns, or across the country or world.

This does not mean that the telephone system is without drawbacks. The performance that can be achieved with regular phone lines is often less than the application really demands. Since the phone system was designed for analog voice signals, it is internally optimized for nondigital signals. Some performance compromises must be made to use it for data. The phone system is an open system, and so special steps must be taken to prevent unauthorized users from "tapping in" to private data. It is certainly possible to prevent them, but doing so costs more money. Finally, since the phone system is shared with other users, it may be difficult to get through to the desired party at the other end if the system begins to overload or someone else is connected to the desired receiver of the data.

The drawbacks of the analog phone system are being overcome as new, digitally oriented links are installed by the various phone companies. Even without these digital links, the telephone system is a powerful and popular vehicle for the transmission of data.

Since a computer or other source of data may need to connect to the telephone system through any telephone line, it is necessary to understand what the telephone line looks like. The voltages, signal timing, and sequence of handshaking used by the telephone lines and systems are very different from those used by computers and other inherently digital devices. Part of the reason for this is historical: the choices made for the telephone system were convenient at the time the system was being formed, beginning about 100 years ago. However, many of the reasons are technical in nature. A low-cost, long-distance system which can handle thousands and millions of users reliably technically requires certain types of signals and equipment, which are very different from the equipment and circuitry which would be chosen for a "close together" group of data communications systems.

The basic telephone is a *dial phone*. The telephone companies support this type of phone (and there are millions of them) with what is referred to as *Plain Old Telephone Service* (POTS). This dial phone does not itself allow features such as multiple lines, automatic redial, or many of the features that new phones have. Yet the electrical interface for the POTS phone is identical to the interface for the more advanced phones.

The Basic Telephone The simple dial or pushbutton telephone unit seems very far away from modern communications. In many ways this is so. At the same time, even the latest systems must interface and be compatible with existing phone lines and offices designed to serve the simple phone. The central phone company office does not know whether it is connected to a telephone or to a computer used for data communications. Therefore, it is helpful to look briefly at the circuitry of a telephone and the signals that appear on the telephone lines.

The simplified schematic of the standard telephone is shown in Fig. 10-1a. When the phone is not in use, called "on-hook," it looks like an open circuit. When the phone receiver is lifted from the cradle switch, the circuit to the local office is completed. A current of about 20 mA flows through this loop, from a battery at the phone office. The presence of this flow tells the office that the phone has gone off the hook.

Dialing involves rapidly opening and closing this loop. The phone office senses and counts the open and close pulses to see what number is being dialed. Time delay circuitry at the phone office makes sure that the opening of the circuit is not misinterpreted as the phone being hung up—a genuine hang-up "gone back on-hook" condition exists only when the phone is down for about one half second. Modern tone dialing phones do not open and close the loop, but instead put tones on the wire pair to the phone office, which has special circuitry to detect the tones. Figure 10-1b shows some of the key voltages and current flows in the phone lines during the telephone operation, including off-hook, on-hook, talking, listening, and ringing.

A special internal transformer is used inside the phone to prevent the users' voices from coming back too loudly to the ear. At the same time, this transformer passes a small amount of voice signal from the microphone element to the speaker so that users can hear themselves talking. Otherwise, it is like trying to talk without the regular audible feedback that guides you as to how your voice sounds, which is very difficult.

The ringing circuit in the phone is electrically ahead of the cradle switch. The phone office ringing signal does not reach and affect the rest of the phone, yet can ring the phone even when it is on-hook. The ringing voltage goes to the coil of a bell ringer. The bell is in series with a capacitor which blocks the normal direct current of the phone line and allows only the alternating current ringing signal to pass through. The ringing voltage is usually a 20-Hz, 90-V rms signal, on for 2 s and off for 3 s. Many modern electronic phones use electronic ringers which do not need this much voltage to ring—but they must be able to withstand it since that is what is supplied. In contrast to the large ringing voltage on the phone lines, the signal levels during talking are only a few volts at about 3 mA.

Any device that is connected to the phone line, either in place of a phone or alongside it, must meet strict standards to ensure the safety and reliability of the overall phone system. These rules determine the amount of power the phone can

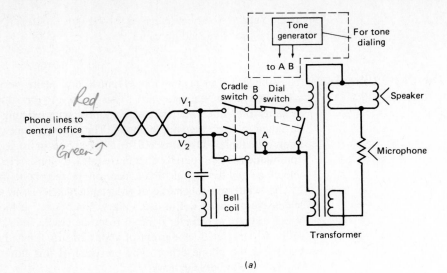

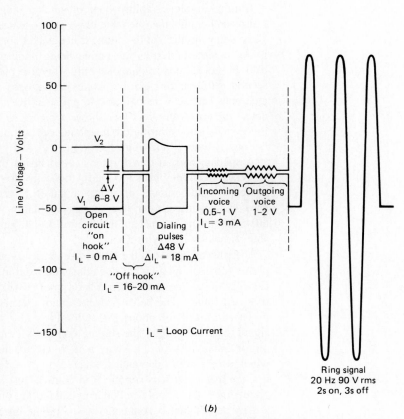

Fig. 10-1 (*a*) The schematic of a standard dial telephone shows the essential elements. (*b*) The signal voltages and currents on the phone lines during various parts of the telephone operation.

draw from the line (a phone is usually powered by the phone company office, not by built-in batteries or an external ac power supply), the kinds of signals the phone will respond to, and the phone's electrical appearance when it is in use (off-hook) or not in use (on-hook). The dialing signals are also carefully defined.

Telephone Office Functions

The pair of wires coming from the phone company local central office to the telephone user are designed to provide the functions needed to support POTS. These functions have their own set of acronyms. The interface circuit to the user's phone is called a *Subscriber Loop Interface Circuit* (SLIC). There is one SLIC required for every phone in the system. The SLIC provides six basic functions, called the *BORSHT functions*. This stands for *battery*, *overvoltage protection*, *ring trip*, *supervision*, *hybrid*, and *test*. In more detail, these are as follows:

Battery: The phone requires some voltage to operate, only when it is off-hook. This voltage is most commonly -48 V dc in the United States.

Overvoltage Protection: The phone lines are subject to many types of electrical disturbance, from misconnection during installation to voltages induced by lightning. The SLIC must be protected against these, so that phone lines and SLIC are not damaged under any foreseeable and most unforeseeable conditions.

Ring trip: The phone company office sends a special voltage to ring the phone. When the phone is picked up, the SLIC must detect this and disconnect the special ringing voltage. The normal talking circuitry must be switched in, to replace the ringing circuitry.

Supervision: Supervision refers to the SLIC monitoring the phone line to detect when it is picked up by a caller who wishes to use the phone. When a phone is picked up, the local office must send out a dial tone, then wait for the user to dial. The SLIC must "keep an ear" on the line to determine when the call is finished and the phone is returned to the on-hook condition.

Hybrid: The signals between the phone and the phone office are carried by a single pair of wires. On a simple loop like this, this single pair forms the complete circuit needed for full-duplex voice conversation. However, for the duplex signals to be passed into the rest of the communications system or even to another user connected to the same office directly, a separate channel must be established in each direction. This requires four wires, two for each direction (Fig. 10-2). A special circuit called a *hybrid* accomplishes this function. It is able to separate the signal coming from this phone and the signal from the other phone into two separate groups. A hybrid is also known as a two-wire-to-four-wire circuit, since it takes the two-wire loop and converts it to four wires and performs the reverse function of combining the four wires (two loops) of the phone office into a single pair of wires to the phone.

Test: The phone company may want to put special test tones and signals on the line, to check the quality of the wires and their overall performance. The SLIC must allow these test signals to go out to the phone and pass the resulting signal back to the office.

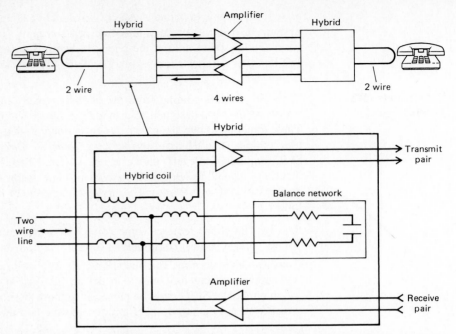

Fig. 10-2 The hybrid circuit at the telephone office performs the required two-wire-to-four-wire conversion on the duplex signals that exist in the phone at the same time.

Of all the SLIC functions, the hybrid function is one that most distinguishes the needs of a voice system from those of a digital data system. Certainly, regardless of what is on the line, there is a need for power, overvoltage protection, and so on. But the two-wire/four-wire case is very different. Two voices (one incoming, one outgoing) can be combined for short distances on the same wire loop, and both parties can have a normal conversation—in fact, the same happens when two people talk to each other face to face. The voices mix in the air, and the person speaking hears his or her own voice with the voice of the other person speaking, if both talk at the same time. In contrast, if a single pair of wires is used for full-duplex communications by having each data source put a signal on the line, the result is a useless mix of signals. The hybrid function is absolutely essential to proper operation of a digital data system using regular phone lines. One of the limiting factors in the performance of standard phone lines for data communications is the ability of the hybrid circuit to perform this two-wire-to-four-wire conversion for digital signals. Some technical limitations cause the hybrid to make a less than perfect separation, and so data errors occur. The problem is more severe at higher baud values. Hybrid imperfections are not so critical in voice-only applications.

The SLIC provides the basis interface functions needed to electrically connect the telephone or data communications device and its two wires to the telephone equipment at the local office. The device connected to the line—either phone or computer—must still have a way to signal to the phone company office in the desired identification number of the intended receiver. This is the *phone number*, and there are several ways to dial it. The next section discusses this in more detail, as well as the effect of the dialing scheme on data communications devices.

An Actual SLIC Circuit

Modern IC technology allows many of the BORSHT functions of a SLIC to be performed within a single IC. Some additional components are needed to complete the overall function. Figure 10-3 shows an SLIC circuit based on the Motorola MC3419 IC. The IC provides most of the required functions, along with some auxiliary circuitry.

There are several points to note about telephone circuits. The two wires from SLIC to the phone are called the *tip* and *ring wires*, from the historical use of plugs and jacks. The phone system is designed so that any phone that is on-hook—not in use—looks like an open circuit to the SLIC. This means that the phone company office does not have to supply any power to a phone, unless the phone is off-hook. Any activity of the phone can be detected by the SLIC since taking the phone off-hook completes the circuit between the tip and ring, and thus some current flows through the completed loop.

The IC itself can operate on any voltage from -20 to -56 V dc, with respect to ground. This is compatible with most phone systems, the majority of which use -48 V dc. The IC itself uses less than 5 milliwatts (mW) when the phone is on-hook. This is an important consideration because a typical phone office has 10,000 lines, and overall power consumption is a critical design factor. Every effort is made to minimize the amount of power that an out-of-use phone consumes, since most phones are on-hook at any instant.

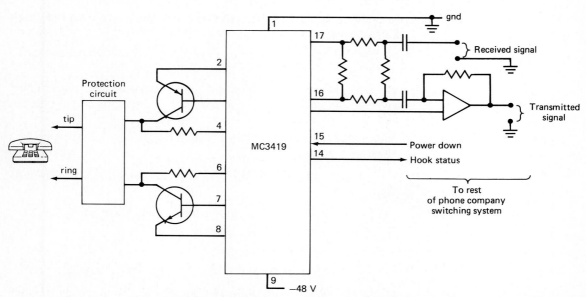

Fig. 10-3 A SLIC IC, such as the Motorola MC3419, performs most of the necessary interface functions between the phone instrument and the phone office circuitry.

The IC has fairly complex internal circuitry to provide the BORSHT functions. One BORSHT function which it cannot fully provide is the *overvoltage protection*. External components are used between the subscriber lines and the IC for this.

In operation, this SLIC circuit provides the tip and ring wires to the outside world. The signals that the SLIC must interconnect with the remainder of the telephone office are more varied. They include the received signal (from the phone office to the phone), the transmitted signal (from the phone to the office), and a digital indication showing whether the phone is on- or off-hook. The phone office can then decide, on the basis of the on- or off-hook condition, what to do next: send a dial tone, look for dialing signals, start the rest of the call sequence, and so on.

The MC3419 also has a pin that allows the phone company office to *power down* the IC and SLIC. This does not shut off the power, but puts the SLIC into an idle state in which it ignores any activity at the phone. This is used to deny service to the phone user without physically disconnecting any circuitry.

QUESTIONS FOR SECTION 10-1

1. Give three reasons why the telephone system is used for data communications.
2. What are some of the limitations of using the telephone system for data?
3. Why are telephone systems different in design from other electronic systems?
4. What is basic telephone service?
5. What is the on-hook condition? The off-hook condition? What does each look like electrically?
6. What does dialing look like electrically? What is the difference in signals when using regular dialing versus more modern tone dialing?
7. Where is the bell ringer located electrically? What does the typical ringing signal from the phone office look like? Why is this important even for modern electronic ringers?
8. What does "SLIC" stand for? What is BORSHT?
9. What are the BORSHT functions, in detail?
10. Why is a hybrid circuit needed? Explain the two-wire-to-four-wire conversion concept.
11. Why is the hybrid function especially critical in digital communications?
12. What are the tip and ring connections?
13. Why is SLIC power usage important?
14. What is the typical range of power supply voltages in a telephone system?
15. What signal wires does a SLIC provide to the outside world? To the telephone office?

10-2 Dialing

The act of dialing a telephone to reach an intended receiver is something that is done by the user without too much thought, because the phone system appears very simple. In fact, from the perspective of the telephone system, dialing is a complicated procedure. Dialing is usually thought of as something a person may do. However, it is also a necessary feature of data communications systems that must automatically, without human assistance, enter into the phone system as part of transferring data to another user or computer.

There are many examples of the use of this *autodial* feature. A retail store that is one of a chain may have computerized cash registers, all recording the day's sales and transactions (what merchandise, how much money). Every night, after the store is closed, the computer calls up the central offices of this store and reports on all the day's activities, so that chain management can get a daily report on store business. The computer in the store must dial up, connect, and converse with the computer at headquarters without any human intervention. Therefore, it must know how to dial, how to recognize whether the other side is busy, whether the phone rings, or whether it has been picked up by the central office computer.

Dialing is accepted only as a response to dial tone. The telephone office sends the user dial tone as an audible handshake signal to say "I'm ready for your dialing." If dialing is done before any dial tone, it is ignored by the phone system.

Pulse Dialing

The simplest form of dialing, and the one that POTS supports, is *pulse dialing*. This is the kind of dialing signal that a telephone with a rotary dial provides. The idea of pulse dialing is to make the phone generate a number of on-off pulses, corresponding to the digit being dialed by the user.

The on-off pulses are generated by the phone dial actually opening and closing the loop circuit from the telephone office, at a fixed rate. Recall that the loop from the telephone office is open when the phone is on-hook. When the phone is off-hook, the loop circuit is closed. The dialed digits are signaled to the telephone office by the dial mechanism opening and closing at the rate of 10 times/s. The number of open-close cycles is equal to the desired digit. The *duty cycle* (ratio of on time to on plus off total time period) is 50 percent, so the on time and the off time are equal (Fig. 10-4a).

The telephone office knows that a dialed digit is complete because of the relatively long time period between digits. It takes at least several tenths of a second, typically 0.5 s, for the person dialing to rotate the dial to the next digit. Therefore, the stream of dialing pulses has a group of 10 pulses/s, then a 0.5-s gap, then another group of pulses at 10 pulses/s, until all the digits have been dialed. It looks like Fig. 10-4b.

Pulse dialing worked effectively for the first 50 years of widespread phone service, from about 1900 through 1950. But it has some drawbacks:

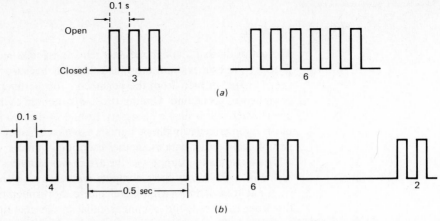

Fig. 10-4 Standard dial pulses have a 50 percent duty cycle (equal on and off times), and each pulse is spaced 0.1 s apart: (*a*) dialing numbers 3 and 6; (*b*) the internumber time period is typically about 0.5 s.

The mechanical make and break of the circuit is difficult and costly to implement via a computer or other electronic device. Some sort of electromechanical relay is needed.

The actual time to dial the complete number is fairly long and in fact depends on the specific digits that are dialed. For example, consider dialing the following two numbers at 10 pulses/s, with 0.5 s between numbers:

345 takes $0.3 + 0.5 + 0.4 + 0.5 + 0.5 = 2.2$ s

789 takes $0.7 + 0.5 + 0.8 + 0.5 + 0.9 = 3.4$ s

This is a problem for two reasons, one being that time is a precious commodity in a communications system. If these digits have to be passed along to a remote office (as they have to be on any call that goes to another phone office such as in the next town or over a long distance) then the interoffice lines are tied up with just passing digits. Second, the length of time is uncertain and depends on the number sequence itself, even if the person or machine doing the dialing maintains exactly 0.5 s between numbers. This means that the receiving circuitry must be prepared for a wide variety of time spans, and this capability necessitates an inefficient and costly design of this circuitry.

Finally, these make/break pulses cannot really be used for any other purpose. Once the numbers are dialed and the call established, any opening or closing of the loop may be confused with hanging up the phone! The use of the dial pulses for any type of signaling, such as to activate equipment at the receiver end, is impractical.

Tone Dialing

The Bell Telephone System recognized the long-term weaknesses of pulse dialing during the 1950s and began to set in place the framework for a more advanced dialing system. This system makes use of tones instead of pulses. Each dialed digit is represented by a unique pair of sine wave tones, and the tone pair is sent to the phone company office as each digit to be dialed is pushed on the phone. The relationship of the phone digits and the tones is shown most clearly by a diagram (Fig. 10-5). The numbers (and two extra keys called * and #) are arranged in a four-row-by-three-column rectangular matrix. (A fourth column is available, but not used on most phones.) Each row and column has a unique tone. When a key corresponding to a digit is pushed, the tones for its row and column are transmitted. The technical name for this scheme is *dual tone multifrequency* (DTMF).

For example, pushing the 1 key on the phone sends a signal at 697 Hz and one at 1209 Hz. The number 8 is represented by 852 Hz and 1336 Hz. Each tone is sent only as long as the key on the telephone is pressed. Practically, this is never less than about 0.1 s, since this is the fastest rate at which an average person can push the buttons on the phone. A typical value is about 0.25 s per key, with about 0.25 s between keys, but many people and all automatic units dial much faster than that.

The frequency values selected by the Bell System for the tones may seem somewhat random and arbitrary. They actually were chosen as a result of careful analysis. The following are the factors that were involved in the choices:

None of the tones is a harmonic of any other, or even a close harmonic. This means that any distortion in a tone, which may produce harmonics of the fundamental frequency (recall the Fourier analysis concepts), does not cause a misleading signal to be received. Example: The harmonics of number 1 are 1394 and 2418 Hz.

The amplitude modulation of one tone by another does not produce a sum or difference frequency that may be confused with a genuine value. This modula-

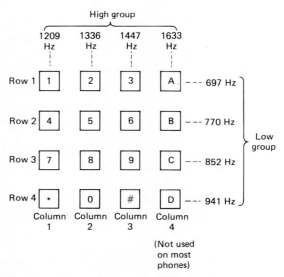

Fig. 10-5 The keypad for DTMF signaling shows the four rows and columns and the way each key generates a unique tone pair consisting of one high-group and one low-group tone. Most phones do not have the last column, but the tones are specified anyway.

tion can result from technical imperfections in the circuitry that are almost unavoidable, except at great cost. If the tones representing number 3 modulate each other, the result is frequencies at 780 and 2174 Hz. These are not near any of the allowed values, so the system can be easily designed to ignore them.

The advantages of using DTMF are in the areas of both the usefulness of the tones and the circuitry which can be used to determine which tones are sent. The usefulness is enhanced in the following ways:

The time to send a complete number is greatly reduced. A three-digit phone number takes only 1.25 s, regardless of the digits, using the 0.25-s value for tone and intertone times. A seven-digit number can be sent in 3.25 s by the average person. This means that the phone circuits are tied up for much less time with the dialing information.

The time to send a number is the same, regardless of the actual digits themselves. All seven-digit numbers take the same time; all long-distance numbers (1 + area code + number) take the same amount of time. The circuitry to receive this is easier to design.

The tones can be used for signaling purposes, once the dialing is over. The phone system is designed to ignore any and all frequencies within the frequency band that the voice uses, since circuitry designed to take some specific action once the call is established may be accidentally tripped by the user's voice. The tones can therefore be used to turn on equipment or send a coded message to the system at the other end.

This is in contrast to the pulse dialing signals, in which the phone system is designed to look, at all times, for the opening of the loop to the phone, since this indicates that the phone has been hung up.

Circuitry to send tones is easier to build with modern ICs than circuitry to send pulses, and the same tone generation circuitry can be more easily controlled by computers or other electronic equipment. This means that tone dialing is more compatible with modern systems and automatic, unattended operation.

The circuitry to receive tones can also be implemented more effectively with present-day electronics than the circuitry to count the pulses. A tone receiver, at the phone company office, is built of seven narrow-bandwidth filter circuits (Fig. 10-6). The incoming dialing from the phone goes through these filters, which effectively sort the incoming frequencies into two of the seven values. If the incoming signal has a frequency component corresponding to the value of the filter, the filter has an output signal. The other filters have no output. Simple digital circuitry can then look at the filter outputs and determine which number is dialed by seeing which two filters have an output. In addition, there should be exactly two outputs from the seven filters. Only one output, or three or more outputs, indicates that some noise or severe signal distortion has corrupted the dialed signal or put a false signal on the line.

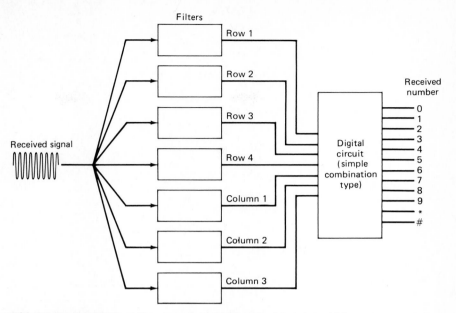

Fig. 10-6 A DTMF receiver can use eight filters to determine which two tones are sent. Simple digital logic can then determine which number on the keypad is pushed on the basis of identity of the two tones.

Circuitry for Tone Generation

When the concept of DTMF signaling was being developed, transistors were still new. The circuitry for generating the tones required seven separate oscillators. These oscillators had to provide good sine wave outputs that would not vary with temperature, power supply voltage variations, or some change in component value over time.

Modern ICs make tone generation very simple and accurate. A typical *tone generation IC* is the Motorola MC14410, shown in Fig. 10-7. This is a digital IC which generates the desired DTMF frequencies by dividing down a master clock, provided from a 1-MHz crystal. Very few additional components are needed for the tone generation. The MC14410 provides the *high-frequency (column) tone* on one pin of its package, and the *low-frequency (row) output* on another pin. They are combined to go onto the phone wire by simple resistors. The IC actually can develop one of four row and one of four column frequencies. The fourth column does not appear on most phones, but frequencies were set aside by the Bell System for this column. The keys of this fourth column can be used for special features in the phone, such as conferencing or interrupting on other calls.

The countdown scheme works like this: the master clock oscillator provides a 1-MHz signal to the IC. The IC uses internal circuitry in the form of counters to divide this 1-MHz value by some number N and puts out a signal at frequency 1 MHz/N. If the divisor N is 1435, then the resulting frequency is 696.8 Hz, very close to the desired value of 697 Hz for row 1.

Telephone Systems and Modems

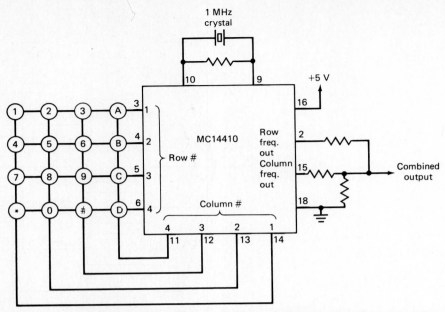

Fig. 10-7 The Motorola MC14410 is designed for DTMF dialing. It senses which key is pushed and generates the correct DTMF tones which it also combines onto a single output to the phone lines.

The overriding advantage of this IC method over the individual oscillators is that the absolute accuracy of the outputs, and their stability with respect to temperature, voltage, and other factors, is determined only by the performance of the master oscillator crystal. It is relatively inexpensive to obtain a high-accuracy, high-stability crystal. Any fine-tuning that has to be done is only on this one component, not individual, independent oscillators.

The Motorola IC is used for a phone that has buttons pushed by a person. Suppose a modern telecommunications system needs to have an autodial feature for unattended dialing. One approach is to have the controlling computer cause the row and column connections to be made, through some sort of computer-controlled switch mechanism. This is costly and unnecessary. Instead, ICs similar to the MC14410 are used, except that these ICs can be controlled directly by a computer. These computer-compatible ICs have internal registers, or *mailboxes*, which the computer can address. The computer puts a binary code corresponding to the desired tone pair into the register. It also indicates to the IC how long the tone output should last. The IC then performs as it has been directed.

QUESTIONS FOR SECTION 10-2

1. What is autodialing?
2. What is pulse dialing? What does it do to the phone office loop?
3. How does the phone system know that a new digit is starting in pulse dialing?

4. What are three drawbacks of pulse dialing?
5. Why can't the dialing pulses be used for any other purpose?
6. What is DTMF dialing? How does it work?
7. How were the DTMF frequencies selected?
8. What are three advantages of DTMF over pulse dialing?
9. Why can DTMF be used for signaling once the dialing is complete and the call connected?
10. How was DTMF tone generation originally achieved? How does a modern IC do it?
11. What are the advantages of DTMF dialing with present-day ICs?
12. How is DTMF dialing performed through computer control?

PROBLEMS FOR SECTION 10-2

1. Using the normal times for pulse dialing, how long does it take to dial the numbers: 163, 784, 223?
2. Repeat Prob. 1 for the following numbers: 784-1213, 461-2001, 687-3489.
3. Repeat Prob. 1 for the following long distance numbers: 1-413-223-3230, 1-916-769-8790.
4. What tone pairs are generated for the following numbers by using DTMF: 3, 8, 6, 4?
5. Repeat Prob. 4 for the following digits: 0, 2, 7, 9.
6. What are the harmonics of the tones used to represent 4 with DTMF? How about 2?
7. What are the frequencies that result from AM of the two tones representing 1? What about 6?
8. Repeat Prob. 7 for tones for 5 and 0.
9. Using standard dialing times, how long does DTMF dialing take for the following numbers: 345, 721, 784?
10. Repeat Prob. 9 for the long-distance numbers of Prob. 3.
11. The divisor of the 1-MHz clock for the DTMF IC must be an integer. What divisors are used to generate tones for digit 4? Digit 9?

10-3 Telephone Lines

Sending data requires bandwidth. The amount of bandwidth needed is directly related to the data rate that is desired. An analog voice signal contains its data in a relatively narrow bandwidth, in proportion to the amount of data it carries. This is because an analog signal is an efficient way to encode data.

The telephone system as it historically developed was designed for voice and analog signals. Many technical decisions and designs were implemented on that basis. Some of those decisions, in retrospect, were not the best for sending digital data. A surprising number, however, were forced by the unavoidable constraints of the ways wires and signals behave over long distances and the need to make a reliable, rugged phone network. These choices were the correct ones initially and still are correct in many ways. The important point is to understand the true nature of telephone lines and learn to work with them to support digital signals properly.

There are two inherent characteristics of a signal line that are very important to carrying signals with minimal distortion or corruption. These are the amplitude response versus frequency (called *frequency response*), and the time delay versus frequency. Various amounts of imperfections in these two aspects can cause widely different amounts of distortion and problems in signals as they are received after being transmitted down a wire.

The *frequency response* shows how each frequency component of a signal fares as it passes through the wire and system. Any wire or cable is electrically equivalent in many ways to a filter circuit. A filter has the property of being able to pass frequencies from its input to its output side, but some frequencies are passed with more or less attenuation than others. This is the purpose of the filter. The frequency response is actually used to characterize virtually any system which must pass signals. Recall from the study of Fourier analysis concepts that the shape of a signal is determined by the frequency components it has and their amounts. A digital signal with sharp corners needs significant high-frequency components to provide those corners. If the high-frequency components are attenuated, there is a rounding of the digital signal. Similarly, the low- and medium-frequency components contribute to the correct shape of the signal. If these frequency components are attenuated, different distortions result. The output waveform does not look like the original waveform.

The other contributing factor is the *delay characteristics*. Any electrical filter not only has the effect of changing the resultant amplitude versus frequency, but also introduces some time delays (which are the same as phase shifts) in the various frequency components of the signal. The *time delay phase shift* (also called *envelope delays*) is not the same for each frequency, however. Depending on the exact nature of the filter or wire, there is some combination of frequency response and phase shift. The effect of these phase shifts is similar to the effect of frequency response characteristics. As Fourier analysis shows, the time and frequency domains are linked together—any change in one is also a change in the other.

The distortions caused by variations in frequency response and time delay make the signal as received differ from the original transmitted signal. The amount of difference depends on the exact amounts of both frequency and time distortions that occur. For voice signals, a relatively large amount of distortion is acceptable, since the human ear can understand voices even with distortion that looks severe to the eye. For digital signals, these distortions may cause the receiver to misinterpret the signal that is sent and so produce an error.

Standard Telephone Lines

The regular telephone loop from the local office to the phone is guaranteed by the phone company to have some specific characteristics. This type of line is the lowest performance line, called *voice grade conditioning*. Figure 10-8 shows

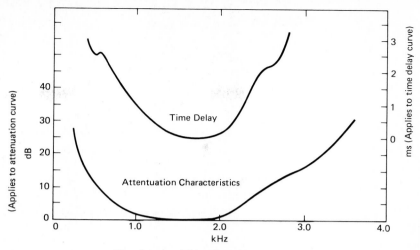

Time (envelope) delay distortion and attenuation

Fig. 10-8 Attenuation and time (envelope) delay characteristics of an unconditioned telephone line, from 300 to 3600 Hz. Note that the left-hand axis is attenuation (decibels) and the right-hand axis is time delay (milliseconds).

frequency and time delay characteristics of this type of line. The least attenuation and time delay occurs from 1.5 to 1.8 kHz. At 1 kHz the signal components are attenuated by about 2 to 3 dB, and at 3 kHz the frequency is attenuated by about 16 dB. The time delays at these points are 1 ms and more than 2 ms, respectively.

Similar line characteristics are offered by telephone companies on the lines that go between phone company offices. These interoffice lines are called *trunks*. Any phone interconnection that goes from one three-digit prefix (called the *exchange*) to another requires such a trunk, since each local office serves only one prefix. The exchanges within a smaller region all have the same area code. Different area codes are used to connect different regions. A system user must be concerned about the signal quality from one end to the other, not only to the local exchange. To meet this need, the interoffice trunk performance is also carefully defined.

Because any phone line can connect one user to another user through the phone system, the user has a line assigned randomly, through the phone offices. This is called the *dial-up* or *switched network*. Any phone can be used to dial up the other end, and the phone company can switch in any suitable line to complete the connection. Lower-performance applications (typically below 2400 or 4800 baud) can use the switched system with acceptable rates of error. Certainly, the public dial-up lines are very convenient and have easy access.

For applications which need greater performance than these dial-up lines can offer, telephone companies offer specially conditioned lines. These lines—both from the phone to the office and between phone offices—provide better frequency response and time delay characteristics. These are shown in Fig. 10-9. This kind of conditioned line is leased by the user, who pays more for it than for a standard line. The terms "dedicated" and "leased" are used when the phone company has set aside a conditioned line for a communications link. Different

Telephone Systems and Modems

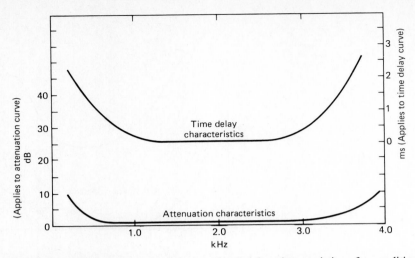

Fig. 10-9 The attenuation and time (envelope) delay characteristics of a conditioned telephone line.

grades of leased lines, with grades of increasingly better conditioning, are usually available. These higher-grade lines can support higher data rates, as high as RS-232 allows (20 kbits/s), or even 50 to 100 kbits/s at a much higher price.

Any receiving circuitry that uses the telephone network must also be designed to cope with signals that arrive with much lower amplitude than they had at the transmitter, regardless of frequency. The long lengths of phone links cause drops in signal amplitude due to cable and wire resistance. Some of this loss can be compensated for by amplifiers along the way, from one office to another. The final loop to the phone is very difficult to manage, unfortunately. The phone may be a few hundred yards from the exchange or several miles. If the phone company boosted the signal at the office, a separate amplifier and adjustment would be needed for every phone location. The amplifier would have to be readjusted whenever phone numbers were reassigned or changed; and this would create a maintenance problem. Simply boosting the signal for all phones would not work, since the phones close by would receive too much signal and be overloaded. Also, simple amplification would boost any noise that had been picked up along the way and make matters worse.

The result is that the actual signal levels that the receiver must accommodate vary considerably, and the receiver must provide error-free (or nearly so) performance in all these cases. How much attenuation occurs? Figure 10-10 shows the amounts. The figure indicates, for three distance ranges, the percentage of calls with received attenuations less than the number on the X axis. For example, on a 130- to 360-mi link, there is an attenuation of 20 dB, but not more, for 60 percent of the calls. The other 40 percent, on average, suffer an attenuation of more than 20 dB. Similar numbers can be obtained from the graph for other cases.

Any designer of systems for communications over the telephone lines has a performance goal that defines what percentage of interconnections should be

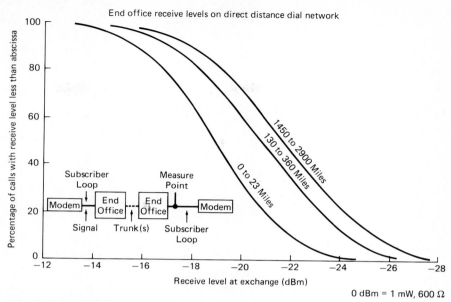

Fig. 10-10 This graph of line attenuation versus distance shows the percentage of calls with a received signal level less than the value of the horizontal axis.

successful. A typical goal may be to make sure that the circuitry works for 90 percent of the long-distance connections. Using the chart, the designer can see that the circuitry must accommodate attenuation values for the top 90 percent of users, and not for the lowest 10 percent. These 10 percent receive signals with attenuations greater than 25 dB, so the designer designs only for attenuations from 0 to -25 dB and ignores those below -25 dB.

Telephone Line Echoes

There is another characteristic of telephone circuits (or any medium) that can cause problems with reliable, error-free communications. This is called *echo*. Echo occurs as a result of unavoidable irregularities and imperfections in the signal path which cause it to have impedance "bumps." A small part of the signal energy coming down the wire reflects back from the bump, just as a wave running into a trough has a small backwave if the trough has a sudden narrowing neck. The transmitted wave reaches the intended receiving end, and a small part of the energy reflects back to the transmitter. This is called the *talker echo* (Fig. 10-11). It normally does not cause a problem for data because the transmitting circuitry is not set up to look for signals in this reverse direction. (It is important to note that the talker echo may be very annoying to a person using the system to talk, if the echo time is greater than a few milliseconds.)

When the talker echo reflects back and reaches another irregularity in the line, it then produces a second small echo back to the intended receiver. This second echo is called the *listener echo*. It looks exactly like the original signal but is

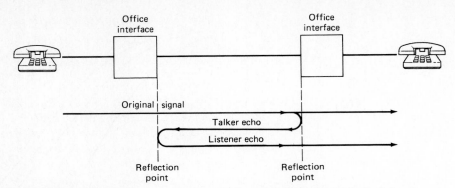

Fig. 10-11 Echo is caused by reflections in the signal path, which result in a secondary, delayed signal after the main signal. Talker echo is heard by the talker, and listener echo is heard by the listener.

lower in amplitude and delayed in time. The intended listener may confuse it with the original signal.

The most common location of these echo-causing irregularities in the signal path is at each end of the link. The hybrid circuitry which performs the two-wire-to-four-wire conversion cannot always be matched perfectly to the lines. The echo distance is therefore usually the entire signal path length, minus the local loop length.

For short-distance communications, the echo time is usually small enough that it does not cause a problem. In addition, the path is essentially the same each time a short-distance connection is made, so the echo can be "tuned out" by some special circuitry. For long distances this echo time can become large enough to cause signal confusion and error. Consider a signal from one coast of the United States to the other. The distance is 3000 mi, and the signal propagates at 70 percent of the speed of light. The total echo time is

$$\frac{3000 \text{ mi} \times 2 \text{ trips}}{186{,}000 \text{ mi/s} \times 70 \text{ percent}} = 0.023 \text{ s}$$

The long-distance path may not be the same each time a connection is completed, however. The New York to Los Angeles path may be cabled through the shortest path (the middle of the United States); it can be cabled from New York to Chicago to Los Angeles; or it can be a satellite link. All provide the identical user function but with very different echo times.

Experience has shown that whenever the echo time is greater than about 45 ms the echo can be a problem. Several solutions are used in communications. One solution is to use a scheme to suppress the echo, by employing circuitry which cuts the signal transmission path from the listener to the talker (Fig. 10-12a). This echo suppression is effective but makes full-duplex communications difficult, since the data signal is allowed to flow in only one direction at a time. The user is restricted to half duplex, which is completely adequate for many applications. The real drawback is that the circuitry to switch the echo path out cannot be switched instantaneously on and off. It takes about 100 ms to give the echo suppressor the instruction to turn on (or off) and to implement it. This 100-ms period is wasted time in the communications system. If the half-duplex operation of this application involves frequent changes, a large percentage of total available time is lost.

Another solution is to use echo cancellation, made possible by modern sophisticated ICs. In *echo cancellation*, a portion of the original transmitted signal is taken, delayed, and then subtracted from the echo signal itself (Fig. 10-12b). If it is done perfectly, the echo is canceled completely. The echo cancellation circuitry must constantly adjust both the time delay and the amplitude of this opposite signal so that it is exactly equal and opposite to the undesired echo. It is possible to do a very good job of this with present-day circuitry.

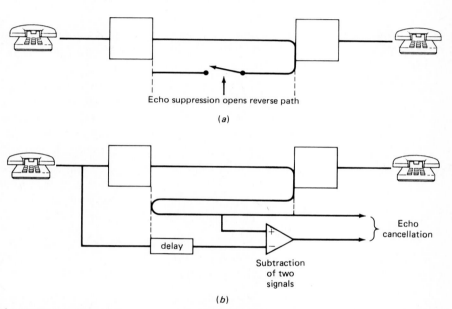

Fig. 10-12 Echo suppression can be achieved two ways. (*a*) The reverse path (from listener to talker) is opened, or (*b*) part of the talker signal is delayed and subtracted from the main signal, to cancel the echo.

QUESTIONS FOR SECTION 10-3

1. What is the meaning of frequency response? What causes the frequency response variations?
2. What is time delay? What causes it?
3. What is the effect of time delay on the received waveform?
4. Why is time delay not a problem to the ear for voice? Why is it a problem for data?
5. What is meant by the frequency response of a voice grade line? Where is the attenuation the least?
6. What is a dial-up or switched line? What baud values can it support?
7. What is a conditioned or leased line? How does it differ from a dial-up line?

8. What causes overall amplitude loss in the signal path? Why can't amplifiers make up the loss entirely?
9. How does the expected loss affect the system design goals?
10. What is echo? What causes it?
11. What is the effect of echo?
12. Compare short- versus long-distance echo.
13. How much echo can typically be tolerated in a system?
14. How does echo suppression work? What are its drawbacks?
15. How does echo cancellation operate? What does it require?

PROBLEMS FOR SECTION 10-3

1. A system uses cable for transmission from Chicago to Florida, approximately 2000 mi. What is the echo time, if the signal speed is 75 percent the speed of light?
2. The same connection is made via satellite that is 22,000 mi above the earth. Signals travel through air at 97 percent the speed of light. What is the echo time?
3. A call is made to a nearby town. The distance is 10 mi. What is the echo time? (Signal speed is 60 percent speed of light.)
4. The call of Prob. 3 is repeated 10 min later. Because of line problems, the phone company automatically routes the call to another center 1500 miles away and back to the desired receiver, using cable. (This does happen; you just never notice it!) What is the echo time?

10-4 Private Exchanges

Many users of data communications systems belong to the same company, factory, or other organization and are located in a single building or group of buildings. The telephones of these users do not connect directly to the phone company central office. Instead, a private exchange is located at the company or factory location. This is called a *private branch exchange* (PBX) or *Private Automatic Branch Exchange* (PABX). The PABX is simply an automated version of the older, manually operated switchboard—the PBX.

The function provided by the PABX and PBX is shown in Fig. 10-13. The important point is that calls between users in the same company no longer have to go to the telephone office, but are routed entirely within the company and its PBX/PABX. The PBX/PABX is essentially like a scaled-down, simpler version of a full telephone central office. The PABX or PBX connects to the actual telephone office via a number of trunk lines. These are used for any incoming or outgoing calls—calls that involve anyone outside the company or factory.

What are the advantages of a PBX/PABX? There are several:

Fewer lines are needed over the distance to the phone company office.

The phone company office is not used for calls over short distances.

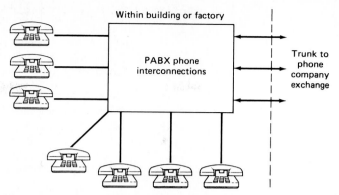

Fig. 10-13 A PABX is a small phone exchange which interconnects all the phones in a location and has trunk interconnect lines to the local phone company exchange.

Special features that this particular company needs can be provided. These may include call restrictions (so unauthorized phones cannot be used for long-distance calls), speed dialing (so remote sales offices can be reached with a two- or three-digit code), and redial for busy numbers. Only an automated PABX can do this, of course.

The internal lines supported by the PABX/PBX and the equipment itself can be designed for high performance in the area of data communications.

The last point is the one of greatest technical interest. Since the lines that the PABX handles are shorter-distance and lines are fully under the control of a local manager, they effectively form a private collection of lines that can be conditioned for data communications and even have special signals and features that data communications users need. It is far easier to provide the line bandwidth and conditioning in a local situation than over long distances. The PABX becomes a group of lines that eliminate, in many cases, the need for expensive leased lines from the telephone company. In many applications, a very high percentage of the data communications flow is entirely within the building and little of it is related to outsiders.

Some of the newest, most advanced PABXs are designed exclusively for digital communications. They assume that the telephone unit is not an analog signal source, but a digital signal source with a well-defined interface. The PABX is optimized for digital performance and can provide very wide bandwidth to users. This bandwidth can support voice, digitized voice, digitized video, and other data which either is inherently digital (financial reports) or can be put into digital form.

The Number of Trunks Needed

The number of internal phone lines that a private exchange must have is easy to determine: all those in the company who need a phone for their job require a phone line. The number of trunks to the local phone company exchange is more difficult to calculate. Using many trunks ensures that any call between outside and inside goes through, but these trunks cost money. Having only a few trunks, conversely, means that many calls are not able to find a line and so receive a busy

Telephone Systems and Modems 337

signal indicating that no lines were available (not that the intended receiver is busy).

The suppliers of exchanges and the phone companies have developed, over the years, various formulas to help determine the number of trunks needed. Two factors determine the number of trunks versus the number of internal lines:

The nature of the business or factory: A real estate business or stock brokerage expects to make and receive many outside calls, and relatively few internal calls. In comparison, a manufacturing plant has many internal calls, but outside calls are mainly those to and from the purchasing and sales groups. The number of internal lines and outside trunks is set according to the needs of the type of business.

The types of calls that are made. Some businesses tend to have shorter calls, and others have longer calls. Once again, stock brokers are a good example of people making short calls; a company that is using phones for discussing engineering designs with its suppliers has longer calls. These longer calls tie up existing trunks, so more trunks are needed. In addition, if the applications involve use of the phones for data communications, the phone lines are tied up for longer periods of time. This is because the data communications link is usually left connected by the user even when there is no data for a short time. When people talk, they hang up as soon as they are finished, or tell the other party they will get back to them later, rather than putting them on hold for more than a minute. With computers and data communications, the user may be working interactively at a terminal connected to a computer, typing a few characters now and then. Although the total amount of data is low, and the data rate is low, the total connection time is very long. The result is that the telephone line is tied up for quite a while.

No calculation can ensure that every user has access to a trunk whenever he or she needs it. The only way to guarantee that is to have a trunk for every internal phone; this is usually impractical. It turns out that experience and analysis can combine to ensure that a certain number of trunks provide a very high percentage of users with a trunk whenever they need it. A PABX with 100 internal users may have 10 trunks and be able to guarantee that 99 percent of the time a trunk is available, depending on the calling patterns of the business. Using 8 trunks beyond the 10 increases the performance from 99 percent to 99.5 percent, but at additional cost. Most users accept a trunk availability of between 97 percent and 99 percent.

QUESTIONS FOR SECTION 10-4

1. What is the role of the PBX? The PABX?
2. Why is the PBX or PABX needed? How do they differ?
3. What are three advantages of using the PABX/PBX versus connecting all calls through the local phone exchange?
4. What factors determine the number of trunks needed? What level of service is usually desired?
5. Why do data communications signals and users occupy the lines for longer periods of time than voice, even though the amount of data and rate are low?

10-5 The Role of Modems

The normal digital signal is not a good match to the technical constraints of the telephone line. Phone lines are designed for analog, voice signals; data is usually represented by binary signals, often of 0 and +5 V. The solution to using phone lines for data communications requires a special communications box, which converts the data signals to those more compatible with the phone line capabilities (Fig. 10-14). This special communications box is called a *modem*. The term "modem" is a contraction of "MOdulator-DEModulator," which describes exactly what a modem does: it uses the data signal to modulate a waveform that is usable with the telephone system. The term "modem" has also come to mean any communications box that makes a digital data signal compatible with any nondigital system and medium besides the phone system. An example is a modem which makes the data bits compatible with a microwave radio system.

In the previous discussion of RS-232, data terminal equipment (DTEs), and data communications equipment (DCEs), the modem was the DCE that the DTE (computer device, terminal, or other) was designed to work with. The theory and application of modems comprise the area in which all the previous concepts of Fourier analysis, modulation, bandwidth, and interfacing come together in a single piece of equipment. The modem acts as the electronic bridge between two worlds—the world of the purely digital signal and the established analog world, such as the telephone system.

Modems are always used in pairs. Any system, whether simplex, half duplex, or full duplex, requires a modem at the transmitting end and the receiving end. The modems do not have to be from the same manufacturer, but as in all aspects of communications, they must use an agreed-to set of rules and signals. Some modem designs are relatively standardized and available from many manufacturers. Other high-performance modems use unique technical schemes to achieve certain levels of performance, and only one manufacturer exists for that specific modem with that performance.

What is the key idea behind a modem? The concept is to use tones (sine waves) of various frequencies, phases, or amplitudes to represent the binary data. *Tones* are what the phone system is designed to handle, since voice is made up of many tones combined. Different modems use frequency modulation (FM),

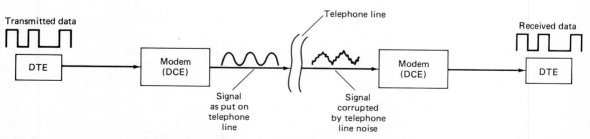

Fig. 10-14 The ordinary noise on phone lines, acceptable for voice, may corrupt data signals represented by simple voltage values.

phase modulation (PM), amplitude modulation (AM), or some combination of these to achieve the desired data rates and low-error performance over different types of telephone lines.

Modem Functions

What are the main functions that a simple modem must perform? Figure 10-15 shows a simplified block diagram of a modem used with RS-232 interconnection. At the transmitting end, the modem must do the following:

Take data from the RS-232 interface

Convert this data (1s and 0s) into the appropriate tones (the modulation process)

Perform line control and signaling to the other end of the phone line

Send dialing signals if this modem is designed to dial without the user present

Have protection against line overvoltage conditions and problems

The receiving modem must perform the corresponding operations:

Receive tones from the phone line

Demodulate these tones into 1s and 0s

Put the demodulated signal into RS-232 format and connect to the RS-232 interface

Perform line control and signaling

Have protection against line overvoltage problems

Adapt its receiving system to variations in the received noise, distortion, signal level, and other imperfections, so that data can be recovered from the received signal with as few errors as possible

Many different modem designs are in use. They range from low-cost, low- to moderate-performance units to much more expensive units whose performance approaches the theoretical maximum that the phone line can provide. Before

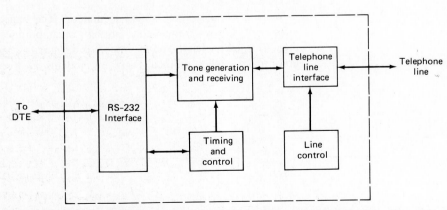

Fig. 10-15 A modem is designed to interface data signals to phone lines. It contains interfaces to the phone line, the DTE (usually via RS-232), tone generation and receiving circuitry, and timing and line control circuits.

studying some actual units commonly used, it is helpful to look at the physical shape of modems and a typical modem application.

Physically, modems are available in two styles. The older and still quite useful arrangement is to have the modem as a separate stand-alone box from the DTE it is connected with, interconnected by the RS-232 cable (or whatever is being used for interface). This stand-alone modem is selected to be the best one for the overall application. It requires its own power cord and takes up some desk space; this is not a problem in cases in which the computer or terminal is staying in one place most of the time. Since the modem size is not restricted, extra circuitry can be put into this modem to provide additional features such as good equalization for fewest errors, indicator lights to show the equipment user what the line conditions are (ring detect, for example), and even lights to show any malfunctions that occur. If a different situation arises that requires a change in the modem (higher speed, more noise resistant modulation, automatic dialing), the user can simply and easily exchange the old modem box for a new one.

However, this external box is not practical for applications in which the DTE is designed to be moved, such as a personal or portable computer. There is also the restriction imposed by the RS-232 interface between the DTE and the modem as DCE. The DTE cannot fully control the modem because any control has to be implemented through the limited RS-232 signal line group. Setting up a list of phone numbers to be automatically dialed by the modem is difficult, for example, since the computer (DTE) somehow has to transfer this list via the RS-232 lines, which were not designed for this kind of transfer. Special sequences of signals have to be employed to separate data the modem is supposed to send.

The rapid improvements in IC technology have made it possible to build entire modems on a small circuit board (Fig. 10-16) and then plug this circuit board into the computer itself. This type of integral modem requires no additional outside power (it takes power from the computer) and is as portable as the computer. Since it connects directly into the computer system, instead of through RS-232, it can be directly controlled by the computer and its program. The computer tells the modem what number to dial, what message to send, what steps to take if the

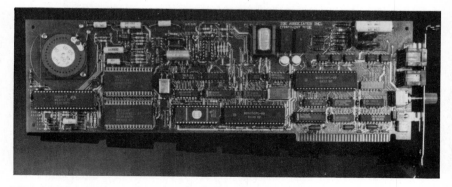

Fig. 10-16 An integral modem is designed to plug into another device, such as a personal computer, and connect directly to the computer bus without using the RS-232 interface (Courtesy IDEAssociates, Inc.).

call doesn't go through, and so on. The relationship between the modem and the DTE is much more intimate, and the modem becomes a part of the computer body rather than a peripheral device to the computer. Present technology limits the capability that can be placed on one of these plug-in cards. The capability of these cards (typically 1200 to 2400 baud) is equivalent to that of a stand-alone modem several years ago, but they are less capable of high-speed performance under difficult line conditions than the newest external modems. The data communications user selects which type of modem to use depending on portability, price, and performance needed.

Operation of a Modem

The main activity of a modem is to send and receive digital data. Consider a case in which the modem uses two frequencies to represent the two binary values. Suppose 1000 Hz is used for binary 0 and 2000 Hz is used for binary 1. A stream of bits then is converted into 1000- and 2000-Hz tones, and these tones go on the phone line. By design, the phone system can handle these tones reasonably well, as compared to putting simple voltages of 0 V and 5 V on the line, for example. (In fact, the phone system cannot handle steady voltages such as 0 and 5 V at all—they are severely attenuated in a few feet.) The rate at which the 1000- and 2000-Hz tones are used to represent the data bits is the baud value in use. In a 300-baud system, there is a new bit every $\frac{1}{300}$ s. If the bit has the same value as the preceding one, the tone is unchanged. If it were different, then the tone would be different. Consider a stream of data bits like the following: 10011010. The phone line receives the following signals as shown in Fig. 10-17: 2000, 1000, 1000, 2000, 2000, 1000, 2000 and 1000 Hz. If two bits in a row are the same, then the tone simply continues for another bit period of $\frac{1}{300}$ s. What the modem has done is perform the digital frequency-modulation (FM) operation. Note that the transition from one frequency to another cannot be abrupt. A sudden change would require high-frequency components, more than the phone line provides.

At the receiving end, the modem must be prepared to look for the 1000- and 2000-Hz signals despite their low signal level (recall that signals are reduced considerably in amplitude by the phone system), continuous variations in the level (fading signals occur especially where satellite and radio links are used by the phone channel), and the addition of corrupting noise. To do this, receiving circuitry in the modem uses designs that amplify the weak signal (and vary the amount of amplification continuously so that final signal level is constant),

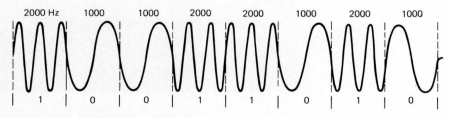

Fig. 10-17 Frequencies used to represent binary 1 and 0. Here a 0 is 1000 Hz and a 1 is 2000 Hz.

special techniques that can reduce the noise and demodulate the signal, and filters tuned to 1000 and 2000 Hz. The output of the filter circuit goes to a signal detector, which sees which filter has an output at any instant of time. If the 1000-Hz filter has an output, then a 0 must have been received; an output at the 2000-Hz filter indicates that a 1 was received. The modem uses its timing circuitry to reformulate a string of 1s and 0s from the received tones that is identical to the string of 1s and 0s that the equipment at the transmitting end generates.

A basic, low-performance modem for RS-232 is intended for use at speeds of up to 300 or 600 baud. A midrange modem can operate reliably at from 1200 to 4800 baud. Highest-performance modems cover the remainder, up to 19.6 kbaud. As the baud value increases, it is technically more difficult to receive and recover the original data bits from the tones, which have been corrupted by noise and have variations in signal levels. Techniques used to allow performance at these higher rates include modulations such as FM, multilevel modulation instead of simpler binary modulation, and fairly sophisticated circuitry to adapt dynamically to line conditions and process the incoming signal in various ways, depending on the type of line corruption.

Originate and Answer

There is a problem that occurs if full-duplex communication is needed. Suppose both sides of the link use the same frequencies, such as 1000 and 2000 Hz. The signals interfere with each other and the result is meaningless. Two solutions can be used. The easier to implement is using half-duplex communication only, so that each side gets a turn, but never simultaneously. Then, a single pair of frequencies can be used. When the receiving channel needs to have its turn, some protocol is used to indicate this. The data line is then "turned around" so data communications can take place in the other direction.

Typically, this turnaround time is anywhere from 100 to 500 ms for both sides to switch over their circuitry and allow any other signal effects to come to an end. For two-way conversation between people, or between a person at a computer terminal and a computer, this turnaround time is very unpleasant. It makes good interaction between the two users very difficult. For many applications, however, such as sending large blocks of data from one location to another, half-duplex communication is very acceptable. There is no need to change direction constantly, since the data flow in one direction continues for a long period of time. A half-duplex design, properly applied, can achieve very high throughput on the channel and gives good performance.

For other applications, full-duplex design is a necessity. There is a technical solution that allows full duplex to be used. Instead of a single pair of tones, two different pairs are used. One modem may use 1000/2000 Hz, and the other uses 3000/4000 Hz, for example. The modem that is transmitting 1000/2000 Hz must have receiving circuitry and filters set for 3000/4000 Hz. Conversely, the modem transmitting on 3000/4000 Hz must be set to receive 1000/2000 Hz (Fig. 10-18).

Of course, each modem must know which pair it should use. This leads to the terms "originate" and "answer." The modem at the end of the link which initiates the call is called the *originate modem* and always uses one pair. This modem that is at the receiving end of the original call always uses the second

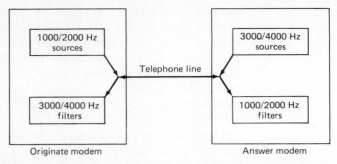

Fig. 10-18 A full-duplex modem needs two pairs of tones. One pair is used by the originate modem, and the other is used by the answer modem to send data. Each must have filters corresponding to the other's frequency for receiving the tones.

pair, which is called the *answer modem*. The modem users must set their modems for either originate mode or answer mode manually. Some newer modems automatically set themselves up for the proper mode. If no signals are coming in, the modem stays in originate mode. Once some tones indicating that someone is trying to call that modem are received, the modem switches over to be an answer modem. The user of the modem never sees any of this, so operation is transparent and easy.

The two-tone-pair-solution allows any phone line to be used for full-duplex communications. However, the bandwidth of the phone line must still be shared by two users, so that the data rate that can be used is lower than it would be if a single direction could make full use of the line. There can still be problems from the reflections, especially at the two- to four-wire hybrid in the phone office. For these reasons, many modem installations use what is called a *four-wire system*. Instead of the regular two-wire loop from the phone company exchange to the phone (and modem), two separate loops are used. One loop is dedicated to signal transmission in each direction. With two separate loops, there can be no interference or interaction between the data flowing to and from the modem. These two loops are not the regular phone loops that serve a standard phone. They are special loops that do not have the hybrid circuit at the phone office, since the hybrid is unneeded and is one of the main sources of line problems. Each loop runs to the phone office, and then onto a wire pair to the next phone office or other user. The connection between the two users is thus four wires—two wire pairs—all the way through (Fig. 10-19).

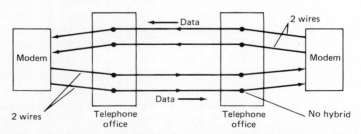

Fig. 10-19 A four-wire modem provides better performance than a two-wire unit, since the hybrid circuitry is eliminated. Two separate pairs of wires are needed, however.

Certainly, this four-wire method is more costly in terms of phone lines than a two-wire connection. It does allow much better performance: higher data rates, full duplex, and no echo. Modems are specially designed for four-wire operation, and some modems can be switched from lower-performance two-wire operation to higher performance four-wire operation by the user. A four-wire design is much easier to implement from a circuitry perspective. One of the most difficult problems in modem design is having the modem receiving circuitry properly extract the relatively low level received tones while the much higher level transmitted tones are also present. Even though the tones use different frequencies, the receiving circuitry is easily overloaded by the other tones, which may be 40 to 60 dB larger. It is like trying to hear someone next to you who is talking at ordinary volume while you yourself are shouting. A four-wire design never has this overload problem since the transmit and receive tones are on entirely different paths the whole way.

Connecting the Modem to the Line

RS-232 or a computer bus connects the modem to the DTE (computer, terminal, or any other device). The modem, acting as the DCE, must somehow connect physically to the telephone line. There are two ways to do this, acoustic coupling and direct connect.

Acoustic coupling is performed by putting the telephone handset into rubber cups in the modem (Fig. 10-20). These cups isolate the telephone handset microphone and speaker from outside noises. The modem itself has a microphone and earphone. The modem-generated tones go through the modem speaker and into the handset microphone. Tones coming in from the phone lines go into the handset speaker and into the modem microphone. In this way, the tones can be transferred to and from the telephone line without any actual wire connection. Any phone can be used—the user simply has to place the handset in the modem.

Acoustic coupling is simple and has the virtue that it can be used with any phone, even a pay phone. There are some problems with it, unfortunately. First, it has to be physically large enough for the handset, so it doesn't fit easily into the design of small computers. In fact, acoustic coupling modems are usually built only as separate boxes, rather than on integral modems. Second, the performance that can be achieved is relatively low. A speed of 300 baud is usually all that can be realized. This is because the coupling between the handset and the modem is never absolutely sound-tight and some extraneous room noise gets in and corrupts the modem tones. Simple nearby activities such as typing on the computer keyboard, rustling papers, or air conditioning can create enough noise with energy in just the right frequency bands to cause very unreliable operation. Finally, acoustic coupling cannot be used for automatic dialing. The tones, in passing through the modem speaker and handset microphone, are distorted too severely to be recognized by the equipment at the phone office. Instead, the user must manually dial the phone, then place the handset in the modem. Completely automatic and unattended operation is not possible with acoustic coupling. Despite these weaknesses, acoustic coupling has a place whenever a very flexible modem that can be used with virtually any phone is needed.

The alternative is *direct connect modem*. This has become the more commonplace modem in the last few years. The direct connect modem actually plugs into

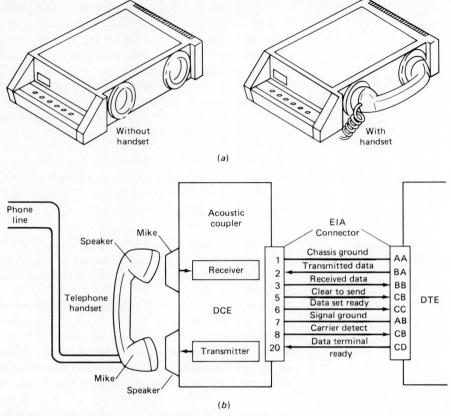

Fig. 10-20 An acoustic coupling modem: (*a*) physical appearance; (*b*) schematic. The phone handset is placed into special cups on the modem, and there is no electrical connection between the phone line and the modem.

the phone line, using the existing plug that normally goes into the phone (Fig. 10-21). Most of these modems also have a connector so that the phone can be connected and still used as long as the modem is not in use. The direct connect modem can be used with any phone line that has the proper plug. If a plug that does not match is on the phone line, or the phone is an older one with a screw connection box, some special cables have to be made, but the principle is the same. The direct connect modem puts its tones, as voltages, directly onto the phone wires, without any microphones, earphones, or acoustic sounds along the way.

Using direct connection, baud values and overall performance are not limited by the audio noise in the areas or microphone or speaker distortion. Direct connect modems can dial automatically, answer, and recognize phone line tones such as the busy signal. The direct connect modem offers high performance at low cost, since the acoustic elements (microphone, earphone, and rubber cups) of the acoustic coupling modem are relatively expensive compared to those of the all-electronic direct connect modem.

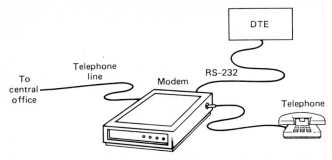

Fig. 10-21 The direct connection modem plugs into the phone line. The modem usually has another jack so that the phone can still be used without disconnecting and reconnecting cables.

This section has examined the concept of a modem and the way it forms a part of the overall data communication system. It has also shown some of the practical aspects of modems and their features as well as limitations. The next section studies some specific modems that are in industrywide use. These modems use various techniques and modulation schemes to achieve the desired performance. Examining these modems in detail provides a good understanding of some practical considerations with which modem designers and users must deal.

QUESTIONS FOR SECTION 10-5

1. What is a modem? What does the word "modem" stand for?
2. Why must modems be used in pairs?
3. What is the concept underlying a modem? How does it combine many concepts of communications?
4. What are the functions of a transmitting modem?
5. What are the functions performed by the receiving modem?
6. What is a stand-alone modem? What are its pros and cons?
7. What is the limitation of the RS-232 interface to a modem?
8. What is an integral modem? How does it differ from a stand-alone unit? What are the advantages and drawbacks of an integral modem?
9. How does an FM modem represent 1s and 0s? What functions must the modem receiving these signals perform?
10. What are the baud value ranges for low-, medium-, and high-performance modems? What is the difference in internal implementation?
11. Why are originate and answer signal pairs needed for full duplex?
12. Why doesn't half-duplex communication need this pair of signals? What is the drawback of half duplex? Where is it useful?
13. What are two limitations to the data rate for two-wire and full-duplex communications?
14. Compare a two-wire and a four-wire full-duplex system.

15. What is technically difficult to achieve in a two-wire modem for full duplex? Explain.
16. What is acoustic coupling?
17. What are the virtues and drawbacks of acoustic coupling for modems?
18. What is a direct connect modem? How does it work? What are its advantages?

PROBLEMS FOR SECTION 10-5

1. Sketch the waveform for a modem using 1000 Hz for 0 and 2000 Hz for 1, when the data is 0101.
2. Repeat Prob. 1 for data 0011.
3. What is the sequence of frequencies for the following data patterns: 01110101, 00001111, and 11001101? (Use 1000 and 2000 Hz.)

10-6 Some Specific Modems

The data communications industry has developed many modems for various applications and performance levels. Some of these are provided only by a single company. In other cases, one company has produced a modem that is so popular that it becomes a standard model, which other companies then also offer. This chapter will study a few of these more common and standard modems, along with some of the ICs that are used in building an integral type of modem for use in a personal computer. These standard modems were originally developed by the Bell Telephone System. They are known in the communications industry by their Bell System model numbers: the Bell 103, the Bell 212, and the Bell 202 units.

The Bell 103

The *Bell 103* unit is one of the simplest modems available and provides full-duplex communications at up to 300 baud. The modulation used is *frequency-shift keying*, with one pair of frequencies for each direction of data communication. Figure 10-22 shows the frequency spectrum for this modem. The outer envelope is the bandwidth of the normal telephone line. The frequencies used must be within this envelope. For the 103-type modem, the frequencies used are also shown in the figure. The originating modem uses 1070 Hz for space, and 1270 Hz for mark. The answer modem transmits with 2025 Hz for space, and 2225 Hz for mark. Even when the receive level is as much as 50 dB below the transmit signal level the modem can still filter out the received tones properly.

Consider the sequence and timing of frequencies that are seen on the telephone line if this modem sends the letter D using one start bit, seven ASCII bits, no parity, and one stop bit. The bit pattern is space (for start), 0010001 (letter D in ASCII, sent with LSB first), and mark (for stop). The 103 modem in originate mode sends the following pattern of frequencies:

1070, 1070, 1070, 1270, 1070, 1070, 1070, 1270, 1270 Hz

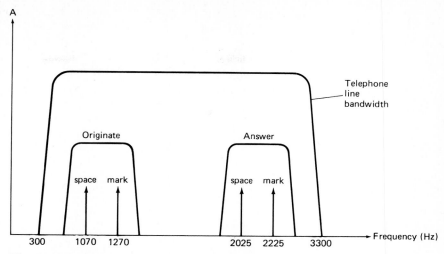

Fig. 10-22 The frequency spectrum for the tones of a Bell 103 modem.

A 103 modem in answer mode sends the following pattern:

2025, 2025, 2025, 2225, 2025, 2025, 2025, 2225, 2225 Hz

The Bell 212

The *Bell 212* full-duplex modem supports data communications with two speed ranges: 300 and 1200 baud. The 300-baud mode is asynchronous only and is used when the telephone line is of lower quality. The 1200-baud mode is either synchronous or asynchronous. The modulation in the 300 baud mode is identical to the modulation for the Bell 103 type of modem, with the same frequency pairs.

For the 1200-baud operation, there are two significant changes in the modulation. Instead of FM, the Bell 212 uses phase shifting between various digital values. The amounts of phase shifting are 0, 90, 180, or 270 degrees. Also, instead of using a single phase shift for each bit, the Bell 212 uses groups of two bits at a time, called *dibits*. For each pair of bits—the dibit—the modem causes the phase of the modulated waveform to shift by some amount as shown in Table 10-1.

Table 10-1

Dibit	Phase Shift, degrees
00	90
01	0
10	180
11	270

Phase modulation is used because at these baud values it is capable of better performance, with fewer errors. The originating and phase modulate a 1200-Hz signal, and the answer and phase modulate a 2400-Hz signal. Dibits are used because they allow the higher rate of bits per second without using more bandwidth than the standard voice bandwidth telephone line can support. If a phase shift were made for every bit, there would be twice as many changes/second. This would require a much greater bandwidth than the dibit scheme. The phase modulation pattern for the letter D, sent the same way as for the Bell 103 modem, would be as shown in Fig. 10-23. The bits are grouped in pairs, and then the phase shifts are applied. (Note: This is really only a 600-baud rate, but 1200 bits/s are transmitted.) Even though the number of bits in this case is odd, the bits after the stop bit are just like more stop bits, since the line is in idle state.

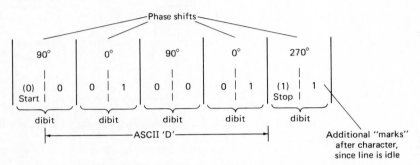

Fig. 10-23 The phase-shifting pattern that occurs when a Bell 212 sends the letter D using asynchronous ASCII characters.

The Bell 202 Modem

The *Bell 202* modem is designed for half-duplex operation only at 1200 baud on the dial-up lines, and 1800 baud on leased, conditioned lines. It uses two frequencies: 1200 Hz for mark and 2200 Hz for space. (Since it is half duplex, the issue of originate frequency pairs versus answer frequency pairs does not arise.) When used on the dial-up lines, the received signal can be anywhere from 0 to -50 dB below the transmitted signal; on leased lines the received signal must be from 0 to -40 dB of the transmitted signal.

Half-duplex operation is acceptable for many applications, but there is often a need for having the receiving end send special signals or information to the transmitting end. The 202 does this with a special reverse channel transmitter and receiver, operating at 387 Hz. This reverse channel can be used at very low rates, up to 5 bits/s, to signal in the reverse direction with simple amplitude modulation. Figure 10-24 shows the frequency spectrum of the 202-type modem.

ICs for Integral Modems

The popularity of personal and portable computers has produced a need for modems built of a relatively few ICs that are small and consume little power. These integral modems are either built into the main circuit board of the computer or on a small circuit card that is plugged into the computer, in an extra card slot. The simplified schematic of such a modem is shown in Fig. 10-25. This modem does not use the RS-232 interconnection to the computer. Instead, it connects directly to the data bus of the processor within the computer. Three main ICs are needed to provide the performance equivalent to the Bell 103 and 202 models discussed previously. A UART IC gets the characters to be transmitted, or those that have been received, and acts as the interface between the data bus, which provides parallel character bits or expects to see character bits as a

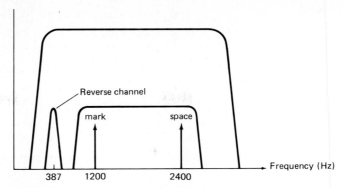

Fig. 10-24 The frequency spectrum of the Bell 202 modem, including the special low-speed reverse channel.

parallel group. The UART connects to a modem IC which provides the necessary modulation and demodulation, filtering of the received frequencies, generation of transmitted tones, and other modem functions. Finally, an interface IC is used to meet the strict telephone line specifications on allowable signal levels, detect the ringing signal on incoming calls, and protect the rest of the modem against excessive voltage on the telephone line.

The three ICs work together to provide a low-cost modem which provides reliable performance in a very small space. The entire modem requires only a few hundred milliwatts of power with supply voltage values of plus and minus 5 V and plus and minus 12 V.

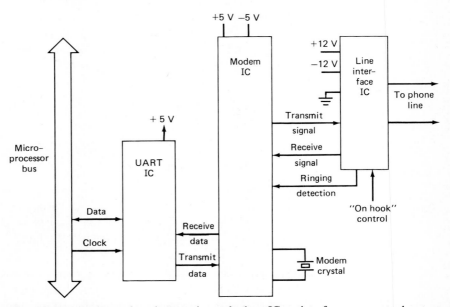

Fig. 10-25 An integral modem requires only three ICs to interface a computer bus to a phone line. A UART connects to the processor bus and in turn connects to a modem IC which provides the required tones. A line interface IC provides the required line interface functions.

Telephone Systems and Modems   **351**

QUESTIONS FOR SECTION 10-6

1. What is the baud value for the Bell 103 type of modem?
2. What modulation is used for the type 103 modem? What are the originate and answer frequencies?
3. What signal level can the type 103 modem handle on the transmitter while it is receiving signals?
4. What modulation is used for 1200-baud operation of the Bell 212 modem? Why?
5. Compare 300-baud and 1200-baud operation of the type 212 modem.
6. How does the Bell 212 modem use pairs of bits? Why does it use this scheme?
7. Why doesn't the Bell 202 modem have an originate and answer pair of signals?
8. What is the reverse channel of the type 202 modem? What is its frequency? What modulation is used in this reverse channel?
9. With present-day IC technology, how many major ICs are needed for an integral modem? What are their functions? What baud value is achieved?

PROBLEMS FOR SECTION 10-6

1. What frequency sequence is sent by the Bell 103 modem when in originate mode, with seven character bits, no parity, and two stop bits for the letter Q in ASCII?
2. Repeat Prob. 1 for the modem in answer mode.
3. Repeat Prob. 1 for sending the digit 8 in ASCII, with one stop bit.
4. What pattern of frequencies is sent by a Bell 103 in answer mode, when sending symbol & in ASCII, with seven characters and one stop bit?
5. Using a Bell 212 modem in originate mode, what frequency is used to send data? What are the phases that are sent for binary 10100100?
6. Use the character conditions of Prob. 1, and show the phases that are sent at 1200 baud by the type 212 modem.
7. Repeat Prob. 4 for the Bell 212 modem at 1200 baud.

10-7 Other Specialized Modems

Not all modems are designed for connection between a DTE and the telephone line. There are other interconnection media that are used in many cases. The same underlying modem concepts are used in these cases. This section will discuss some of these other important modem types.

Fiber-Optic Modems Like any electrical wire, the telephone loop is very susceptible to electrical noise from machinery, lights, motors, and radio transmitters. It can also be tapped by knowledgeable individuals. Optical fibers offer absolute immunity to electrical noise, interference, tapping, and harsh environmental conditions such as water and dirt and provide electrical isolation.

In order to use data communications equipment with fiber optics, special modems are used which replace the line driver and receiver circuitry that normally interface to the telephone wires with optical fiber interfaces. The data signal from the DTE is used to on/off modulate a light source such as an LED, which puts light waves onto the optical fiber cable. At the receiving end, the on-off light causes two different amounts of current to flow in an optically sensitive phototransistor, which then acts as the demodulator from light energy to electrical energy. It is possible to buy fiber optic modems which are identical to telephone modems in performance (in terms of baud values and full-duplex operation) but use fiber optics instead of the phone interconnection. These modems are transparent to the communications user, since the DTE cannot even tell which is being used.

The potential bandwidth of the optical fiber is many times greater than the bandwidth of the telephone line. Some modems use non-RS-232-type interfaces to allow data rates from the DTE to the DCE of hundreds of kilobits per second or even megabits per second, and can support these rates on the wide bandwidth optical fiber. Phone companies are now installing local loops of fiber optics on a preliminary basis. These will require a fiber-optic modem and offer many performance enhancements.

Direct Connection (Short Haul) Modems

In many applications the two DTEs are relatively fixed in location, within a single building, for example a computer and its terminal in an office building. It may be unwise to use a telephone line as the medium, since the line is more useful for connections that have to be moved around, or for people who are talking. In other cases, the two DTEs may have to connect in areas where there are few phone lines, such as to and from computer terminals scattered around a large refinery at distances of several miles.

A special modem called a *direct connect*, *short haul*, or *metallic modem* can be used. Technically, this modem is very similar to a telephone line modem, with one exception. This modem is designed for use where there is an absolute, direct wire path, without breaks or switching, between users (which results in the term ''metallic''). Since the interconnection medium is dedicated to this user pair, its characteristics do not change each time the connection is made. Unlike long-distance telephone connections, there is no radio, satellite, or unknown link that can be part of the channel. This modem is designed for a more limited distance and is capable of high performance within that distance. The design of this modem is simpler than that of a telephone modem because there is a direct, continuous electrical path. In fact, this modem can check whether the other modem is connected simply by sending a current through the loop—if current flows, the loop must be complete and the other modem is connected. Telephone line modems cannot do this for two reasons. First, it is not feasible for technical reasons to try to force current into the telephone loop. Second, it would not do much good to do so, since there are interposed elements in the telephone loop that can give false readings, such as transformers and signal couplers. The telephone system ensures that signals get through, but does not promise that solid wire is used end to end.

Digital Modems

The normal telephone line is designed for analog signals. In recent years, the phone system has been installing special digital data lines which are intended for digital signals only. The bandwidth, frequency response curve, and time delay characteristics of this type of line are optimized for digital signals.

However, common digital voltages of 0 and 5 V cannot be put on the digital line. Usually, the voltages used must be centered around 0 V, to ensure that some types of inevitable distortion are minimized. Also, the digital lines are usually used in synchronous mode, so that highest performance and throughput are achieved. This means that some sort of formatting which includes clocking information is desired.

The digital modem meets these needs. It takes the digital voltages that the data source uses and converts these to *bipolar* (two voltages one on each side of 0 V) *signals*. It also performs some form of encoding, such as the Manchester encoding discussed earlier in the book, to provide a clocking signal that can be used for synchronization at the receiver.

Multiplexers and Concentrator Modems

Suppose an office has 10 terminals that need to connect to a central computer in another building, either nearby or distant. Ten telephone lines and modem pairs can be used and can provide the needed performance. As the number of terminals increases beyond 10, additional individual loops and modems are required. This may be costly and tie up many telephone lines that are needed for other purposes.

A *multiplexer* (mux) combines several lower-bandwidth data communications channels into one or several higher-bandwidth channels, using time or frequency multiplexing. The total bandwidth needed is approximately equal to the sum of the individual bandwidths. The company then rents, buys, or installs the higher-performance links and can serve all the terminal users. The multiplexer not only provides the combination function, but also the modulation and demodulation function that a modem must provide. The mux is a multichannel modem and provides the following advantages:

Often many lower-bandwidth links can be replaced at less cost with one much higher-bandwidth link or several moderate-bandwidth links.

If the mux and channel are selected with some excess capacity, then adding a few more users is easy and inexpensive. No new DCEs or phone lines are needed.

Only one piece of equipment is needed, instead of many smaller ones. This applies at both the computer terminal and the computer end of the system. This saves a great deal of hardware cost and makes repairs easier.

There is also another aspect of the mux that results in further savings. Most of the terminal installations are not active all the time. A typical computer terminal may be in actual use only 5 to 25 percent of the time. This means that even a single line can be shared among several users. A special mux called a *concentrator* is used in these applications. The concentrator is able to "make do with less" by actively sharing some of the idle bandwidth among the terminals that need it. The goal is to have the wide-bandwidth lines used as close to 100 percent as possible. This is achieved by the concentrator, which interleaves the signals of active terminals with any unused bandwidth it can find. Here is the way this works in a system with 10 terminals but bandwidth for only 5 (assume the individual terminals are only used for about 10 percent of the time each): The concentrator is managing a link that can support up to 5 terminals full time. As each terminal needs service, the concentrator assigns it to the next available spot. When a terminal is finished, the concentrator internally notes that the spot previously occupied by the terminal is now vacant (Fig. 10-26). It is somewhat like having parking spaces for cars. New cars entering the parking lot take the place of some cars that have left. It is not absolutely necessary to have a spot reserved for each car that may come.

Of course, there is a chance that all spots will be taken and a terminal will have to wait. If the concentrator is selected on the basis of a careful study of the number of terminals and the amount of data they send, this should happen very rarely. The concentrator has internal buffering for each terminal, so that even if a spot is not available immediately, the terminal can still go ahead and does not have to come to a complete stop. The concentrator also allocates its slots on a very rapid basis (typically, every 10 or 100 ms) so that it can assign slots and reassign them as soon as they are available. What the concentrator often does is use some of the "dead time" between characters when someone is typing at the terminal keyboard, for example. Because a concentrator design is based on probabilities that not all users are active all the time, it is also known as a *statistical multiplexer*.

What is the difference between a multiplexer and a concentrator? A multiplexer is a relatively simple device, which takes a large group of low-speed channels and combines them into one or several higher-speed channels. Each of the low-speed users has a unique spot assigned, and that spot is unused when that user is idle. There is no chance that a user will have to wait. In contrast, the concentrator constantly reassigns available spots to various users, on the basis of the continuous change in demand. This results in a more efficient system, because less bandwidth and a lower data rate are needed at the concentrator output to serve the same number of users. The concentrator has two drawbacks: It cannot guarantee that every user will be served immediately (but it can come very close, with 98 or 99 percent of users never having to wait). It also is a more complicated device than a mux and so costs more to buy, install, and service. The increased cost of the concentrator may outweigh the savings in channel costs.

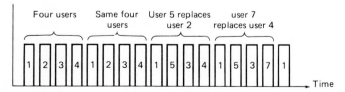

Fig. 10-26 A data concentrator allows many users to share fewer lines, by assigning unused, available time slots to new users. As users drop off, new ones can replace them, so no time slots are left unused.

QUESTIONS FOR SECTION 10-7

1. Why is a fiber-optic modem used? How is it similar to a telephone line modem? How does it differ?
2. What data rates can be achieved with a fiber-optic modem?
3. What is a transparent fiber-optic modem?
4. Where is a direct connect modem used? Why? Why is it sometimes called *metallic*?
5. Why is a direct connect modem simpler than a telephone line modem?
6. Where are digital modems needed? What function do they perform?
7. What is a data communications multiplexer (mux)? What savings does it offer?
8. What is a concentrator? How does it operate? What advantage does it have compared to a mux? What disadvantage?
9. Compare the operation of a mux versus a concentrator.

SUMMARY

The phone system was designed for analog signals but is a very important medium for the transmission of data signals. The local connection between the phone company exchange and the phone is called a *loop*, and the phone office supports this loop with special circuitry called an *SLIC*. The SLIC provides six basic interface functions, summarized by the acronym "BORSHT." Dialing to the SLIC is either make/break pulses or tone pairs. The tone pairs are more versatile, more rapid, and more compatible with computer-based equipment.

Telephone lines can support various data rates, depending on their frequency response and time delay characteristics. Phone lines which are perfectly adequate for voice may not be best for data, however. The modem (modulator/demodulator) is the basic interface between the data signals and the phone lines and system. Modems send tones (sine waves) through the phone system, and they

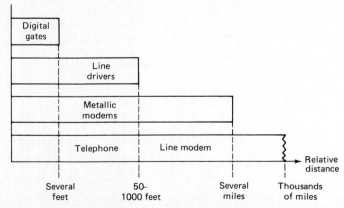

Fig. 10-27 The distances achieved with different types of data communications interfaces, from digital gates (short distance) to telephone lines and modems (worldwide).

modulate these tones to represent the data 1s and 0s. Different modulation schemes provide differing performance levels, at several baud values. The four-wire modem provides better performance than the regular two-wire version in full duplex, because the inevitable echo of a duplex link is avoided and signal overload is much less of a problem.

The modem is the longest-distance data communications interface. As shown in Fig. 10-27, there are four groups of interfaces. The shortest distance comes from ordinary digital ICs such as gates. Line drivers provide greater distance, and modems provide much greater distance. The metallic direct connect modem is used over long distances, and the telephone line modem can provide the longest distance of all, sending data signals around the world via the telephone system.

END-OF-CHAPTER QUESTIONS

1. Give two advantages and two disadvantages of using the standard telephone system for data communications.
2. What does the loop from the local phone exchange see when the phone is on-hook? Off-hook?
3. Why does each phone in use need an SLIC? What does the SLIC do?
4. Explain, briefly, each of the BORSHT functions: battery, overvoltage protection, ring trip, supervision, hybrid, and test.
5. Explain the need for a two-wire-to-four-wire hybrid circuit. What problems can a hybrid cause?
6. What are the names often used for the two wires from the phone office to the phone? What power supply value is most commonly used for standard telephones?
7. Compare pulse dialing and tone dialing in these areas:
   a. The way the dialing is done
   b. The appearance of the dialing signal
   c. The time to dial as compared to the actual specific digits
   d. The advantages and disadvantages of each
   e. The circuitry used to dial
8. Explain the concept of DTMF dialing. Why were those tone frequencies selected?
9. What else can the DTMF tones be used for? Why can this be done, although pulses from pulse dialing cannot be used for other activities?
10. What is the purpose of the fourth column in DTMF signals?
11. What is frequency response? Why is it an important consideration for determining the baud value that a link can support?
12. What is the time delay characteristic? How does it affect achievable performance?
13. Why are the frequency response and time delay needed for data communications different from those for voice?

14. Compare voice grade conditioning on a phone line with leased line conditioning.
15. What is a dial-up or switched line?
16. What causes echo? What problems does echo cause for communications?
17. What is the difference between echo suppression and echo cancellation? How is each achieved? What are the pros and cons of each?
18. For what types of distances is echo a problem? How can echo be a problem even when the two users are located relatively close to each other?
19. Why would a business or factory want its own phone office? What are the benefits?
20. What can the PABX do for the phone user that the regular phone company exchange may not be able to do?
21. What determines the number of outside lines versus inside lines needed on a PABX/PBX? What kinds of users need more lines?
22. What is the function of the modem? Why is it necessary?
23. Explain why modems must be used in matched pairs.
24. Explain the way a modem takes the data signal, in RS-232 format, and puts it onto the telephone system.
25. Explain in what way a receiving modem performs the reverse operation of the transmitting unit.
26. Compare the physical appearance and performance of the stand-alone modem and the integral modem.
27. Explain the way that a modem may use FM to represent signals.
28. Why must pairs of signals that differ be used for the originate modem and the answer modem? Explain why a half-duplex modem does not need more than one pair of signals.
29. Why is the two-wire, full-duplex modem limited in performance potential? How does a four-wire or a half-duplex modem overcome this?
30. Compare the virtues and problems of acoustic coupling for modems versus direct connect modems.
31. Compare the baud rates, modulation techniques, and operation (half versus full-duplex) for the Bell 103, 212, and 202 modems.
32. What is the reverse channel on the type 202 modem? How is it implemented?
33. What are the three main functional blocks in an integral modem built from present-day ICs?
34. Where would a fiber-optic modem be used? Compare its similarities, differences, and possible advantages to those of the standard telephone modem.
35. Where would a metallic modem be used? How is it related to a telephone line modem?
36. Why can't digital signals from electronic circuitry be plugged directly into the telephone line if the line is designed for digital signals?

37. Where and why is a mux used? Why is a concentrator an improvement over a mux? Why is a simpler mux sometimes a better choice?
38. Explain the principle behind the operation of a concentrator. Compare this to a mux operation.

END-OF-CHAPTER PROBLEMS

1. How long is the start to finish dialing time for the following numbers, using pulse dialing: 123, 111, 777, 789? (Use the dialing rates given in the chapter.)
2. Repeat Prob. 1 for tone dialing, using the dialing rates in the chapter.
3. What is the quickest long-distance number to dial with pulse dialing (a total of 11 digits)? How long does it take?
4. What is the slowest long-distance number to dial? How long?
5. Repeat Probs. 3 and 4 for DTMF dialing.
6. What DTMF pairs are used to represent the following digits: 1, 5, 9, 0?
7. What harmonics can be generated as a result of signal or circuit problems with the digits 2, 3, 6, and 7?
8. What frequencies may result from modulation of the tone pairs for the digits 1, 2, 5, 7, 8, and 9?
9. The 1-MHz master clock of the DTMF IC is used to generate the tones for digits 1, 2, 5, and 8. What divisor (integer) should be used by the IC?
10. The master clock of 1 MHz is going to be used to generate a special tone at 600 Hz. What would be a good divisor? How about for a 950-Hz tone?
11. A system designer is making some echo time calculations. The worst-case distance from New York to Paris is by satellite, orbiting at 22,500 mi above the earth. What is the approximate time for the talker echo? The listener echo? (Assume signal speed is 97 percent that of light.)
12. The same New York to Paris link can be made with an undersea cable. The signal travels through the cable at 60 percent the speed of light. The distance is 6000 mi. What is the listener echo time? The talker echo time?
13. A New York to Los Angeles link uses cable, with propagation speed of 70 percent of light. The main source of echo is a repeater station exactly in the middle of the 3000-mi distance. What is the listener echo time?
14. A modem uses 1000 Hz for 0, and 2000 Hz for 1. Sketch the waveform seen on the signal line for the following pattern of data bits: 11001010. The data is being sent at 1000 baud. Show the timing of a single bit.
15. Using the frequencies of Prob. 14, show the observed signal for the letter T in ASCII. (Remember that the LSB is sent first.)
16. Repeat Prob. 15 for the case in which one start bit, seven character bits, no parity bit, and one stop bit is used with the character.
17. Two Bell 103 modems are connected and communicating in full duplex. Sketch the waveform observed when each is sending characters with one start bit, seven character bits, no parity, and two stop bits for the originate

modem sending letter Y and the answer modem sending letter S simultaneously.

18. The Bell 212 modem is used to send this bit stream: 11001101. What is the pattern of phase shifts?
19. The Bell 212 modem is used to send the ASCII representation of the symbol %, using one start bit, seven character bits, parity bit set to 1, and one stop bit. What is the pattern of phase shifts?
20. Repeat Prob. 19 for the parity bit set to 0.

11 Networks

Networks allow many users to share a common pathway and communicate with each other. The subject of networks is complicated, yet the use of networks is growing at a fast rate. More people have direct access to computers at their desks or factory equipment, and larger computers are transferring enormous amounts of data from building to building, from city to city, and even around the world. This can be done successfully only with a carefully laid out plan and definitions of the network rules and regulations. The design and installation of a network involve many considerations. What data rates, costs, data reliability, and numbers of users must be supported? All of these issues have different answers, depending on the application of the network.

In order to meet the needs of these applications, networks are available with different interconnection layouts and plans, methods of access, protocols, and media. A local area network is used for a limited geographical area. Wide area networks are used over longer distances and usually involve sharing the network equipment with other users who have messages of their own. The wide area networks often group messages going to the same place and then split them off to the intended receivers at the far end.

Cellular networks increase the number of mobile phone and terminal users that can be supported by a fixed amount of frequency bandwidth. Instead of one large central base station, many smaller base stations split the overall area into smaller cells. The local base stations of the cells must communicate with adjacent cell base stations and coordinate the operation of the cellular system.

11-1 What Is a Network?

A communications system that supports many users can be called a *network*. The term "network" means many different things in various applications, and there is no single "most correct" definition. By describing where networks are used, and under what conditions, however, the general meaning of the word as used in communications will be made clear.

A network serves many users, but not necessarily at the same time. The phone system is the largest network in existence, and many of the potential users—those who own phones—use it at any instant, but many other users are on-hook (inactive). The many potential users of the phone system network do not share a common application. This means that most of the users are independent of each other and have no interest in the messages and activities of any other users, except the one they are talking to. The phone network is designed to let any user talk to any other user, and to allow many such conversations to take place at the same time. It also allows many users to call in to a central point, such as a toll-free number, central computer, or prerecorded message center.

Many networks are not related to the telephone system. These networks bring together users who do have a common purpose, application, or organization. The users can be spread across wide distances or located in a single building. What are some typical examples of these networks, which tie together many users who have something in common?

A company may need to link together all of the executives with access to some critical company data, such as sales figures, production plans, or personnel information. The executives can each call up the needed information on their terminals (after issuing the right password). The data is located in a central computer, which may be miles away from some of the users.

An electronic mail system for a large corporation can link together many offices and factories spread across the country or even the world. Instead of sending memos on paper, which may take days to arrive, the electronic mail system makes any typed messages instantly available to the intended readers. Each reader can then save or discard messages. The sender of the message can retrieve any memo without having to shuffle through paper files.

A manufacturing factory can tie all aspects of the manufacturing operation together so that the entire production activity is coordinated and organized efficiently. Orders are entered into the system, and the correct parts are requested at the stockroom. The production equipment on the shop floor manufactures the product and keeps the rest of the factory informed about the manufacturing cycle. When the manufacturing is complete, the shipping department and the customer are notified that the item is ready to go. The accounting department is also informed that the finished product has been shipped, so it can send a bill to the customer.

One of the largest networks in the world is run by the National Aeronautics and Space Agency (NASA). This network combines all of the worldwide space tracking stations, the NASA research facilities, and the NASA operational offices. Data from any location can be instantly sent to any other, as can memos, messages, and conversation. The network is so advanced that it is as if all the employees were under one roof, in close physical proximity with each other, rather than spread all over the earth (including some very remote locations).

Networks can be generally grouped into open networks and closed networks. An *open network* is public and is available to virtually anyone. The telephone system is a good example of this, of course. In an open network, special care must be taken that unauthorized users do not access information, and that authorized users do not damage information that the network can provide and cannot physically damage the network, except perhaps where they physically connect. In a *closed* or *private network* the situation is much easier. Only users who have a direct, close relationship to the application have connections. A typical closed network is one which links together people and machinery in a manufacturing factory so that products and the data needed to produce them can flow wherever needed. Outsiders cannot connect to this network—only company employees with the special network interface assigned to them can.

Networks can be used between people, between people and computers, between computers and equipment or machinery, or between computers only. The modern corporation, either the office or factory, uses networks in all of these ways. Depending on the application, the network needs different speeds, degree of reliability, and cost.

These are just a few of the applications of networks. What is not generally considered a network? A dedicated channel between two users usually is not considered a network. The system which uses a direct, metallic modem to connect a terminal to a computer several miles away is not a network. A system designed for only a few users, from two to perhaps five, is really a very small subgroup of what is usually considered a large network.

A network is much more than just a physical interconnection of all the users. There are many issues to resolve and plan. Can one user talk to more than one other at a time? What if several users try to talk to another one at the same time? How does the network handle noise and errors? What happens if there are any network failures, in either a user's circuitry or the interconnection circuitry—does the whole system stop, or only the failed part? How does the network ensure that messages reach the intended user in a timely fashion, without excessive delay due to the network operation?

When these network issues are analyzed technically, the answers and solutions involve many aspects of the network. These aspects are as follows:

Topology, which is the interconnection plan for many users. Various topologies can be used, and each has performance or cost advantages and drawbacks.

Protocols chosen.

Types of layers in the protocol, ranging from the lowest level (such as ASCII serial) to the highest level (the way the overall message is presented, along with preamble and postambles).

Electrical interface: RS-232, RS-422, or others.

Type of modulation used. The carrier used, if any, is also a factor.

Medium chosen. The many types of wire and cable each have a role in speed, distance, reliability, ease of adding new users, and security. Fiber optics is also a choice. Closely associated with the medium are the physical connector and the way that new users are attached or detached from the network.

Network procedures for handling problems, errors, or noise.

This chapter will examine the various aspects of networks in detail, as well as some actual networks in use. For each issue studied, there is no single best choice. There are many valid ways to achieve the network goal, and each of these has reasons why it is desirable and reasons why it may be less than ideal. It depends on how the network will be used, by whom, for what applications, with what expected performance, for what distances, and at what allowed cost. All of this must always be considered with the basic goals of the network in mind: to allow users to send messages to other users and to allow users to share information and data on the network.

QUESTIONS FOR SECTION 11-1

1. What is a network? What are some of the general characteristics of a network?
2. Are networks capable of letting only two users communicate at the same time? Explain.
3. Do network users all have something in common with each other? Give examples of when they do and don't.
4. Give three examples of networks in use today.
5. What is an open network? Give an example.
6. What is a closed network? Give two examples.
7. Give an example of communication that is not achieved via network in the usual meaning of the term.
8. What are some technical issues regarding development of networks?
9. Explain why there is no best single network choice available.

11-2 Topology

"Topology" is a mathematical term which refers to the way the interconnection paths between the many users, or *nodes*, are arranged. There are many ways to run the channels that allow data to go from one user node to another user node. The simplest way, at first, is to have a path from each user of the network to every other (Fig. 11-1). Although this is effective, the number of links increases very quickly as the number of users increases. The number of such links needed is given by the following formula, for N users.

$$\sum_{n=1}^{N} (n-1) = \frac{N(N-1)}{2}$$

Using it, it is easy to calculate the number of links for a desired number of users, as shown in Table 11-1.

Table 11-1

Number of Users	Number of Links Needed
2	1
3	3
5	10
10	45
50	1225
100	4950 $= \dfrac{100(100-1)}{2}$

This topology can be effective for very small numbers of users, but in practice it is hardly ever used. Even if the performance and number of links is acceptable, it is very expensive to add one more user. Most network designers must allow for

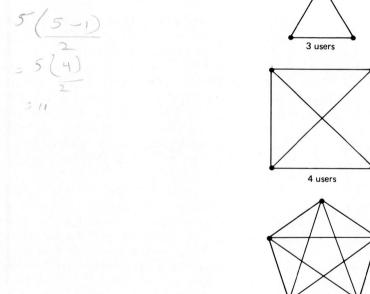

Fig. 11-1 A network can use a link to connect each user to every other user directly, but the number of links needed grows quickly with the number of users.

Networks 365

the simple fact that new users will probably have to be added and that the cost and aggravation of adding them should not be excessive. Consider going from a 99-user system to a 100-user system—99 new links are needed! Adding the 101st user requires another 100 links. In the real world, this is not a workable scheme.

The networks in data communications are usually one of the following three topologies: star, bus, or ring. These are the fundamental network topologies, and they can serve as building blocks for even more complicated topologies in turn. These complex topologies are found in very large systems, such as the phone system of the worldwide NASA network. The analysis of the large systems is performed by dividing the overall network into these smaller, basic network topologies. Equally importantly, many networks in use today are made up of one simple, fundamental topology itself. For these reasons, it is very useful to study the star, bus, and ring in detail.

Each of these is studied in terms of its flexibility, ease and cost of adding another user, reliability, and key features and drawbacks.

Star Topology

The *star topology* uses a single central node, and all the users are connected directly to this point (Fig. 11-2). The number of users can be as large as the central node can handle. This can be anywhere from a few to dozens or even several thousand users. The local telephone exchange is a star network designed to handle 10,000 users; this quantity is as large as is practical in most cases.

The reliability of this topology is high, since a problem with any single user does not affect other users, if the interface circuitry at the central node is properly designed. There is really only one weak spot in this design: the reliability of the central node itself. Every effort is made to design the highest reliability into this part of the network, often with extra circuitry double-checking and triple-checking performance of the primary circuit.

In operation, the data from any user passes through this central point on its way to the intended receiver. In some star systems, there can only be one data message passing through at a time. In more complicated systems, many paths through the central node can be in use at the same time. The telephone office, again, is an example of the type of star network which allows many connections between pairs of users at the same time. The single-message system is, of course, simpler than the many-message system.

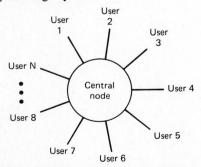

Fig. 11-2 The star network topology has a central node to which each user is connected, and the path from one user to the other is made at this central node.

What does the central node do? In the simplest approach, it merely acts as a large switch to direct the data from the sender to the intended receiver. The central node is the key point of the network, but beyond switching the data onto the correct path it adds nothing to the performance of the overall system. More advanced designs put some computer power into the central node itself. Suppose the sender wishes to send the same data to several other users. In the simple design, the user has to send the message once for each intended receiver and direct it to each receiver in turn. The more advanced star configuration allows the sender to tell the central node who the intended receivers are, and the central node resends the message to each of the desired receivers. This capability frees the sender from this basically repetitive and time-consuming task, so it can do other things such as process new data and make decisions. The central node acts to relieve part of the communications burden from the sender.

Expanding this star configuration to handle another user requires just one more link from the user to the central node. This is easy in concept. In practice, its cost may be high, if the distance from the new user to the central point is large. The star topology, for these reasons, is said to have ease of expansion with high cost. "Just one more wire" may necessitate running several miles of new cable from the new user to the central node, and there may already be a cable from an existing user nearby, to add to the frustration.

Where is the star topology used in data communications? Besides the telephone system local exchange, the star is used mainly where the users are located within a clearly defined physical area. Often, the system is installed with many extra wires to handle new users—the wires are in place but not connected until needed. This is because it is relatively inexpensive to put wires in place all at the same time. The wire itself is not the major cost; instead, the expense is going back later to add just one more wire. A building under construction may be prewired with cables going to every room and section on each floor, with all the wires running back to a central computer which maintains and controls the heating and cooling. The parts of the building which are unoccupied are sealed off and the wires not used. When those parts of the building are finally rented, the wires are put to use.

The star topology is also used when more than one pair of users must send data at the same time. If the central node of the star is designed for multiple pathways, many user pairs can be connected to each other simultaneously. This is what the central office exchange at the phone company does. Since there is already in place a wire from each user to this central node, the only cost associated with allowing multiple conversations is the additional circuitry within the central node itself. The rest of the network remains unchanged.

Bus Topology

In *bus topology*, a common pathway called a *bus* is shared by many users (Fig. 11-3). The bus is an elaboration of the multidrop studied earlier in the text. Users can be connected to the bus at any point. Adding a new user is mainly a matter of physically connecting the user to the nearest point on the bus. No new links are needed, although the bus may have to be extended physically. In data communications, the bus length can be anywhere from a few inches or feet (for example,

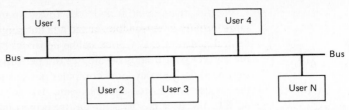

11-3 In a bus topology, a common pathway called a *bus* is shared by all the users. Only this one pathway is needed to interconnect all users.

within a computer chassis) to several thousand feet (around an office or a small building).

The bus is a very flexible structure because users can be put anywhere along its length and added for very low cost and with little difficulty. If the bus interface circuitry is designed properly, most failures of a user will not affect other users or the bus operation. However, it is possible to have failures which electrically short-circuit the bus and make it unusable by anyone. This is a weakness of the bus—a single user can bring the system down. There are design precautions that can be taken to minimize the chance of this occurrence, such as extra rugged interface circuitry that fails in a way that does not cause a short circuit. Optoisolators are often used for this purpose.

The cost of the bus itself is fairly low. There is no central node which has to be built and installed, as there is in the star topology. For these reasons, the bus is a popular configuration.

The bus does have some weaknesses, besides the possibility that a failed user can stop the entire bus. Most important, since all the users share a common path, only one conversation or data message can be passing on the bus at a time. If two users are passing data, all other users must wait before they can have access to the bus. This disadvantage may be an advantage in some applications, though, in which a user needs to send data to many other users. The bus then is a common broadcast medium with many users listening simultaneously.

The second major weakness of the bus is that it requires more management than the star. Since the bus is shared, there has to be a protocol that is used to decide whose turn it is, what to do when another user is active, and how to prevent data from various users from colliding on the bus. There are many protocols that can accomplish this, but all are more complicated than the simple star, which really needs very little management (since each user has its own path to the central node). This protocol is more difficult to design, to keep working, and to fix. Even if it works perfectly, the back-and-forth messages of the protocol use the precious communications element of time—some of the bus traffic is not the actual user data, but setup and completion messages of the protocol. This limitation reduces the efficiency and throughput that the bus can achieve, even if the actual data rate is high.

Despite these limitations, the bus topology is one of the most popular. Many commercial networks in use are based on it. Virtually all computer systems, in fact, use the bus to interconnect the various circuit cards within the system because of its advantages. Buses are popular for applications in which there is a need to add users or move them around frequently, and the cost of running the bus initially is reasonable. If the bus is being used at a high bit rate, and the bus management is not very complicated, it can provide good performance for many users.

Ring Topology

The *ring topology* connects all the users in a large ring (Fig. 11-4). Data from any user must pass through the other users along the way until it reaches the intended receiver. Adding a new user means breaking into the ring, and this is sometimes not convenient. It also means that besides the physical break, the network is temporarily out of service while the break is made, the new user is installed, and the network is reconnected. However, adding new users does not require running any additional wire or cable since the new user simply goes in the existing ring (which may require some extension).

How does the ring topology operate? Messages are passed from user to user, sort of like a bucket brigade. Just as with the bus topology, the ring has many users sharing the same physical path, and management is needed. For the bus this involves complicated protocols. For the ring, the protocols may be much simpler. The function of each user is to take an incoming message, check whether the message is for it, and pass it along if it is not. If it is, the user keeps the message and instead puts an "accepted" message onto the ring. The "accepted" message is passed along until it reaches the originator of the message, who then knows it has been received. The key to this operation is that each user performs a relatively simple operation: receive message, check whose it is, accept it, or pass it along. The overhead in the protocol to do this is less than in the bus protocol. Specialized circuitry can be designed which does this efficiently, and not much of the ring time is taken by this protocol. Thus, the efficiency of the ring can be high. A message intended for several users is also simple—each user reads the message and then passes it along unchanged.

The ring topology is especially useful when the interconnection medium is not wire, but optical fiber. This is because present-day technology does not have an easy way to put taps into a fiber-optic cable. These taps are needed for the bus type of interconnection, in which the user is connected into the common bus.

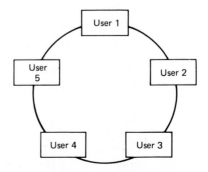

Fig. 11-4 A ring topology has each user as part of the signal path which interconnects all users.

Instead, fiber-optic connections are more easily made with an in-line connection, in which the user circuitry has two connections—one for signal in, and the other for signal out. The ring topology is a good fit with this connection, since adding a new user means opening the ring and inserting the new user, with the in and out connections.

One drawback to the ring topology is that any failure at any user or interuser link may cause the entire ring network to stop. If a user is unable to receive and pass along the data it receives from the ring, the chain of events is stopped. Any break in the interuser link has the same effect. A ring is like a chain, and the entire ring is no more reliable than any single link (user). In order to overcome this limitation, the ring interface circuitry must be designed for highest reliability. Often, two rings are used. One serves as a spare, or backup, unit for the other ring. In this way, there is some insurance that allows the ring to continue to operate if there is a problem with either the links or a user.

QUESTIONS FOR SECTION 11-2

1. What is the simplest network topology, at least initially? What are its disadvantages?
2. What is the formula for the number of interconnections for this simple topology?
3. What is the effect of adding just one more user?
4. What is the star topology? Give an example of it. How many links for N users?
5. What is the overall reliability of the star? What is its weakest link? What is done to improve the reliability?
6. What does the star central node do, in the simple case? In the more advanced case?
7. What is the "ease of expansion" of a star? The cost? Explain.
8. Why is the star preferred when more than one pair of users must send data over the network at exactly the same time?
9. What is the bus topology? How are new users added?
10. What is the point in a bus network that is critical to reliability?
11. What is the weakness of a bus, in performance?
12. Why does a bus need more protocol than a star?
13. Where are buses a good choice?
14. What is a ring topology? How are new users added?
15. What are the advantages and disadvantages of adding new users in a ring?
16. How does a ring operate to send messages around to the various nodes?
17. Why is a ring efficient, in terms of the amount of data it can handle?
18. Why is the ring topology good for fiber-optic systems?
19. What is the main drawback of the ring? How is it overcome? At what cost?

PROBLEMS FOR SECTION 11-2

1. How many links are needed for the simple network in which every node is connected directly to every other node, for 60 users?
2. How many links does the simple network of Prob. 1 need for 75 users?
3. How many additional links are needed as the number of users of the simple network increases from 25 to 26? From 25 to 30?
4. How many links does a star network need for 10 users? For 25 users? When the number of users increases from 25 to 26? From 25 to 30?
5. A bus topology has 10 nodes. Two new users are added. How many new connections must be added?
6. A ring topology has 10 users. Two additional users must be added to the ring. How many splices must be made? (A splice is a cut in an existing wire.)
7. A ring topology has a loss of 1 dB of signal at each connection. How many decibels are lost in a 12-node ring when a signal passes through all nodes, back to itself?
8. A bus also has 1-dB signal loss at each connection. How many decibels of signal are lost from one end to the other, for 12 nodes? (Hint: not the same answer as in Prob. 7.)
9. Draw a network for seven users which connects every user to every other.
10. Draw a seven-user star network.
11. Draw a seven-node bus network topology.
12. Draw a seven-node ring network topology.
13. For the drawings of Probs. 9, 10, 11 and 12, how many links are needed in each case to support seven user nodes?

11-3 Basic Network Protocols and Access

The topology used is only one part of what is needed for a successful, properly functioning network. Because there are many users, some protocols must be established to define the rules of conversation: who speaks, under what conditions, what happens when two users need to send data, and similar issues must be carefully defined. There are many levels to the complete protocol, or *access*, as it is often called in networks. The most basic level is the interaction at the network topology itself. There are four common protocols in use: command/response, interrupt-driven, token passing, and collision detection. Each of these will be examined in detail in this section. Any of the four can be used with any of the three topologies that were discussed in the previous section.

Command/Response In a *command/response system*, one user is called the *master*, and all the others are *slaves*. A *slave* can send its message only when the master sends it a command to do so. The *master* sends the command; the slave responds. In theory, this is a simple protocol.

When a user needs to send a message to the master, it must wait for its turn, indicated by the master's command. If the master has a message to pass to a slave, it incorporates the message into the command and waits for a response indicating that the message has been received properly.

For example, look at a network with one master and five slaves (Fig. 11-5). In order to receive data from the third slave, the master says, "Slave 3, give me your data." The slave then responds with the data. To send data to a slave, the master says, "Slave 4, here is some data," followed by the data. The slave responds, "Slave 4 has received the data."

A slave may also send data to another slave, using command/response protocol, but it is done indirectly. The slave must first send the data to the master, and then the master retransmits it to the other slave. The sequence would be as follows for slave 2 sending data to slave 1:

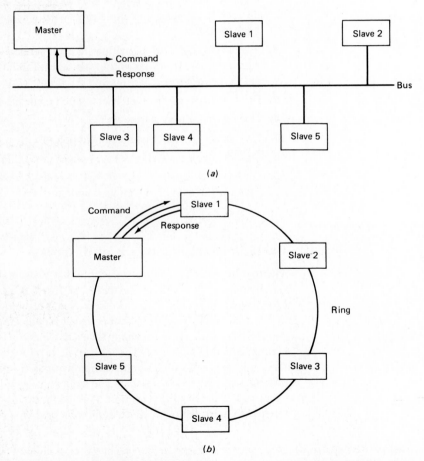

Fig. 11-5 In command/response protocol, one unit is the slave and the others are masters: *(a)* a master on a bus topology sending a command and receiving a response; *(b)* the way a master on a ring sends a command and receives a response.

Master: "Slave 2, give me your data."
Slave 2: "Here is my data, intended for slave 1" (followed by the data).
Master: "Slave 1, here is some data" (followed by data).
Slave 1: "Slave 1 has received data."

Although the protocol is relatively simple, it is not efficient. The master must spend a lot of its time checking with all the slaves. Even sending a small amount of data from the slave to the master takes two messages (the command and the response), so the network is tied up for a relatively long period of time. Sending a data message from one slave to another takes four messages, and this occupies the system for even longer. The response time of the overall system to sudden events is slow, since the master cannot receive any data, even critical emergency messages, unless it asks for it. This command/response protocol is not suitable for applications in which data must be passed quickly, without delay. It is also not a good choice when there are large amounts of data to be sent, since there are many protocol messages needed for each block of data.

The term "polling" is often used, especially in connection with this sort of protocol. In polling, the master periodically checks with each slave to find out whether any slave has data to send. In a communications network, this may mean checking with each slave several times a second. As the number of slaves becomes large, this polling takes up most of the master's time and network time, yet there is no data to transmit in most cases. It is like trying to find out exactly when people arrive home by calling their houses every few minutes. This may be a nuisance for checking on one person, but imagine trying to check on ten people every five minutes. The person doing the checking would be constantly dialing and listening for the phone to be answered. The person's phone, in turn, would be occupied (busy) for a large portion of the time. Overall polling is practical only if there are a few slaves, or if the slaves only have to be checked relatively infrequently. A good example of this is a building heating and cooling control, which needs to check each room in the building every half-hour or so. In contrast, polling to see which person is typing at one of many terminals connected to a computer, and to see what he or she is typing, would probably result in a very slow response.

In a simple case, consider the performance if there are 10 users and each message, from any slave to the master, or master to any slave, takes 1 ms. A simple command/response cycle takes 2 ms. Passing a message from one slave to another takes $2 + 2 = 4$ ms. Polling all 10 users, to find out whether any has any message, takes 10×2 ms, or 20 ms. This means that the entire polling cycle can be completed at a rate of 1/20 ms, or 50 times per second, at best. This time may be adequate for building heating and cooling management, but too slow for handling 10 terminals connected to a single computer.

Interrupt-Driven

The *interrupt-driven protocol* is a refinement on the command/response protocol. The purpose of the interrupt-driven scheme is to improve the efficiency of the network by overcoming the two major weaknesses of the command/response method: its low efficiency due to the extent of protocol needed to transfer data and its long, slow response for immediate events and occurrences.

In the interrupt-driven protocol, there is a master just as in the command/response protocol. Instead of the master polling each slave for data, the master instead waits for a special signal, called an *interrupt*, which comes from any slave when the slave has data to send (Fig. 11-6a). Only then does the master try to get the data from the slave that has caused the interrupt. There are several approaches to doing this.

One way is to have the master perform the polling operation only when this interrupt occurs. The master says to each slave in turn, "Did you have any data for me?" Each slave responds with either "No, I did not" or the message. This is more efficient than the repetitive polling of the previously discussed command/response, since there is no activity when there is nothing to transfer.

A further improvement is possible by using more signal lines for interrupts in the network (Fig. 11-6b). Each slave can be assigned its own interrupt line or interrupt code. Then, when the interrupt occurs, the slave not only interrupts the master, but tells the master which slave it is that has some data or a message. The master then immediately responds to the correct slave. This method minimizes the time wasted polling through slaves that have nothing to report. It costs a little more in terms of signal lines but improves the performance a great deal.

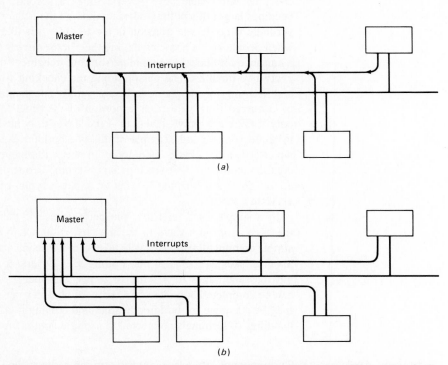

Fig. 11-6 Interrupts allow a slave to get the attention of a master quickly: (*a*) a single interrupt line shared by all slaves in a bus system; (*b*) a single interrupt line per slave (multiple interrupt signal lines).

By using the same message times discussed in the command/response example (each message takes 1 ms), the effect of interrupt-driven access can be calculated. When there is no interrupt, the master performs its other activities, such as calculations on numbers it may have collected previously. When the interrupt occurs, the master begins polling. The polling takes 2 ms per slave. It turns out that the maximum polling time is 10×2 ms, since there are 10 slaves, but the average polling time is only half of that. This is because half the time the slave which interrupts is slave number 1, 2, 3, 4, or 5, and half the time it is slave 6, 7, 8, 9, or 10. Once the master finds out which slave has interrupted, it can stop checking all the slaves. This has the effect of reducing by one-half, on average, the number of slaves that must be checked. The master only uses 5×2 ms to check, or 10 ms, compared to 20 ms for a complete poll through all slaves. To this is added the saving in time that results since the master checks only when it is interrupted, so there are no wasted polling cycles through all 10 users only to find out there are no messages. This interrupt-driven method is much more efficient in use of the network and master time.

Token Passing

The master/slave approach, whether used in the command/response or the interrupt-driven protocol, has weaknesses that make it inefficient for some types of network applications. The large number of protocol messages causes low efficiency, and the problem is worse when many of the messages need to go to other slaves, not just the master. A typical example is a network which links together all of the manufacturing equipment, planners, and support operations in a factory. Messages may have to go from the area where orders are received to the stockroom to get the parts for the production, from the stockroom to production to set up the production equipment, from production back to order entry to provide the status of the production (when the finished item will be ready to ship), and many other ways. The messages crisscross each other continually to form the total production-related system.

A command/response and master/slave system requires very high network speed to support all these messages going from virtually every slave to every other slave or from one slave to several others. An alternative to the master/slave system is the token-passing protocol. In *token passing*, there does not have to be a master, with the rest of the users as slaves. All users can be relatively equal in stature, although one unit may act as the overall coordinator in some implementations.

Token passing works as follows: A message, in a special format called a *token*, is sent from one user node to another, in one direction (Fig. 11-7). At the next receiving node, the user examines this token. The user has a specific amount of time during which it can remove the token for its own use or add a message to it. During that time, all the other network nodes can only listen to the network. The token passing continues until the original sending node receives the token from the last network node, which thereby acknowledges that the intended receiver has seen the message.

The effect of token passing is that each user node has the control of the transmitting on the network for a period of time. The node with the token is in

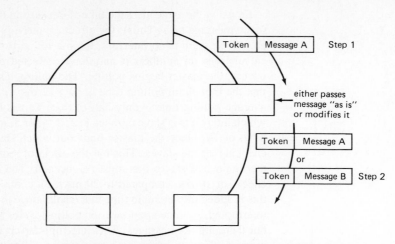

Fig. 11-7 A protocol called *token passing* does not have a single master. Instead, the network control is passed from unit to unit with a token. Only the token holder can control the network.

temporary control. The control token is passed from node to node, sequentially, so that every user has a turn.

One great advantage of token passing is that each user has access to the network within a predictable amount of time, since each node has the token for only a specific period and no longer. This is important in some time-critical applications, in which the delivery time of the message must be short and guaranteed. The total time for the longest cycle on the network (the time when the user last had a turn and has the turn again) is equal to the time that each node has the token:

Number of nodes × time per node

The time for the message to propagate between nodes adds to this total but is usually much shorter than the time that each token has the node. If the propagation time is not so short, it is added to the calculation:

Number of nodes × (time per node + propagation time between nodes)

[**Note:** In some applications the nodes are spaced at very uneven distances, so the propagation time between nodes is not the same for each pass of the token. In this case, the time for each node must be calculated separately.]

A network with 8 nodes, each handling the token for 0.5 ms, guarantees that the message reaches any user in 4 ms (or that a user has a turn to get on). If the propagation time of 0.01 ms is added to the calculation, the total time is 0.48 ms.

Token passing guarantees that there are no conflicts on the network, since a user transmits only when it has the token. All other users must listen when they do not have the token and must be prepared to accept the token when it is passed to them. There is a weakness in the basic concept of token passing: what happens when any node unit malfunctions and does not pass the token on? Or some other problem occurs and causes a malfunctioning unit to transmit when it does not have the token? This can be a real problem in the practical implementation of the network. The solution is to include in the token-passing scheme some special rules and enhancements which cause a special recovery plan to be put into operation automatically when a unit has not received the token within some maximum amount of time. If a node knows that it should receive the token every 100 ms, and it does not get it after 100 ms, then special restart and recovery procedures are used. These do work and get the network restarted, but they also add to the overall complexity of the protocol.

CSMA/CD

The final network protocol examined is *carrier sense multiple access/collision detect* (CSMA/CD). Because it is such a long term the acronym is always used. The concept behind CSMA/CD is very different from the concept of token passing. The rule in CSMA/CD is that any network node can transmit its message onto the network, after listening to the network to make sure that it is not busy. If the network is busy with another user that is transmitting, the node must wait until it is clear before transmitting.

This seems to be a simple and straightforward way of accomplishing the desired goal of giving every user a chance to transmit whenever it has something to say, except for one problem. There is a small but definite chance that another unit will also be trying to do the same thing at exactly the same time—checking the network to see whether it is clear and then putting a message onto the system. Between the time that the user finishes checking whether the network is busy, makes the decision that all is clear, and begins transmitting, another user can come on and also begin transmitting. The result is a collision of the two messages.

The CSMA/CD scheme makes provision in its design for this occurrence. Whenever a collision occurs, the signal levels on the network change as a result. Each node has circuitry built into it to detect this sudden, undesired change in level. Thus, each of the transmitting nodes knows that a collision has occurred (hence the "collision detect" [CD]). Then, each user follows this rule: stop transmitting immediately, back off, and wait a random amount of time before trying again. The circuitry in each node is designed to use a varying random amount of time each time it detects a collision. The idea is that the two users who collide then have a different back-off waiting time, and one comes back on sooner than the other. The protocol then can begin again without collision.

Let's go through an example to show exactly how this works. Suppose the network has three users, called *A, B,* and *C*. User A has a message it wants to put on the network for user C. User B also has a message to put on the network, for user A. A checks, sees the network is clear, and begins transmitting. However, user B was checking at the same time and also begins transmitting because B thinks the network is clear, too. They detect the collision, and both stop. User A waits 0.1 s, and user B waits 0.2 s, before trying again (Fig. 11-8). Of course, since user A waits less time, A gets onto the network. If A is still on after B has waited the 0.2 s, B has to wait until A finishes. If A is finished, then B has its turn. If there is another collision, A, B, and C have different waiting times than they had the first time.

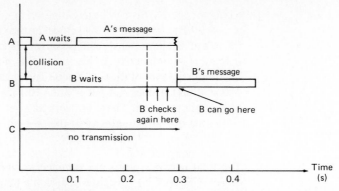

Fig. 11-8 In the CSMA/CD protocol each user checks the network before transmitting. If there is a collision in trying to access the network, the users back off and try again later, after a short random time period. Here, A and B collide; A waits a shorter time and thus can go first; B then has to wait until A is finished. In this example, C never has any message to send and so is off the whole time. Note: Here, A's message is long and B cannot send at end of the 0.2 s wait.

Why are random times used, instead of fixing different wait times for each user? The reason is to ensure that on the average, each user has roughly equal access to the network. If the waiting times were fixed instead of random, then A would always have a higher priority than B, and B would have a higher one than C. As the amount of data being passed increases, it is very possible that C will be "locked out" and never have a chance to get onto the network, like a younger brother trying to get a word in among many other brothers and sisters.

The concept behind CSMA/CD may appear very unnatural, but in fact it is one that is often used when people are talking. Of course, sometimes collisions do occur when two people start to speak at the same instant. Usually, both stop, wait some small, random amount of time and then the one that waits the shorter amount of time begins to speak. CSMA/CD is really an electronic version of the rules of personal conversation in some ways.

The CSMA scheme is very practical but does have some characteristics in operation about which users must be concerned. The main problem is that as the number of users in the network increases, the overall message traffic can increase to the point where the network is busy a large part of the time. The result is that a node trying to send a message has no guarantee of how long it will have to wait. There may be repeated collisions with other users: Suppose a network has 10 users, A through J, and A wishes to send a message. First, A collides with B, so both wait. On the first retry, A collides with D, who also now has a message to send. They back off, and A waits. After the waiting period, A finds that the network is busy while D uses it. A waits until D is finished, and just as A begins, J starts. They collide, and once again A waits. It is impossible to determine what the total waiting time will be until A finally gets onto the network. If the probabilities of the random waiting periods and the random messages on the network do not "roll" in favor of A, the wait may be very long. In some applications, a few hundred milliseconds of extra waiting may be acceptable. In others, such as critical data used to control production machinery, the wait may be too long. Worse, unpredictability of the wait may make it impossible to guarantee the overall performance of the various pieces of machinery and computers as they are linked.

Consider the case of a three-user network for users A, B, and C, as shown below.

	User A	User B	User C
First wait time, ms	2	1	4
Second wait time	4	3	1
Third wait time	1	5	3

The messages, as before, are always 1 ms long (this may not be the case, for in most applications the message lengths may be shorter or longer, but it does simplify the example). Suppose A and B have messages that they start to transmit at the same time, and so they detect a collision and back off. In Fig. 11-9a, the

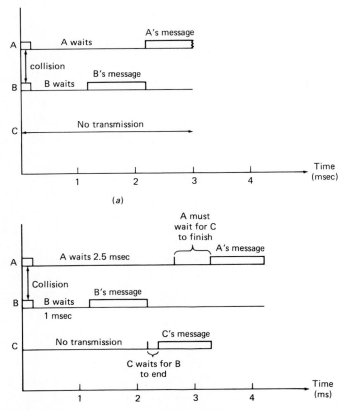

Fig. 11-9 (a) Both A and B have messages and collide. B's wait time is less, so B is able to go first. (b) Now, C has a message, too. B and C's wait times cause B to go first, then C, and finally A, since here A must wait for C to finish.

Networks 379

	Command/Response	Interrupt	Token Passing	CSMA/CD
Concept simplicity	Simple	Simple	Moderate	Complex
Efficiency	Low	Low–medium	Medium	Medium–high (depending on amount of data)
Response time predictability	Predictable	Fairly predictable	Somewhat predictable	Hard to predict
Reliability/ruggedness	High	High	Medium	Medium

Fig. 11-10 Summary of the characteristics of the various protocols in the areas of simplicity, efficiency, the predictability of the response time and of reliability. No single network is "best" in all categories.

sequence of events that occurs as A tries to get its message on the network is shown. A gets through after B.

Now, repeat the same exercise, but in this case user C has a message also. User C knows nothing about A or B, but C tries to send its message 2 ms after the first time that A and B tried and collided. Figure 11-9b shows that A is forced to wait even longer, since C has gotten in ahead of it! Yet, A started trying much before C.

This is in contrast to token passing. In token passing, the worst-case (longest) waiting time can be calculated knowing the number of nodes and the time each node is allowed to hold the token. The engineering term used to describe this situation is "deterministic" for token passing, because it can be determined in advance. The corresponding term for CSMA/CD is "probabilistic" or "statistical," since the performance depends on probabilities and overall statistics over many repeated tries.

Each protocol used to access the network has cost and performance tradeoffs in various aspects of the operation. These are summarized in Fig. 11-10. Understanding the network topology and the method of access provides two key pieces to establishing a complete and functioning network that can support many nodes. Another essential piece is the nature of the electrical signals used to put the data signals onto the network medium at the nodes. This is the topic of the next chapter.

QUESTIONS FOR SECTION 11-3

1. What is access? Why do networks need it?
2. What is command/response? What is the role of the master? Of the slave?
3. How is data sent from master to slave? How is data sent from slave to master?
4. How does a slave-to-slave data transfer occur?
5. Give three weaknesses of command/response.

6. What is polling? Why is it inefficient?
7. What kind of application is suitable for command/response?
8. What is interrupt-driven access? How is it a derivative of command/response?
9. What is the advantage of interrupt-driven access over command/response?
10. How does an interrupt work? What action does it cause?
11. How can the interrupt scheme be made more efficient? Explain. What is the cost?
12. Why is the average polling time one-half of the maximum?
13. Why is the command/response scheme, or the interrupt scheme, with its master and slaves, not good in some applications?
14. What is token passing? How does it work? Where are the master and slaves?
15. Why, in token passing, do all users know they will have a turn and what the maximum wait will be?
16. How is the total waiting time calculated?
17. How does token passing prevent conflict over access to the network?
18. How are token passing failures handled?
19. What is the idea of CSMA/CD? How does it aim toward network efficiency, with low overhead?
20. What is a collision? How does CSMA/CD detect it? How does CSMA/CD handle it?
21. Why is the back-off waiting time for each node random under CSMA/CD?
22. What is the concern of many user applications about CSMA/CD?
23. How can a user node become locked out? What two factors determine this possibility?

PROBLEMS FOR SECTION 11-3

1. Show the sequence of messages, using command/response, for a six-node network which needs to transfer data from slaves 3 and 4 to the master, who is user 1.
2. For the case of Prob. 1, show the sequence when a message must go from user 4 to user 2, and from user 2 to user 6.
3. Show the sequence, for the case of Prob. 1, when user 4 needs to transfer data to user 3 and the same data must also go to user 5. What about when the master must send data to all slaves?
4. A 10-node network is using interrupts for access. There is a single interrupt indication which any slave can send to the master (node 1) which then checks for the interrupt source. Show the sequence of messages when user 4 has a message and causes the interrupt. When user 9 causes it.
5. If each message—interrupt or data—takes 1 ms in Prob. 4, how long does it take for data from node 3 to reach the master? From node 8?

6. For the case of Prob. 4, what are the shortest and longest times it can take for a critical message to reach the master, after the master has received the interrupt? What is the average time? (All messages are 1 ms long.)

7. A network is using token passing access, among its 15 nodes. How long does a message take to reach node 5 from node 2? (Assume all transfers take 0.1 ms.) How long does it take for confirmation of message received to get back to the sender?

8. For the network of Prob. 7, what is the shortest time to get a message to the intended receiver? The longest?

9. The network of Probs. 7 and 8 is expanded to serve 20 nodes. Repeat Probs. 7 and 8 for this case.

10. A network with 10 nodes, called A through J, uses CSMA/CD access. All messages are 1 ms long, for simplicity. Node A wants to send a message to node J; at the same time node D wants to send a message to node E. Node A has a back-off time of 2 ms, and node D has a back-off time of 4 ms. Sketch the network activity versus time, for the first 10 ms.

11. Repeat Prob. 10 when node D has a back-off time of 2.5 ms. Compare this to the result of Prob. 10.

12. The network of Prob. 10 has been modified so that each node has a different back-off time each time there is a collision. For the first three back-offs, the times are as follows: node A, 1, 2, and 3 ms; node B, 2, 3, and 1 ms; node C, 3, 1, and 2 ms. Sketch the network activity when node A and node C each try to send a 2-ms message at the same time.

13. Repeat Prob. 12 with nodes A, B, and C trying to send 2-ms messages at the same time. Whose message gets through first? Whose is last?

11-4 Media, Modulation, and Physical Interconnection

Whenever a network is installed, the system designer must decide the best medium to use. There are many considerations that are involved in the decision. These include bandwidth, number of nodes that can be handled, distance achievable, ease of installation, cost, and noise resistance. The system designer may not have full control over the choice of medium. If the network is covering a large distance, between buildings or across several miles or more, the network may make use of already installed links such as the telephone system. In that case, the designer may be prevented from choosing the technically best medium, in order to save on installation cost and time. If the network is intended to cover a very wide area, such as the entire state, country, or globe, most of the links will be using media already in place: telephone, satellite, or special wide-band cables that are installed by communications companies.

For the case in which all or a large part of the medium is within a single location, the choices with modern circuitry and technology usually involve one of four possibilities. These are twisted wire pairs, coaxial cable used in base band mode, coaxial cable used in broadband mode, or fiber optics. Each of these has areas in which it is superior to the others and weaker than the others.

The twisted pair wire is the lowest-cost medium. It is very widely available, and since it is among the older interconnection technologies, all the pieces needed for a complete system—connectors, splices, test equipment—are widely available. The drawbacks of twisted pair are in its bandwidth and the number of nodes it can handle. The physical nature of twisted pair gives it a bandwidth maximum of anywhere from 0.1 to 1.0 MHz, depending on the exact type of twisted pair used (conductor size, insulation, shielding). This may not be enough for an application which is sending lots of data in a short time span. At the same time, the twisted pair is best suited for networks in which there are relatively few nodes, up to about 20 or 30. Usually, the total amount of data on a network with so few nodes is low enough that the bandwidth of twisted pair is sufficient. In this sense, there is a good fit between the node capacity and the bandwidth of twisted pair. A typical user of twisted pair is the small office, with several users sharing terminals, computers, and printers. The distance that twisted pair can serve is also on the lower side—on the order of 1000 ft—so the office is probably a good fit. The noise immunity of twisted pair is relatively low, so it is not suitable for electrical noise environments such as manufacturing plants with large motors which generate noise.

Coaxial cable is another medium that is often used for networks. It can be employed in either of two ways: base band or broadband. In base band mode, the digital signals directly connect to the coaxial cable. The voltages of the binary 1s and 0s can be placed on the coax by using the correct kind of electrical circuitry to drive the cable. This base band mode has a bandwidth of between 10 and 50 MHz, depending on the exact coax used, and can run for moderate distances, up to several miles or even tens of miles. The number of nodes that can be handled is higher than with twisted pair, up to about 100. Of course, coaxial cable is more costly than twisted pair, and the connectors used are more expensive and complicated than the simple plugs and sockets which twisted pair can accept. Since the coaxial cable is shielded, it has much better noise immunity than twisted pair.

The base band coax does not have any carrier associated with it. This means that only one signal can be on a single coax link at any time. Even though the bandwidth of the coax may be high enough for the application, the system performance may still be too low. There is often a need for full-duplex operation, which requires either two separate links (one for each direction) or some modulation of carriers. The two-link method is costly and not flexible enough for many applications, so modulated signals are used. This technique is called *broadband coax*.

In *broadband coax*, one or more high-frequency carriers are used. Each of these carriers is modulated by the data. One carrier may be at 5 MHz, and the other at 10 MHz. Data flowing in one direction modulates the 5-MHz carrier, and data going in the other direction modulates the 10-MHz carrier. Frequency shift is the most common form of modulation. The use of broadband coax requires signal amplifiers at each node, and boosters if the distance is long. This is not so expensive as it may at first seem, since the most common amplifiers used are the same as the amplifiers used for cable TV. The cable TV equipment is low in cost and available, along with the miscellaneous items such as connectors and splices. The data communications industry makes use, in this way, of the very high-volume/low-cost electronic equipment made for the cable TV business.

Broadband coax has very wide bandwidth, up to 300 to 500 MHz. The distance can be dozens of miles, since the amplifiers increase the signal level as needed. Several hundred nodes can be supported, as well. The major drawback is that the cost of each node is higher than that of base band coax, and the people installing, maintaining, and repairing the equipment must understand very high frequencies. Using the regular base band techniques for troubleshooting, for example, is not enough, because some of the signals are at several hundred megahertz, and regular oscilloscope may not operate in that region.

There is another advantage to broadband coax. More than two carriers can share the cable. It is technically practical to have many carriers spaced along the entire frequency bandwidth, so that multiple groups of users can share the same physical medium. This can allow, for example, a data processing network to share the cable with a manufacturing network, and yet each operates independently.

Fiber optics is better than any of the previous media in many respects. The bandwidth is the highest, in principle, and the noise immunity is very high, nearly absolute. Any other cable can be run alongside high-voltage or high-frequency wires and still have no noise effects. The drawbacks to fiber optics lie more in the practical implementation. Good low-cost, high-bandwidth interconnections are not fully available for widespread use. The light-emitting diode which can take several hundred-megahertz electrical signal and convert it to a light wave is expensive, and the phototransistor that must be used at the receiver is also less common than a purely electronic component. The physical connection is also difficult. Adding a node is complicated, because the fiber must be cut, a new connector added, and the cable respliced, using very special tools and techniques. Even then, a fiber loses anywhere from 0.1 to 1 dB of its light signal at each splice. This signal loss can affect the number of nodes and the distance the original signal can travel effectively before it loses so much of the light intensity that it can no longer be used. Fiber optics is best suited to applications in which the number and location of nodes are known in advance, and there will not be constant additions or rearrangements. If used in the proper situation, it can provide the bandwidth and noise immunity that no wire — twisted pair or coax — has.

The Physical Connection

One of the key practical considerations in a network is the way new nodes are added. Suppose a coaxial cable runs throughout an office building or factory. How can a node be installed? One way is to install, in advance, connectors that tap into the cable every few feet. This is costly, of course. A better way is to allow connectors to be added to the cable as needed. A technician can come, cut the cable, and add new connectors, and then the new node is ready. A better way is to use a tap that can be installed with the cable intact. This sort of tap is now in

Fig. 11-11 A self-tapping connector allows a tap to be added to a coaxial cable without having to cut and splice the cable or stop operation in any way (Courtesy AMP Incorporated).

use (Fig. 11-11). It has special cable piercing points that can pierce the outer layer of the coax, connect the inner connector, and also connect to the outer braid, all without short-circuiting the inner to outer connector while the tap is made. This type of connector is expensive (about $40) but quick to install by almost any user, whenever and wherever needed.

Some networks use simple plugs, similar to the plugs on telephone systems. These cannot support the wide bandwidth of coaxial cable but are suitable for the simpler twisted pair type of medium. The user node usually has two jacks. One jack is for connecting to the existing network wires. The second allows the addition of a new user, in what is called a *daisy chain*. The new user is looped off any existing user (Fig. 11-12).

QUESTIONS FOR SECTION 11-4

1. What are the four likely media that can be used when the network developer and designer have a choice?
2. What are the characteristics of twisted pair with respect to cost, familiarity, bandwidth, and number of nodes?
3. Repeat Ques. 2 for coax.

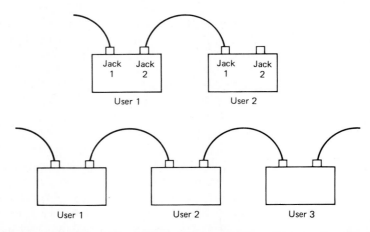

Fig. 11-12 The daisy chain approach to interconnection. Each user has two connectors, one to connect to the existing network and one to connect to the next user.

Networks 385

4. What is base band versus broadband mode? How many channels on each can be supported?
5. What is used for full duplex?
6. What circuitry is used for broadband coax?
7. What are the characteristics of broadband coax?
8. How can two separate networks share the same physical cable?
9. Compare fiber optics for networks to twisted pair and coax. Where is it a better choice? A poorer choice?
10. What is the effect of splicing in fiber-optic cable?
11. What is one easy way to tap into coax? Why is installing connectors every few feet not a good idea?

11-5 Local Area Networks

A *local area network* (LAN) is a network that is specifically designed for operation over a relatively small physical area, such as an office, factory, or group of close buildings. LANs are very widely used, and their applications are increasing rapidly. They are easier to design and troubleshoot than a wide area network, which can span the country or the entire globe. They are also interconnected to wide area networks to span the larger distances, in some cases.

In recent years, the need for LANs has increased because of the growth in use of distributed computing power, which has been made possible by the microprocessor and low-cost computers. In offices, this means that such equipment as word processors is on the desks of many office workers, and the office also has high-speed printers, along with data storage units such as tape systems which store large volumes of documents. There is a need to link these together, so any word processor or personal computer can send messages to others, use the printer, or store and retrieve memos and reports. In engineering departments, sophisticated workstations are used in computer-aided design of circuitry and mechanical pieces. These workstations are linked to powerful computers which perform detailed calculations, graphic plotters which produce engineering drawings, and some of the equipment used in actual production. Factories also have small, computer-based controllers to ensure best performance of machinery such as punches, plastic extruders, and heat treating ovens. All of these operate independently, but they have to be connected to each other to pass critical information, and to a central computer which provides overall direction (such as indicating the number of pieces to manufacture of a given item). In plants which manufacture products such as plastic, chemicals, or glass, continuous control of the process is needed. The flow rates and temperatures, for example, must be controlled precisely and repeatedly so the end product has highest quality and lowest cost. A computer-based controller is used to maintain the process at the required temperature, pressure, and flow values. These controllers, in turn, need to link to the plant computer to get overall direction (the color of paint, for example) and to report any problems (inability to achieve desired temperature, as a case).

All of these four applications—office, engineering, factory, or industrial process—have different needs in terms of the amount of data that must be passed on the network, speed, reliability of the communications, and physical ruggedness of the communications equipment. In practical terms, the total amount of data is not the only critical parameter. There is the response speed required to consider, and the way the data is grouped. It can be short bursts, such as when a computer workstation sends a cluster of data points to a printer, or it can be more of a spread-out flow, which occurs when production equipment sends back a steady stream of data on the status of the production process.

The four applications groups for LANs and their characteristics are summarized and compared in Fig. 11-13. These are carefully compared to the many LAN possibilities that are available by the systems analyst responsible for the network or the company that provides most of the equipment that will be connected to the network. The choices are varied. They include token passing ring using base band coax or CSMA/CD bus on twisted pair, for example. Many commercially available networks which meet the requirements can be purchased. The various applications have different needs and priorities on these needs. A system used to link low-cost office computers wants low cost per node, and a short delay or hard-to-predict time to reach a printer is not critical. In contrast, a network used in linking process control machinery costing tens or even hundreds of thousands of dollars is less concerned about cost, but short and guaranteed response time is important.

Feature	Office	Engineering	Factory	Process
Amount of data	Messages, memos (5 kbyte), reports (50 kbyte)	Engineering files (1 Mbyte)	Instructions and status messages (1 kbyte)	Same as factory
Nature of data	Infrequent bursts	Infrequent bursts	Continuous	Same as factory
Required response time	2–10 s	1–5 s	0.5–2 s	0.1–0.5 s
Data reliability need	Low–moderate	Low–moderate	Occasional	Frequent
Packaging	Office	Office	Semirugged	Rugged
Cost/node (1986)	$300–500	$500–2000	$800-2000	$200–2000

Fig. 11-13 The various applications for LANs have differing needs in terms of the amount of data to transfer, the nature of the data flow (burst or continuous), the response time required, the reliability and integrity of the data, the physical packaging of the electronics, and the acceptable cost.

Networks and Standards

Just as with any communications activity, standards are needed to ensure that the various aspects can communicate successfully. This is especially true for networks, in which many users are involved. Network standards fall into two general categories, called *proprietary* and *open*. A *proprietary standard* is one which a single manufacturer develops and sells. It can offer excellent performance, low cost, and many other good features. However, only that one company, or a small group of companies, makes equipment that is compatible with the standard.

An *open standard* is one which some industrywide organization has developed. This is done by gathering a large group of users, manufacturers, and other interested parties and formulating a standard to which all can agree. This process usually takes several years, since there are many technical and business-related issues to work out. Companies can then design and build network equipment which meets the standards and tell their customers that the network meets standard number 367, for example. Users do not have to become experts in the subtleties of the various aspects of the standard, since they can have a successful network simply by buying equipment that meets standard 367.

Standards are especially important with networks because the electronic equipment that is connected to the network can be supplied by many different companies. The computer may be from company A, the printer from company B, the plastic extruder from company C, and so on. The issue is not whether the standard is proprietary versus open, but whether it is clearly defined so users know what other equipment is available and compatible. Many open standards begin as proprietary ones. They become very popular and then a standards committee is formed to make them open standards that others can also use.

Many of these standards are made final and published by the Institute of Electrical and Electronics Engineers (IEEE), a nonprofit organization devoted to advancing electronic technology. Publishing standards is one of the many activities of the IEEE. Many of the common networking standards commercially available are defined by an IEEE specification. Examples include the following:

IEEE 802.3 CSMA/CD for base band and broadband systems

IEEE 802.4 token passing for base band and broadband bus

IEEE 802.5 token passing for base band rings

A standard does not necessarily define everything about the network. It may leave the choice of medium to the user, as long as the medium supports the desired data rate, bandwidth, and topology. The user can decide whether coaxial cable or fiber-optic cable is the better choice in the particular application. The network operation is not affected, but the physical interface box has to be designed for one or the other. The network standard may be medium-transparent and independent in these cases. In other cases, the standard specifically calls for several features, such as lower speed with coaxial cable.

Many networks are available today. They offer a wide variety of performance and are designed for different applications and needs. The next section examines a few of these in detail.

QUESTIONS FOR SECTION 11-5

1. What is the local area network? Where is it used?
2. Why has the need for LANs increased in the past few years?
3. What are the four main areas of application which use LANs?
4. What are three key areas of concern in each application of LANs? Why are there different priorities in different applications?
5. What is a proprietary standard? An open standard?
6. How are open standards defined and developed? Why do they take a long time to become official?
7. What are the IEEE 802.3, 802.4, and 802.5 standards?
8. What technical points may a standard leave undetermined, for the user to specify?

11-6 Some Network Examples

There are many networks in commercial use. Over 100 different proprietary networks are available presently. Some of these are intended primarily to connect equipment from the same manufacturer together. For example, WANGNET from WANG Laboratories Inc. is designed to interconnect all of the office automation products that WANG Inc. makes, and AppleTalk from Apple Computer, Inc., is for the Macintosh line of products. Besides these proprietary networks primarily intended for use among the products of a single company, there are networks sold by companies that do nothing but design and manufacture networks. These are then used to interconnect computers and related equipment from other companies. One example of this is Ethernet, initially developed by Xerox Corp., and now sold by companies such as Ungermann-Bass.

Many open networks also are in use, defined by industry group specifications. Some of these open networks began as proprietary ones and became very popular, and a formal industry standard developed from them. Others have been set up in advance by many users meeting together and deciding what they would like to see in a network, before the first one is ever built. The *Manufacturing Automation Protocol* (MAP) was originally proposed by General Motors for use in manufacturing plants, and many potential users and network suppliers have joined together to write the final specifications.

In this section, four existing networks are examined. For each one, some of the key features and technical attributes are discussed. None of these networks is better than the others. Each is designed for a different application and gives good performance in that application. Using the network in another type of application would result in disappointed and frustrated network users.

AppleTalk

The AppleTalk network is made by Apple Computer, Inc., to interconnect the Macintosh series of products. These include the Macintosh personal computers, printers that can be used with the Macintosh (both conventional printers and high-performance laser printers), and centralized mass-storage devices such as disk drives. Using the AppleTalk network, any Macintosh can send messages to any other device in the network. One Macintosh may send a memo to the printer, while another may retrieve, from the disk, a long document placed there by another Macintosh.

One of the goals of the AppleTalk was low cost per node since users who spend only $1000 for a device on the network are not willing to spend another $1000 per node of network. The AppleTalk price is approximately $50 per node. The cable is also very inexpensive twisted pair. Of course, the speed and ruggedness are not the highest, but they are well suited to the office environment where a few seconds of delay in receiving a message is not critical, and no catastrophe results if an error in the message occurs, as long as the system can determine that there has been an error (more on error detection and correction in Chapter 12).

The specifications of the AppleTalk network are relatively simple:

Topology: bus, with serial data

Wiring: shield twisted pair

Signal standard: modified RS-422

Signal speed: 230,400 bits/s

Signal modulation: frequency modulation between two values

Frame format: synchronous data-link control (SDLC)

Maximum number of nodes: 32

Maximum distance total: 300 m (1000 ft)

The cable is also carefully specified:

Conductors: No. 22 AWG, 17 Ω/300 m resistance

Shield: braid with 85 percent coverage over the conductors

Impedance: 78 Ω

Capacitance: 68 pF/m

Rise time: 175 ns, maximum, from 0 to 50 percent of full value

Users of the AppleTalk connect onto the last node or can break into an existing node. Each user tap has two connections, so that users are added in a daisy chain style. This makes it relatively easy to add, remove, or reconfigure the nodes.

The actual interconnection box to the cable must be low in cost and simple. The AppleTalk uses transformer coupling for this feature. *Transformer coupling* uses a very small, high-performance transformer to couple the energy from the cable of the network to the user equipment, and vice versa (Fig. 11-14). This may seem an outmoded mechanism, but it has some advantages. Transformers are inexpensive and reliable—they rarely fail. The transformer provides electrical isolation, so any miswiring in the user equipment does not affect the network, and any large voltages that get onto the network do not couple into the user's computer or printer. (The AppleTalk allows up to 500 V of common mode noise to be present and still does not damage the user equipment.) The transformer makes it easy to hook in new units and remove old ones.

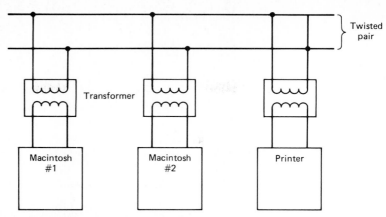

Fig. 11-14 The AppleTalk network uses low-cost isolation transformers and twisted pair wire to link personal computers and peripheral devices such as printers.

AppleTalk is a proprietary network. The integrated circuits (ICs) and software used to drive and control it, as well as manage the network, are under the control of Apple Computer. If other companies wish to develop products that are AppleTalk-compatible, they must negotiate with Apple for rights to do so, and Apple provides the additional technical information needed.

IEEE-488

For many years, a piece of test equipment was used as an isolated instrument. A signal generator put out signals, a frequency meter measured their frequency, and a voltmeter measured their voltage. As microprocessors became common, designers started adding them to instruments to make the instruments more capable and versatile. The user of the equipment did not have to perform detailed calculations based on the numbers from the instrument, because the "intelligence" of the instrument could do this. Many of the manual adjustments that were very time-consuming and error-prone, such as selecting the correct range of operation, could be made automatically under the control of the microprocessor. However, each instrument still acted independently from all the others that might be used in a test setup.

In many test and instrumentation applications, several different instruments are used, each performing a specific specialized function. Consider the testing of a radio (Fig. 11-15). The radio needs a signal source to generate signals into the antenna connection. It also requires a voltmeter to measure the voltages at certain points within the radio, a spectrum analyzer to check frequencies at some key internal points, and a distortion analyzer to monitor the audio quality at the output. Each of these instruments has to be set and varied during the test, and the results have to be recorded manually.

It would be much better if there were a way to link all of these instruments together, along with a computer acting as a controller. This controller would send commands to the various instruments which would take the place of manually setting ranges, output levels, and so on. In addition, the controller could take readings from each instrument automatically, minimizing the need for

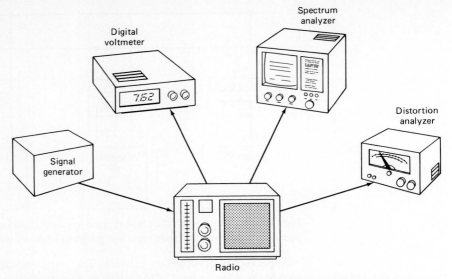

Fig. 11-15 Testing an electronic instrument such as a radio involves simultaneous use of several pieces of specialized test equipment which must be manually operated by the test technician.

someone to copy down all the values. Results and summaries could be printed or drawn on a plotter. The ability to link instruments and make them communicate would allow powerful testing systems to be built, using instruments from various manufacturers.

The key to this link is a standard that defines the rules for the link. This standard specifies the signals on the link, the type of connector to be used, and the way that messages are passed back and forth. Such a standard was first put into place by the IEEE in 1975, and it was given the number 488. This IEEE-488 is often referred to as the *488* standard or the *general-purpose interface bus* (GPIB). More than 300 manufacturers now make equipment of all types that can communicate on the IEEE-488 bus.

Some of the rules specified by the IEEE-488 bus standard are as follows:

Up to 15 devices may be on the bus.

Signals on the bus must have defined voltage levels and shapes.

Signals are base band with a maximum rate of 1 MHz.

A total cable length of 20 m (60 ft) and specific cable and connectors must be used. The cable uses 24 wires: 8 for data, 8 for control, and 8 for ground. Data is sent in parallel form, 1 byte at a time.

Each instrument on the bus has a unique address, which is set by the user with small switches built into the instrument. These addresses are used to direct messages to an instrument or to indicate to the receiving instrument who the sender is. It is possible to have two identical units on the bus at different addresses.

Devices on the bus are categorized as listeners, talkers, or controllers. The *controller* sends messages and receives them. There can be only one controller in a system. The controller is usually a small computer, with an interface to the IEEE-488 bus. *Listeners* are devices which can only receive messages, and *talkers* can only send them. An instrument can be a talker or listener, depending on its function.

A 24-pin connector is used. Connector mismatch never occurs.

Examples of Talkers, Listeners, and Talker-Listeners

Talker: a voltmeter which makes and reports readings but has no settings that can be made by the controller

Listener: a power supply which can be set to any output voltage by the controller but can report nothing back

Talker-listener: a voltmeter which not only makes readings but also receives settings and ranges from the controller

Both command/response (or polling) and interrupt-driven modes can be used, with the controller as master and the talker or listener as slave.

Use of IEEE-488 based equipment to test radios is shown in Fig. 11-16. The controller is programmed to run the specific tests required. It puts prompting messages, such as where to put the test probes, what settings to use on the radio itself (not an IEEE-488 device, of course), and what results to look for, on the screen for the test technician. The controller then sets the output of the signal source to the proper level and instructs the voltmeter, spectrum analyzer, and distortion meter to go to certain ranges and make the proper number and types of readings. It receives these readings and compares them against a list of what is acceptable and what is not and determines whether the radio under test has passed or failed. The controller may also print summaries of the test results on a printer and draw graphs of the results on a plotter, also connected via IEEE-488.

The standard rules for intercommunications between instruments defined by IEEE-488 are quite complicated. There are approximately 100 groups of tables,

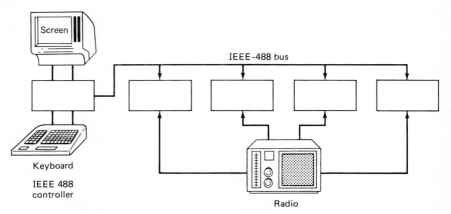

Fig. 11-16 Using IEEE-488, the various test instruments can be linked together and operated, and data analyzed by a single computer acting as the IEEE-488 controller.

Networks 393

charts, and diagrams which spell out what happens under certain conditions. This is because the various instruments may be in different states of operation, and each instrument needs to be able to handle all applicable states. It is similar to driving a car—there are a few basic rules, but there are also rules and guidelines to follow under special circumstances, such as intersections, cars stalling, cars meeting at signs, and so on. The IEEE-488 standard covers every possible circumstance in order that everything be clear to all instruments, but the rule book is thick. Any attempt to write a computer program to perform according to these rules would be complicated, would probably be filled with errors, and would take much of the available processor memory and time to execute. Soon after the IEEE standard was adopted, manufacturers of the ICs began to develop support ICs that actually implement the charts and diagrams of the standard. The processor of the system only has to indicate the message it wants sent, and the instrument which is to receive it, and the support IC takes care of the actual communications and message handling in both directions. The same support IC can be used in a controller, talker, listener, or talker-listener.

Ethernet

Ethernet was originally developed by the Xerox Corp. It is now available from several companies and is covered by IEEE standard 802.3. Ethernet was intended primarily for use in office environments, to link many types of office equipment. Ethernet is also used to link engineering computers used for circuit design with the associated plotting and drafting equipment.

The Ethernet network uses the CSMA/CD access protocol in a bus topology. Up to 1024 users can be supported on the bus. The bus medium is usually coax, with base band signals at a 10-Mbit/s rate. The maximum distance that the coax can be used is 500 m (about 1500 ft), although special repeaters can increase this. Fiber optics is also available as medium.

Since Ethernet is one of the older established standards for networks, there are ICs available from several manufacturers which are designed to implement various aspects of the specification. This makes the design of an Ethernet system and Ethernet-compatible equipment much easier than designing a complete circuit using small, lesser-function ICs and programming the Ethernet protocol into a microprocessor. As with any communications system, there are many subtleties in the complete protocol. The Ethernet ICs, examined later in this chapter, provide a quicker, less expensive, more reliable, and error-free way to get an Ethernet network built and operational.

Ethernet is suitable for applications in which very short response times are not absolutely necessary and the response time does not have to be guaranteed. This is typical of office equipment linked together: a few more tenths of a second to get a response or a slightly longer delay because the system is busy is not a problem. This performance is a result of the CSMA/CD access. Unfortunately, factory floor and production equipment cannot accept this—they need guaranteed, short response times, even if the cost is higher.

Manufacturing Automation Protocol (MAP)

The *manufacturing automation protocol* (MAP) is designed to meet the needs of factory and production plant automation. It is a token passing bus, running at either 5 or 10 Mbits/s. Frequency shift keying is used as the modulation method.

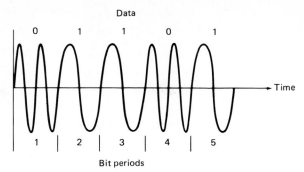

Fig. 11-17 The MAP network standard uses FM. A data bit of 1 is represented by one complete cycle of the carrier; a 0 bit is represented by two cycles within the bit period. Shown are five bit periods with the data pattern 01101.

The MAP specification is being formalized by the IEEE, and it will have the number 802.4. Even within the specification, there will be variations allowed in type of cable, connectors, and other issues.

The MAP standard uses token passing to guarantee that the longest response time is still within a desired value, and that no user is "shut out" as the traffic on the network increases. Experience and mathematical analysis have shown that for the CSMA/CD method, the system response becomes very sluggish and hard to predict when the amount of data traffic exceeds 30 percent of the theoretical maximum. With token passing, this limitation does not occur.

The MAP standard provides either 5- or 10-Mbits/s performance. The modulation used is called *phase coherent frequency modulation*. The digital data directly controls a carrier which goes onto the cable. If the data bit is a 1, one cycle of this carrier is sent in the bit time period. If the data bit is a 0, two complete cycles are sent (Fig. 11-17). For a bit rate of 5 Mbits/s, this means that the carrier is either at 5 or 10 MHz. (Why?) For the 10-Mbits/s bit rate, the carriers are 10 and 20 MHz.

One other critical consideration in the factory or production plant is the presence of large amounts of electrical noise. The MAP standard requires that receivers of the network nodes be able to operate without error as long as the signal-to-noise ratio (SNR) is at least 20 dB. This means that a noise signal relatively close to the amplitude of the actual data signal can be tolerated.

QUESTIONS FOR SECTION 11-6

1. What are the two ways that an open network standard can be developed?
2. What is an AppleTalk intended for?
3. What is a key goal of the AppleTalk network? Why?
4. What is the AppleTalk topology? Protocol? Bit rate? Modulation? Cable?
5. How does an AppleTalk node couple into the network electrically? Why is it done this way?
6. What is the IEEE-488 specification intended for? How is it used?
7. What are six key features of IEEE-488?

8. What is a talker? A listener? A controller? How do they differ?
9. Which access protocols does IEEE-488 support?
10. For what application is Ethernet intended?
11. What access protocol, bit topology, bit rate, and number of nodes are used or supported by Ethernet?
12. What is the response of an Ethernet system? Where is this satisfactory?
13. What media does Ethernet use?
14. For whom is MAP intended? What access, topology, bit rate, and modulation are used?
15. Where does the performance of CSMA/CD become sluggish?
16. What are the bit representations for 1 and 0 with MAP? What modulation is this?
17. What SNR is allowed for MAP? Why?

11-7 Wide Area Networks, Packet Switching, and Gateways

Whenever a network must span a large distance or widely separated locations, a local area network cannot be used. Instead, a wide area network must be installed. The main communications links in a *wide area network* are provided by telephone company facilities, including leased lines, satellite links, and similar channels. In some cases, the network user actually owns and maintains its own facilities, but only a few major companies have the resources or desire to do this. It is cheaper and more efficient to make use of the phone network (or other common suppliers) for the links.

Because of the distances in a wide area network, propagation delay and the uncertainties of the signal travel time become major factors. Most wide area networks are not used for time critical applications. It is unrealistic to expect to control the position of a motor on a drilling machine from 1000 mi away. Practical problems such as the network reliability and time delays would make this a difficult and even dangerous approach to control. It would also be very expensive, since dedicated communications equipment would be required to ensure that the user had the network available at the many instants he or she needed it.

Instead of using a wide area network for this sort of highly time critical application, most wide area networks are used for transferring large blocks of data between users. The exact times are not critical, because the data is usually from existing records or files, such as banks, or memos from people in different divisions of the same company. Another type of wide area network that many people are familiar with is an airline reservations system. Terminals are located around the country (and even other countries) through which the ticket agents can see which flights have seats and can make reservations. All the terminals use the same common data, provided by the central reservations computer. If an agent in Los Angeles books a seat, an agent in New York knows that the seat is taken and should not be booked by another person. The wide area network makes this possible.

Even though the system response time at the agent terminal seems very quick, it is sometimes a few seconds before the terminal screen updates with the correct information. This is because the wide area network provides service that is fast enough for this noncritical application, but still not on the order of milliseconds. The wide area network is usually a packet switched network.

A *packet switched network* allows many users, even from completely different sources and with different applications, to share many of the communications facilities. To transmit a message from point A to point B via packet switching, the message is broken into smaller pieces called *packets*. These packets are delivered to the network via local links. Once on the main network, they are routed through it more or less independently of one another. The packets may take different pathways through the network and arrive out of order at the destination (Fig. 11-18). (The network chooses the path for each packet on the basis of the channels and resources it has available at any instant.) The packets must then be put back into order and delivered to the final user. If a packet cannot proceed to the next node in the wide area network because there is no link available, it is temporarily stored at the node. Packet switched systems are also called *store and forward systems* since each node must store messages, then forward them when there are enough or a clear path has been found.

Managing a packet switched network is an enormously complex task. There are many users to deal with at both ends, sending messages from one to another. The resources available are constantly changing, or there may be sudden increases and decreases in the amount of data between some users. The links themselves may suddenly have technical problems, such as increases in noise on a satellite link that make it unusable. The protocol that is often used is called *X.25*, supported by an international standards committee.

The *X.25 protocol* provides for packet addressing, allows nodes that are next to each other in the network to send and receive packets between them, and

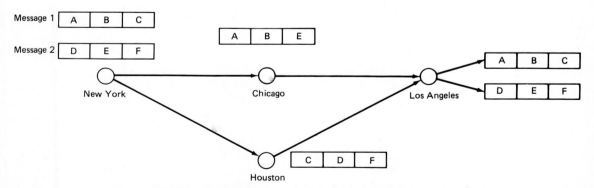

Fig. 11-18 In packet switched networks, the user messages are divided into standard size packets. They may travel over different paths to the final destination and must be reassembled when they reach the far end of the system. Many users share the network facilities.

provides a way to move packets between nodes that are not next to each other on the network. It also provides a means to detect errors, or recover packets that may have developed errors, between two nodes. The use of protocols such as X.25 is transparent to the user who originates the message. That user is responsible merely for delivering the initial message to the nearest node. Once there, the X.25 protocol takes over.

Wide area networks often connect to LANs. A special interface circuit is used, called a gateway. The gateway allows the LAN protocol to interface to the wide area network protocol, Fig. 11-19. Certainly, this is an extremely complicated function. Gateways must be able to support two protocols at once, and insure that any problem in the LAN does not interfere with overall operation of the wide area network, or vice versa. The gateway can also be used to connect two widely separated LANs together (either of the same type or different) via the wide area network.

An example of this is a manufacturing company, such as an automobile maker. The orders for cars and all the related paperwork are handled at a large office, which is internally linked via a LAN such as Ethernet. This office directs

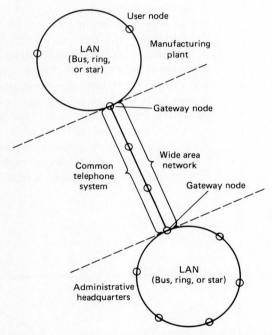

Fig. 11-19 A *gateway* is a special network node that acts as the interface between two networks. A manufacturing plant using MAP is shown connecting via a wide area network to the administrative office using another LAN. The MAP network has a gateway node to the wide area network, and the administrative office LAN has a gateway to the wide area network. In this way, the two different LANs can pass messages and data to and from each other.

the various manufacturing plants on how many car parts to make. One plant stamps the sheet metal, another builds engines, others manufacture other car components, and a final assembly plant puts the vehicle together. Each plant receives overall guidance of what to make from the central office via a wide area network and reports its daily production results by using the same network. Within each plant, the equipment used in production and the plant management and supervisor use a network such as MAP. A gateway from Ethernet to the wide area network, and from each MAP network to the wide area network, is required to make everything work smoothly and efficiently. Packet switching from the central office to the individual plants is acceptable since the data that moves between them is not required to the second, but represents only general instructions and reports.

QUESTIONS FOR SECTION 11-7

1. What is a wide area network? Where is it used? Give two examples.
2. What is a packet switched network?
3. How is a long message transferred in a packet switched network?
4. Why are various paths employed, even for the same message?
5. How is a message, which has been received as a collection of out-of-sequence segments, reassembled into the original sequence?
6. What are some of the management issues in packet switching?
7. What is the X.25 protocol? What does it do?
8. What is a gateway? Where is it used? Give an example.

11-8 Network Layers

When computers are used to develop data communications systems that serve many users, located over a wide physical range and with many needs, the protocols used in the network require more than a network standard for interconnection and formatting. Even a protocol such as SDLC is not sufficient for many of these advanced applications. To help ensure that nationwide and worldwide data communications systems can be developed and be compatible, an international group of standards is being developed.

These standards fit into a framework that has been established by the International Organization for Standardization (ISO) in Geneva, Switzerland. The framework is called a model for *open systems interconnection* (OSI) and is usually referred to simply as the *OSI reference model*. It defines seven levels or layers in a complete communications system. These range from the lowest layer, where the physical and electrical connection to the network is made, up to the highest level, where the actual user of the data that has been passed resides.

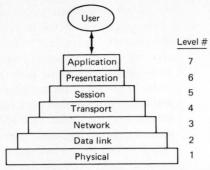

Fig. 11-20 The open systems interconnection model for network layers of protocol and structure has seven layers. The user is at the highest level 7; the physical interconnection to the network is made at level 1.

The seven layers in the ISO model are as follows (Fig. 11-20):

Level 7 (highest): application

Level 6: presentation

Level 5: session

Level 4: transport

Level 3: network

Level 2: data link

Level 1: physical

Not all applications need or use all seven layers for a successful, practical working system. The lowest three layers are enough for many applications. A full data communications system designed to serve many types of users at each end, however, may need all seven layers. Each layer is built from electronic circuitry and/or software and exists separate from the other layers. The function of each layer is to handle data or messages from the layer immediately above or below, using a carefully defined set of rules. Each layer has no idea of what really occurs in the layers above and below—only that it takes data from the adjacent layer, handles it according to these rules, and then passes the result onto the next layer on the other side.

First, let's study the definition of each layer and its role (Fig. 11-21), then use this in a practical example which will make clear how the various layers work together to form a fully integrated and complete data communications system. Finally, let's examine how the OSI model fits into the networks studied so far.

The Physical Layer Layer 1 includes the functions needed to activate, maintain, and deactivate the physical connection. It defines what voltages and signal rates, for example, are needed, but does not say anything about the medium itself, or the modulation.

The Data Link Layer Layer 2 covers the synchronization and error control for information which is to be transmitted over the physical link, regardless of the content.

	Layer	Functions
7	Application	User programs
6	Presentation	Transformation, selection of syntax
5	Session	Dialog management and synchronization
4	Transport	Reliable end-to-end data transfer
3	Network	Message routing and switching
2	Data link	Framing, error control
1	Physical	Isolation, encoding, modulation

Fig. 11-21 The name and functions of the seven levels of the OSI network model show the action and responsibility of each layer.

The Network Layer Layer 3 acts to route the communications through the various communications resources (such as different channels) to the other end. It acts as the network controller by deciding which route data should take—either out along a physical network path or up to an applications process (the user). Data which is routed between networks or from node to node within a network requires only the functions of layers 1 through 3. As shown in Fig. 11-22 a network may act as a relay link, and the message goes up from layer 1 to 3 and then back down again.

The Transport Layer Layer 4 is responsible for multiplexing a group of independent messages over a single connection (when this is appropriate), as when many users have messages going in the same direction. It also breaks the data groups into smaller-sized units so that they can be handled more efficiently by the network layer. It guarantees data from one end of the system to the other.

The Session Layer Layer 5 must manage and synchronize conversations between two different applications. For example, streams of data are properly marked and resynchronized so that ends of messages are not cut off prematurely.

The Presentation Layer Layer 6 makes sure that the information is delivered in a form that the receiving system can understand and use. The form and syntax

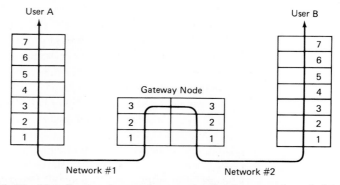

Fig. 11-22 Not all network operations require that data pass through the seven levels. A gateway node only has to pass the data from level 1 up to 3 and then back down again.

(language) of the message are decided by each of the two communicating parties. The function of this layer is to translate, if needed. For example, if one end of the system is transferring a file that uses ASCII code, and the other uses IBM's EBCDIC, this layer provides the translation.

The Application Layer The top of the model is the application layer, layer 7, which provides for the manipulation of information in various ways. It provides the resources for transferring the files of information, distributing the results, and transferring the complete message to another program that the user (who sits above this layer) may need.

All this seems very complicated, and it is. An illustrative example will show how a real-world data communications need is filled by this seven-layer OSI model. Suppose a manager of a Paris branch office of a New York company wishes to send a thank you note to the sales director of another company in Chicago. The following example illustrates the way the OSI model functions (Fig. 11-23).

The Paris branch manager represents the application process that initiates the conversation. He deals in the general meaning of the communication (the semantics) and merely tells the office public relations (PR) manager to send a thank you note. The public relations manager is the application layer. She calls on the services of the layers below to meet the needs in getting this message transmitted. The PR manager dictates the thank you note in French into a dictation machine and then gives the tape to her secretary, who acts as the presentation layer. The secretary translates the message into English and types it out as a formal business letter.

After the secretary has typed out this letter, she hands it to her administrative assistant—the session layer. The assistant records the letter in the Paris office's file on correspondence with the Chicago company, making sure that the right person has been addressed, that the title and spelling are correct, and that the address is correct. This type of checking allows both ends of the communications to synchronize and organize their dialogue, by noting where the message goes and when it was sent. If there is going to be a back-and-forth exchange, then this session layer manages the dialogue.

The next layer, the transport layer, is provided by the manager of shipping and receiving. His job is to negotiate the quality of service available from the network layer below (how will the message go? what delivery is promised?), approve the connection, and provide for receipt and delivery. The manager of shipping and receiving is really guaranteeing end-to-end transmission. If something unexpected happens during the transmission, such as a breakdown in the truck carrying the message, the manager recovers by sending another copy of the letter since he always copies a letter before sending it.

After he copies the letter, he assigns it a number sequence in whatever group in which it is going (such as number 2 of 10). He then passes the entire shipment, tagged with both the destination address and sequence numbers, to a shipping clerk. He tells the clerk to establish the route over which the note will be sent to Chicago. The shipping clerk acts as the network layer and selects the route, but also advises her boss, the transport layer, about it.

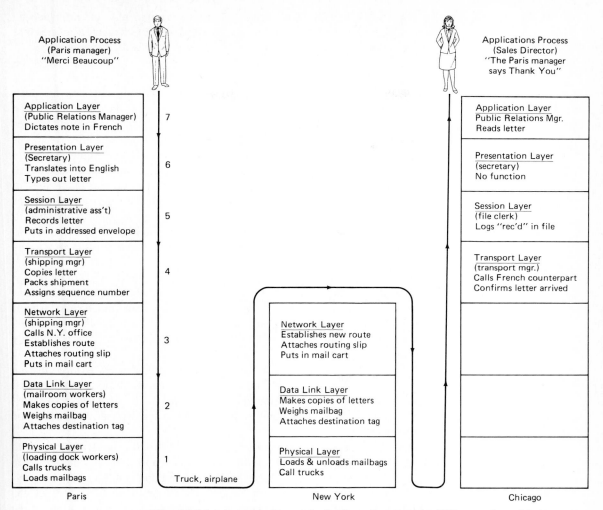

Fig. 11-23 A practical example that is analogous to the OSI network model shows the way a message from a Paris office manager is sent to a sales director in Chicago. The person at each level has a specific task to perform in order to make the whole process work efficiently and correctly.

The shipping clerk calls her counterpart in the New York office. From this call, she learns that her company's internal mail service can take the shipment to the New York office, without using a stamp on the envelope for the public postal system. The New York clerk also tells her that the message can then go from the New York office to Chicago by an express delivery service and be there the next day. This seems fine to the clerk, who attaches a routing slip to the letter and puts it into a cart with others, all labeled for New York. She then sends the cart to the mail room, which acts as the data link layer.

The workers in the mail room weigh the whole mailbag for New York on a very accurate scale, record the weight and destination, and then move the bag to the shipping dock, which acts as the physical layer. This physical layer acts as the interface to the medium, which is the combination of trucks and airplanes that take the mailbag to the receiving dock at the New York end.

When the mailbag arrives in New York, the workers at the dock there—the physical layer—pass the bag up to the workers in their mail room—the data link layer. Here, the bag is weighed again to check whether even one letter has been lost. If the weights do not match, the entire shipment is rejected and the mailroom in France is notified to send replacement copies. (This weight discrepancy corresponds to some error in the transmission of data. Other error possibilities within the message are checked at higher levels.) If the weights agree, the mailbag received goes to the routing clerk in New York, who acts as the network layer. This clerk opens the mailbag and sorts the mail. Mail for employees in New York offices is passed along to the transport manager (transport layer) for processing within the New York offices. Other mail stays at the network layer to be rerouted. The routing clerk recognizes the thank you letter as one that should be sent through an express service, so she tags it as such and sends it back to the mailroom.

The mailroom groups together all mail for express delivery to the Chicago company (multiplexing), as there may be a large quantity of regular correspondence between this company in New York and the company in Chicago. Again, the contents are copied, weighed, sealed in a bag, and tagged with a new shipment number and address. The bags go out onto the loading dock and away in express service trucks.

As the mailbag and message are routed further by the express service only the lower three layers are involved. This is because they are the only ones needed when a message is routed via intermediate networks. The upper layers—transport and above—are involved only at the origination and destination points of the communication.

When the express package arrives in Chicago, the routing clerk passes it up to the transport manager, who checks the packing slip and calls her counterpart in Paris to let him know that the letter has arrived in good condition. In this way, the transport layer acknowledges "end-to-end" communcations. All of the previous acknowledgments have been at the data link layer, from one leg of the trip to the previous leg. This final acknowledgment connects the end of the trip to the beginning, no matter what carriers or media (reliable or not so reliable) were used in between.

Once the communication has been received and acknowledged by the transport layer, it is passed along to the session layer. A file clerk logs the letter in the file of the Chicago company and takes it to the presentation layer—the secretary of the sales director. She reads the letter and determines that it is in English, and therefore no translation from the French is necessary. She gives the letter to the PR manager of the Chicago company, who acts as the applications layer. At a regular staff meeting, the PR manager informs the sales director that a "thank you" has been received from the head of the Paris office of the New York company. The receiving application process—in the form of the sales director of the Chicago company—receives the semantics of the message but not the original message itself, which was in French (saying "Merci beaucoup").

Certainly, this reference model is a framework for an elaborate structure. Is it needed? It is, if the goal is an easy-to-use system. Note that the application itself—the head of the French office saying thank you in French—did not have to know any of these details. He merely had to describe his application need, and the system automatically took care of the rest for him. When viewed as a total system, the OSI model can be broken into two groups. The bottom three layers (physical, data link, and network) cover the components of the network used to transmit the message. The top three layers (session, presentation, and application) reflect the technical characteristics and, needs of the end users. Their functions take place regardless of the physical medium actually used, whether a satellite link or a local area network. The transport layer acts as the liaison between the end user layers and the network layers. The complexities in the model occur because the model is designed to support many types of users who may be using the system to reach many other types of users. These include engineering design computers, process control computers on the factory floor, robots, graphics, document and memo transfer, and electronic mail. These users are sending numeric data, graphical data, and financial data, using various formats, such as ASCII, EBCDIC, or binary, to give a few examples.

The OSI Model and Local Area Networks

The LAN standards discussed previously were put in place with this OSI model in mind. Standards 802.3 (CSMA/CD), 802.4 (token passing bus), and 802.5 (token passing ring) define a physical layer (layer 1) and a data link layer (layer 2). These are enough to build a local area network. When the user of the network is separated from the application by intermediate equipment such as computers and terminals, more layers are needed for a complete system. If the user is located directly at a LAN node, then the LAN specification in the 802 series may be sufficient for sending messages to other users located at other nodes. It depends on the nature of the application and whether the overall data communications system is designed to be a general-purpose system. If the system is designed for a limited set of circumstances, then there is no need for a seven-layer OSI system. The two layers implemented by the LAN are sufficient. Note that a simple description of the interface (simply saying RS-232 or even SDLC) is not sufficient to ensure that both the sender and receiver can transmit and receive correctly, and that the receiver can make sense of the meaning of the message.

Are all seven layers necessary? For many applications, the answer is no. When equipment using the same language, or all designed for one application (such as running a machine shop) is interconnected, there may be only the lower group of the seven. However, in communications, success is achieved only when both users agree to a set of rules, terms, and conditions and have a common language. The goal of the seven-layer OSI model is to ensure that the user who understands only English is able to receive and use messages written in French. Otherwise, the message can be perfectly sent and received, and still not be usable. A practical example of this is an engineer who is using a computer design workstation to design a new mechanical strut for a car. This design system allows the engineer to complete the entire design and analysis on the screen of the computer workstation. Somehow, this must be translated into the details to control the machine tools (lathe, milling machine) that actually produce the part. The way that the part is described by the engineer, however, may be quite different from the machining instructions needed in production. The engineer may describe the part by its corners, using X and Y coordinates against a graph. The machine tools need these coordinates, but they also need a set of directions for the correct order of machining, since the cutting head of the tool cannot jump around in any order or grind an internal spot before the outside metal has been cut away. The engineer and the machining tools have the same goal and are talking about the same thing, but using different frames of reference. The seven-layer OSI model, if fully implemented for both users, lets the engineer design a mechanical strut on a screen and then translates the design specifics into a set of directions that will control the activities of the machining tools.

QUESTIONS FOR SECTION 11-8

1. What is the OSI reference model? Why is it needed?
2. How many layers are there in the OSI model? What are their names?
3. Explain where all the layers are needed, or not needed.
4. How do the layers of the OSI model interact with each other?
5. What is the role of each layer?
6. Why is this multilayered model easy to use, despite its design complexity and elaborateness?
7. What is the role of the bottom three layers, in summary? The top three? The middle one?
8. What is the relation of the OSI model layers to IEEE 802.3, 802.4, and 802.5?

PROBLEMS FOR SECTION 11-8

1. A message is going from a user who implements all seven layers of the OSI model to another similar user. Along the way, a LAN is used for part of the channel. Sketch the passage of the data message through the layers of the two users and the LAN. How many layers are used in the LAN?
2. A message in English is sent to a Spanish-speaking user, with both ends using the seven-layer ISO model. At what layer is the translation from English to Spanish made? At which end? How about the translation for the response from the Spanish-speaking person to the English-speaking side?
3. Two users are connected through the ISO model. One is describing the position of a machine tool by using ASCII characters, while the other uses binary numbers. Where is the translation made (which end and layer)?
4. A network uses the seven layers of the OSI model. The Ethernet standard is used for the LAN part. Which layers are defined by the Ethernet specification? The network also connects to the MAP LAN, so that the plant chief executive can communicate with production machinery and supervisor. Which layers of the OSI model are affected?

5. Sketch the seven-layer OSI model, which allows the plant chief executive to communicate to both the office staff (using Ethernet) and the production staff (using MAP).

11-9 State Diagrams and Network ICs

The complicated rules and protocols of a network could be implemented by each manufacturer of network equipment. This would be time consuming if each manufacturer did it individually. In addition, the protocols are so complex that the many specific implementations would probably have errors and bugs.

Another possibility would be for some integrated circuit manufacturers to provide microprocessor systems preprogrammed with the rules of various networks. These would then be available to other equipment manufacturers. In fact, this is what is done, but with a technical variation. An ordinary microprocessor is not fast enough to keep up with the fast data rates and events that occur on a network. This is because a microprocessor (or any computer) must execute several software program steps for each decision and action it takes, and each step takes time (on the order of 10 to 100 μs, typically). Each network action requires many such decisions, so the overall decision-making process would be slower than acceptable.

The alternative is an IC designed to implement sets of rules and events, in what is called a *state machine*. The state machine implements what is called the *state transition diagram*, or simply *state diagram*, of the network protocols. Exactly what is a state machine? What are state transition diagrams? An example will illustrate.

Suppose a person can be in one of several places: at home, at the baseball stadium, or in a restaurant. These are the states. The state transition diagram defines the events that direct the person to go from one state to another. A person at home, desiring to go to the baseball game, checks the weather. If it is clear, go to the stadium; if raining, stay home. At the stadium, if there are seats, go in; if all seats are taken, go back home. From the stadium, if the person runs into a friend, then go to a restaurant; if the person meets no one, go home. This process is illustrated in Fig. 11-24. The state transition diagram shows each state (home, stadium, restaurant) as a circle. The lines connecting the states show what events occur and how these events cause a person to go from one state to another. For each state, there are several possibilities that can occur, depending on what event occurs.

The state diagram is used extensively in defining what states and possibilities exist and how transitions from one state to another occur. The state changes are initiated by events such as message received ok, error detected, collision, or no signal present. Manufacturers of network ICs implement these state diagrams with logic circuits on the IC. Network equipment manufacturers can buy these ICs and have a large part of the complicated network interface problem greatly simplified for them. State diagrams are also troubleshooting maps, guiding the repair person on correct operation.

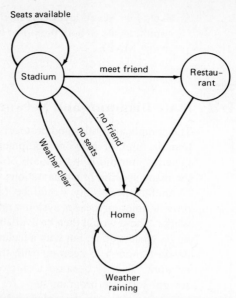

Fig. 11-24 A state diagram shows the possible conditions or states of a network and the paths for making the transition from one state to another. The example shows someone who can be at home, at a stadium, or in a restaurant, and the possible events that would occur to make the transition from one state (location) to the next.

There are many ICs available, in order to handle the many networks in commercial use. The IEEE-488 network and Ethernet network ICs show some of the similarities and differences in the networks and the features that are needed in an IC that supports and implements them.

Integrated Circuits for IEEE-488

Many manufacturers provide ICs which can be used to implement the IEEE-488 circuits network standard. These ICs are not identical from manufacturer to manufacturer, but all have many similarities and provide the same functions (although sometimes with subtle differences). The TMS9914A is made by Texas Instruments and is designed to function as the interface circuitry for a GPIB controller, talker, or listener (Fig. 11-25). The IC communicates with the microprocessor of the equipment in which it is used through registers. These registers are used to instruct the 9914 and set it up for the desired modes of operation. They are also used to allow the 9914 to send status and indications of IEEE-488 bus activity back to the microprocessor.

The 9914 also contains other registers for the actual data that is to be passed from the system to the network bus and is received from the network bus intended for the microprocessor.

The 9914 actually is a state machine with gates, flip-flops, and other digital circuitry interconnected so that each of the state diagrams of the IEEE-488 standard is implemented. When IEEE-488 bus activities occur (such as new data, no new data, interrupts, error on received signals), the 9914 uses the state diagram of the mode it is in and reacts according to the rules of that diagram.

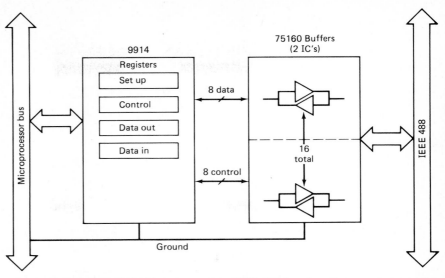

Fig. 11-25 The TMS9914A is a complete IEEE-488 interface that goes between the system processor and the IEEE-488 buffers. It implements the many state diagrams of the standard and allows data to be sent and received after the IEEE-488 protocol.

Figure 11-26 shows one such example. The figure illustrates the state transitions for the talker mode of IEEE-488 operation. There are only four possible states, called by the mnemonics "ENIS," "ENRS," "ENAS," and "ERAS." The operation of the talker model starts with a system reset, which puts the IEEE-488 bus in ENIS state. Then, the interface can follow the arrows only to the other states, depending on what event occurred. If the 9914 is in ENRS, either a source delay SDYS occurs, causing a return to ENIS, or a write to the data register WDOT occurs, forcing a jump to the ERAS state.

The 9914 is only part of what is required for a complete IEEE-488 interface. The specification defines the precise voltages, signal shapes, and protection levels needed for the physical connection to the IEEE-488 bus. To meet this need, there are line driver and receiver ICs which are specifically designed for IEEE-488 operation. One such type is the 75160 series, also from Texas Instruments. The electrical connection to the IEEE-488 bus is not isolated, so the bus interface ICs can connect directly to it. There are a total of 24 wires used. Of these, 8 are ground, 8 are for the data bits themselves, and the remaining 8 are for bus management. Most of these lines carry bits in either direction (from IEEE-488 bus to the 9914, or vice versa) so the 75160 series of buffers is also bidirectional.

The connector used for IEEE-488 is called out very clearly by the standard. It is a 24-pin connector, and the signals of the interface occupy one pin each. There is no ambiguity about which signal is on which pin, and the standard is written in such a way that there is never a need for adapter cables or similar cable modifications. There may be operational problems, due to software problems in layers above the 9914, in making an IEEE-488 system function, but the physical and electrical interface is never the source of the problem.

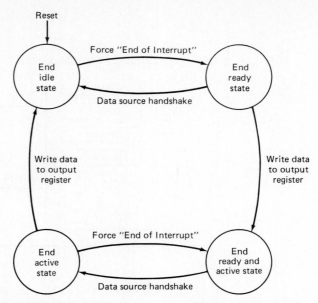

Fig. 11-26 The state diagram in IEEE-488 for transitions between idle, ready, and active states. Operation begins with a reset signal, which causes the system to end the idle state. From idle, various interrupts or handshake signals cause the transitions to other states; writing data to the output registers can also cause transitions.

Ethernet ICs

As in the IEEE-488 interface, two IC groups—the state diagram implementer and the bus interface—are needed to build an Ethernet node. However, the Ethernet standard is much more complicated than the IEEE-488 standard, so the ICs used are also more complex. The Intel Corporation makes ICs for Ethernet networks (Fig. 11-27). The 82586 IC is called a *LAN coprocessor*, because it is a highly advanced component which acts in close relationship with the processor of the user's system. It is an intelligent peripheral IC that completely manages the process of transmitting and receiving frames over the network. It off-loads the main processor of virtually all tasks associated with data communications and does not depend on the main processor for time-critical functions, which can quickly absorb much of the capability of the processor. It sends data, receives it, manages the framing of incoming and outgoing data, and keeps track of various error conditions (such as framing and overrun) that may occur. It also has diagnostic capability. It can send messages for test purposes, and even send itself messages to determine whether there are any problems within the IC.

The physical and electrical interface to the Ethernet cable requires more than the simple drivers and receivers of IEEE-488. The Intel 82501 Ethernet Serial Interface is intended specifically for this purpose. It connects between the 82586 and the Ethernet cable connection. The 82501 has built-in 10-MHz clock circuitry to generate the required clock signal, performs Manchester encoding and decoding (since that is the way Ethernet signals are encoded), and has electrical interface circuitry for the Ethernet cable and some diagnostic capability. It also has special circuitry which continuously checks to make sure that the transmitter

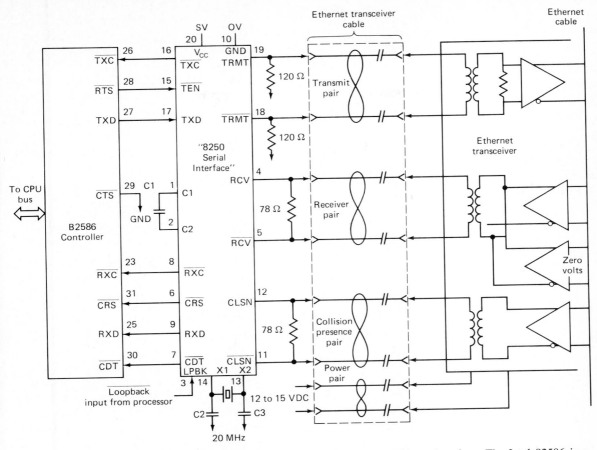

Fig. 11-27 Several ICs are required for an Ethernet interface. The Intel 82586 is a microprocessor dedicated to handling LAN protocols and is called a *LAN coprocessor*; the Intel 82501 is an Ethernet serial interface which provides the clocking and Manchester encoding/decoding that Ethernet requires (and the diagnostics which help test the interface) and the special interface circuitry which actually connects to the Ethernet cable.

of the Ethernet node does not stick in a transmit mode and "lock up" the bus. This function is called a *watchdog timer*. If the 82501 watchdog detects that a transmission from the IC has lasted for more time than the maximum allowed, it generates a special signal which forces a reset.

Finally, the connection to the Ethernet cable is made through the Ethernet transceiver cable, which connects the 82586/82501 pair to the Ethernet cable connection point. The 82501 has the correct voltage levels, current drive, and short-circuit protection for the specification requirements. There are three signal pairs which come out of the 82501: transmit data, receive data, and collision detect (which indicates that a collision has occurred). These signal pairs can be transformer-isolated and then combined at the Ethernet coaxial cable itself.

Any company that wants to manufacture Ethernet-compatible systems can buy these ICs and build the circuitry, as well as add some software that binds the

82586/82501 pair to the main system processor. An alternative choice is a circuit board, from Intel and others, which already does this. These circuit boards plug into the bus of the system and combine not only the Ethernet chip pair but also the various miscellaneous components (resistors, crystal, smaller ICs, and capacitors) needed to provide a properly functioning Ethernet interface. They reduce the risk that manufacturers face in designing their own and also save considerable design and debugging time.

QUESTIONS FOR SECTION 11-9

1. Why is a microprocessor running a software program usually not fast enough for a network manager?
2. What is a state transition machine? How is it implemented?
3. What is a state diagram? What does it show? Why is it useful?
4. How does the TMS9914 communicate with the processor of its circuit?
5. Why are line drivers and receivers needed for the IEEE-488 specification?
6. How many wires are there in an IEEE-488 cable? What are their functions?
7. What interface and support are needed for Ethernet?
8. What is the role of the 82586? The 82501?
9. How does the 82501 manage a lockup situation?
10. What Ethernet signals must the 82501 provide?

PROBLEMS FOR SECTION 11-9

1. A student has three possible states: at home, at school, or at the library studying. The events that can occur and cause transitions from one state to the other are new school day, no school today, term paper due, school day is over, library closed because of holiday. Draw a state transition diagram for this.
2. Draw a three-state transition diagram in which some of the states can be reached from two other states. Are there any real-life situations that are like this?
3. Draw a state transition diagram for a relatively simple RS-232 interface which is sending data, with a handshake line from the receiver indicating whether it can/cannot accept more data.
4. Sketch an IEEE-488 network which has four nodes. Show how many 9914 type ICs and 75160 ICs are needed, and where they are located.

11-10 Cellular Networks and Systems

A network that is very different from any others studied in this chapter is now in use for a specialized application. Cellular networks are designed for mobile telephones, such as those used in cars. The mobile phones of cellular systems can also be used for data communications, and some special vehicles (police, repair crews) are now fitted with a small computer which is linked to the main office not by wire, but by radio link. This allows users to be part of the larger wide area networks and telephone system without a physical connection to the system.

Ordinarily, radio-based telephones (and two-way radios) use a powerful transmitter and sensitive receiver at the base station to cover the total geographic area (Fig. 11-28a). A typical range is 50 mi. Each phone is assigned a unique frequency within the band reserved for this service, and this allows many phones to be in use at the same time without interference. The base station then links into the rest of the regular phone system, so a mobile phone can be connected to any telephone.

The problem is that the available bandwidth in the spectrum is relatively small, and only a few phones can be accommodated. In major cities, there are only a few hundred channels for the general public, since large blocks of the spectrum are reserved for police, fire, medical, and other government users, as well as taxis and businesses. There seemed to be no way to allow more users onto the system within the spectrum that was available.

To overcome this apparent unsolvable problem, a new concept of cellular systems was proposed and investigated by the Bell Telephone System. This concept has been verified as practical and effective and is now available in many parts of the United States. In a cellular system, the single high-power base station is replaced by many low-power stations. Each low-power cellular base station covers a small area, and there is some overlap at the edges so that no area is uncovered (Fig. 11-28b). The base stations themselves are linked together by a private network, so each base station can communicate with the others.

The cellular system works as follows: each mobile phone is initially assigned a frequency and communicates with the closest base station. Within the zone of each base station, there is only one phone per frequency, but an adjacent base

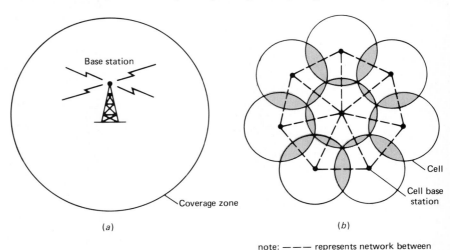

Fig. 11-28 (a) A single base station can be used to link many mobile phones, with one frequency channel assigned to each phone. (b) A cellular system uses many smaller base stations with minimal overlap to provide frequencies to many users in the same larger area. The smaller base stations coordinate their actions.

station can have a phone at the same frequency as its neighbor. There is no interference because the base station signals are low-power and do not spill over into adjacent zones. As a phone user crosses from one zone to the next, the cellular base station does two things. It passes control of the phone to the base station of the new zone, and the base station of the new zone also instructs the mobile phone to switch to a new frequency, one that is unoccupied in the new zone. The frequency that the phone had in the old zone is now available for any other phone in the old zone.

The result of this is that all the frequency assignments are available to many more users, since the overall geographic area has been divided into so many smaller zones and all the frequency slots can be reused in each zone by others. A simple numerical calculation gives an idea of how powerful this can be. Suppose the single base station covers 100 square miles (mi^2) and there are frequency assignments for 1000 users in the spectrum. The cellular approach can divide this into 30 zones of about 33 mi^2 each. Within each 33-mi^2 zone, there can be this same 1000-user potential. The total number of potential users is now 30×1000 or 30,000. If the zones were made smaller, there would be even more users than could be handled.

Technically, the keys to cellular systems and networks reside in the local base stations. These stations must handle each mobile unit within the zone and also communicate over their own private network to other adjacent zone base stations. The base stations, working together and directed by a computer, must solve the following problems:

Identify which base station is closest to the mobile phone.

Determine whether the signal from the mobile unit is starting to fade at one base station, but increasing at another, meaning that the user is moving in a certain direction. If so, preparations for switching control of that phone to the new base station must be made.

Keep track of which phones are assigned which frequencies in each zone.

Reassign unused frequencies to any phone that enters the new zone and direct the phone to change its frequency.

Free any frequency that was used by a phone that just left the zone.

Make sure that the coverage area of each zone overlaps enough with adjacent zones so there are no dead spots.

At the same time, make sure that the overlap is as little as needed, since two mobile units cannot share the same frequency assignment in the overlap area.

Coordinate all these activities with adjacent zones and the central system computer which is managing the overall cellular network.

The concept of cellular phones and systems is elegant, but considerable technology and network capability among the zones are necessary to make it work reliably. It may seem that many users can be added simply by making the zones smaller. This does not work out, in practice. As the zones become smaller, their coverage circles overlap more and more (Fig. 11-29) so that there is more

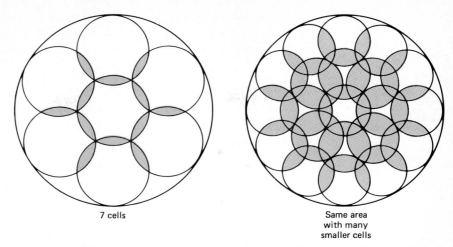

7 cells

Same area with many smaller cells

Fig. 11-29 The amount of overlap in coverage (in which only one phone can have a single frequency) is related to the number of cells in an area. Smaller cells, which can accommodate more users overall, also have more overlap, which corresponds to wasted areas.

wasted overlap area. Second, the cost of building and installing many base stations makes the overall system too expensive. Third, as the number of base stations increases, the internal network by which the base stations communicate with each other grows very complicated and difficult to manage. The studies done on the technical and cost issues related to cellular networks, primarily by the Bell Telephone System, give some guidelines as to what size and number of zones give the best performance and cost compromise for an overall larger area. The cellular zone size tends to be about 10 mi, on average. It may vary, depending on the nature of the area: heavily populated cities have smaller zones than the more open suburbs.

An example will show the effect of varying some size. Suppose a base station for 100 mi^2 costs $1,000,000, but a base station for a zone only costs $100,000. Then, 10 zones can be built for the cost of a single large area base station. However, the cost of the intrazone network must be added, regardless of the number of zones. This may be another $500,000. Of course, the number of users in a 10-zone system is about 10 times the number of those in the single large zone, so the income from telephone charges will be about 10 times greater. A 20-zone system can support 20 times the number of users, in theory, but the internal network cost increases to $1,500,000. Now, compare the items shown in Table 11-2.

Table 11-2

Zones, Number	Cost, Million Dollars
1	1
10	$(10 \times 0.1) + 0.5 = 1.5$
20	$(20 \times 0.1) + 1.5 = 3.5$

Networks 415

Finally, the number of users is the theoretical maximum. Because the zones overlap, the actual number is less than in theory. As the overlap areas increase, the number achievable is much less than it is for smaller amounts of overlap.

QUESTIONS FOR SECTION 11-10

1. How is mobile communications ordinarily achieved?
2. What range is supported by the typical mobile base station?
3. How many channels can usually be handled?
4. How does the cellular concept work?
5. How does cellular design increase the number of users? By what amount?
6. Identify five problems that each cellular base station must deal with and handle.
7. Why can't cellular zones be made smaller and smaller, to support more users? Give both technical and cost reasons.

PROBLEMS FOR SECTION 11-10

1. A cellular system is using 20 base stations to replace a single large base station. The single base station has 200 user frequencies within the spectrum. In theory, what is the maximum number of users that the cellular system can support? What if 20 percent of the increase cannot be used because of overlap in cell coverage?
2. Sketch a large circle representing the area of coverage of a single large base station. Then, overlay this with a sketch of the total area covered by seven cellular areas. In order to cover the total original area how much overlap occurs with the smaller cells? (Estimate; don't measure.)
3. Show how three cells could cover approximately the same area as one large cell. If the cell areas just touch (use circles), but do not overlap, approximately how much area is left uncovered by the cells?
4. A single large base station costs $5,000,000 and can support 1000 users of mobile phone, over a 100-mi^2 area. It will be replaced by a cellular system which has 10 cells, in which each base station costs $1,000,000. The cellular base station interconnection network costs another $100,000 per cell.
    a. What is the total cost of the cellular interconnection network?
    b. What is the total cost of the cellular base stations?
    c. What is the total cost of the cellular system? Compare this to the cost of the noncellular system.
    d. How many users can the cellular system support, in theory?
    e. Each user pays $1000/yr for the phone. What is the maximum amount of income from the single base station system? From the cellular system?
    f. Because of cell overlap, 25 percent of the cellular frequencies cannot be used across cells. What is the income, in this situation?
5. A single base station system has 500 users and costs $3,000,000. There are plans to replace it with a cellular system, but the number of cells is not yet decided. Each cellular base station costs $750,000, and the interconnection network among cells costs $100,000 per cell. As more cells are used, the

overlap among cells reduces the number of users that can be supported by some percentage below the theory. Fill in the following chart:

Cells, Number	Overlap, Percentage	Total Cost	Users (Theory), Number	Users (Actual), Number	Cost/User
1	0				
7	15				
15	25				
30	35				

11-11 The Integrated Services Digital Network

The numerous advantages of digital communications over traditional analog systems are spurring the development of all-digital systems installed and serviced by the phone company and available to the public. In these systems, the signals either originate in digital format (such as from a computer) or are converted to digital via an analog to digital converter at the signal source (the telephone, for example). Modems which use analog tones and channels which are designed for analog signals can be replaced by systems that were designed for handling digital signals only. A new set of standards called the *Integrated Services Digital Network* (ISDN) is now being developed for worldwide use by the International Telegraph and Telephone Consultative Committee (CCITT, using its French initials), which is an international body operating under the guidance of the United Nations. The CCITT develops many European and worldwide communications standards.

The ISDN proposed standards are still in the development stages. The plan is to have standards which cover a wide range of digital communications requirements, including the transmission of signals, switching from one line to another, channel control and signaling, equipment interfaces, and overall service. The long-range plan is for ISDN-compatible systems to allow direct digital communications between any systems that can transmit digital information, such as digitized voice, computer data, digitized video (television) signals, and information which controls the system operation.

The structure of the ISDN system is shown in Fig. 11-30. There are three interface points, called S, T, and U, with S closest to the system user and U farthest away. The S interface ensures that the digital signal from the user device has format, signal levels, and timing compatible with the system. If the signal is analog in nature, such as voice, it is converted to digital at the S interface. The T interface combines multiple user signals into a higher-bit-rate digital signal. The signals then go to the U interface, which is the major interface point between the local loops of the user and the central office.

The current ISDN specifications allow two different types of access to the system. These are called *basic rate* and *primary rate*; they differ in the amount of data capacity they offer. Both basic and primary rate services support both user data channels and separate channels which have signaling and controlling infor-

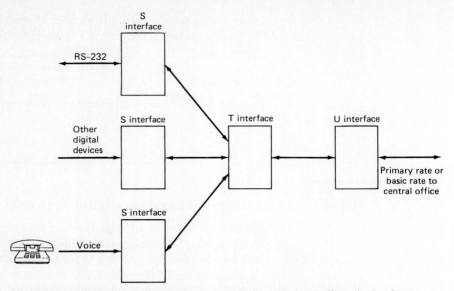

Fig. 11.30 ISDN has three interface points S, T, and U and allows both a lower rate basic digital service or higher rate primary operation.

mation about what the user data channels are doing and what they need. This frees up the data portions to carry user data rather than data which is used for system control and management.

With *basic rate* service, there are two 64-kbit/s user data channels (called *B channels*) plus one 16-kbit/s D channel for control (called a *D channel*). Therefore, the basic rate provides what is called *2B + D service* with an overall rate of (2 × 64) + 16 = 144 kbits/s. This is only a moderate-bandwidth requirement for the ISDN system, so the amount of data and number of users is low.

The *primary rate* uses much more bandwidth. It supports 23 B channels and one 64-kbit/s D channel for signaling and control (23B + D). The overall data rate is therefore (23 × 64) + (1 × 64) = 1.536 Mbits/s. This rate is suitable for use with the T1 multiplexed data transmission system which operates in North America and Japan at 1.544 Mbits/s. In Europe, the equivalent of the T1 system is called *CEPT*; it operates at 2.048 Mbits/s. Under the ISDN system, this would support 30 user B channels and 1 D channel (30B + D) or 30 user B channels and 2 D channels (30B + 2D). The ISDN standards proposed by CCITT allow for both the North American/Japanese and European systems.

QUESTIONS FOR SECTION 11-11

1. What is the goal of ISDN?
2. What rule-making body is developing the ISDN standards?
3. What are the S, T, and U interfaces of the ISDN standard? What is the function of each?
4. What is basic service? How many user data channels does it support? At what bit rates? What type of signaling and control channel does it support?

5. What is primary service? How many user channels and signaling and control channels does it allow? What is the overall bit rate?
6. With what data rate is the overall ISDN primary rate compatible?

SUMMARY

Networks provide a way for many users to share data with each other easily and conveniently. From a technical perspective, networks are complicated interconnections of circuitry and functions. Some networks are open to all users, and others are for private use within a company or factory. Many network types are available. Each serves some type of application better than other network designs.

Network topology describes the way in which the many user nodes are interconnected. Star topology has a single central node, to which all users connect. This is the same as the local telephone company exchange. A bus topology has the individual nodes connecting anywhere along the network, using the network as a common highway. The ring topology connects each node with a link from node to node, in a large circle.

The topology alone does not define the network. Each user node must have a means to access the network. Command/response and interrupt-driven access are used when a single user can be called the master, and all others are slaves. The slaves do not speak unless they are addressed. This access method works but requires many overhead messages to transfer data, especially from slave to slave. Other alternatives include token passing and CSMA/CD. In token passing, each node has a turn at transmitting when it has the token message which puts it in charge. It passes the token to the next node, within a specified time period. The CSMA/CD approach allows any node to transmit, but only if it first checks to make sure the network is unoccupied. Even then, two nodes may begin to transmit at the same time. When this collision occurs, both nodes stop and wait for a random amount of time before trying again. The token passing scheme guarantees node access times; CSMA/CD does not.

The network medium can be twisted pair, base band coax, broadband coax, or fiber optics. The twisted pair is the least costly and easiest to work with, but it has the lowest performance potential. The other media have various characteristics in the areas of distance, speed, noise immunity, and cost.

Local area networks provide intercommunication for many users within a smaller area, such as an office or factory. They allow users to transfer memos, manufacturing information, or data from experiments among each other. They are now available under several industry standards, including Ethernet, IEEE-488 (for test instrument control), MAP (for interconnection in factory production applications), and AppleTalk, provided by Apple Computer for the Macintosh family of computers. Each of these networks attempts to balance cost, ease of use, reliability, and guaranteed performance needs of the intended type of user.

Of course, many data communication systems span large areas. For these, wide area networks which make use of telephone facilities or other common communications channels (such as satellites) are used. The wide area networks usually act as packet switching networks, whether messages are collected at one node or grouped together and sent to a distant node, where they are separated for the many users. Within a group of messages, the overall collection is broken into smaller packets, and each packet is sent by the best channel available at the instant it is transmitted. One packet may go by cable, and the next by satellite. The wide area network must reassemble the packets, in the correct order, at the far end.

A total communications network involves many layers of protocol to ensure that reliable, meaningful communication can be achieved between users who have different protocols, languages, and syntax. The seven-layer OSI model provides a framework for achieving this. The lowest layers deal with the transmission of the message, and the highest layers make sure that the message can be understood by the actual receiver. Not all networks need all these layers. A local area network, for example, may need only the lowest two or three.

The ICs which are used by data communications system designers to build networks provide much of the network protocol and access. These rules are summarized in state transition diagrams, which define what state a network can be in and how it can get to other states. State diagrams clearly show this and form the basis for the design of specialized ICs which are optimized for network management. Buffer ICs are also needed, to connect the management IC with the physical network medium.

The need for mobile communications, for both phones and small computers, has created a demand for many more channels of radio link to the telephone system. The single powerful transmitter does not make the best use of the limited frequency spectrum. By subdividing the area to be covered into smaller cells, more users can be handled. Each cell has its own base station, and the base stations interconnect to each other to coordinate activity when a mobile phone crosses from one cell to the next.

The subject of networks is a complicated one. However, the need for networks on both a local and wide area is increasing at a dramatic rate, as more users in offices and factories have computers and computer-based devices at their desks or plant machinery.

END-OF-CHAPTER QUESTIONS

1. What are some of the applications of a network? What are some of the characteristics of a network?
2. Give an example of two data communications applications that are not considered networks.
3. Compare open and closed networks. What is similar about them? What is different?
4. What are some of the features with which a network designer or implementer must be concerned?
5. Is there a single best network for all applications? Explain why not.

6. What does network topology mean? What are the implications of having different topologies?
7. What is the star topology? The bus? The ring?
8. Compare the number of links needed in each of the three topologies for five users. How many new links for the sixth user?
9. What are the critical reliability issues for each topology?
10. What are the costs and advantages for each topology?
11. What types of media are supported by the various topologies?
12. What is network access? How is it related to protocol?
13. Compare the operation of command/response, interrupt-driven, token passing, and CSMA/CD access, with respect to efficiency, reliability, predictability, and cost.
14. How does interrupt-driven access offer an improvement over command/response?
15. What is the weakness in having a master and several slaves? What is the strength of such a scheme?
16. How do token passing and CSMA/CD operate without any node fixed as the master?
17. What is the waiting time for any user to have his turn in CSMA/CD? What factors affect the total wait?
18. Why does CSMA/CD use random back-off times when collisions occur? What is the potential drawback to this?
19. How does token passing ensure that a node will receive a message within a known amount of time? How is this time calculated in advance?
20. Compare twisted pair and coax (base band and broadband) with respect to cost, ease of use, bandwidth, and number of nodes.
21. What is the technical difference between base band and broadband coax?
22. Compare fiber optics to coax, in terms of cost, ease of use, noise immunity, and number of nodes.
23. Explain how different networks can share the same cable in some systems.
24. How are connections made into a twisted pair network? A coax network? A fiber-optic network?
25. Where are local area networks used, geographically? What are four typical application areas for LANs?
26. What are the three important technical concerns in choosing a LAN for each of the application areas?
27. Which IEEE standards represent the topologies and access schemes discussed previously?
28. Explain applications for which a standard may leave the selection of a network issue to the user.

29. Compare AppleTalk, IEEE-488, Ethernet, and MAP, in terms of performance, cost per node, number of users, interconnection medium and connector, and intended application.
30. Why does the AppleTalk network use transformer coupling?
31. What does the interconnection of an IEEE-488 system look like?
32. What access scheme does Ethernet use? Compare this to MAP.
33. What is the intended function of a packet switched network?
34. How are data messages handled by the packet switched network? What are the media? What channels may be used?
35. Why must a packet switched network often accommodate different channels, even for the same message?
36. Explain how a long message is split and reassembled by a packet switched network.
37. How does a gateway function between wide and local area networks?
38. What are the functions of the seven layers on the OSI reference model?
39. Why are so many layers needed? Explain applications for which only a few may be needed.
40. Summarize the function of the lowest three layers (numbers 1 through 3). Compare these to the IEEE-802 standards for networks.
41. Why can a network protocol be characterized by a state transition model? How does the model show the rules of the network?
42. How does the state diagram show the transition?
43. What kinds of ICs are needed for a complete IEEE-488 interface? What is the function of each IC in the group?
44. Why does Ethernet require such advanced ICs, compared to IEEE-488? What are the functions of these ICs?
45. Why does a cellular phone system allow many more users within the same bandwidth?
46. How does a cellular system work?
47. Why can the number of users be increased more and more by shrinking the cell size? What causes the difference between the theoretical and actual increase?
48. What are three of the functions of each base station in a cellular system?
49. What is ISDN? What is the intention of the ISDN standards?
50. How does ISDN provide for user data channels as well as signaling and control channels?
51. What is the difference between an ISDN basic interface and a primary interface, in terms of data rules, number of user data channels, and signaling and control channels?
52. Compare the primary rate ISDN system for North America and Japan versus Europe.

END-OF-CHAPTER PROBLEMS

1. A network which uses links from each node to each other node is being built for 60 users. How many links are needed? How many new links are needed when the number of users increases to 65? From 65 to 66?

2. A network is being built to support 10 users. The nodes are located on a circle of about 1000-ft diameter.
    a. For a star network, how many feet of cable are needed?
    b. For a bus topology, how many feet of cable are needed? (Assume each node is located on a 5-ft cable off the bus.)
    c. How many feet of cable are needed for a ring topology?

3. In Prob. 2, another node is added on the same circle. How many additional feet of cable must be added in the star, bus, and ring topologies?

4. In Prob. 3, how many new cable splices must be added in each case? (A splice is a cut into an existing cable, not a new connection.)

5. A command/response system is used to send messages from slave 1 to slave 2, then from slave 2 to slave 3, and from slave 3 to slave 4. Each message takes 0.5 ms. How long does the entire sequence of messages take?

6. The command/response of Prob. 5 is replaced by an interrupt-driven access. The interrupt signal also takes 0.5 ms. What is the total message time? Compare this to the answer in Prob. 5. What is the amount of time saved (or lost) by command/response versus interrupt?

7. What would be the effect on response time in Prob. 6 if the slaves were numbers 7, 8 and 9 instead of 1, 2, and 3, for both command/response and interrupt-driven access?

8. A network uses token passing access. Each message transfer takes 0.2 ms. How long does a message take to go from node 4 to node 10? From node 4 to node 3, assuming there are 10 nodes in the system?

9. A token passing network is in use for a large computer communications system, with 200 nodes. What is the longest time that it takes for a message and its response to be sent and received at user 7, if each transfer of the token takes 0.13 ms? Sent and received at user 85?

10. A CSMA/CD system has two users, A and B, who have data to transmit at the same time. The back-off times are 1 ms for A, 2 ms for B. The messages are 3 ms long. How long does A wait to transmit? How about B?

11. Sketch the situation of Prob. 10.

12. The system of Prob. 10 now has three users who wish to transmit 3-ms messages. A has back-off times of 1 and 2 ms for each successive collision, and there are back-off times of 2 and 1 ms for B and 3 and 2 ms for C. Sketch the network activity as all three try to begin at the same time. Who waits longest before beginning? How long?

13. Repeat Prob. 12 for the case in which the back-off time is always 2 ms, for all users and every collision. What happens?

14. Draw a state transition diagram for a full-duplex communications port, with handshaking in each direction. The possible states are receiving data, waiting for new data, transmitting data, waiting for OK to transmit. What are the events that cause state transitions to occur?

15. An IEEE device is designed to operate as either a talker or a listener, but not both, at any instant. It goes from one state to the other by command from the controller if it has data to send (talker) or expects data so that it can complete task (listener). Draw the state transition diagram.

16. A cellular system is replacing a single base station system that has 100 users. The cellular system is a dense urban area, so the cells are small, and the overlap is high because of the number of cells (and the local receiving conditions with tall building interference). The overlap is about 40 percent.

    a. What is the increase in the number of users if the number of cells is 10?
    b. Because of the urban area, the local cell base stations are low-powered and inexpensive, only $200,000 each. The interconnection between the base stations, however, is expensive at $500,000 each. How much does the cellular system cost?
    c. The number of cells is doubled to 20. Repeat parts **a** and **b**.

17. Thought problem: Normally, the coverage areas of cells are thought of as circles. This causes overlap. What coverage is available if the areas of coverage are:

    a. Triangles?
    b. Squares?
    c. Pentagons?
    d. Hexagons?

    Does there have to be overlap? Sketch a large area covered by triangles, squares, pentagons, and hexagons. How many of each would give good coverage of approximately a circle? Note: This shows the key role of simple plane geometry in communications and cellular systems.

12 Error Detection, Correction, and Data Security

No customers would tolerate a bank that made errors in their accounts once a month, as a result of occasional errors in transferring account-related numbers. A data communications system must provide error-free data to the user at the receiving end. Errors can result from external noise added in the link, noise in the transmitting or receiving electronics, or marginal circuit performance due to power supply, temperature, or other fluctuations. A good system design attempts to minimize the sources of these errors, but also assumes that some will occur.

To handle the inevitable errors, special bits are added to the data to allow the detection of errors. A simple parity bit allows some types of errors to be detected, and more complex schemes allow virtually all errors—single or multiple bits—to be detected.

Many applications require that errors also be corrected. This can be done by going back to the transmitter and executing a retransmission, once the error has been detected. It also can be done without requiring a retransmission by adding special error-correction bits to the original data. Requesting the data again requires a reverse direction channel which may not always be practical or efficient. Error detection and correction, also based on algebra theory, allows many types of errors to be corrected as the bits are received but requires additional bits along with the data bits.

A closely related subject is the need to make data transmissions secure from unwanted listeners. The transmission scheme can try to prevent the unauthorized eavesdropper from even receiving the signal. For cases in which this is impossible, as in regular telephone wires, the data bits can be encoded in such a way that even if the bits are perfectly received, they are meaningless to anyone without the proper decoding key. Random bit sequences generated by simple shift register circuitry make this practical. Once again, algebra theory supplies the basis for the method. The data security application can use one of many variations, and a standard method has been published by the U.S. government. Several integrated circuit (IC) manufacturers implement this method in advanced ICs which are used in data security equipment at low cost. The encryption can also be performed, more slowly, by software programs on many computers.

12-1 The Nature of Errors

In the ideal world, there would be no need to worry about errors in data communications. Unfortunately, errors do happen. There are many causes and types, but even a properly designed data communications system must make some provision for dealing with errors. The impact of an error can be as simple as one letter of a message changed or, more critically, numbers changed from their original values. Errors can also cause the system to lose synchronization or cause the wrong receiver to respond to a message.

Errors can be caused by *noise*, which is any unwanted signal that occurs with the desired signal and adds (or subtracts) from the original value of the signal. The sources of noise range from spectacular lightning strikes to dirty contacts on signal switching equipment. An important characteristic of noise in data communications is the length of the noise disturbance. A short noise burst, say 0.01 s, occurs commonly during phone conversations and sounds like a click. It doesn't prevent the two parties on the phone from understanding each other, however. Compare this to data being sent at 4800 baud. A noise-caused disturbance of 0.01 s means that 48 successive bits have been corrupted. This means that a whole group of characters or block of data is suspect. If the errors occur in bursts, but the bursts only occur infrequently, then the average number of errors per hours is low—but this is deceptive. The overall average must be low, but there are clusters of severe errors. Averages can be deceiving. The average of 1, 1, 2, 1, and 10 errors in each of 5 seconds is only 3 errors/s but 10 errors may be unacceptable in any second.

This "bursty" aspect of noise is very important when deciding on a way to handle the detection and correction of errors.

What, specifically, are some of the causes of errors and noise? They come not only from external electrical disturbances such as motors switching on and off, but from legitimate signals of other parts of the circuitry, operating normally. They can be either external or internal disturbances in its source. The following are some examples:

- If the power supply in the system is not exactly at the specified voltage, the components of the system may not operate perfectly. A typical system using transistor-transistor logic (TTL) ICs requires a power supply voltage of 5 V, with a tolerance of 10 percent. An IC that is located far from the supply may be operating at only 4.5 V or even 4.4 V as a result of voltage drop in the supply wires. Its operation is not guaranteed at these levels.

- A data signal which is traveling along a wire is weakened by the resistance, inductance, and capacitance in the wire. It may start out as a perfect 0- and 5-V signal, but arrive as a 0- and 4-V signal. The receiver may be designed to handle 0- and 5-V signals and operate less than perfectly on the 4-V signals.

- The same signal can arrive, after its trip through the wire, with the correct voltage values, but distorted. This distortion in the signal shape means that the receiver is not able to make a decision on a signal of the proper shape, but is trying to judge on the basis of an improperly shaped signal. The timing of the receiver circuit may cause it to look at the voltage value 0.001 s after the signal edge. For a perfect signal, this is the correct point; for a distorted signal, it may be too early, and the signal may not be up to its full value yet.

The circuitry may be operating at its high or low temperature limits, so the performance of some digital gates may not be perfect.

Crosstalk from adjacent signals can corrupt the signal. This crosstalk can occur on the circuit boards of a system, where a signal in one part of the circuit couples some of its energy into another wire. There is no physical path needed—the two wires act as miniature transmitter and receiver antennas for the electromagnetic energy. Crosstalk is especially a problem when the signal paths run parallel to each other, as in wires in a cable. The parallel signal paths allow the maximum amount of energy coupling.

Finally, errors occur in modern systems by the laws of probability, especially with computer memory circuits and ICs. A computer system used with a data communications channel may have millions of bits of storage. Even a very low chance of error, such as 0.0001 percent, means that some bits are in error (in this case, 1 bit per million data bits). The designers of memories for computers and communications realize that although it is good to strive for absolute perfection in the memory, it is foolish to expect it. Even a stray amount of electromagnetic energy in the system can cause one of the millions of bits to change from a 1 to a 0 or vice versa. In any large collection or group, there are always some bits in memory that are more resistant to error, and some that are "weaker" and more error-prone. These error-prone bits are affected first by low voltage, crosstalk, or signal noise.

The case in which a single bit of data is changed because of noise is the easiest to deal with. If these single-bit errors occur at widely spaced intervals of time, there are some very effective techniques that cost little to implement. If the noise occurs in bursts—even short ones of only 3, 4, or 5 bits—then the problem is more difficult to solve, and the solution costs more circuitry, power use, or transmission time. It is almost impossible to say that no error can occur, since this would take extraordinary designs in the communications circuitry, protocols, and formats, but the probability of error can be made very low, for example, 0.000000001 percent, or a *bit error rate* (BER) of 1 in 10^{11}. Another approach is to design a system that ensures that all errors of a certain type, such as single-bit errors and bursts up to 10 bits, will be recognized with absolute certainty. Any errors above 10 bits in a row will be detected with very high probability. This is a more practical approach, because it defines the nature of the errors expected and handles them without any possibility that errors will be missed.

Although noise is a random event, this randomness is not impossible to characterize. Noise from nearby motors, for example, has a strong 60-Hz component. Noise from crosstalk takes on the aspects of the signal wire that is causing the crosstalk—when it is not in use, the noise is less. Noise from external events that have sudden on-off characteristics has a wide bandwidth of frequency components (recall Fourier analysis). The more that is known about the type of noise, the better the circuit and protocol design can be to defeat it.

A large portion of the design and engineering effort that goes into a data communications system revolves around noise and errors. The first line of defense is to try to minimize the noise that enters the system. To do this, cables may be shielded against crosstalk, or power supply wires may be made thicker to minimize any voltage drop that may occur in them as they supply current to the system. But noise and errors are inescapable realities, and they cannot be kept out—especially if the cause is a "weak" memory bit in the many megabits in a computer system. The effort then must shift to dealing with it.

The first objective is to detect any errors that the noise may cause. A system can be designed to do any of several things once it knows that an error has occurred: ignore the data, request the data again, or notify the operator, for example. A second objective may be to actually correct the error. This can be done either by a request that the data be sent again or by a technique called *error correction*. In *error correction*, the data somehow is formatted so that it contains extra information. This extra information is used to correct most errors that are detected. There are several techniques for performing error detection and correction and for dealing with errors. These will be studied in detail in sections of this chapter, along with the technical advantages and drawbacks to each.

Another issue is *data security* when the data must somehow be protected against unauthorized receivers. In this chapter, the term "data security" does not mean physical security, such as using passwords to keep others away from the data on the computer terminal. It does mean having the data set up in such a way that even if every bit were intercepted or read without error, the eavesdropper still could not figure out the message. (Physical security is another problem entirely, and it is often difficult to achieve, especially when data travels through the public phone system or by radio.) The reason for studying data security here is simple: many of the theories and techniques used to detect and correct errors are also used to make data meaningless to anyone except the desired receiver. Developments in the area of error detection and correction are often applied soon thereafter to data security design, and vice versa.

Finally, it is important to understand that most larger data communications systems do not use a single method of achieving error detection and correction. If the data communications system has several levels, then each level can apply error-detection and -correction techniques of its own. The actions at each level are transparent to the other levels, and each level does what it is designed to do with any group of data bits. The levels do not have to know what the actual bits represent (actual information, headers, error-checking fields) in order to apply these additional methods. Each additional error-detection and -correction scheme simply adds to the capabilities of the preceding schemes and makes the system even more reliable. Specific examples of this will be studied in this chapter, as well.

QUESTIONS FOR CHAPTER 12-1

1. Why is noise a concern even for perfect systems?
2. Compare the length of a typical noise signal to the number of data bits affected.
3. Explain the concept of average error rate versus burst errors.
4. What are some of the causes of noise and errors, both internal and external?

5. What are two general approaches to dealing with noise? Explain.
6. What is the first step in minimizing errors from noise?
7. What can happen when an error is detected? What does error correction require beyond error detection?
8. What are the objectives of data security?
9. What is the relationship of data security to error detection and correction?
10. Does having an error-detection feature preclude having data security? Does having security prevent having error detection and correction? Explain.

12-2 Parity

The simplest and oldest form of error detection is called *parity checking*, or simply *parity*. In a *parity check*, one extra bit is used as a checking bit on the correctness of a group of bits. A typical case would be in the ASCII representation of letters, numbers, and symbols, in which 7 or 8 bits are used to encode the character. An additional bit for parity is added before the stop bit to allow the receiver to determine, in some cases, whether the character bits are correct as received. Parity can be either odd or even.

Exactly what is parity? The parity bit is calculated at the transmitter to make sure that the total group of data bits, including the added-on parity bits, has an even number of 1s (for even parity) or an odd number of 1s (for odd parity). This is regardless of what the data itself is. A few examples will show this, in the case of odd parity first (Table 12-1).

Table 12-1

Data Bits	Parity Bit for Odd Parity
0101	1
0100	0
1101	0
1111	1
0000	1
0001	0
00101000	1
01010110	1
01110000	0
10100101	1

Note that if the original group of bits has an even number of 1s, then the parity bit is set to 1 to make the total odd. If the original group has an odd number of 1s, then the parity bit is set to 0 so the total remains odd. For even parity, the opposite rule applies. The total number of 1s must be even, as shown in Table 12-2.

Table 12-2

Data Bits	Parity Bit for Even Parity
1001	0
0100	1
0111	1
00101001	1
00010111	0
10010000	0

The number of bits that a parity bit is used with is not restricted to only 7 or 8. Parity has been used with as few as 4 bits or as many as 16 or 32.

Parity is a fairly simple scheme, which has the following benefits:

It is very low in cost, and simple, to implement with digital circuitry.

It provides a quick check of the correctness of the data.

It can be easily calculated mentally; this is important if the engineer or technician needs to verify the operation of the parity circuitry.

The difference between odd and even parity is a matter of custom and preference, not of performance. Odd parity does not have any performance advantages over even, and the reverse is true. Some system designers prefer odd parity. The reason for this is that a group of all 0 data bits then has a 1 for parity. This makes it easier to troubleshoot the system with an oscilloscope, since there is something to see on the scope screen. If even parity is used, then all 0 data bits have 0 parity, and it shows up as a constant 0 value. This is easily confused in systems which use 0 V for binary 0 and some nonzero voltage for binary 1. The presence of at least one 1 makes it easier to see the data bits and synchronize the oscilloscope.

How is the parity bit used? The receiver gets the data and parity bits. It knows where the parity bit is in the field of received bits (the last bit in the case of asynchronous ASCII, for example). The receiver recalculates the parity bit,

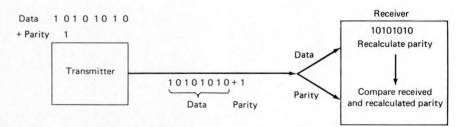

Fig. 12-1 The parity bit is generated at the transmitter and sent along with the data. The receiver recalculates the parity bit value on the basis of the received data bits and then compares this to the received parity value.

on the basis of the received data bits, and compares it to the received parity bit (Fig. 12-1). If they differ, there is a parity error, and one or more of the data bits has changed because of noise or other problems. The error can also be related to the parity bit itself, of course. Table 12-3 gives some examples of parity errors.

Table 12-3

Data Bits	Parity (Odd)	Received Bits
0110	1	0111 1
0111	0	0011 0
1000	0	1001 0
11000101	1	11000001 1
10101111	1	10001111 1
00010000	0	10010000 0

Weakness of Parity Parity used for error detection is simple to implement and use, but it does have weaknesses. The obvious weakness is that it "costs" 1 bit for every group of data bits. In a 7-bit ASCII character, this extra bit means an additional overhead of $1/7 = 14$ percent.

This weakness is not the major one, since one extra bit is a reasonable price to pay for effective error detection. The problem is that parity bits are relatively ineffective at detecting many types of errors. A simple example illustrates this. Suppose the system is using odd parity, and the data is an 8-bit group:

    10110101   Parity bit is 0

If one bit changes, the parity bit will show the following:

    10010101   Parity bit should be 1

However, if two bits are in error, the parity bit won't show the error:

    10000101   Parity bit is still 0

In fact, a parity bit (whether odd or even) only indicates when an odd number of bits are in error. If an even number of bits change, the parity bit remains unchanged. Figure 12-2 shows examples of this. The usefulness of parity in detecting only odd numbers of errors, not even, applies regardless of the number of bits the parity bit is checking. Parity bits are most commonly used on up to 8 or 16 bits, because the chances of a multiple-bit error are much greater (and the effectiveness of parity reduced or eliminated) beyond that number of bits.

Since parity has this technical weakness in dealing with even numbers of bit errors, where and why is parity used? It is useful when the application is not very critical, for example, a local nearby printer. It is also useful in a relatively non-noisy environment, where the chances of a bit error are very low and a burst or cluster of errors is not expected to occur. An example of this is communications between two sections of a single, larger system chassis. Because of the limited

		Odd Parity
Original data	1 0 1 0 0 1 1 0	1
Single-bit error	1 0 1 $\underline{1}$ 0 1 1 0	0
Double-bit error	1 0 $\underline{0}$ $\underline{1}$ 0 1 1 0	1
Double-bit error	1 $\underline{1}$ $\underline{1}$ 0 0 1 $\underline{0}$ $\underline{0}$	1
3-bit error	1 $\underline{0}$ 0 1 1 1 $\underline{1}$ 0	0
3-bit error	0 0 $\underline{1}$ 0 0 1 0 $\underline{1}$	1
6-bit error	$\underline{0}$ 1 0 $\underline{1}$ $\underline{1}$ 0 $\underline{1}$ 0	1
7-bit error	$\underline{0}$ 1 0 $\underline{1}$ $\underline{1}$ 0 $\underline{0}$ $\underline{0}$	0
8-bit error	$\underline{0}$ $\underline{1}$ $\underline{0}$ $\underline{1}$ $\underline{1}$ $\underline{0}$ $\underline{0}$ $\underline{1}$	1

Fig. 12-2 Parity, whether even or odd, can only show when an odd number of bit errors has occurred. Even numbers of errors cause no change in the parity bit. The underlined bits are in error from the original data bits, but the parity bit, as shown, cannot indicate this in all cases.

distance and relatively good electrical environment, errors are very rare. Such error may occur when a voltage at one IC is too close to the allowable lower limit, and the IC is not electrically robust or is slowed down by low ambient temperature. Since parity is very easy and low in cost to implement, it is often included as an essentially effortless safety factor.

The performance of parity bits for error detection can be improved by adding additional parity bits. As shown in Fig. 12-3a, 4 bytes of data are sent by the communications system. With each byte, there is an associated parity bit on the same horizontal line. The communications system can also develop a parity bit in the vertical direction, for each bit position in the byte. These are called *horizontal* and *vertical parity bits*, respectively. The vertical bits provide cross-checking and can be appended to the data bits as an error-check group (called a *checksum*). The cost of eight additional parity bits provides extra error-detection capability. As indicated in Fig. 12-3b, some errors have occurred. Note that one of the horizontal parity bits has changed, even though an even number of errors exist, as illustrated by a horizontal byte.

Unfortunately, even this vertical parity cannot detect all error possibilities, as shown in Fig. 12-3c. To detect more error types, some new type of error-detection scheme is needed. This is discussed in the subsequent sections, after some actual parity bit generation and parity error-detection circuitry is studied.

Parity Circuitry

The digital circuitry needed to generate the parity bit is made from a relatively simple combination of boolean logic gates. One such circuit for four data bits is shown in Fig. 12-4. A single control line is used to select generation of even or odd parity. The logic gates take the input data bits and produce the correct parity bit. The data bits and the parity bit are then transmitted by whatever scheme is used (serial, parallel, and so on) to the receiver.

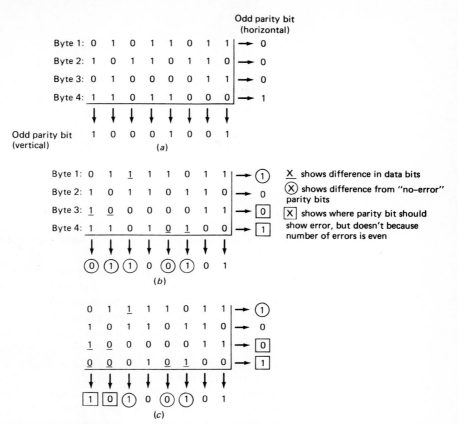

Fig. 12-3 Vertical parity bits are a cross-check on the standard horizontal parity: (*a*) the original data bits and the resulting horizontal and vertical parity bits (for odd parity). In (*b*) the data bits have errors. The horizontal parity bits don't show all of these, but the vertical ones do. In (*c*), even the vertical bits may not see some error patterns.

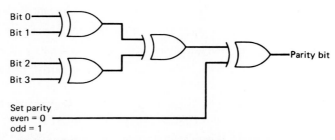

Fig. 12-4 Digital gates can be used to generate parity bits. The circuit shown develops the parity bit for a 4-bit data group; even or odd parity can be selected by a single control line. Exclusive OR gates are used.

Error Detection, Correction, and Data Security

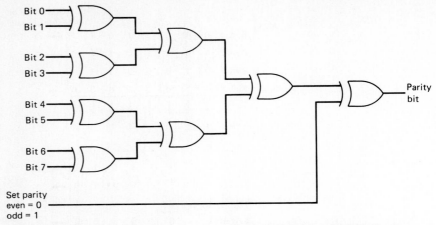

Fig. 12-5 The circuit of Fig. 12-4 can be easily expanded to handle 8-bit data bytes.

At the receiver, the circuitry must take the data bits, develop the parity bit again, and then compare the new parity bit to the received parity bit. If they agree, there is no parity error; if they differ, a parity error exists. The circuitry at the receiver is designed to generate a parity error bit for the receiving system. An error bit equal to 1 indicates parity error, and a 0 means no error.

The simple 4-bit circuit concept is easily expanded to handle 8 bits (Fig. 12-5). It can be further expanded if needed. Since there is a great demand for parity circuitry, the manufacturers of digital ICs have developed ICs which include the complete parity circuitry. These ICs can be interconnected (*cascaded*) to provide parity on any number of bits (Fig. 12-6). Similarly, the manufacturers of universal synchronous receiver-transmitters (UARTs) embed parity generation and checking circuitry into the UART IC, so no additional ICs are required to add parity to the serial receiver and transmitter function.

QUESTIONS FOR SECTION 12-2

1. What is parity checking? How is parity generated?
2. How is parity used for error detection?
3. What is the difference between odd and even parity, in terms of generating the parity bit and in terms of technical performance?
4. What are some of the benefits of parity?
5. Why do many systems use odd parity instead of even?
6. What is the "cost" of adding parity? Is it a function of the number of bits? Of the number of messages?
7. Why is parity a relatively weak error check? What types of errors can it detect? What types does it miss?
8. Where is parity commonly used? Why? Why is it often used even though it doesn't provide strong error-detection capabilities?
9. Where is horizontal parity? Vertical parity? What are the function and capability of each?

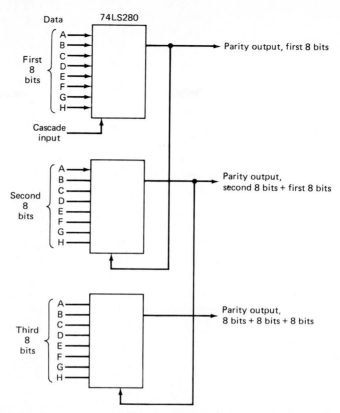

Fig. 12-6 ICs which incorporate all of the parity circuitry are available. The 74LS280 provides parity for 8 bits and can be combined (cascaded) to handle larger groups such as the 24-bit group shown.

PROBLEMS FOR SECTION 12-2

1. Calculate the even parity bit for the following groups of data bits:
   0110, 1100, 0001, 0000, 1111, 1110
2. Calculate odd parity for the following 6-bit groups:
   011001, 110011, 010001, 110000
3. Find the even parity bit for the following groups of 16 bits:
   00001101 10100100, 01010111 00110101, 11111111 00000010
4. Calculate the vertical parity bits for the data groups in Prob. 2, using odd parity.
5. Calculate the even vertical parity for the data groups in Prob. 3.
6. The following pairs of data bits represent the data as sent and the data as received. Calculate the even parity bit for both before and after cases, when an error may have occurred during the transmission:
   01101001 — 01111001
   10100001 — 10100010

Error Detection, Correction, and Data Security

00000000—00000001
11111111—11111100

Where are the errors, as indicated by looking at the parity bits? Where are the errors really located?

7. Use odd parity bits, and see which errors can be detected for the following pairs of sent and received data bits:
10101001 00000001—10101010 11000000
00010001 10010010—00011110 10010010
11111111 11110000—00000000 11110000
01010101 01010101—01010111 01010101
Where are the actual errors in the received data groups?

8. Use the parity circuit of Fig. 12-4. Show the way it determines parity by marking the binary value of each point in the circuit for the bit pattern 1001. Use even parity.

9. Using the parity circuit Fig. 12-4 and odd parity, note the binary value at each point in the circuit for data 1110.

10. (Advanced Problem) Find the parity of 0110 and of 1000. Next, find the odd parity value of the two parity bits just calculated. Compare this to the odd parity value if parity is calculated across all eight bits as one group.

11. (Advanced Problem) Find the even parity bit of 01011000 and of 11110000. Find the even parity bit of the two 8-bit units as a single 16-bit unit. Compare this to the even parity value calculated on the two separate parity bits of the 8-bit bytes. Can you draw any conclusion about parity of subgroups versus the parity of a larger group of bits from this?

12-3 Cyclic Redundancy Codes

The weakness of the parity method of error detection is clear—only odd numbers of errors can be detected. The need for an error-detection scheme that provides much more confidence to the user led to the development of more effective error-detection technology. The most common method is based on the use of *cyclic redundancy codes* (CRCs).

The impetus for the development of these codes was the need to somehow generate a unique fingerprint based on the data itself. This fingerprint can be sent along with the data. The receiver uses the same scheme, generates the fingerprint again, and then compares the two. If the fingerprints agree, there is no error; if they differ, an error has occurred. The problem of developing an effective method of detecting errors is then a problem of figuring out how to set up a formula or algorithm that generates this unique fingerprint based on the data bits.

This is the environment in which the CRC error detection is used. The theory of CRC is derived from a branch of mathematics called *algebra theory*. The theory itself is quite complicated, but the circuitry to physically implement the theory and make use of it is very simple (and will be shown in this chapter). The algebra theory shows the type of formula that should be used and the effectiveness of the formula. In a CRC-based scheme, all the data bits of the message are put through circuitry which embodies this formula. The result is a *CRC checksum*, which is a group of bits which is the fingerprint of the entire group of data bits. Unlike parity, which uses a single bit for every relatively small group of data bits (typically 48 or 16), the CRC checksum can fingerprint hundreds or even thousands of bits of data. The checksum itself is several bits long. Typical checksum lengths that are used are 16 and 32 bits. The result is that the CRC error detection "costs" only a very small percentage of the total number of bits to be sent.

The Concept of CRC Checksums

The theory of CRC checksums is developed by using algebra and polynomials. These polynomials are equations which have the form of powers of X:

$$X^N \ldots + X^3 + X^2 + X^1 + X^0$$
[**Note:** $X^0 = 1$]

For binary data, each term has a 1 in front as its coefficient if the data bit is 1, and a 0 coefficient if the bit is a 0. For example, the polynomial for the binary pattern 10011 is $X^4 + X + 1$ (since the third and second terms are missing) with X^4 as the most significant bit (MSB) and 1 representing the least significant bit (LSB).

The polynomial for binary 11001010 is

$$X^7 + X^6 + X^3 + X^1$$

The polynomial which represents the data bits is called the *message polynomial*, usually shown as *G(X)*. There is a second polynomial, called the *generator polynomial P(X)*. It has the same format as *G(X)*. The goal is somehow to combine these two polynomials *P(X)* and *G(X)* to produce the CRC checksum polynomial *F(X)*. It is accomplished as follows:

1. Multiply the message $G(X)$ by X^{n-k} where $n-k$ is the number of bits in the CRC checksum (4, 8, 16, or 32 typically). This has the effect of increasing each power of X in $G(X)$ by $n-k$ so X^3 becomes X^7 if $n-k = 4$, for example.
2. Divide the resulting product $X^{n-k}[G(X)]$ by the generator polynomial $P(X)$.
3. Disregard the quotient. Add the remainder $C(X)$ to the product to give the code message polynomial $F(X)$, which is represented as $X^{n-k}[G(X)] + C(X)$. $F(X)$ is the original data followed immediately by the CRC checksum $C(X)$.
4. The division is performed in binary without carrying or borrowing. In this case, the remainder is always 1 bit less than the divisor. The remainder is the CRC, and the divisor is the generator polynomial; therefore, theory shows that the bit length of the CRC is always one less than the number of bits in the generator polynomial.

A simple example is explained next.

1. Given:

   Message polynomial $G(X) = 110011$ ($X^5 + X^4 + X + X^0$)

Generator polynomial $P(X) = 11001\ (X^4 + X^3 + 1)$

$G(X)$ contains 6 data bits

$P(X)$ contains 5 bits and yields a CRC checksum with 4 bits; therefore $n - k = 4$.

2. Multiplying the message $G(X)$ by X^{n-k} gives
$$X^{n-k}\,[G(X)] = X^4(X^5 + X^4 + X + X^0) = X^9 + X^8 + X^5 + X^4$$
The binary equivalent of this product contains 10 bits and is 1100110000.

3. This product is divided by $P(X)$.

4. The remainder $C(X)$ is added to $X^{n-k}\,[G(X)]$ to give $F(X) = 1100111001$. Note that the six MSBs are the original data $G(X)$, followed by the CRC checksum $C(X)$.

Circuitry to Implement CRC Checksums

The division needed to perform the CRC checksum would take a long time as the message $G(X)$ became longer and included more data bits. A processor programmed to perform the operation would spend many tenths of seconds or even seconds to do it. It is not acceptable in communications to have an error-detection scheme whose implementation takes a long time and actually has the effect of delaying the messages. In fact, the delay would occur twice: once at the transmitting end to calculate the CRC checksum value and again at the receiving end, where it is recalculated and compared to the received CRC checksum value.

Fortunately, there is a very simple and essentially instantaneous way to develop the CRC checksum using digital circuitry, by using shift registers and boolean logic gates. The data bits of the message are shifted into the shift register. Some of the shift register locations are fed back into the register, using *exclusive OR* (XOR) *gates*. The *shift register* is the physical implementation of the generator polynomial $P(X)$. After all the bits of the message have been shifted into the register, what is left in the shift register is the remainder $C(X)$ of the polynomial division. Figure 12-7 shows one such example. Here the generator

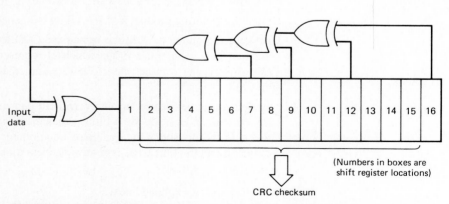

Fig. 12-7 Simple circuitry can implement the complex theory of the CRC checksum. A shift register and XOR gates take the data bits and feed them back in the required sequence. The resulting bit pattern in the shift register after all the data bits have passed through in the CRC checksum, is shown.

polynomial $P(X)$ is $P(X) = X^{16} + X^9 + X^7 + X^4 + 1$. A 16-bit shift register is used, with feedback through XOR gates at shift register locations 7, 9, 12 and 16 going back to shift register location 1. The XOR gate has a simple boolean rule: the output is 0 if both inputs are 1 or both inputs are 0; the output is 1 if either input is 1 and the other input is 0.

(The relationship between the location of the feedback taps and the desired polynomial $P(X)$ is mathematical and simple and will be discussed after the operation of the circuit is covered. Assume these taps are correct for this example.)

Before the data enters the shift register, the register is cleared to 0. The data bits are then shifted into the register. As each bit is shifted in, it is first combined with the result of the XOR combination, which itself was based on the previous contents of the shift register. Assume a simple case of 16 data bits in which the first bit of the data is a 1, and the remaining 15 bits are 0. The contents of the shift register at each cycle, as each bit is shifted, are shown in Fig. 12-8a. After 16 shifts and XOR steps, the register has the remainder. This remainder is the CRC checksum value, which is 0100 0000 1010 0101 in this case.

Now, change the data so that the first two bits are 1 and the remaining 14 bits are 0. Figure 12-8b shows the effect as these data bits pass through the shift register and XOR circuitry. The binary value in the shift register after all the data bits have been shifted through is the remainder, exactly as would be calculated by performing the division of a polynomial $G(X)$ into polynomial $P(X)$. The single-bit change in data results in a completely different CRC checksum. This is a very simple, inexpensive, and quick solution to the problem of performing the complicated division that algebra and CRC theory show must be done.

In practice, the transmitter passes all the data bits through a shift register and XOR circuit (Fig. 12-9). The data bits are sent, followed by the CRC checksum. There is no time lost in developing the CRC value, since it can be done by the shift register and XOR gates as the data bits are actually being transmitted. At the receiver, the bits are received and passed through the identical circuitry that is used at the transmitter. When the last data bit is received, the value in the shift register should agree exactly with the CRC value received. Once again, there is no time lost calculating the CRC value because it can be done as the bits arrive. The only time cost is the time to send the CRC bits themselves.

Location of Feedback Taps

The location of the feedback taps for the XOR gates is simple to determine. The feedback tap locations are the inverse of the polynomial $P(X)$. This inverse is easy to find. Take the powers of X that are in $P(X)$. Subtract each power value from the number of locations in the shift register, which is the same as the highest possible power in $P(X)$. The result of each subtraction is the tap number. Do this for each of the powers of X in $P(X)$. Therefore, for $P(X) = X^{16} + X^9 + X^7 + X^4 + 1$ the powers of X that exist are 16, 9, 7, 4, and 0 (since 1 is X^0). The powers of X are subtracted from 16 (the length of the shift register) with the result 0, 7, 9, 12, and 16, and so the inverse of $P(X)$ is $X^{16} + X^{12} + X^9 + X^7 + 1$. The taps are placed at shift register locations which correspond to the powers of X, except at $X^0 = 1$, which is where the data enters the shift register.

a) CRC for Data 0000 0000 0000 0001

Bit Number	Input Bit Value	1	2	3	4	5	6	7	8	9	10	11	12	13	14	15	16
1	1	0	0	0	0	0	0	0	0	0	0	0	0	0	0	0	0
2	0	0	0	0	0	0	0	0	0	0	0	0	0	0	0	0	1
3	0	0	0	0	0	0	0	0	0	0	0	0	0	0	0	1	0
4	0	0	0	0	0	0	0	0	0	0	0	0	0	0	1	0	0
5	0	0	0	0	0	0	0	0	0	0	0	0	0	1	0	0	0
6	0	0	0	0	0	0	0	0	0	0	0	0	1	0	0	0	0
7	0	0	0	0	0	0	0	0	0	0	0	1	0	0	0	0	0
8	0	0	0	0	0	0	0	0	0	0	1	0	0	0	0	0	0
9	0	0	0	0	0	0	0	0	0	1	0	0	0	0	0	0	1
10	0	0	0	0	0	0	0	0	1	0	0	0	0	0	0	1	0
11	0	0	0	0	0	0	0	1	0	0	0	0	0	0	1	0	1
12	0	0	0	0	0	0	1	0	0	0	0	0	0	1	0	1	0
13	0	0	0	0	0	1	0	0	0	0	0	0	1	0	1	0	0
14	0	0	0	0	1	0	0	0	0	0	0	1	0	1	0	0	1
15	0	0	0	1	0	0	0	0	0	0	1	0	1	0	0	1	0
16	0	0	1	0	0	0	0	0	0	1	0	1	0	0	1	0	1

—Remainder—

b) CRC for Data 0000 0000 0000 0011

Bit Number	Input Bit Value	1	2	3	4	5	6	7	8	9	10	11	12	13	14	15	16
1	1	0	0	0	0	0	0	0	0	0	0	0	0	0	0	0	0
2	1	0	0	0	0	0	0	0	0	0	0	0	0	0	0	0	1
3	0	0	0	0	0	0	0	0	0	0	0	0	0	0	0	1	1
4	0	0	0	0	0	0	0	0	0	0	0	0	0	0	1	1	0
5	0	0	0	0	0	0	0	0	0	0	0	0	0	1	1	0	0
6	0	0	0	0	0	0	0	0	0	0	0	0	1	1	0	0	0
7	0	0	0	0	0	0	0	0	0	0	0	1	1	0	0	0	0
8	0	0	0	0	0	0	0	0	0	0	1	1	0	0	0	0	0
9	0	0	0	0	0	0	0	0	0	1	1	0	0	0	0	0	1
10	0	0	0	0	0	0	0	0	1	1	0	0	0	0	0	1	1
11	0	0	0	0	0	0	0	1	1	0	0	0	0	0	1	1	1
12	0	0	0	0	0	0	1	1	0	0	0	0	0	1	1	1	1
13	0	0	0	0	0	1	1	0	0	0	0	0	1	1	1	1	0
14	0	0	0	0	1	1	0	0	0	0	0	1	1	1	1	0	1
15	0	0	0	1	1	0	0	0	0	0	1	1	1	1	0	1	1
16	0	0	1	1	0	0	0	0	0	1	1	1	1	0	1	1	1

—Remainder—

Fig. 12-8 The CRC checksum (*a*) for data bits 0000 0000 0000 0001, using the shift register and feedback taps of Fig. 12-7; (*b*) for data bits 0000 0000 0000 0011. Note that a slight difference in data bits results in a big difference in checksums in this case.

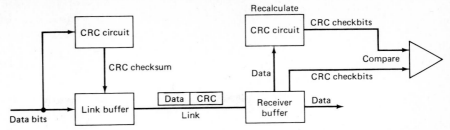

Fig. 12-9 A system using the CRC checksum computes the checksum at the transmitter and sends it along with the data bits. The receiver recalculates the checksum, comparing it to the received value. If they differ, there has been an error.

There is an important point about the CRC checksum: the "meaning" of the checksum itself. As the previous figures showed, a single-bit change causes a very different checksum. Is it possible to say by looking at the checksum what bit or bits change? The answer is no. It is impossible to go in the reverse direction, from the checksum, to figure out what the digital bit pattern that caused it was. A single CRC checksum can be caused by many data patterns. (If it were possible, an efficient system would send only the CRC checksum and let the receiver figure out what the data bits were. Thousands of bits would not have to be sent.) More important, the difference between the correct checksum and the checksum that resulted from an error can be small or large. A single-bit error does not cause the checksum to change by only one bit, and in fact, many bits of the checksum can change. The converse is also true: a single-bit difference in the checksum does not mean that only one data bit is in error. The only conclusion that can be drawn from the checksum is whether the received data is correct or incorrect. It is not possible to determine, with the CRC checksum, how "incorrect" the data is.

Certainly, there are many possibilities for the polynomial $P(X)$. The mathematicians and communications specialists have studied many of these and found that some $P(X)$ polynomials have practical advantages over others. Two of the common ones in industry use are

CRC-16: $P(X) = X^{16} + X^{15} + X^2 + 1$

CCITT: $P(X) = X^{16} + X^{12} + X^5 + 1$

The types of errors that these can detect are both individual errors and clusters that occur from longer noise burst. Various segments of the data communications industry have standardized on one of these (or some other) types of CRC polynomial $P(X)$. Both sender and receiver must agree to the same one for the CRC checksum method to work, of course.

Benefits of CRC Checksum

The CRC checksum is a vast improvement over simple parity checking, at a low cost. It takes only some simple circuitry to generate or check the checksum for error and does not take transmission time except for actually sending the CRC bits. In fact, it takes less time to send one 16-bit CRC checksum for 1000 bits than it does to send one parity bit for every 8-bit byte. A data block consisting of 1024 bits (128 bytes) would require 128 parity bits, or an extra 12.5 percent. It would require only a 16-bit CRC checksum, or $16/024 = 1.6$ percent.

However, the CRC is not absolutely perfect at detecting errors. It can be looked at this way: a 16-bit checksum can take on one of $2^{16} = 65{,}536$ unique values. Therefore, the CRC scheme has produced a fingerprint of the data, but the fingerprint is 1 of 65,536 types. It is possible to have several different bit patterns that produce the same fingerprint, since only these 2^{16} fingerprints are allowed (in contrast to people, for whom every fingerprint is absolutely different), although a string of N bits has 2^N possibilities. The odds—that the original good data and the data with error—will both produce the same fingerprint are very low. A more likely occurrence is that the original data produces one fingerprint, and the data with errors produces another one. A typical CRC polynomial $P(X)$ detects all single-bit errors, 99.8 percent of the double-bit errors, and 99 percent of 3- to 6-bit errors. Different CRC polynomials can detect different percentages of the various types of errors, and this is what leads to the selection of the various polynomials. Longer $P(X)$ polynomials are more effective, since there are more "fingerprints" and less chance of two data bit sequences yielding the same CRC checksum.

The circuitry which generates the CRC checksum can operate on whatever bits go into it. This means that the message can consist of data and parity bits, for example. It makes no difference to the circuitry, and the operation is transparent to the bits. In a multilevel data communications system, the CRC operation is often performed on bits that are other checking bits. The parity bits are unaffected by going through the CRC circuitry, and the CRC circuitry doesn't know which are parity bits and which are actual data bits. Consider a computer connected to a nearby printer and using only parity for error detection. The cable to the printer is then connected to a modem, which connects to a packet switching network to another printer in another state. The packet switching network uses CRC checksums to check for errors. This CRC checksum is developed on both the ASCII bits that represent the character to the printer and the parity bit associated with each character. The receiving end of the network checks for errors by using CRC, strips off the CRC bits, and passes the received characters (data bits and parity) to the new printer. Two levels of error detection occur. In fact, if an error occurs between the computer and the packet network prior to the calculation of the CRC checksum, it is not detected unless the parity bit indicates it. This is because the CRC generation circuitry makes no determination on the errors in the bits that enter into it. A parity error and parity bit go through the CRC process, get the CRC checksum, and are transmitted. The receiver CRC process proves only that no errors have occurred between the transmitter and the receiver, not that the bits entering into the transmitter are error-free.

QUESTIONS FOR SECTION 12-3

1. What is the concept behind CRC? From what is the theory developed?
2. What is the CRC checksum?
3. How many bits are there in the CRC checksum? What determines this number?

4. Over how many bits is the CRC operation performed?
5. What is the message polynomial? The generator polynomial?
6. What is the relationship between CRC checksums, $P(X)$, and $G(X)$?
7. How is the CRC checksum implemented in circuitry?
8. Where do the feedback taps go? How are they determined? What do they represent?
9. Explain the operation of a CRC checksum circuit.
10. What does the CRC checksum "mean"? What does a 1-bit difference in CRC checksums tell about the nature of the data error?
11. Is it possible to go from the CRC checksum to the original data bits? Explain.
12. What are two common $P(X)$ polynomials?
13. Why is CRC more efficient than simple parity?
14. Why is CRC not perfect at error detection? How many "fingerprints" does a CRC code support? How many possible data streams are there?
15. Explain how both parity and CRC can be used on the same data bits.

PROBLEMS FOR SECTION 12-3

1. Express the following bit groups in polynomial form: 1010, 1111, 0010.
2. What is the polynomial form for the following bytes?

    10010000, 01010111, 11001010, 01110100.
3. For the polynomials shown, what is the bit pattern?

    $X^3 + X + 1$
    $X^4 + X^2$
    $X^7 + X^5 + X^3 + X^2$
4. Express the polynomials shown as bytes of data bits:

    $X^5 + X^4 + X^3 + 1$
    $X^6 + X^4 + X$
    $X^7 + X^6 + X^3 + X^2 + 1$
5. What are the inverse equations for the polynomials of Prob. 3?
6. What are the inverse equations for the polynomials of Prob. 4?
7. Draw a shift register with feedback taps to implement the first polynomial of Prob. 3.
8. Draw the shift register and feedback taps which represent the first polynomial of Prob. 4.
9. A CRC checksum is used with 16 bits of CRC per 128 bits of data. What percentage of overhead does the CRC add to the system? Repeat for a parity bit used with the same 128 bits divided into 16 bytes.
10. What is the additional percentage of bits that a 16-bit CRC adds to a 512-bit data block? What would it be if the CRC were replaced by a parity bit for every 8 bits of data?

11. A simple 4-bit shift register has feedback taps at locations 2 and 4. The register is set to all 0s initially. The data bits, entered serially, are 1010 (rightmost bit first). Show the contents of the shift register for each bit entered. What is the CRC checksum?

12. Use the same situation and answer the same questions of Prob. 11, with the serial data 1011.

13. Use pattern 1011 on a shift register with taps at locations 3 and 4. Compare the CRC checksum to the result of Prob. 12.

12-4 Dealing with Errors

A good protocol must deal with any detected errors, regardless of the way they are detected. Most protocols support a method for automatically requesting a repeat of a transmission of data that has an error. This is called *automatic request for repeat* (ARQ). There are several ways to implement the ARQ function. One method involves waiting for a "received OK" or "not received OK" message from the receiver before sending the next block of data bits. The other method uses a time relationship between the blocks of data and a "not received OK" message to determine which block of data was received incorrectly to initiate retransmission beginning at error block only.

The first method is called "stop and wait." The transmitter sends the data block and waits for a message of ACK (acknowledgment—data received without error) or NAK (negative acknowledgment—data received with error). If the transmitter gets a NAK, it simply resends the block of data. This is a very simple and straightforward method of dealing with errors that are detected. Unfortunately, it is also very inefficient.

There is the obvious dead time while the transmitter waits for the ACK or NAK message to come back. During this time, the transmitting channel is idle (Fig. 12-10a). Suppose 1000-bit blocks are sent at 1000 bits/s. After each block, the transmitter waits 0.1 s for the NAK or ACK. Even if there are no errors, and all data is received perfectly, the channel is wasting about 10 percent of the time. In practice, the efficiency that this system can achieve is closer to 70 to 80 percent, but no greater.

The problem is worse when a half-duplex channel is used. Then, the channel must be "turned around" for the NAK or ACK to be sent back. Typically, in a large system, this takes 100 to 200 ms. The data bits are sent, the channel is turned around, the ACK or NAK is sent by the data receiver, the channel is turned around again, and transmission can resume (Fig. 12-10b). The wasted time is equal to twice the turnaround time, plus the time to send the ACK or NAK. For the case with 1000 bits sent at 1000 bits/s this means that every second of transmission requires 400 ms of turnaround time (if the turnarounds are 200 ms each). Even if there are no errors, 400 ms of every 1.4 s is wasted. A typical half-duplex system using stop and wait usually is no better than 50 percent efficient. If there are errors, of course, this percentage decreases further as blocks with errors are sent again.

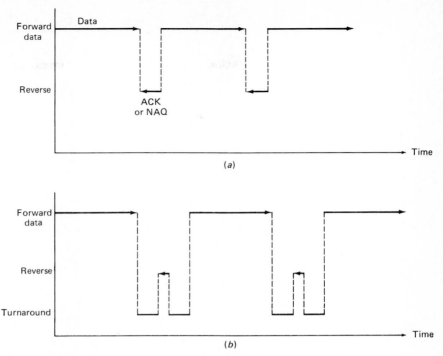

Fig. 12-10 (a) In a full-duplex ARQ system, the transmitter stops and waits after each data block for an acknowledgment that the data is received correctly or incorrectly. In the half-duplex system shown in (b), the link must be "turned around" after each block; this procedure wastes time.

To overcome the disadvantages of "stop and wait," another scheme is often used. This second method is called "go back n," where n is some number of blocks. In many applications, n is set equal to 2. The "go back 2" approach to dealing with errors requires a full-duplex channel and works as follows: After each block is sent, the reverse channel from the receiver to the transmitter carries either a NAK or an ACK message. While this ACK or NAK is coming back, the transmitter continues and sends the next block. The timing of events is shown in Fig. 12-11. After finishing the transmission of block 10, the transmitter immediately sends block 11. During the transmission of block 11, the ACK or NAK reply to block 10 is sent back. Therefore, the transmitter knows during the sending of block 11 whether all is well with block 10 and it can go on to send block 12, or whether it should go back and send block 10 again. When there is an error, the transmitter backs up two blocks and begins again from block 10. For this reason, this is called the "go back 2" method. It is important to note that after the transmitter has gone back and sent block 10 again, it then continues automatically from block 11. This is done even though block 11 is originally received correctly. The effect of a received error in a block is to cause the transmitter simply to back up and start again from the block with the error.

There is some inefficiency in having good blocks, such as block 11, sent again. However, there are considerable benefits because both transmitter and

Error Detection, Correction, and Data Security

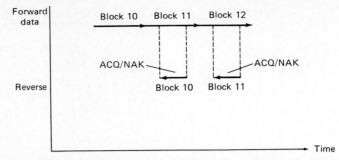

Fig. 12-11 In the "go back 2" error protocol, the receiver continually sends status about received blocks. The status about block 10 is returned while block 11 is being sent.

receiver cannot get out of synchronization with each other with the "go back" scheme. When the error is detected, the receiver knows that the transmitter will back up and start again. Otherwise, there is a possibility that receiver and transmitter will lose track of which blocks they are referring to in the go back scheme, and then all the remaining blocks will be affected.

The advantage of the go back 2 approach is that the transmitter can go at full speed, without dead time, unless an error occurs. When an error does occur, only the block with the error and the one immediately subsequent block are sent again. The efficiency of this is high, but a full-duplex channel is needed. In some communications system, a half-duplex or even simplex-only channel is all that is available. The need for error-free communications with channels that are not full duplex, combined with some mathematical theories that extend the CRC concept, led to the development of checking codes that detect errors and allow them to be corrected as well, without retransmission. The ability to both detect and correct errors, on the basis of received data itself, is called a *forward error correcting* (FEC) scheme, in contrast to the "stop and wait" or go back n, which requires retransmission of any blocks that have errors.

QUESTIONS FOR SECTION 12-4

1. What is ARQ? Why is it needed?
2. What is "stop and wait"? How is it done?
3. What is required in the data link?
4. Why is "stop and wait" inefficient? What is the effect of half duplex on efficiency?
5. What is "go back n"? How does it work? What type of link does it require?
6. Why does the "go back n" protocol require that not only the bad block, but all subsequent ones, be sent again? What are the pros and cons of this?
7. What are the advantages and disadvantages of stop and wait versus go back n?
8. What is FEC? Compare it to an ARQ procedure.

PROBLEMS FOR SECTION 12-4

1. A "stop and wait" ARQ system is using a full-duplex channel. Each block of data uses 200 ms, and the ARQ message takes 50 ms. How many blocks of data are sent in 5 s? How much time is lost to the ARQ message? What percentage of the total time is this?

2. The system of Prob. 1 is now using a half-duplex channel, which takes 100 ms to turn around from forward to reverse direction and to go from reverse to forward direction. For this case, how many blocks of data are sent in 5 s? What percentage of time is actual data?

3. Draw the sketch of line activity versus time for the first second for the case of Prob. 1.

4. Draw line activity versus time for the half-duplex channel of Prob. 2.

5. An ARQ scheme uses "go back 2." It is sending out 10 blocks of data on a full-duplex channel, and block 6 is in error. Sketch the line activity versus time. Number the blocks as they are sent.

6. In the situation of Prob. 5, there are errors in blocks 3 and 7. Sketch the activity on the line and show the block numbers. What percentage of the total time is lost in resending blocks that have already been sent?

12-5 Forward Error Correction

In *forward error correction* (FEC) the transmitted bits contain the actual data along with checking bits which allow both the detection and correction of many errors. The mathematics theory that shows how to do this is developed from the same algebra theory that is used for CRC analysis. FEC is used whenever a simplex-only channel exists, or when switching a half-duplex channel direction is too time-consuming and inefficient. Whenever the highest level of data integrity and confidence is needed, FEC is considered. In some applications, the problem is not simply that the reverse channel doesn't exist or the half-duplex system would be inefficient. The problem may be that the data is unrecoverable if it has an error, when the source of the data has no copy of it available. Consider a computer memory IC which is storing data that must be sent to another system. If the data bit in the memory changes for whatever reason, there is no way that the receiver can recover and correct that bit, even if the data communications channel is error-free. The data is changed and thus lost at the source, before it enters the communications system. It is equivalent to an original document going into a copying machine. If the original is of good readable quality, but the copier is not making clear copies, then the user can take the original to another copier or wait until the first copier is adjusted. If the original is smudged or illegible, no copier can make it better. The information that is unreadable on the original is lost forever. The same holds true with data in communications. Sometimes, FEC is used because the original source of data may not be able to recreate or resend it.

FEC is not without cost, unfortunately. Unlike parity (which requires 1 parity bit per group of bits) or CRC (which uses a single group of checking bits for hundreds or thousands of bits), FEC schemes need several bits per small number of data bits. For example, it takes 6 checking bits per 16 data bits to provide the following capability:

Detection and correction of any single-bit error within the data bits

Detection, but not correction, of all 2-bit errors and some errors of more than 2 bits

Detection and correction of any single-bit error within the checking bits themselves

Adding more checking bits allows the detection and correction of all 2-bit errors, but the cost in additional bits is high. Some applications do require it, however, and pay the price in extra bits and transmission inefficiency for the additional data integrity.

Hamming Codes

There are many ways to implement an *error-detection and -correction* (EDC) system, as this implementation of FEC does. One of the most common ways is based on work done in 1950 by the mathematician R. W. Hamming, so the EDC system is called a *Hamming EDC circuit*. Other similar approaches, based on mathematical work by Golay and others, can be used. Hamming is the most common, and "Hamming circuit" is sometimes used as an equivalent term for "EDC circuit," even when Hamming theory is not the exact one used.

The best way to understand EDC and Hamming concepts is to discuss some of the background and then the way an actual EDC system is constructed, using advanced digital ICs. To reduce the complexity of the discussion, consider a group of eight bits. These require five Hamming bits for EDC. The function of the EDC system is to do the following:

1. Take the eight data bits as a group.
2. Form the five check bits, called *CB0* through *CB4*. Each of these is formed by simple parity (odd or even) determination across some of the eight bits of data. The theory indicates which combinations of bits to use, as shown in Fig. 12-12.

```
                    Data bits
Check bits  7   6   5   4   3   2   1   0   (LSB)
CB0                     X   X       X   X
CB1             X   X       X   X       X
CB2         X       X   X       X   X
CB3         X   X                   X   X   X
CB4         X   X   X   X   X
```
X means perform parity (odd or even) on these bits only

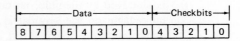

Fig. 12-12 Hamming bits for error correction. The check bits CB0 through CB4 are formed by taking the parity of certain subgroups, as shown, of the data bits. The checkbits are then sent after the data bits.

3. Transmit the eight data bits and the five check bits.
4. At the receiver, take the eight received data bits and perform the same check bit generation action as in step 2 (using the same parity series).
5. Compare the check bits just generated at the receiver with the received check bits. If all the check bits agree, there is no error.
6. If check bits do not agree, note which check bits differ. This pattern of where the check bits agree and disagree is called the *error syndrome*. A difference is shown by a 0.
7. Using Fig. 12-13, match the error syndrome. This figure then shows which bit is in error. The error is corrected by simply inverting the bit. (This is an advantage of binary systems: if the received bit is known to be wrong, then the correct value must be the opposite state.)

A specific example will show the way this works. Suppose the data bits are 1100 0101. In the figure, the check bits are determined with odd parity, in this case, to be

$CB0 = 0$, $CB1 = 0$, $CB2 = 1$, $CB3 = 1$, and $CB4 = 1$.

Errors cause the received data bits to be 1100 0100. (Note that the rightmost bit is in error.) Go through the parity operation again, and the check bits are

$CB0 = 1$, $CB1 = 1$, $CB2 = 1$, $CB3 = 0$, and $CB4 = 1$.

If the check bit groups are lined up, it is easy to see where they differ:

Transmitted check bit group	1 1 1 0 0
Received check bit group	1 0 1 1 1
Error syndrome	1 0 1 0 0

Error Location		Error Syndrome Bits				
		CB0	CB1	CB2	CB3	CB4
Data Bit	0	0	0	1	0	1
	1	0	1	0	0	1
	2	1	0	0	0	1
	3	0	0	1	1	0
	4	0	1	0	1	0
	5	1	0	0	1	0
	6	1	0	1	0	0
	7	1	1	0	0	0
Check Bit	0	0	1	1	1	1
	1	1	0	1	1	1
	2	1	1	0	1	1
	3	1	1	1	0	1
	4	1	1	1	1	0

Fig. 12-13 The error syndrome bits show where the checkbits differ between the received values and the recalculated values. The pattern of syndrome bits shows where the error is.

The difference is in CB0, CB1, and CB3. In the table on page 449, this indicates that the data bit 0 is in error. Simply invert this bit, and the error is corrected.

The construction of the Hamming code necessitates the following rules:

1. Any single error in the eight data bits causes a difference in exactly three of the five check bits.
2. It is possible that a check bit is changed by a transmission error. This causes only that single check bit to change. No other bits (received data or other received check bits) change.
3. An error in 2 bits causes an even number of check bits to differ, and therefore the 2-bit error is detected but cannot be corrected. This 2-bit error can be either entirely in the 8 bits of data, or entirely in the 5 check bits, or in both the data bits and the check bits.
4. Three or more simultaneous bit errors can fool this system into thinking that no error at all has occurred, a correctable single-bit error has occurred, or an uncorrectable 2-bit error has occurred, all of which are incorrect.

In order to detect and correct more error bits, the Hamming code must be expanded. A field of 7 check bits allows all 2-bit errors to be detected and corrected, and all 3-bit errors to be detected. There are two problems with adding more check bits. First, it increases transmission time and lowers efficiency. Second, the ICs needed at both ends to perform the necessary boolean operation that produces the check bits and the syndrome bits become much more complex and thus need more circuit board space, power, and testing. Nevertheless, applications such as deep-space satellites sending back critical data from the temperature of Jupiter, for example, use the extra bits. In such applications, the reverse channel is very unreliable (transmitting to a satellite 500 million miles away) and the signal propagation time is so long (45 min) that requesting a retransmission is very impractical. On a smaller scale, the popular compact disk audio players use Hamming-type codes and FEC. The compact disk is a miniature data transmission system. The surface of the disk has a binary pattern corresponding to the digitized values of the audio signal. In playing the disk, the bits are passed through a digital to analog converter, which reconstructs the analog value from the digital one. The nature of the disk surface is such that a small speck of dust or dirt can "hide" a burst of bits, which are then read in error. Since it is impossible to prevent this dirt or dust from affecting the reading, the designers of compact disk systems instead store the data on disk with many EDC bits. As each group of digital bits is read from the disk, it goes through a circuit which performs EDC on it and its associated check bits. In this way, most of the errors are corrected and the music sounds perfect—because it is.

This example has discussed Hamming codes and check bits for 8 bits of data. The theory shows the number of Hamming bits that are needed for longer data fields to detect and correct all single-bit errors and detect all 2-bit errors (Table 12-4). From this, it seems that the most efficient approach is to use longer data fields. This is certainly the case, since a 64-bit data field and its check bits only use 11 percent of the total 72 bits for check bits, versus 38 percent for an 8-bit data field. The system designer must weigh this against two factors: the circuitry

Table 12-4

Number of Data Bits	Number of Check Bits	Total	Check Bits as Percentage of Total
8	5	13	38
16	6	22	27
32	7	39	18
64	8	72	11

cost for larger numbers of bits (data and check bits) is greater, and the chance of a multiple-bit error in the data bits and check bits is also higher. If the odds are that a 2-bit error will rarely occur in 16 data bits plus 6 check bits, but a 2- or 3-bit error will more likely occur in a field of 64 data bits plus 8 check bits, there is no point in using 64 + 8 bits since it won't detect or correct the errors that do occur. The money spent on EDC will be wasted.

An EDC IC

The need for EDC and the logical nature of EDC circuitry mean that digital ICs can be used to provide the EDC function. Several manufacturers offer EDC ICs which combine all the logic gates needed for the EDC system. This means that the hardware circuitry needed for EDC is available via one large-scale IC rather than many smaller ones that have to be interconnected. One such IC is the 74LS636 from Texas Instruments. The 74LS636 was designed initially for EDC on memory circuits. It can be adapted, with a few external ICs, to be used for serial communications EDC. (This IC was chosen because EDC for data communications is usually embedded in a much more complicated protocol IC.) It can do the following:

Generate the check bits for an 8-bit data field

Read in 8 data bits and the 5 check bits

Develop the syndrome bits on the basis of 8 data bits received, the check bits received, and the newly generated check bits

Correct errors if possible and indicate whether uncorrectable 2-bit errors have occurred

A block diagram of this 20-pin IC is shown in Fig. 12-14. The IC requires +5 V for operation. The mode of operation is controlled by two control lines S0 and S1, as shown in the control function table (Fig. 12-15a). When both S0 and S1 are low (binary 0) the 74LS636 takes in the data bits D0 through D7 and provides a 5-bit check bit field, which can then be transmitted with the data bits. The receiver uses the same IC, but with the other modes of operation. The S0 and S1 bits cause the IC to read in the received data bits and check bits (S0 = 1, S1 = 0), indicate to the system the nature of the error via error flag bits (S0 = 1, S1 = 1), and then output the corrected 8-bit data field and the syndrome bits (S0 = 0, S1 = 1).

The two error flag bits are used to show the rest of the receiver system precisely what the 74LS636 has determined. As shown in Fig. 12-15b, the

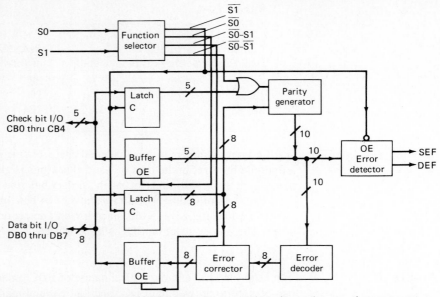

Fig. 12-14 The 74LS636 IC is a complete error-detection and -correction system on an IC. It accepts data and computes checksum bits, error syndromes, and the corrected bits.

(a) Control Function Table

Memory Cycle	Control S1	Control S0	EDAC Function	Data I/O	Check Word I/O	Error Flags SEF	Error Flags DEF
Write	0	0	Generate check word	Input data	Output check word	0	0
Read	0	1	Read data and check word	Input data	Input check word	0	0
Read	1	1	Latch and flag errors	Latch data	Latch check word	Enabled	
Read	1	0	Correct data word and generate syndrome bits	Output corrected data	Output syndrome bits	Enabled	

(b) Error Function Table

Total Number of Errors 8-Bit Data	Total Number of Errors 5-Bit Check Word	Error Flags SEF	Error Flags DEF	Data Correction
0	0	0	0	Not Applicable
1	0	1	0	Correction
0	1	1	0	Correction
1	1	1	1	Interrupt
2	0	1	1	Interrupt
0	2	1	1	Interrupt

Fig. 12-15 (a) Two control lines, S0 and S1, direct the mode of operation of the 74LS636 when it is reading or writing data. (b) The error flags from the IC, SEF, and DEF, show what kind of error (if any) has been detected and corrected, if possible.

possibilities are no errors at all, a single error in either the data bits or the check bits, or two errors, spread across the data and check bits. For example, if flag SEF = 1 and flag DEF = 0, a single-bit error has been detected and corrected. In the case of 2-bit errors, the receiver needs to know that an uncorrectable error has occurred, so it can choose to ignore the data or mark it as suspect. Perhaps at a later time it can use a reverse channel to ask for retransmission, if the channel allows for this.

The TI 74LS636 is designed for use on 8-bit bytes and cannot be applied to 16-bit data groups. Some other manufacturers provide more complex ICs which can be cascaded. In this way, two 8-bit EDC ICs, each of which produces 5-bit checking fields, have special input and output interconnections so they can be combined to produce a single 6-bit checking field across 16 data bits. The IC itself is more complex and costly because of this cascade capability.

QUESTIONS FOR SECTION 12-5

1. What is FEC? Where is it needed and useful? Where is it essential?
2. What is the cost, in bits, of FEC?
3. Why is FEC also called *error detection and correction*?
4. What levels of EDC can be achieved with 6 bits per 16 data bits?
5. How are the check bits generated?
6. How are the check bits used to determine where the error is located?
7. What is the error syndrome?
8. What is done to a bit known to be in error?
9. What do the syndrome bits show when one data bit is in error?
10. What do they show when a check bit is in error?
11. What do they show for a 2-bit error, either in data bits only, in one data and one check bit, or in check bits only?
12. What do they show for three or more error bits?
13. Why not simply add more check bits to detect and correct more errors?
14. What is a Hamming code? Where are Hamming code and EDC absolutely necessary even in applications that are not critical?
15. What number of Hamming bits is required for the simplest EDC for data fields of 8, 16, 32, and 64 bits?
16. What influences the decision on the size of data bit blocks on which to perform EDC?
17. How does a typical EDC IC get its instructions? How does it indicate the outcome of the EDC cycle?

PROBLEMS FOR SECTION 12-5

Note: Use the Hamming check bit tables in this section for the problems.

1. Data bytes are transmitted and received with Hamming bits for EDC. For the data bits shown, calculate the check bits and the error syndrome bits, and show which bits are in error, where possible:

Transmitted	Received
a. 0110 1001	0010 1001
b. 0010 1111	0001 1111
c. 1001 1001	1011 1001
d. 0011 1100	0000 1100
e. 0011 1100	1101 1100

2. The error may be in either the data or check bits. A transmitter uses Hamming codes to generate the check bits and send them along with the data. The receiver takes the data and check bits, recalculates the check bits, and finds the error syndrome bits. For the following data and check bits, determine the error syndrome bits:

Received Data	Received Check Bits
a. 0010 1001	00011 (CB4)
b. 0010 1111	00110
c. 1001 1000	11100
d. 1001 1000	11101

What do they indicate has occurred?

3. Using the same situation as in Prob. 2, find the error syndrome bits and indicate what they mean.

Received Data	Received Check Bits
a. 1100 0001	10100 (CB4)
b. 1100 0011	10001
c. 0011 1100	10111
d. 0011 1100	01100

4. The TI 74LS636 IC is used to implement the Hamming code. Using the results of Prob. 2, what error flags are set in each case?
5. Use the results of Prob. 3 and show which error flags are set for each case.
6. Data is sent as 0110 1100 but is received as 0101 0000. What are the check bits at the transmitter and receiver? What is the error syndrome?
7. Severe noise causes the first 7 bits of a transmitted byte to look like binary 1s. What are the check bits if the data transmitted is 01000001? What are the check bits after the data is received? What is the error syndrome?
8. What do the following error syndrome bits indicate?

    a. $CB0 = CB1 = CB2 = 1$ $CB3 = CB4 = 0$
    b. $CB0 = 0$ $CB1 = CB2 = CB3 = CB4 = 1$
    c. $CB0 = CB3 = CB4 = 0$ $CB1 = CB2 = 1$

12-6 The Nature of Data Security

There are many cases in which a user does not want any unauthorized listeners to receive and understand the data being transmitted. In military situations, any information that the enemy can gain by understanding communications can of course create a serious problem. In commercial applications, a bank may want to make sure that data which represents the names and amounts of money in various accounts is kept confidential, even though this information must be transmitted between bank branches. Companies do not want their rivals to eavesdrop and learn of any new products, plans, or problems as they are transmitted and discussed among various company locations.

Another aspect of data security is to prevent unauthorized persons from tapping into a channel and putting false information into the system. This is critical in military applications and commercial ones. A skillful thief can cause a bank to transfer money from one account to another, for example. Regardless of the major concern—listening in or adding false information—the methods used are the same. The goal of data security is twofold: to prevent eavesdropping (or false transmissions) and to make it difficult to understand the meaning of the stolen data (or to put data that makes sense onto the channel). Any data security system design must also consider the amount of time the data must be protected. A system transmitting stock prices only has to make the data secure long enough to prevent someone from figuring out the prices before the next day's paper is published. On a longer scale, a system transmitting military information may want the data to be incomprehensible to the enemy for months or years. Therefore, the data security scheme must be much more difficult to break. It is like a car theft or house burglar alarm. In some cases, the alarm is designed to attract attention in the first few minutes, so that the potential thief becomes discouraged and worried and gives up quickly. In other cases, the homeowner may want to make sure that the house cannot be entered even after hours of trying, since the house is in a secluded area and the owner is on vacation. Different applications have different amounts of ruggedness in their security needs.

Approaches to Data Security

What are some of the ways to make data secure so that only the intended receiver can receive it or make sense of it? There are three basic techniques.

1. The data can be encoded simply by using some nonstandard format. Instead of ASCII representation of characters, use some other set of bit patterns, for example. Alternatively, use letter Q for A, P for B, and so on. These simple schemes work only for a short period. If the unauthorized receiver studies the data, the new encoding pattern can be figured out. If the characters are mostly English text, the letter E is most common, so the most common bit pattern is probably E. The letter T is second in frequency. A further weakness here is that if the undesired eavesdropper ever gets a copy of the unencoded text (called the *plaintext*) and can line it up with the received data bits that were

specifically encoded, the entire pattern becomes clear. This simple scheme is too easy to break in nearly any real application and is rarely used.

2. The data bits can be left in the usual encoding (such as ASCII), but the transmission can be done so that the frequencies used or time periods used keep changing during the transmission. The eavesdropper then has a problem in finding where the bits are. It is like trying to follow a radio station that randomly jumps from one frequency to another every second. Even if you know the language, it is difficult to pick up the signal. The general name used for this technique is "spread spectrum."

3. Modify the actual data bits by a complicated encoding scheme which is very difficult to break. Both the sender and intended receiver need some kind of key. The sender uses this key to encode the data, and the receiver uses it to decipher it. Depending on the length of the key and the exact steps used to turn plaintext into encoded text with the key, this can be a moderately difficult to nearly impossible code to break. This encoding operation is called *encryption*.

In actual data communications, only the last two methods are used. They can be employed separately or even combined for additional security. If the data is encoded by using some complex scheme, and the transmission frequencies or times also vary, the real message is extremely hard to capture and decipher. Many applications are well served by just one of the techniques, however, if it is done with enough complexity.

The heart of both the spread spectrum and encryption techniques is the use of random and near-random sequences to control the frequency or time jumps or the encryption method. The theory behind the random sequences, in turn, is also derived from algebra theory used for CRC and Hamming codes. In error detection and correction, the goal is to preserve correctness even if errors occur. In data security, the intention is to put a cover of confusion on the data, so that it appears to be erroneous. Only a receiver with the right keys can remove the "errors" and restore the correctness. The subsequent sections of this chapter discuss aspects of the theory and ways it is put into practice to provide data security. Some actual ICs that provide data security through encryption will also be discussed.

QUESTIONS FOR SECTION 12-6

1. What are two aspects of data communications security from the listener or false signal perspective?
2. Why is the time element a factor in security?
3. What are the pros and cons of security using encoding in a nonstandard format?
4. What is the purpose of spread spectrum?
5. What is the intention of complex encoding? What element prevents eavesdroppers from understanding?

6. What is plaintext? What is encryption?
7. What is at the heart of encryption techniques?
8. What is the similarity between EDC and encryption?

12-7 Random Sequences

One of the elements used to make the data bits unreceivable or meaningless is some type of random sequence along with the data. An example will show the way this works for a message of English letters. The sender wants to use an encoding scheme in which each letter is represented by another letter. The simplest approach is to offset the real characters from the coded ones, by letting each letter be represented by the fourth letter beyond in the alphabet, for example. Thus, A would be sent as E, B as F, and so on. This is a very easy code scheme for the undesired listener to break.

A major improvement is to vary the offset value with each character. Both the sender and receiver, of course, have to know the offset amount and the variations. The first letter of the message would have a $+4$ offset, the second a -2 offset, and so on, all through the message. Consider the situation as shown in Table 12-5:

Table 12-5

Message	H	E	L	L	O	T	O	Y	O	U
Offsets	+2	−3	+6	−1	−4	0	−5	+5	+2	−2
Encoded message	J	B	R	K	K	T	J	D	Q	S

The eavesdropper receiving this gets a jumble of letters that have no apparent underlying pattern. If the offsets are randomly selected then the eavesdropper has no way to determine what the real characters are. This is because any received character gives no clue whatsoever to the meaning of any other received character. There is no relationship or correlation between any characters in the message. An E may stand for Q one time and for F the next. It may even stand for E.

The use of this random pattern to encode the data is very effective. As long as the same random offset pattern is not used again, the receiver has no way of backtracking to the real message. In fact, this is the only scheme that has ever been proved unbreakable. If the sender and receiver never reuse the same random string of offsets, the message is in an unbreakable code. Even if the eavesdropper somehow gets a copy of plaintext message and the encrypted text, they are of no help the next time because the relationship between the original message and the encrypted version is changed each time, on a letter-by-letter basis.

Random Sequences and Binary Data

In a data system which uses only binary signals, the situation is the same but simpler. Each bit can be either a 1 or a 0. The random sequence in the key used causes a bit either to remain at its existing value or to change to the opposite value. There are no other choices. The random string of offsets is replaced by a random string saying "change, don't change, don't change," and so on. The receiver has the same string of change values and applies this to every received bit. An outsider, without the key string, is hopelessly lost in trying to make sense of the message.

The method works in the following way for a message of 16 bits. Let the random string either be a 1, meaning change the bit value, or a 0, meaning make no change.

Original data:	0110	1101	0001	1110
Random string:	1001	1011	1010	0100
Encrypted result:	1111	0110	1011	1010

The receiver gets the result and reverses the encrypting operation; this results in the original data.

How should the random pattern of bits, used to decide whether to change or not change each bit of message, be determined? Any random source of binary values can be used. Flipping a coin (heads = 1, tails = 0) is a legitimate possibility. The energy levels of electrons have also been used. It doesn't matter as long as the random key string meets tests for true randomness: equal numbers of 1s and 0s, no repeating patterns, only a certain number of runs of 1s or 0s, and similar features. Statisticians have many tests to determine whether something is truly random or whether it appears random.

The One-Time Key

The use of random strings to encrypt and then decipher the message is called a *one-time key*. Since the key is unbreakable, it would seem that there is nothing further to say about obtaining data security. Unfortunately, the practical limitations of using this scheme make it very impractical in many cases.

- The key must be guarded and safe, but somehow the intended receiver must have a copy. No third copy should exist, since it may fall into the wrong hands.

- The length of the random string has to be equal to the length of the message. A bank may send millions of bits daily from one location to another. This means that it has to generate random strings of millions of characters each day and somehow transfer one copy to the receiving branch. Certainly this is impractical.

- There has to be some way to physically incorporate the random string into the electronic equipment at the two ends of the link. If there are millions of bits, these may have to be put onto magnetic tape. The tape then must be read into the transmitting circuitry while the data is sent, and into the receiver while the received data is being deciphered.

Clearly, these complications make the one-time key impractical for nearly all commercial and military applications. This does not mean that it is never used. Spies have been captured with booklets of random numbers, to be used as one-time keys. But this system does not work in fast-moving high-data-volume applications or in areas where the intended receiver is not physically reachable easily (such as soldiers in the field under battle conditions). Fortunately, algebra theory provides a way to achieve nearly the same amount of data security with simple circuitry, by using pseudorandom sequences.

Pseudorandom Sequences

A *pseudorandom sequence* (PRSQ) is a group of bits that has the appearance of being random, but is not generated by a true random process such as flipping a perfect coin. A digital circuit is used to generate a string of 1s and 0s which meets nearly all the tests of a true random sequence. The same circuit can be installed in both the transmitter and receiver systems and can provide the random pattern needed to encrypt and decipher the data. It is called *pseudorandom* because it is not a true random sequence from a mathematical view, but it can be very close. An outsider looking at the sequence would find it nearly impossible to figure out that the pattern is not random and to determine the nature of that underlying nonrandomness. Trying to break a code that is developed by using PRSQ groups is not impossible, as it is with a true random sequence of a one-time key. However, the PRSQ can be made so like a random one that it will take computers years of effort to decode a message without knowing the PRSQ.

Very simple circuitry is used to generate binary PRSQs. Take essentially the same circuit with shift registers and feedback gates as is used for generating the CRC checksums. Instead of initially setting the shift register to all 0s, put in some binary pattern. This is called the *key*. Then take the output of the shift register and feed it back into the input (Fig. 12-16). (The data input that is used for the CRC circuit can be omitted.) The result is that the circuit will generate a pattern of 1s and 0s forever, and this will be a PRSQ pattern. With certain feedback tap combinations, the pattern only repeats after 2^N bits, where N is the number of locations in the shift register. (Shorter repeats occur with some tap locations.) A small 16-bit shift register circuit repeats after 65,535 bits in the maximum case. A larger register, with hundreds or even thousands of shift register locations provides many more bits before repeating and is easy to build with today's ICs. (Note: $2^{128} = 3.4 \times 10^{38}$.)

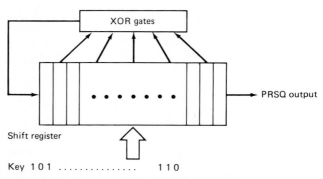

Fig. 12-16 A shift register with feedback taps and XOR gates can generate infinite strings of pseudorandom sequences.

An example will show the way this works. For simplicity, use a very short shift register of only four locations, connected as shown in Fig. 12-17a. Set the initial value to the key of 1101. The PRSQ generated is shown in the figure. If a new key is used, such as that in Fig. 12-17b, a different PRSQ results. Note how the PRSQ repeats only after 16 bits of the sequence have been generated.

Can someone look at a portion of the PRSQ and figure out the key? Yes, but it is very difficult, almost as difficult as breaking the message code. The reason is that although a key generates a unique PRSQ, the reverse is not true. The PRSQ cannot be worked back to a single key. There may be several keys that provide the same PRSQ, especially if the eavesdropper does not know where to start in the sequence.

The use of PRSQs to provide a foundation for data security is now common. It does have one technical difficulty (besides distribution of the keys to all users and the remote possibility that it can be broken): synchronizing both the sender and receiver PRSQs to the same start, not merely to the same bit rate. They must both use the same element of the random string to operate on the same bit position of the message for this scheme to work. If the receiver is decoding bit 7 received with bit 8 of the PRSQ, then meaningless results and errors occur. This

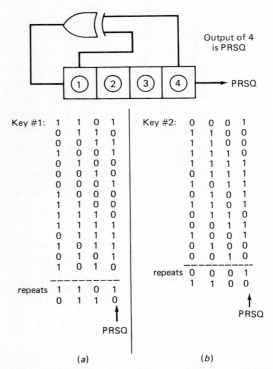

Fig. 12-17 A 4-bit shift register can generate a short PRSQ, with feedback taps shown. Different PRSQs result from different starting keys. The rightmost bit of the shift register, after each shifting step is complete, is the next bit of the PRSQ. The rest of the bits within the shift register are not part of the PRSQ until they shift into the last position. Here, keys 1101(a) and 0001(b) are used.

```
Data                 1 0 1 0 1 0 1
PRSQ (in sync)       1 0 0 1 1 0 0    } Encryption
Result-transmitted   0 0 1 1 0 1 1

PRSQ                 1 0 0 1 1 0 0          } Deciphering in sync
Result               1 0 1 0 1 0 1    OK    } OK

PRSQ                 0 0 1 1 1 0 0 X        } Deciphering    PRSQ shifted left
Result               1 0 0 1 0 0 1 X  Not OK} Result

PRSQ                 X 1 0 0 1 1 1 0        } Deciphering    PRSQ shifted right
Result               X 1 1 0 0 1 0 0  Not OK} Result
```

Fig. 12-18 The PRSQ must be lined up, or synchronized, with the encrypted data. The first grouping shows the original data, the PRSQ, and the result of encryption. The second group shows the way the original data is restored when the transmitted result is combined with the PRSQ in proper synchronization. The last two groups show that the result of the deciphering is completely wrong when the PRSQ is just 1 bit to the left or right of the position where it should be.

problem of getting both units synchronized is not easy to solve. Most applications use a known pattern of data bits, such as sending 101010. . . .

During this time, the receiver takes the encoded, received data bits, decodes them with the PRSQ, and shifts the PRSQ back and forth versus the received bits until the known pattern 101010. . . emerges (Fig. 12-18). When this happens, it means that the PRSQ circuit in the transmitter and the one in the receiver are synchronized. Various enhancements to this basic idea are used in real PRSQ units to make the synchronization quicker and more reliable, but the overall technique is the same.

The problem has now changed from one of generating and distributing random sequences as long as the message to a much simpler one. Only a relatively short key, equal to the shift register length, has to be made up and sent around. It is entirely practical to change keys frequently, since the keys are short and can be carried easily. There is no large mechanism needed to enter the key into the system. The key can be typed in (using binary, hexadecimal, or other characters), or a small-memory IC can be programmed with the key bits and then put into the circuitry. This is how a PRSQ makes an unbreakable but generally impractical key into a nearly unbreakable but highly practical key.

Once a system has been set up to generate and synchronize identical PRSQs, they can be applied to various aspects of data security. These involve frequency changing or time slot changing and are discussed in the next section. They can also be used in modifying the data bits themselves (discussed in Section 12-9).

QUESTIONS FOR SECTION 12-7

1. How is the simple offset encoding scheme improved for better data security?
2. Why is the random variation very hard or impossible to break?
3. What is a one-time key? Why is having a copy of the plaintext and encrypted version not helpful in breaking a one-time key?
4. How is binary data encrypted in a one-time key? What happens to the 1s and 0s?
5. How can a random key string be generated?

6. Give three reasons why a one-time key is impractical. Is it ever used?
7. What is PRSQ? Compare it to a genuinely random sequence. How is it similar? What is the main difference?
8. How is a PRSQ generated with circuitry? After how many bits does the pattern repeat in the maximum case?
9. What determines the key in a PRSQ?
10. Can the key be determined just by looking at part of the PRSQ?
11. Why is synchronization a problem with PRSQ-based systems? How is it achieved?

PROBLEMS FOR SECTION 12-7

1. A user is sending the message COME AT THREE, using fixed offset of $+3$. What is the encrypted version of this message?
2. An eavesdropper knows that the sender is using fixed offset. The eavesdropper intercepts the following message: KP PGY AQTM. What was the original message?
3. A user is encrypting the message with a one-time key. The key sequence is $+2, -2, +1, -4, +6, +6, -1, 0, +1, +3$. The message in plaintext is NOT SAFE YET. What is the encrypted text?
4. Repeat Prob. 3, using the one-time key $0, +1, -2, +2, -4, -1, 0, +3, -4, -2$.
5. A random pattern is used to encrypt data bits. What is the encrypted result for the following data pattern and random pattern?

   Data   0110 1001 1010 0101
   Random bits 1110 0101 0100 1001

6. Repeat Prob. 5 for data bits 1111 1010 0000 1001.
7. Repeat Prob. 5 for random bits 1010 0101 1100 1001.
8. Received data after encryption is 1001 1010 0011 1000, using the random bits of Prob. 5. What was the original data?
9. A 4-bit shift register is used with feedback taps at locations 2 and 4. The initial key is 0011. What PRSQ is generated for the first 16 bits? For the first 20 bits?
10. The same 4-bit shift register and taps are used with a key of 0101. What is the resulting PRSQ (first 16 bits)?
11. An 8-bit shift register with taps at locations 3 and 7 is used to generate a PRSQ, with key 0011 1000. What are the first 10 bits?
12. The shift register of Prob. 9 is used with a key of 1101 1001. What are the first 16 bits of the resulting PRSQ?
13. A 101010... data pattern is used to synchronize transmitter and receiver PRSQ circuitry. The random PRSQ pattern is 1001 1010 0011 0001. What is the encrypted pattern? What is the deciphered pattern if the receiving PRSQ is shifted one position to the left of where it should be? What is it if the PRSQ is one position to the right?

12-8 Spread Spectrum Systems and PRSQ Encryption

Whenever a modulated signal is used to send data, the data acts to vary some characteristics of the carrier used for transmission. The amplitude, frequency, or phase of the carrier can be shifted according to the value of the data bits. But there is no inherent technical reason why the carrier frequency must itself remain unchanged. One way to make it hard for unauthorized listeners to eavesdrop is to vary this carrier between bits.

There are other problems that can occur when the carrier frequency is fixed. In military applications the location of the transmitter is often critical and should be kept secret; otherwise, the enemy at least knows where troops are, even if the message itself cannot be deciphered. Also, if the carrier frequency is known, the signal can be jammed by transmitting a powerful interfering signal on the same frequency. Both of these drawbacks led to the development of what is called *spread spectrum data communications systems*, used primarily in military applications.

In *spread spectrum*, the frequency of the carrier is varied according to the pseudorandom pattern. Even though the PRSQ is binary, a formula is used to link it to many possible carrier frequencies. A typical spread spectrum system uses 25 to 100 possible carrier frequencies, among which the data signal carrier can hop. The transmitter carrier jumps successively to the many different carrier frequencies in use, transmitting short blocks of data bits. The receiver knows this pattern since it has previously synchronized its PRSQ circuit and has the same formula to link the value of PRSQ to the carrier frequency. It can therefore jump around in synchronization and receive the data signals as the carrier changes. The system is called *spread spectrum* because the data signals occupy a much wider bandwidth than if the carrier frequency were fixed during the transmission.

The two advantages of this system to the military concern locating the transmitter and jamming. An ordinary receiver turned to any single carrier hears only very short bursts of signal, spread apart widely in time. It is nearly impossible to locate this signal, as a result, without the correct receiver. A special wideband receiver can be used, but then the received signal also includes a large amount of noise. The effect is that the eavesdropping receiver is looking for a needle in a haystack. Since the signal cannot be received (and the location of the transmitter is also therefore hidden) the signal cannot be decoded. Unauthorized listeners cannot break the code of a signal they can't receive.

With regard to jamming, the reverse situation applies. A successful jamming station must transmit a great deal of power at the right frequency. In a spread spectrum system, there is no "right" frequency, since the carrier is never in at any one value for anything more than a short amount of time. A jamming station may wipe out a few bits, when its frequency and that of the spread spectrum system happen to have the same value, but most of the bits get through. A wideband jamming transmitter is much harder to design and is much less effective. It also tends to wipe out many signals, including those of the unauthorized listeners' radios.

The advantages of spread spectrum appeal to the military, which requires the antijamming and location-hiding aspects. Most commercial installations, in contrast, do not use spread spectrum for data security. The reasons are as follows:

The complexity of frequency hopping transmitters and receivers is greater than that of fixed ones, and these systems cost much more.

Most commercial applications have no need to hide the transmitter location, and jamming is not a concern.

The receiver must be synchronized with the sender circuitry at the beginning of each transmission. This is a problem in half-duplex channels, which are common in many commercial installations. In addition, any long gap in the received signal, produced by poor radio conditions, can cause the two users to lose synchronization with each other. The systems then go back to a special initialization routine to reestablish synchronization. This is time-consuming and inefficient.

Many systems do not have available the wide bandwidths of radio frequencies that spread spectrum requires. Bandwidth is a precious resource, and commercial installations prefer to use it to send, rather than conceal, data.

Encrypting with PRSQ

In cases in which spread spectrum is not practical or convenient, the PRSQ can be used to encode the data itself. The circuitry to do this is simple and effective. A simple addition to the shift register circuitry is all that is needed (Fig. 12-19a). The output of the PRSQ circuit is added, on a bit-by-bit basis during the transmission to the data bits to be sent. Both the PRSQ circuit and the data bits use the same clock, so each data bit has a corresponding bit in the PRSQ. This

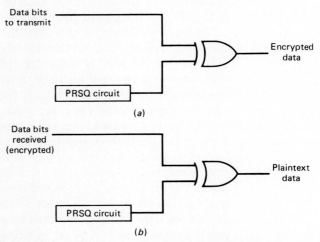

Fig. 12-19 Circuitry for using PRSQs for encryption and deciphering is relatively simple: (a) the way the PRSQ generator circuit output is XORed with the data and the result is transmitted. In (b) the identical PRSQ circuit output is XORed with the received data bits. The output of the XOR gate is the original, unencrypted data.

addition can be performed by a simple exclusive OR gate, since the boolean truth table for the XOR is the same as for adding 2 bits. Note that if the bits are both 1, the sum is 10, and the addition is performed without a carry bit. Thus, the XOR provides the correct answer. The result of the addition is the actual transmitted signal, which now consists of data bits that are combined with the PRSQ bits. To anyone eavesdropping, the sequence is a meaningless pattern of random bits.

The receiving circuitry is similar. The PRSQ generator must of course be synchronized with the one at the transmitter. Once this is done, the received data bits are added to the output of the receiver's PRSQ output, through an XOR gate (Fig. 12-19b). The result is the original data. This method works because the XOR addition has the effect of cancelling the effect of the previous XOR addition, which is one of the mathematical simplifications that binary values allow.

Table 12-6 provides an example which illustrates this:

Table 12-6

Original data sequence	1001 0111 0001 1101
PRSQ	1000 1010 0110 1010
XOR of the two streams (signal sent)	0001 1101 0111 0111
XOR of PRSQ and received signal	1001 0111 0001 1101

One of the nicest features of this method of providing data security is that the level of security is high, and the circuitry is inexpensive with modern ICs. This approach of encoding data to provide security is very practical when the sender and receiver can maintain synchronization and when the resynchronization time at the beginning of every transmission is not a burden. The encryption by combining with PRSQs requires no more bandwidth than the original data, but a synchronous transmission channel is required for this to work effectively. In applications in which there is a half-duplex channel that must be frequently reversed, or a lower-quality, lower-bandwidth channel (such as the standard phone lines), this method may not be a good choice.

QUESTIONS FOR SECTION 12-8

1. What are two drawbacks to a fixed frequency carrier system, in terms of data security?
2. What is spread spectrum? What is it called spread spectrum? How many carrier frequencies are used, typically?
3. How and why does spread spectrum avoid location determination and jamming by the enemy?
4. Give three reasons why commercial users don't need spread spectrum.
5. How can data be encrypted with a PRSQ? How can it be deciphered?
6. Where is this a good method? Where is it not a good choice?

12-9 The Data Encryption Standard

The circuitry for implementing PRSQs, encryption, and deciphering data bits as they are transmitted and received is relatively simple, but the overall system performance may not be sufficient or matched to the application. The need to synchronize the PRSQs and use a synchronous link makes the scheme a poor to fair choice for many applications. This is especially true when dial-up lines, RS-232 modems, and other lower-performance channels are used. A typical user may be a terminal connected to a computer via modems, running at 9600 baud, with the need to make the data on the phone line secure from any eavesdropping or prevent someone from tapping in and entering false data. Fortunately, there is a solution for these users. As the need for data security became more widespread and IC technology advanced, groups of mathematicians, commercial users, engineers, and equipment manufacturers met with the National Security Agency. The result was a series of *data encryption standards* (DES) published by the U.S. government. These standards can be used by anyone to provide a framework for data security. The algorithm is public and published.

The DES standard is formally known as *Federal Information Processing Data Encryption Standard 46*. It supports three levels of data encryption, each with more security and complexity than the previous one. The standard is intended to be used for data sent between banks, company offices, federal agencies, and anyone who feels a need to keep data secure and prevent false data from being entered in place of real data. The standards of DES operate on groups of bits, and it does not matter where these bits come from or how they arrive. It is possible to build an RS-232 modem which connects to the normal computer terminal and immediately encrypts the data as the bits are typed at the keyboard, for example Fig. 12-20. The theory that was used to develop the DES came from number and algebra theory, related to the algebra theory used for CRC.

The DES Concept

The concept used in the DES operation can be explained with a very simple example. Assume that the sender and receiver both have the same key, and this key is a very large number such as 1200. When the sender wishes to send data, such as the number 3, the sender divides 3 into 1200 and sends the result, which is 400. The receiver performs the opposite operation: it gets the 400 and divides it into the key, and the result is the original value that was sent.

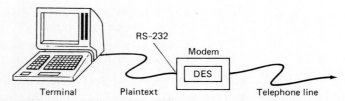

Fig. 12-20 Data encryption circuitry, such as that for the DES, can be built into the modem which connects to the terminal, so that all data bits are encrypted as the characters on the keyboard are typed.

How does this provide data security? An unauthorized eavesdropper does not have the key and so must try to figure out what the real, unencrypted number is by dividing the encrypted value (400) into any of the possible key values. If the allowable number of keys is large, for example, up to 1 million, this process of trial and error takes a long time, since 400 divides evenly into many values, such as 800, 1200, 1600, and 2000. The eavesdropper has to assume that the 400 represents a 1, 2, 3, 4, or other integer and then repeat this trial and error process for the next received value, to try to see what series of deciphered values make sense and contain a meaningful message.

What the DES does is make the relationship between the original plaintext and the resulting encrypted text much more complicated. Each group of bits that goes through the DES procedure is shifted, added, multiplied, inverted, and manipulated in many ways, so that there are an enormous number of possibilities that have to be tried if the key is not known (Fig. 12-21). In the DES model, the key itself is set to 56 bits ($2^{56} = 7.2 \times 10^{16}$ possibilities!). This value was chosen as a good compromise between any key length (for making the system secure) and the ability to design ICs which can efficiently and effectively handle the DES procedure. The DES procedure can be used on any bits: data bits, parity bits, CRCs bits, headers, and so on. It makes no judgment on what the bits mean, but it takes groups of bits and processes them to produce another group which can be deciphered only by someone with the key.

The idea behind the DES procedure is the same as for any way of making data secure. Set up an algorithm, or set of rules, which defines a unique, complicated path from plaintext bits to encrypted text bits. This relationship must be one in which there is only one possible encrypted result that can occur when the plaintext is put through the process using the key. At the same time, make sure that the reverse direction path has many possibilities and many answers (Fig. 12-22). Only having the key points out the correct one. It is exactly the same as

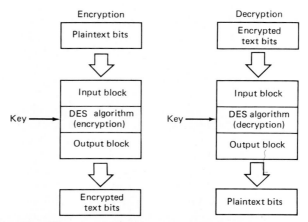

Fig. 12-21 The DES scheme for encryption works as shown in the left-hand figure. The original bits are taken in blocks and mathematically manipulated, and the resulting block of manipulated bits is the encrypted version of the original block of data bits. The decryption process is the reverse operation, as shown in the right-hand figure.

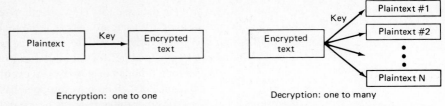

Encryption: one to one   Decryption: one to many

Fig. 12-22 The concept underlying the DES scheme is the unique relationship (using the key) between the original plaintext and the encrypted text. In the reverse direction, for decryption, there are many possible original plaintext results that can occur, without the key to point to the correct one. Without the key, a one-to-many mapping of data bits exists.

someone saying, "The answer is 100; what two numbers when multiplied can give this answer?" If you have the key, say 20, then the answer is 5. If you don't have the key, the answers can be 1×100, 2×50, 4×25, 5×20, 10×10, and so on (using integers only). From a security view, it is said that the process of encrypting the data has a one-to-one mapping; the process of deciphering the encrypted value has a one-to-many mapping without the key.

DES Modes

DES 46 allows three methods of making the data secure:

Electronic Code Book   In the *electronic code book* algorithm, groups of 64 data bits are processed. Every time the same 64 bits are entered, the same encrypted output appears. Each group of 64 bits is totally independent of any other group. This method is used when the data bits are stored as blocks on tape or magnetic disk and may be retrieved in a sequence different from the one in which they were stored. For example, each 64-bit block may be data with a name, address, and personal information. The data is initially stored by someone typing in all the information, perhaps alphabetically. It may be retrieved in any order, as the user needs to look up the file of Jones, then of Adams, then Smith.

     This method is effective, but one guideline to better data security is not to use the same encryption pattern more than once, because it makes it easier for someone to break the code by studying the recurrent sequences.

Chain Block Cipher (CBC)   *Chain block cipher* improves on the first method, because the result of encrypting any 64-bit block depends on the previous block. Thus, the same data bits produce different results if encrypted twice in a row. This is useful in cases in which large numbers of bits must be handled, but they must be sent and received in the same order. This is the way most data communications systems work: a memo is sent as one long string of bits. Even if the bits go through a packet switching network, which rearranges the order of blocks during the transmission, the receiving end of the packet switching network must reassemble the blocks into the correct order (whether the data is sent encrypted or not). After the blocks have been put into the original order, they can go through the DES circuit for decryption using CBC. Anyone eavesdropping along any route of the packet switching network only gets some of the blocks (recall that various parts of the message may take different routes) and therefore have a much harder time breaking the code, without all the blocks, and in the correct order.

Another advantage of CBC is that anyone who inserts data, or substitutes new data for the real ones in an attempt to change the message, affects not only the 64-bit block in which the change is made, but also all the following blocks. The legitimate user at the receiving end notices this, because the data becomes gibberish. If only one block were affected, someone's age might be changed from 47 to 51, but the reader of the printout would not notice anything peculiar.

8-Bit Cipher Feedback (CFB)   The *8-bit cipher feedback* is similar in principle to the CBC, but uses shorter blocks to make it more compatible with medium-speed, character-based communications such as ASCII terminals and modems.

Regardless of which method is used, it is still important to ensure that there are no errors in the received data. The CRC checksum or Hamming code can be used on data which is encrypted, since neither looks at the meaning of the data bits. Normally, the encryption process takes place as physically close to the sender as possible, for maximum security. The encrypted data can then have EDC applied. The reverse is also true—the DES algorithm can be applied to bits which represent both data and CRC or EDC bits. It is possible to have a communications system which has multiple layers, as shown in Fig. 12-23: the data source adds EDC bits, the bits are put into the DES procedure, and the result is then transmitted over ordinary channels with EDC added by the transmitter system. At the receiver, the channel EDC is removed, any transmission errors are corrected, and the encrypted bits are passed to the DES circuit, which decrypts the bits. The plaintext, including EDC bits, then goes to the end user circuitry, which removes the EDC bits that were added at the data source. Although this

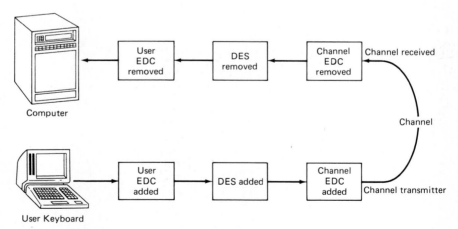

Fig. 12-23   A communications system can have layers of error detection and correction as well as encryption. The original data bits can be encrypted at the terminal, and then the channel itself may add further encryption and EDC. At the received end, the reverse operations take place to restore the original data bits. Each layer does not examine the meaning of the bits it receives, but performs a carefully defined operation on whatever bits it has.

seems inefficient, it is a natural outcome of the fact that in a layered system, each layer operates in a transparent manner to the other layers. Each layer takes the group of bits it receives, operates on them, and perhaps adds some bits. The receiving layers perform the opposite process. In the preceding case, the channel on the next call may not provide EDC between transmitter and receiver, but the EDC that was put on when the data was generated is still useful. If the channel does provide EDC on its own, this is just another level of data integrity.

A DES Integrated Circuit

Many manufacturers of *very large scale integration* (VLSI) ICs have designed ICs which implement the complex DES algorithms. For the system designer or technician who needs to work with DES and data security, these ICs make the job relatively easy. The DES IC designs themselves are verified by the National Bureau of Standards of the United States government to make sure that the IC properly performs the DES procedures. This means that two pieces of equipment can each use ICs from different suppliers and still communicate, as long as they both agree on the DES mode and use the same key. The internal design of the ICs may be very different for each manufacturer, but that is not a problem.

Western Digital Corporation is one maker of DES ICs. Two models are available: the *WD2001*, which uses a 28-pin package, and the *WD2002*, which has 40 pins. The difference is not in the DES functions, but in the way the key is loaded. For highest-security classified information, the key access path to the IC must be entirely separate from the access path for data. The WD2002 has extra pins to provide a separate port for the key to be loaded; the key and data share a port in the WD2001. Regardless of model, once the key is written into the IC it cannot be read back out (so someone cannot come by later and find out what the key is).

A block diagram of the IC is shown in Fig. 12-24. The WD2001 and WD2002 are designed to interface to a microprocessor bus and be controlled by a micro-

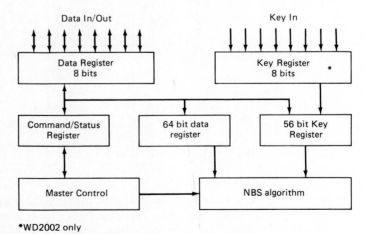

Fig. 12-24 The Western Digital WD2001 and WD2002 block diagram. This IC performs the complete DES sequence on data bits. Note: The two ICs are essentially identical, except that the WD2002 has a separate group of IC pins for loading in the key.

processor dedicated to this function. It has address lines for the microprocessor to address various registers within the IC, bus control lines to direct the read and write cycles between the IC and the microprocessor, and special status and control lines which indicate some special operating conditions or situations of the IC. There are also data lines which are used for loading the key bits, writing to special setup registers, reading back the status registers, and transferring plaintext from the microprocessor for encryption or the results of decryption back to the microprocessor (Fig. 12-25).

Bit	Name	Function
0	Not used	
1	Activate	This bit must be set from 0 to 1 to initiate loading the key register. This bit must be 1 for encrypt/decrypt operation. This is a read/write bit.
2	Key error output enable ($\overline{KEOE}$)	When 0, the $\overline{key\ parity\ error\ output\ pin}$ ($\overline{KPE}$) remains inactive regardless of the status of the $\overline{key\ parity\ error\ bit}$ (bit 5). When 1, the $\overline{key\ parity\ error\ output\ pin}$ is active when the $\overline{KPE}$ bit (bit 5) is 1. This bit is set to 1 upon a master reset. This is a read/write bit.
3	$\overline{Encrypt}$/decrypt ($\overline{E}$/D)	When 0 data is to be encrypted. When 1 data is to be decrypted. This is a read/write bit.
4	Key request (KR)	This bit is set one clock period after the activate bit is set (from 0 to 1). It is reset upon loading of the 8th and final byte of the key register. This is a read only bit.
5	Key parity error (KPE)	This bit is set internally upon detection of a parity error during loading of the key register. It is reset when the activate bit is programmed from 1 to 0 (i.e., chip is deactivated). This is a read only bit.
6	Data-in request (DIR)	This bit is set upon either: a) Completion of key register loading or b) Completion of data register reading (i.e., the last data-out request has been serviced by an 8-byte read and the data register is now empty and ready to be loaded with the next data word). It is reset upon loading of the 8th and final byte of the data register. This is a read only bit.
7	Data-out request (DOR)	This bit is set upon completion of the internal encrypt/decrypt calculation of a data word. It is reset upon reading of the 8th and final byte of the data register. This is a read only bit.

Fig. 12-25 A command and status register is used to control the WD2001 and WD2002. The command bits allow the system microprocessor to direct the IC operation and set the required modes. The status bits allow the DES IC to indicate to the microprocessor when each operation is completed and whether it is successful or not.

Operation has four distinct cycles (Fig. 12-26):

1. *Load key:* The microprocessor tells the WD2001/2002 that a new key is about to be loaded, by writing a 1 to bit position 1 of the command register. The microprocessor then loads the key as 8 successive bytes. Each byte contains 7 bits of the key plus a parity bit, used to make sure the key is loaded properly but not actually part of the key.
2. *Load data:* The microprocessor sets the WD2001/2002 to either encrypt data or decrypt data by setting bit 3 of the command register to 0 or 1, respectively. The data bits are then loaded into the IC as 8 successive bytes of data with write cycles from the processor.
3. *DES:* In this phase, the DES algorithm (encrypt or decrypt) is performed on the 64 bits that were loaded in the previous step.
4. *Unload data:* The microprocessor reads 8 successive bytes from the WD2001/2002. These represent the result of the operation of step 3.

At this point, the 64 bits have been encrypted or decrypted, depending on what was requested in step 2, and passed back to the microprocessor. The IC can

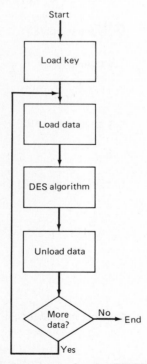

Fig. 12-26 The operational sequence for the WD2001 and WD2002. The key is loaded, followed by data. The encryption process is performed, and the encrypted data is unloaded (sent to the communications link). If there is more data, the sequence continues without loading the key again.

operate on the next group of 64 bits by immediately going to step 2; there is no need to reload the same key. The encrypted bits can be sent by the processor to a regular serial port and modem, if that is the way the connection to the link is made. Similarly, the encrypted bits from a modem can come into the processor, be passed to WD2001/2002 for decryption, and then be returned to the processor as decrypted plaintext. Figure 12-27 shows the path of data during these operations.

The bits of the command and status register are used for telling the WD2001/2002 that a new key is about to be loaded and whether encryption or decryption should be done. It also indicates to the processor and the rest of the system or user (indicator lights if applicable) that a parity error in loading the key has occurred, that the key has been successfully loaded, that the 8 bytes of data have been loaded, or that the encryption/decryption cycle is completed.

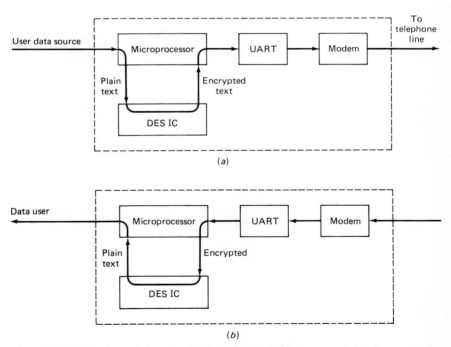

Fig. 12-27 The flow of data in a DES modem. (*a*) When encrypted before transmission, data flows from the source (a keyboard or computer, typically) to the modem microprocessor, which directs the DES IC and passes the data to the DES IC. After the DES operation, the encrypted bits are passed back to the microprocessor, which sends them, like any other bits, to the UART and modem circuitry. (*b*) For decryption, the received data goes into the modem circuitry and UART and then to the system microprocessor, which in turn passes the data to the DES IC while controlling it. The decrypted data bits go back to the microprocessor, which then passes them to the user, such as, a computer or terminal screen.

QUESTIONS FOR SECTION 12-9

1. Why is another encryption scheme needed in addition to PRSQ?
2. What is DES 46? Who publishes it? Is it a secret?
3. What is the concept of using DES and keys to encrypt data? How does the DES provide data security?
4. What do the bits on which the DES operate represent?
5. Why does a large key provide more security? How does the eavesdropper attempt to break it?
6. What is the security in one-to-one encryption and one-to-many deciphering without the key?
7. What is the electronic code book of DES 46? Where is it used? Why is it not a good choice?
8. What is channel block cipher mode? Where is it needed? What are its advantages over ECB?
9. What is cipher block feedback? Where is it best suited?
10. Who verifies DES ICs? Why is this helpful to the users of these ICs?
11. Why do DES ICs have two versions of the key path into the IC?
12. How is the key loaded into the Western Digital WD2001/2002 IC? Why does a 56-bit key require 8 bytes?

PROBLEMS FOR SECTION 12-9

1. An eavesdropper knows that the number 10 was sent as the encrypted value of the data but does not know the key used. He also knows that the number sent divides into the key without any fractional remainder, and that the key can be any number up to 1000. How many possibilities are there for the actual value of the data? What about the number of possibilities if the data is limited to two digits?
2. The plaintext data value is 15. A two-step approach is used to encrypt it. First, the data is divided into the key value of 150. This result, in turn, is divided into a second key value of 100. What is the resulting encrypted value?
3. The eavesdropper only knows the encrypted value which was intercepted in Prob. 2, and that the key can have only values up to 999. What values does the eavesdropper think are possibly the plaintext value? Is the eavesdropper correct? Explain.

SUMMARY

Data communications must function in a harsh and hostile electrical world. Noise from sources both within and outside the system can cause bits to be received or interpreted as a 1 instead of 0 (or vice versa). Even marginal operation of complicated circuits can cause an occasional bit error, from temperature, power supply, or other variations in circuit performance. To deal with noise, the design must first try to eliminate or minimize it wherever possible, with shielding, filtering, and careful design techniques.

Still, error-free performance cannot be promised. The data communications system must assume that some bits will be wrong. This reality does not have to be accepted without any further action. It is possible and common to design with the assumption that there will be some errors. The first step is to put into place a method to determine that an error has occurred. A single, simple parity bit can show whenever a single or odd number of errors exists in a group of data bits. Additional parity bits can be used across several groups of data bits and show some, but not all, even-number bit errors. The cost of the parity bit is low—1 bit per group of data bits—but it may miss many errors. Still, it is often included where very few errors are expected, for example, between a computer and nearby printer.

More advanced methods of error detection are based on advanced algebra theory. This theory shows that a long stream of data bits can be summarized with a nearly unique fingerprint of a few bits. This cyclic redundancy checking provides a way to detect nearly every possible error or group of errors. Although the theory is quite involved, it can be put into actual circuitry with some shift registers and exclusive OR gates.

Regardless of whether simple parity or more advanced CRC codes are used, error detection is achieved by adding the parity bits or CRC bits to the transmitted data and sending both data and error-detection bits to the receiver. The receiver takes all the data bits and uses the same scheme algorithm that is used at the transmitter to recreate the error-detection bits. These recreated bits are compared to the received error-detection bits. If they agree, no error has occurred; if they differ, an error exists.

Once the error has been detected, the receiving system must do something with this information. It can mark the received message as being in error. It can also go back and request a retransmission of the data. Two methods are commonly used for this. The simpler, but less efficient, is to have the transmitter wait after each block of data is transmitted for a message from the receiver that the data is received with or without error. Then, the receiver either resends the block or goes onto the next block of data. A better scheme is to have a "data OK/not OK" message sent while the next block is being transmitted. If there is no error, the transmitter keeps going. If there is an error, the transmitter backs up and begins sending again with the block that has an error.

Both of these methods require some reverse channel. In some applications, such as remote data acquisition from space satellites or weather stations, this reverse channel doesn't exist or is very slow. In other cases, the transmitter may not be able to send the data again, because the data is not stored internally. Finally, the data can be stored in memory at the transmitter, but a bit in the memory may have been changed. Sending the data with that memory bit again simply resends the error. A solution to this dilemma is forward error correction, in which the transmitted data bits contain special additional bits to allow the receiver to detect and correct many types of errors. Algebra theory, such as that developed by R.W. Hamming, shows which types of encoding schemes allow all single-bit errors to be detected and corrected, and all 2-bit errors to be detected. More Hamming bits can be used to allow even more error groups to be detected and corrected, but the cost in additional bits is high. The simplest error-detection and -correction pattern for 8 bits requires an extra 5 checking bits. Nevertheless, some applications require this level of data integrity and use Hamming and other similar codes to provide it.

There is a problem in data communications which is technically related to error detection and correction. Users need to ensure that no unwanted listeners are receiving the signal, or, if they are, that the bits as received make no sense. These users also may want to guarantee that unwanted users don't put misleading or false information on the link. Data security is achieved by using random bit patterns to modify either the signal modulation or encoding. A perfectly random sequence can be used, but there are problems in generating and carrying such long sequences from user to user with absolute security. Once again, algebra theory provides a simple solution. The same shift register and exclusive OR circuit used for CRC can generate pseudorandom sequences which are very close to being perfectly random bit patterns. The initial value in the shift register is called the *key* and uniquely defines all the bits that the pseudorandom generator circuit develops. If the transmitter and receiver have the same circuit and key, they can use these random bits to encrypt and decipher the data bits. An outsider has a very difficult time figuring out the random pattern and data bits without the key.

The same random sequence can also be used to vary the carrier frequency in radio transmission. This is called *spread spectrum* because the single-user signal hops from carrier to carrier and occupies a much wider bandwidth than the data itself needs. A spread spectrum signal is very hard to intercept if the hopping pattern is not known. It is also hard to jam with an interfering transmitter. Spread spectrum is used extensively in military radio applications, in which even the location of the transmitter should be a secret.

The pseudorandom sequences provide a low-cost way to encrypt data bits, but the transmitter and receiver must synchronize their circuitry. This is difficult and uses channel time, especially in half-duplex systems, in which a resynchronization occurs every time the channel is reversed. To overcome this problem, a national standard for encryption was developed, called the *data encryption standard* (DES). The DES algorithm uses a 56-bit key and combines the user data with the key in many operations, such as addition, inversion, shifting, and other combinations. Plaintext data which enters the DES circuitry comes out in a unique pattern, which a receiver can decipher easily if it has the key. Without the key, however, there are a nearly infinite number of possibilities of original data that may have caused the encrypted pattern.

The basic idea is to establish a unique one-to-one relationship between the original data and its encrypted version, using the key, but a one-to-many relationship in the encrypted-to-plaintext direction without the key. The key directs the decryption operation to provide the correct answer. In many ways, the idea of error detection and correction is similar to data security. In the first, the data must be encoded so that errors can be identified and removed. In the second, deliberate errors are added, and only a receiver with the key can separate these deliberate errors from the original data. Related branches of algebra theory, and similar circuitry, are used for error detection, correction, and the pseudorandom sequences of data security.

END-OF-CHAPTER QUESTIONS

1. What are the causes of data bit errors, even in a well-designed system? How can marginal design make more errors occur? Under what conditions?
2. Why are users concerned with both average error rates and burst errors?
3. What is the first part of an error-reduction strategy? Why is this not enough?
4. What is data security? How is it related to error detection and error correction?
5. Explain how parity checking is done? How is the parity bit generated?
6. What is done with the parity bit at the transmitter? The receiver?
7. Explain odd versus even parity in terms of technical difference, error-detection difference, and usefulness to a technician or engineer.
8. How does parity decrease the efficiency of a system? By what amount?
9. What kinds of errors does parity detect? What kinds does it miss?
10. What is horizontal parity? What is vertical parity? Why is vertical parity added to horizontal?
11. Where is parity a good choice for error detection? A poor choice?
12. How does CRC provide error detection? How is the CRC checksum generated, in theory?
13. What circuitry is used to develop the CRC checksum?
14. Is a single CRC checksum used over only a single byte? Over how many bits typically?
15. Why is the theory of CRC checksum related to polynomials and algebra theory?
16. What is used to decide where the CRC circuitry feedback taps are located?
17. What is the meaning of the CRC checksum value? Does it explain where or what the error is?
18. How does the efficiency of CRC compare to that of parity?
19. Explain why CRC checksums are not perfect at error detection, but are very good.
20. Does use of CRC rule out use of parity? Explain.
21. What is done when an error is detected, by whatever means?
22. Compare "stop and wait" to "go back *n*." How does each work? What does each require? What are relative good and bad points of each?
23. What may happen if a "go back *n*" method only sends the bad block again, but not any subsequent good ones?
24. Compare FEC to ARQ. What are the advantages and drawbacks of each?

25. What theory is used to provide EDC? What is done to provide the EDC capability?
26. Where is EDC needed? What penalty is there for it?
27. In a simple EDC structure, with 8 data bits and 5 check bits, what happens when there is a single-bit error? A double-bit error? Three bits in error? What does the EDC reveal? What do the syndrome bits show?
28. Why isn't Hamming EDC made more efficient simply by making the data bit field longer, which requires a smaller percentage of check bits?
29. How does an EDC IC operate? Why is a memory-oriented EDC IC studied, rather than a communications IC?
30. How is the error location determined with an EDC IC? What is done to the bit in error?
31. What does the EDC IC show to the rest of the system?
32. Why is data security closely related in a technical sense to error detection and EDC?
33. What situations must data security protect against?
34. What is the simple offset encoding scheme? What is its weakness?
35. What is a one-time key? How does it make offset encoding unbreakable?
36. What are the problems with one-time keys?
37. How can a one-time key be closely simulated, in theory? How is the theory implemented with circuitry?
38. When does a PRSQ repeat, in the maximum case? Can an eavesdropper determine the key from observing the PRSQ?
39. Why must PRSQs used for security have synchronization at both ends?
40. What is the idea of spread spectrum? What is the reason for the name?
41. What two problems does spread spectrum solve in military applications? Why is spread spectrum less used in commercial situations?
42. Explain the circuitry used to encrypt or decipher data with PRSQs.
43. What is the idea of a data encryption standard? If the standard is published and not secret, how can it provide data security?
44. What is the role of the key in the DES?
45. What are the three modes supported by the DES? Where is each a good choice?
46. What are the four steps in a DES operation with a DES IC? Are all the steps repeated for each group of bits to be encrypted?
47. Explain why the key cannot be read back out of some DES ICs.
48. Explain how a DES IC can be used to build a modem which provides data security for bits coming from a terminal.
49. Does the use of parity, CRC, or data security preclude the use of any other of the three? Explain why or why not.

END-OF-CHAPTER PROBLEMS

1. Calculate the odd parity for the following 4-bit groups:
   1001, 1010, 1100, 0000

2. Find the even parity bit for the following bytes:
   10010111, 00001111, 10100101, 00110011

3. Find the odd vertical parity for the data groups of Prob. 1.

4. Find the even vertical parity for the bytes of Prob. 2.

5. The following bytes show what is sent and received. Calculate the odd parity for each pair. Which errors are detected? Which are missed?

   a. 01010101—01100101
   b. 00001111—11110000
   c. 11000011—01000011
   d. 01011100—10011110

6. Show how the parity circuit of Fig. 12-4 finds the even parity for input 0000, by marking the boolean value at each point in the circuit.

7. Repeat Prob. 6 for inputs 1111 and 1100, with even parity.

8. Express the following data bit group as polynomials:
   0111, 1001, 1000, 1101

9. Express the following data bytes as polynomials:
   10101010, 00110011, 11010001, 00001001

10. What data bytes are represented by the following polynomials?

    $X^5 + X^4 + X^2$
    $X^7 + X^3 + 1$
    $X^7 + X^2 + 1$
    $X^6 + X^5 + X^4 + 1$

11. What are the inverse polynomials for the polynomials of Prob. 10?

12. What are the polynomial and its inverse for the data group 110011? What are they for 011010?

13. An 8-bit CRC checksum is used with every 32 bits of data. What percentage of the data is this? What if parity were used instead, with one parity bit per byte of data?

14. A single 32-bit checksum is used for a large block of 2048 bits. How much more efficient is this than using one parity bit per byte?

15. How many possible CRC checksums are there for the 32-bit checksum of Prob. 16? For the 8-bit checksum of Prob. 15?

16. By what factor is the number of CRC checksum possibilities increased when the CRC checksum is increased by 1 bit? Explain. What about increasing the CRC checksum by 2 bits?

17. What CRC checksum results from data bits 1111 going into a 4-bit register with feedback taps at locations 2, 3, and 4?

18. What CRC checksum is produced by data bits 0000? Is it a function of the length of the register or the taps?

19. Compare the efficiency of a full-duplex "stop and wait" ARQ system to one which is half duplex. The data blocks are 100 ms, the ARQ message is 100 ms, and the line turnaround time is 150 ms for the half-duplex channel.
20. Draw a sketch of the line activity for the systems of Prob. 22.
21. The "go back 2" ARQ method is used for a very poor quality channel. Ten blocks are sent, with errors in blocks 2, 6, and 9. How many blocks are sent in total? Draw a sketch of the line activity with block numbers.
22. Calculate the Hamming checkbits (odd) for the following data bytes:

    1111 1111; 0100 1111; 1100 0001; 0000 0000
23. Using the bytes of Prob. 22, the rightmost bit is made by noise to be a 1 if it already is not. What are the new Hamming check bits? What is the error syndrome in each case?
24. For the data byte 10101001, what is the indication for each of the following groups of check bits?

    11001, 10001, 10111, 10101
25. A user attempts to encrypt a message with a fixed offset of -1. What is the encrypted message for the plaintext message DANGER STAY AWAY.?
26. A message encrypted with a fixed offset is intercepted. The message is LDDS KZSDQ. What is the plaintext?
27. A random series of offsets is used for the message of Prob. 25. These offsets are $+2, -4, +1, +2, -2, +0, -1, -3, -1, +3, 0, +2, +1, -1$. What is the encrypted message?
28. A random pattern is used to encrypt data bits. What is the encrypted result?

    Data   1001 0010 1010 1000
    Random bits   0110 1010 0011 1001
29. Repeat Prob. 27 for data bits 1010 0110 0011 1001.
30. Repeat Prob. 28 for random bits 1001 0100 0001 0101.
31. The data received after encryption is 1001 0110. The random sequence used is 0101 1100. What is the original data?
32. A 5-bit shift register is used with feedback taps at locations 3 and 4. What are the first 16 bits of the PRSQ generated with a key of 11001?
33. Repeat Prob. 32 with a key of 10001.
34. The taps on the 5-bit shift register are moved to locations 2 and 5. What are the first 10 bits of the PRSQ for a key of 01001?
35. Repeat Prob. 34 for the key 11001.

13 Test Techniques and Instrumentation

The testing of data communications systems requires instruments more advanced than the traditional voltmeter and oscilloscope. Specialized instrumentation is required to determine what the system performance is and where it may be faulty. These instruments are not all expensive or complex. They range from simple signal breakout boxes, used for making a basic interconnection work, up through sophisticated equipment for testing and assessing networks. Loopbacks at various points allow testing to be done from one end of the system.

Even for digital data communications, the analog characteristics of the communications link are critical. Transmission system test sets measure the bandwidth, noise, and jitter of the link itself. Time domain reflectometry, which is very simple in theory and use (but requires some advanced electronics), allows a technician to locate cable faults to high accuracy while working from the end of the cable.

The increasing use of fiber-optic cables to replace wire cables means that these optical fibers must be tested as well. There are new types of equipment and methods used for measuring the properties of an optical system. The physical connector and cable splice, as well as the light wavelength in use, are very important factors when checking fiber-optic cables.

13-1 Basic Tests

The traditional tools of electronics, the voltmeter and oscilloscope, are used in parts of the test and troubleshooting aspects of data communications. They are useful, but additional specialized instrumentation is needed in order to properly check out a malfunctioning system or a new design. In addition, data communications systems involve some unique issues. The system is often spread out over a large distance, and the test technician may not be able to access the far end of the system easily, because of the distance or location. Special techniques are used in communications systems to overcome this problem.

The goal of testing is to make sure that the data communications system meets the performance needs and specifications. This is not as trivial as it may sound. Unlike many systems, which either work properly or hardly work at all, a data communications system may work to varying degrees. If noise is a problem, the system may function but with more errors than it should have. If the system has advanced error-detection and -correction capabilities (either through error-detection and retransmission or through error-correction codes) the actual end user may never see the errors. The system appears to be operating fairly well. However, a closer look may show that the throughput of the system is less than it should be since so many data blocks are being retransmitted. The essential points are the following:

- Data communications systems can be operating anywhere from poorly to perfectly and still pass some data.
- The actual user of the data may not see many of the problems the system is having or may see symptoms that don't relate directly to the problems. A system which produces many retransmissions may have slow response time to characters typed at the keyboard, without the users realizing the actual cause.

One of the issues to determine before beginning any work on a system is what the specifications for performance really are. The industry uses many measures of this, such as *bit error rate* (BER). Instrumentation is available to measure BER in many useful ways. Once the specification is known, and the performance doesn't meet the specifications, further investigation is begun. There may be problems that are within the control of the system designer and technician, such as a defective component. There may also be problems that are basically beyond the control of anyone, such as severe noise on the frequencies used by a radio link. It is important to identify the source of the problem so that correction, if possible, is made and, if it is not possible, alternatives are evaluated.

Fiber optics is playing an increasing role in communications, in both long-distance and local links. There are many performance and cost advantages to fiber optics, but new types of test equipment are needed to implement it. The connectors are different from those for electrical signals, and regular electronic equipment such as signal generators, oscilloscopes, and voltmeters is not usable. Also, the parameters that have to be measured are not always the same as those with wire links or radio links. Part of this chapter will discuss test equipment designed for fiber optics.

Role of the Voltmeter and Oscilloscope

The *voltmeter* plays a limited role in checking the performance of a communications system. Since the signals are changing so quickly, and the exact sequence of signals and their shape and timing are critical, the voltmeter is useful only for the first level of investigation. When there seems to be any type of problem in operation, a voltmeter is used to check the value of the power supply at various points and components in the system. This value should be within the tolerance specified for the circuitry. A typical system uses a 5-V supply with a 10 percent tolerance, so the direct current (dc) supply should be between 4.5 and 5.5 V everywhere.

Most voltmeters have a relatively limited frequency response range, so they cannot be used for checking the values of wide-bandwidth signals such as modulated carriers, unless they are designed for this type of operation. There is another aspect of digital communications that makes obtaining accurate meaningful voltmeter readings difficult. This is the fact that although the signals may be digital and rise to some specific voltage each time a binary 1 is sent, the voltmeter tends to average out the ratio of 1s and 0s and show the average value. If the system is sending one 1 bit and nine 0 bits (using 5 V and 0 V, respectively) the voltmeter shows the average, or 0.5 V. This number is meaningless because the average value of a stream of data bits tells nothing about the system performance. The technician really needs to see that the 1s and 0s reach their correct values.

The *oscilloscope* can show much more than the voltmeter. It is used for the following types of investigation:

The power supply should have the correct voltage, and there should be no noise on the power signals.

The data signals should reach the correct voltage values and also have little or no noise. If there is noise, it should be a small oscillation on the steady value of the digital signal, rather than a large spike (Fig. 13-1). The spike causes errors and is very difficult to filter out of the system. It may even cause some of the clocking circuitry not directly related to the data signal to malfunction or lose synchronization.

The data transitions from one binary state to the other should be fairly sharp and clean. Here, judgment has to be applied by the technician. The transitions are sharpest at the data source. At the receiving end of the data, these transitions are affected by any cable bandwidth limitations, capacitance, and inductance, so the rise time is longer and there may be ringing (Fig. 13-2). A properly designed receiver circuit takes these changes into account before deciding whether a 1 or 0 has been received, but the signals should still be checked to make sure that the degradations are not excessive.

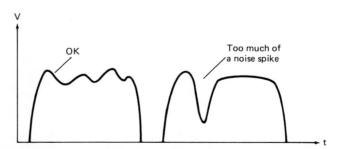

Fig. 13-1 Some amount of noise is always present on digital signals. A small variation or oscillation is usually acceptable, but a large noise spike (either positive or negative) can cause a bit error and system problems.

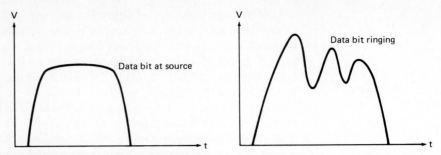

Fig. 13-2 The sharp bit transitions and signal at the data source are usually corrupted by cable capacitance and inductance at the receiving end, often with some "ringing."

The scope can be used to study the overall pattern of signals as they are received by using the eye pattern (Fig. 13-3). This single pattern summarizes the overall condition of the data bits. It shows the amount of distortion, noise, and timing variation in the data bits. It also shows in what way the 1 and 0 values are separated. Most important, it allows the effects of any adjustments in the system to be quickly observed. If there are fine-tuning adjustments, the eye pattern shows the effect of the adjustments. The point where the eye is most open is the normal point for deciding whether the received signal is a 1 or a 0. If this point is not steady, the receiver has trouble consistently looking at the data at the best time within each data bit period.

There is a definite limit to what can be accomplished with the voltmeter and scope. Other types of equipment are needed to continue the test and troubleshooting activity. Some of these are relatively simple, and others are quite complicated. Two of the simplest and most useful are the breakout box and line monitor.

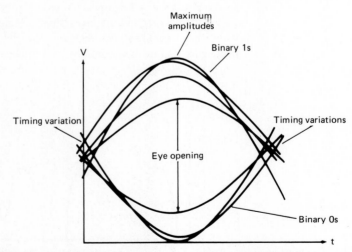

Fig. 13-3 The eye pattern is a good summary of the quality of the received signals. The eye opening shows the difference between 1 and 0 values, even with the noise, and the point of maximum eye opening is the best place to determine whether a 1 or 0 is received. The spread of signals at the crossover points indicates the amount of jitter that exists.

Successfully connecting an RS-232 terminal to a local computer usually requires only a few simple items. Checking out a long-distance link using synchronous data-link control (SDLC) protocol requires much more, of course.

QUESTIONS FOR SECTION 13-1

1. What is a goal of data communications system testing? How does it differ in some ways from testing of other electronic systems?
2. What facts should first be determined about the system?
3. How is the standard voltmeter used? What are its limitations?
4. What should the oscilloscope show for the power supply of the communications system?
5. What are the effects of noise spikes?
6. What should data bit transitions look like at various points?
7. What does the eye pattern show? Why is it very useful when making adjustments?

13-2 The Breakout Box and Line Monitors

One of the most common and sometimes frustrating activities in data communications is interconnecting two devices, both of which should be able to talk to each other, only to find that they don't. This occurs most commonly with RS-232 devices, as was discussed in detail in Chapter 8. Fortunately, a simple, low-cost device called a *breakout box* makes the job of making two RS-232 devices work together much easier.

The *breakout box* takes all 25 wires in the usual RS-232 cable and breaks them out so they are all accessible to probes (Fig. 13-4). It also has a simple on-off switch for each of the wires (except ground) so the path of the signal can be made or broken. Indicator light-emitting diodes (LEDs) show the status of key RS-232 lines, such as transmitted data, received data, CTS (clear to send), RTS (return to send), and a few others that are essential for handshake. These data and handshake lines are available on both sides of the on-off switch as test points which can also be used for jumper wires. The breakout box itself is passive and puts no signals on any of the lines.

How is the breakout box used in a typical situation? It is used to resolve conflicts between pin assignments, disconnect lines that are interfering with each other, and make desired handshake connections. It is often used with a *gender changer*, which also overcomes the pin/receptacle mismatch on the connectors themselves.

Suppose the technician has to connect two RS-232 devices, such as a computer and a terminal. Both are normally data terminal equipment (DTE) devices, so there is a good possibility that the connection will not work the first time. The first step is to make the connectors mate, and also make sure the baud values, bits per character, and parity setting of each unit are the same as those of the other. Once this is done, the technician should study each device and see what pin each puts its data on and what handshake lines each expects to see before it transmits.

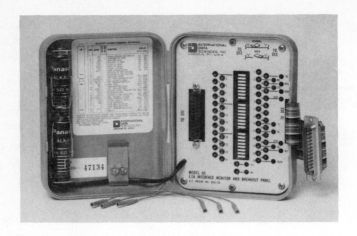

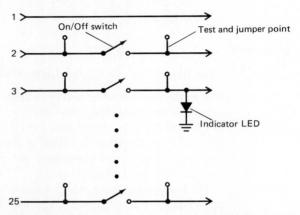

Fig. 13-4 A breakout box (top photo) is very helpful in accessing the signals of the RS-232 cable. The schematic shows the function of the breakout box: to bring out test points for all lines, indicate their state with LEDs, and allow any line to be opened or jumpered to another line (photo courtesy of International Data Sciences, Inc.).

The goal is to make at least one side send data. This requires making at least pins 2 (transmit) and 3 (receive) of the normal 25-pin connector compatible, if both devices are DTEs. The switches of the breakout box are all opened, and then transmit data and receive data can be jumpered by crisscrossing, if needed. The switches for pins 2 and 3 are closed if jumpering crisscross is not needed.

The breakout box is then used to make sure that the terminal, for example, sees whatever handshake lines it needs. This is done by opening all the handshake lines in the cable, and forcing the needed lines to a high with jumpers. Once the terminal sends characters, the computer must be able to accept them. Once again, the handshake lines are the key. The LEDs are used to monitor the handshake line from the computer which indicates that the computer is ready and able to accept characters. If this line is low, the problem is back in the computer serial interface. Perhaps its buffer is full, or the computer software has not set the proper initial conditions into the interface circuitry. If the handshake line from the computer shows that it will accept the character, the character from the terminal should be received.

It may turn out that the character is sent by the terminal, but the computer refuses to accept it properly even though the handshake line indicates it is willing. This indicates a problem with the baud value, bits per character, parity setting, or computer software that actually processes the characters as received. Once the computer accepts the terminal characters, the actual handshake line pair match from the terminal to the computer can be worked out, by jumpering the ''I'm ready to accept'' line from the computer to the ''Can I send you data?'' line at the terminal. Exactly which lines are used by each side is determined by study of the data sheets or manuals of each device and by some trial and error.

For the common case in which both devices are DTEs, the breakout box is used to cross lines 2 and 3 of the normal RS-232 interface, so that both units don't send data on the same wire.

By a combination of carefully reading the data sheets, thinking about what is supposed to happen, observing what actually occurs (via the LEDs), and making things happen with the switches and jumpers, a working interface is achievable. The breakout box cannot solve problems, of course, when the fault lies in initialization or setup within the interface circuitry itself. These must be done at the terminal or computer. Once a working interface is established, a proper cable which replaces the breakout box can be made up. This can put the data and handshake wires in the correct place and substitute for any missing handshake lines, as discussed in Chapter 8.

Line Monitor

The breakout box enables the technician or engineer to establish communications on a character-by-character basis. This does not mean that the messages are understood, however. Perhaps the characters are not the ones the computer needs to see, or they cause the computer to take peculiar actions because of differences in protocol, headers, and character fields within the message. This is a situation in which some type of line monitor is needed. A *line monitor* is connected to the data lines and shows the characters as they appear on the link. The person trying to make the system work looks at the characters, determines whether the intended message has been sent, and ascertains whether this agrees with what the receiver wants to see.

The simplest line monitor for RS-232 communications is an ordinary cathode-ray tube (CRT) terminal. It can be attached to the data in one direction (not both) and show in its screen what characters are appearing on the line (Fig. 13-5*a*). In order to make the terminal work in this way, it must have its baud value, and so on, set the same as the expected data. More important, it must have its transmit data (which goes from the keyboard outward) disconnected, so it doesn't conflict with a data line. The handshake lines of the terminal must also be disconnected so they don't cause interference. Usually, this also requires that the monitor handshake lines are jumpered to fool the monitor into always being ready to accept data.

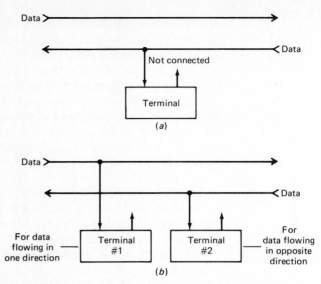

Fig. 13-5 (*a*) An ordinary terminal can be used as a line monitor for one direction of data flow. The line carrying data from the terminal to the RS-232 interface must be unconnected. (*b*) Two terminals can be used to monitor both directions of data, but the procedure is very awkward.

Although this simple line monitor works, it does have some drawbacks. First, it cannot monitor data in both directions at the same time. Its input can only be connected to one direction of data flow on the interface. (Two CRT monitors can be used, one for each line, but it is nearly impossible to watch two screens at once [Fig. 13-5*b*]). Second, the monitor does not show the nonprintable control characters that may be sent from the real terminal to the computer, or vice versa, or it may be confused by these. Characters which represent line feed (LF), carriage return (CR), or special functions such as S0 and S1 do not appear on the screen, since they represent control functions, not alphanumeric characters.

To overcome these limitations, special monitor terminals are available. These are designed to read the signals on the data lines passively, in both directions. The monitor screen uses regular video display for one direction (light characters on a dark background) and reverse video display for the other direction (dark characters on a light background). These two displays appear simultaneously on the screen, so the user can actually see the data flow in both directions at the same time. The control characters are shown with special symbols on the screen (Fig. 13-6*a*), but cause no action by the monitor. The monitor is absolutely passive and transparent to two devices that are connected—it is never seen by either of them.

Some more advanced monitors for RS-232 also allow the status of the various RS-232 control lines to be shown on the screen at the same time, in timing diagram fashion (Fig. 13-6*b*). This allows the technician or engineer to study not only the data characters themselves, but the interaction of handshaking. This is very helpful when the characters are coming across the interface correctly but buffers are filling too quickly, or handshake problems are causing characters to

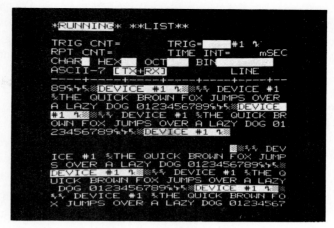

Screen showing both data directions and control characters
(a)

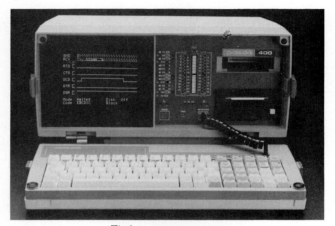

Timing as seen on screen
(b)

Fig. 13-6 *(a)* A line monitor can be designed to show data flow in both directions, with regular video display for one and reverse video display for the other. Note that control characters (such as line feed) are also shown, using special characters (such as a small LF symbol). (Photo courtesy Hewlett-Packard Co.) *(b)* Some line monitors can also show the various signal lines and their timing, which is used for troubleshooting some problems. This unit combines breakout box features (middle of the top panel) with the ability to show both characters and timing on the CRT screen, while a keyboard lets the user type in commands on what signals should be displayed and in what format (Photo courtesy Digilog Corp.)

be sent but not received. For example, a device sending data may not respond to an "I can't take any more" handshake from the receiver quickly enough and may send out one more character. The transmitting unit sends out the full message, but the receiver shows that the receiver has said, "Please stop" via the hand-

Test Techniques and Instrumentation 489

shake lines, and the sender gets there too late to prevent it from sending out that next character. The monitor shows the flow of characters and the relationship, in time, to the handshake line activity. Using this information, the circuitry in the transmitter can be modified to respond more quickly, or the transmitter can be modified to send out the indication sooner.

QUESTIONS FOR SECTION 13-2

1. What is a breakout box? What features does it have? What does it do to the various signal lines?
2. What are the first steps in using a breakout box to connect two RS-232 devices? What is the initial position for the switches?
3. What are the next steps when using the breakout box?
4. What does the breakout box accomplish? What can it not do?
5. What is a line monitor? How can a simple terminal be used as a line monitor?
6. What are the limitations of using a terminal as a line monitor?
7. How does a specially designed line monitor instrument overcome these limitations? What does it do?
8. Why is observing the relative timing of the data and control lines important?

13-3 Loopbacks

In a data communications system there is often a need to perform all the testing and diagnostics from one end of the link. The other end may be unreachable, far away, or unattended. Even if both ends are accessible, it may be difficult to coordinate activities and test equipment operation. The solution is to use a *loopback*, which is simply a way to route the data signals in a "U-turn" at the receiving end. The loop-back is a very powerful tool in a test and diagnostic situation. It allows a single test operator to run a series of data diagnostics and determine what is working and how well the link is performing.

In order for a loop-back to operate, there must be a path in the reverse direction. A simplex system does not support loopback, although a half-duplex or full-duplex system does. In general, a full-duplex link is preferable, but a half-duplex link can work. In a loopback situation, a data sequence is sent and then returned. By comparing what is sent with the returned data, the number and types of errors can be measured.

Of course, many applications have only a simplex link. The loopback cannot be used in these cases. However, it turns out that although the individual user may have only a simplex link, the total communications system often has at least one link in the reverse direction. Otherwise, even installing and setting up the system, and verifying that it is working, is very difficult. How can the sender have any idea that data has been received? How can the sender know that the data signals are not simply being sent into "thin air"? Some reverse channel, even if available only on a part-time or shared basis, is usually installed.

Another advantage of the loopback is that it allows automatic, unattended tests to be performed. Since there is no need for an individual at the remote end, a technician or engineer can run the tests alone. If there is a computer or programmable device at the sending end, this computer can be programmed to run a series of tests on its own and report on the results. The tests do not have to be individually run and supervised by any person. Most large data communications systems have built into the design specific programs which perform these tests on a regular basis during idle times, when the system is not being used for data. In the telephone system, this commonly occurs at night, when the amount of data falls off tremendously. The computer that controls a local exchange, for example, sets up a loopback with another exchange and then sends signals down every trunk line between the two exchanges. The quality of the signals is recorded; it may point to trunks that are not working up to the desired specification but haven't failed completely. This is preventive action, since the trunk can be checked and fixed before its condition degrades to a point at which it must be taken out of service. In the telephone system, a single reverse channel may be used and shared, in sequence, among all the regular trunks. These reverse channels are designed for test use only and so may have wider bandwidth, lower noise, or some other characteristics that make it especially good for testing purposes. The cost of this one special channel is shared by the many lines it serves, so it adds little to the overall system cost.

Loopback Locations A total communications system usually has many loopback locations. Consider a simple case of a DTE device connected to a data communications equipment (DCE) unit, with this DCE unit communicating to another DCE some distance away. The second DCE is in turn connected to its DTE. Where are the loopbacks?

The first one is located at the first DTE (Fig. 13-7). Using this, the DTE can send data and get it back without going through any DCE device or link at all. This loopback verifies that the DTE is actually capable of sending and receiving data on its own. If the DTE cannot do this, then there is a basic problem that must be fixed at the DTE itself.

The second loopback is at the DCE device of the first DTE system. This loopback allows the data to go from the source DTE, through the first DTE to DCE cable, through the DCE, and back to the DTE. By using this loopback, the

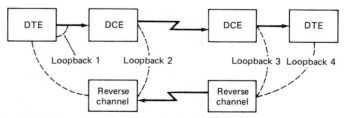

Fig. 13-7 A full data link, with two DTE/DCE pairs, has four possible loopback locations. Each loopback indicates additional information about system performance.

signal quality and system performance at the transmitting location (such as an office) are verified and measured. It also checks the operation of the first DCE device. If the data is returned without error or with few errors, then the DCE is probably operating properly.

A third loopback is performed at the remote DCE. Here, the sending system has the opportunity to verify the data link and the remote DCE device. Unfortunately, this doesn't allow the testing end to determine whether the problem is with the link or the remote DCE. Many large systems, with multiple DCE devices, allow the same link to switch to one of several DCEs. This permits a substitution for what may be a defective DCE. If the third loopback is tried with several remote DCEs, the local unit can decide whether the link is a problem or the DCE hardware at the remote end is suspect.

The fourth and final loopback is through the remote DTE. The DTE unit accepts the data as received and then returns it. If this works, the complete user-to-user system is verified. Even if it does not work, the problem is now more clearly located, since it must be somewhere between the remote DCE and the remote DTE, or in the remote DTE itself. This has narrowed down the area to check to a much smaller section. In many cases, the communications support personnel believe that the communications system has been checked out successfully, since the DCE/link/DCE part of the system is working. The DTE units are really just the individual pieces of electronic equipment that a single user is operating, such as a terminal or computer, and not a part of the data communications system itself, which is defined by the two DCEs and whatever is between them.

The loopback is a very powerful and useful tool. If any data at all can be sent through a system and returned, a great deal about the operation and quality of performance can be learned. If no data at all can be looped back, then a major problem exists and the loopback can only show where the problem is. The flow of data through a loopback must be established so that testing beyond the basic "It seems to work OK" can be performed and specialized equipment for measuring the quality of the system can be used.

Implementing the Loopback

The loopback can be implemented through special cables, switches, system commands, or some combination of these. Each method has some benefits and drawbacks and also some implications for the completeness of the loopback check itself. The simplest method is simply to use a cable, with the correct connections and writing, to connect an output port of the DTE or DCE back to its input port.

Consider the way this is done in the case of the basic RS-232 interface, which exists at loopback locations 1 and 3 of the four identified previously. A specially wired 25-pin RS-232 connector is used in place of the regular connector and cable which connects the DTE to the DCE (Fig. 13-8a). This connector is wired to send the transmit data signal to the receive data pin and also to connect the handshake lines so they match. Any data sent by the DTE automatically comes back to the same DTE. As far as the DTE unit itself sees, the data has gone to another device and is now coming back. This sort of simple connector method

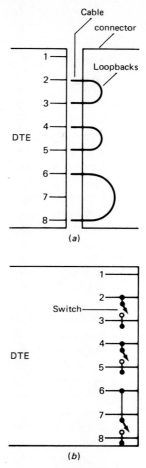

Fig. 13-8 (*a*) A loopback can be implemented with a special external connector (not all wires shown), or (*b*) with internal circuitry at the interface and some switches.

verifies all the circuitry in the DTE, all the way through to the connector itself. The confidence after this sort of loopback is performed is very high.

The drawback to this method is that the existing cable must be disconnected; the cable is often difficult to reach without disturbing an existing equipment setup. Disconnection also takes time, which is important in a field service situation. To achieve the same loopback effect, some devices have special switches which can be simply flipped. These switches disconnect the external connector and instead route the signals within the DCE in the same way that the external connector does (Fig. 13-8*b*). The advantage is that the changeover from regular operation to loopback operation is quick and easy, and no existing cables have to be disturbed. This is an important feature in a real installation, where every cable changeover takes time and has the risk that it may not go back the way it should.

The mechanical switchover using ordinary, built-in switches does not provide the same 100 percent confidence as the external loopback connector. There is a small possibility that the problem lies at the DTE connector itself, and the switch effectively bypasses that. However, this is a small risk and is outweighed by the convenience of the easier switchover. Also, the technician can still use the external loopback connector, if there is a suspicion that the problem is between the loopback switches and the DTE connector.

A larger problem with the loopback switchover using switches is that it still must be carried out by a person. For remote units or unattended units, this is very impractical. The solution to this problem is a remotely initiated switchover mode using system commands. The electronic system contains the same type of internal loopback switchover function, but one which can be set to switch over by using a special system command, sent through the communications link. This allows a remote unit to be put into loopback mode conveniently, either by a person at the sending end or by a system which is going through a set of computer-driven tests. A loopback command might be the ASCII group "#@LPBK," which never occurs in the regular data.

The remotely initiated loopback is a very powerful tool in testing large, spread-out systems. It does require more circuitry in the unit performing the loopback; this adds to the cost slightly. It also means that there is a larger possibility that the problem is with the loopback circuitry and not the unit itself, although in a carefully designed circuit this is unlikely. More important, it means that the basic communication to the device must be working, in order to receive and understand the loopback command. If the communication is poor or nonexistent, then a command cannot reach the unit properly. If the unit is malfunctioning severely, it may receive the loopback message but not be able to implement it. For these reasons, most units which have remotely initiated loopbacks also have a set of manual switches so a person can perform the procedure if necessary. Despite these minor drawbacks, most higher-performance systems have remotely initiated loopback commands as part of their design, because they are very useful and powerful.

Once the flow of data through the system has been established, either with or without loopbacks, the specialized data communications equipment can be used. This equipment includes error rate testers, pattern generators, and protocol analyzers, which are discussed in subsequent sections.

QUESTIONS FOR SECTION 13-3

1. What is a loopback? Why is it useful in data communications systems?
2. What must a system have in order to support the loopback function? Where can this be a problem? Why is it often not a problem in a large system?
3. How does the loopback allow unattended testing? How is this testing performed on large, multi-channel systems?
4. Where are the loopbacks located in a typical system? What is tested by each loopback location?
5. How is a loopback implemented in a cable? What are the advantages and drawbacks to this approach?

6. How can the disadvantages of a special loopback cable be overcome? What is lost when the disadvantage is worked around?
7. How is a loopback implemented remotely? What does this require? What are the pros and cons of this approach?

13-4 Pattern Generators and Bit Error Rate Analyzers

Testing a system, either with or without loopbacks, means sending known data and determining whether this known data is actually received. The instrument used to develop this known data is the pattern generator. It sends a large block of data bits and usually has several modes of operation:

It can send the same short bit pattern again and again (1100 1100 1100. . .).

It can send a long pattern, of thousands of bits, once.

It can send the longer pattern repeatedly.

It can send bit patterns of character patterns. A typical character pattern is the ASCII representation (with start and stop bits) for the FOX message THE QUICK BROWN FOX JUMPS OVER THE LAZY DOG, which includes all the letters of the alphabet and is easy to recognize.

The pattern generator is used to put data into the system, which can then be followed through the data communications system. As the known data is modified by any encoding, modulation, protocol, and formatting, the person running the tests can determine whether these transformations are occurring properly. The pattern generator sets up the repetition, if needed, so that stable patterns can be established and seen on the oscilloscope or other measuring devices.

In operation, the pattern generator is often connected in place of the actual source of the data. It is used to replace the terminal DTE device, for example, that is connected to the modem DCE. The pattern generator then replaces a person trying to type at the keyboard. The modem itself can also be replaced by a special pattern generator modem, which transmits a test pattern. Some modems actually have some special pattern generation capability built into their design. By setting a special switch or sending a control command, the modem acts as the source of the data, and no terminal is needed. This is useful when the link and receiving end must be checked, and any extra equipment at the transmitting end (such as the terminal) can interfere with test procedures.

The next step is to use the pattern generator to measure the performance of the data communications system accurately. This is done with a corresponding unit at the other end, which receives the data pattern. The receiving unit is also a pattern generator but is set up to receive and compare patterns. In this way, it can help assess the quality of the received data versus the transmitted data. When a known data pattern is used and compared, the instrumentation is called a *bit error rate* (BER) *analyzer*. If a loopback is used, a single physical instrument can act as the pattern source and receiver at one end.

BER Analyzer Functions

A BER analyzer can do much more than simple testing of the error rate, which involves basic comparisons of bits. There are many versions and variations of BER tests that are useful in data communications. A typical BER analyzer unit is shown in Fig. 13-9. This instrument is designed for use with telephone standard multiplexed T1 data transmission lines, which are used at 1.544 Mbits/s. The types of tests, though, are typical of any high-performance system checkout procedure. These tests use various patterns and include the following:

Generating various bit patterns, including all 1s, all 0s, alternating 1s and 0s, and certain ratios of 1s to 0s.

Generating different frames which are compatible with the T1 frame format. (A *frame* is a group of bits in the protocol, with their own header, data, checksum, and ending sequence.)

Using *nonreturn to zero* (NRZ) and *bipolar encoding* (in which a binary 1 is one voltage and a binary zero is the same magnitude, but opposite sign.)

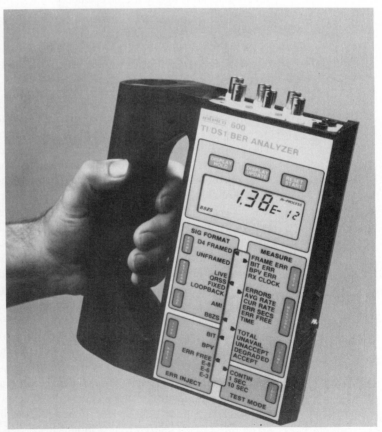

Fig. 13-9 A typical BER analyzer can perform many data communications tests and provide analysis of the test results in easy-to-use format (photo courtesy of Intelco Corp.).

Using these patterns, the instrument makes measurements and provides analysis in the following areas:

Bit errors, as an average over many thousands of bits
Percentage of seconds that are absolutely error-free
Percentage of frames that have errors, compared to error-free frames
Loss of signal, framing synchronization, or character synchronization

The meaning of these error measures, and the implications, are shown by an example (Fig. 13-10a). Suppose data is being sent at the rate of 1000 bits/s, in 100-bit frames. In the first second, there is one error in each frame. In the next second, all frames are error-free. The BER over the 2 s is 10 errors/2000 bits, or 1 error/200 bits. The percentage of error-free seconds is 50 percent. The percentage of frames that have no errors is 10 out of 20, also 50 percent.

Next, use the same total number of errors (10) but spread them differently: 5 errors in the first frame of second number 1, and 5 errors in the first frame of second number 2 (Fig. 13-10b). The BER remains 1 error/200 bits. The percentage of error-free seconds is 0 percent, and the percentage of frames with errors is 2 out of 20, or 10 percent.

The reason for different measures of BER is this: if the BER and the frame error rate are high, and the percentage of error-free seconds is low, then the errors are scattered randomly throughout the data transmission and may be due to

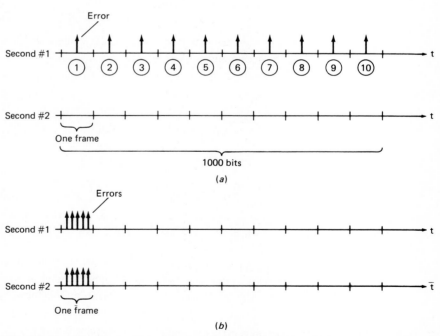

Fig. 13-10 There are various measures of error performance. In (a) errors are scattered throughout the frames of the first second. In (b) the same number of errors are concentrated within the first frame of each second.

Test Techniques and Instrumentation

many types of noise or circuit malfunctions. If, however, the BER is high but the frame error rate and the percentage of error-free seconds are low, then the errors are most likely due to some impulse type of noise that occurs infrequently or to problems in synchronizing the system. This distinction helps the technician understand possible causes of problems. Typical BER values range from 1 bit in 10^5 up to 1 bit in 10^{12} (for very good system performance).

The unit shown also has the capability of taking the known repeating pattern and deliberately putting known amounts of errors into the data bit sequence. It can inject these errors at the rate of 1 error per 10^3, 10^6, or 10^8 bits. This procedure is used to verify the operation of the unit and to determine whether the data communications system can cope with errors by using its error-detection or error-detection and -correction circuitry. If errors are included but not detected, the system may be operating, except for the error-related functions. It is important to test out that capability as well, since the system may be working error-free today but suffer from errors another time. All the tests and patterns can be run continuously or repeated at the rate of 1 test/s or 10 tests/s.

QUESTIONS FOR SECTION 13-4

1. What is a pattern generator? Why is it needed?
2. What are some of the various modes of pattern generator operation?
3. What is the FOX message? Why is it used?
4. Why are repetitive patterns needed?
5. Where are the places that a pattern generator can be connected in operation?
6. How is the pattern generator used to measure BER? What is needed at the receiving end?
7. What kinds of digital patterns does a typical pattern generator provide?
8. What are some of the measures of error that the BER instrument provides? Why are they needed? What do they imply?
9. Why are errors deliberately put into some of the patterns?

PROBLEMS FOR SECTION 13-4

1. A communications system sends data at the rate of 10,000 bits/s. The data is divided into frames of 200 bits each. Single-bit errors occur over a 4-s period at the following time intervals: 0.11, 0.12, 0.23, 1.11, and 2.05 s. What is the bit error rate? The percentage of error free-seconds? The percentage of error-free frames?
2. The same system is used as in Prob. 1. Two errors occur in the first frame, and then the next 25 frames each has a burst of 50 errors. What are the BER, the percentage of error-free seconds, and the percentage of error-free frames?
3. A system is sending 1000 bits/s, with frames of 100 bits. A total of 100 frames is sent. Compare the three measures of error for the case in which each frame has exactly one error, to the case in which the same number of errors are all in the first frame.

4. Repeat Prob. 3 for the situation in which each frame has two errors, versus the case in which the errors are evenly spread out in the frames of the first second of data only and all subsequent frames are error-free.

13-5 Protocol Analyzers

A pattern generator and BER analyzer can help test bit patterns and characters on a communications link. In some cases, the patterns are made to resemble standard protocol frames and the BER analyzer can begin to help check whether the header, data bits, checksum, and similar parts of the message are received correctly. However, as communications networks become more complicated, the protocols involve handshake messages, formats, and signal timings and relationships that must be maintained exactly. In order to help diagnose and test these types of systems, a much more advanced type of instrument is required.

The *protocol analyzer* is one such instrument. It is a data line monitor, like the one discussed previously, but with many features added that make it especially useful for working with systems that use known protocols. A data communications system which is using the SDLC protocol, for example, uses the analyzer to receive messages and break the frame of received bits into the pieces that make it an SDLC message. These parts—the message number, the data itself, the checksum, the ending tag—are displayed separately on the screen of the analyzer. The technician or engineer can then examine each part separately, to determine whether the overall SDLC format is being used properly. The message numbers would allow the flow of communications to be followed, as well. A protocol such as SDLC has a certain required back-and-forth pattern of acknowledgments and handshake messages, and the analyzer shows these on its screen. Finally, the analyzer is aware of the timing requirements of the protocol or can be told, via a keyboard, what the time periods and rates for this particular installation are. If a return message should appear within 0.1 s and no response is seen within the specified period, the analyzer signals the test operator.

A protocol analyzer also has a large amount of built-in memory capable of storing many hundreds or thousands of data bits. This allows the analyzer both to capture messages as they occur and to hold many messages of the recent past. The reason for this is that often understanding what is occurring in a system requires looking at what happened before the apparent problem. Many times, the observed problem is really caused by a message or hidden problem that occurred several messages previously. The operator can look back into the past, from the point of the observed problem, to see what events preceded it. For example, suppose one user node in the system is having trouble transmitting and receiving. The problem may be in that node, or it may be that another user is sending messages to the wrong address and inadvertently jamming other users with messages. The analyzer can capture all the messages that have occurred over several seconds and then present them to the operator. By looking at these, the technician or engineer can see that some messages are misaddressed and confusing other users of the system.

LAN Protocol Analyzer

The increase in use of local area networks (LANs) has been a problem for most test situations. These LANs have many users, each sending messages at almost any time. It is very difficult to obtain a clear and faithful picture of what is happening on the LAN without the appropriate equipment, designed for the LAN. One such type of instrument is the 4971S LAN Protocol Analyzer from Hewlett Packard Company (Fig. 13-11). This instrument is designed for use with IEEE 802.3 LANs, such as Ethernet. It can monitor and capture message frames in a 1-Mbyte buffer and actually be set in advance to indicate to the test operator when certain LAN events occur. These events are anything that the test operator wants to see and study in detail and are used to trigger further data analysis of the LAN activities.

The 4971S is a very powerful and sophisticated instrument, with many features. Many of these are really necessary for proper analysis of a LAN that may be having problems or one that has to have its performance analyzed and measured. Among the functions the instrument can perform are the following:

Be set up as a monitor on the LAN, invisible to all other nodes.

Be configured as an actual node and user of the LAN.

Transmit messages onto the network used to test other users.

Store every message that occurs on the LAN, along with the exact time, with a time stamp resolution of 32 μs. This is used to establish timings and sequences and to measure network speed and response time.

Act as a source of messages on the network, to allow measurements of how well the network performs when the message traffic is heavy. The 4971S can load the LAN to 90 percent of its theoretical capacity, which will show how other user nodes are affected as the LAN becomes busier.

Fig. 13-11 A protocol analyzer for local area networks has many features designed specifically for the complex operation and protocol of an LAN. The HP 4971S can act as a passive monitor for Ethernet, or as an active node in the network (courtesy of Hewlett-Packard Co.).

Generate messages that have deliberate errors or are incomplete, to observe whether if the network handles these properly without hanging up or crashing.

Store only those messages that occur between operator-defined messages, or *markers*. A typical application is to store only messages directed to user 7, or messages from user 2, or messages transmitted between certain time periods.

Vary the timing and response time to messages and requests, to see how the other users react to messages that take too long to come back, even though they are otherwise correct.

Equipment like the Hewlett Packard 4971S is relatively expensive and represents the highest performance in a LAN protocol analyzer. It is almost a necessity for detailed study of a LAN such as the Ethernet, although there are some types of tests that can be performed with simpler, less expensive equipment. Protocol analyzers which have fewer capabilities are still very useful for many situations and are about one-half the price of the 4971S unit. They may not have all the features needed for detailed analysis or finding subtle, stubborn problems, however.

QUESTIONS FOR SECTION 13-5

1. Why is a protocol analyzer needed? What weaknesses of the pattern generator and BER unit does it overcome?
2. What does a protocol analyzer do?
3. How does a protocol analyzer help analyze timing relationships?
4. Why does a protocol analyzer need memory? What is this memory used for?
5. Why is the apparent observed problem only a starting point for analysis? How does the analyzer help in these problems?
6. What features does a LAN protocol analyzer provide? Why are they needed?
7. What does the LAN protocol analyzer do in the monitor-only mode?
8. What can the protocol analyzer do on the LAN when it is used as an active node?

13-6 Transmission Impairment Measurement Sets

The instrumentation discussed so far is designed for analyzing and testing the digital data performance of the communications system. It uses the binary data as a starting point, to check for bit errors, handshaking, and protocol performance. Even a digital system, though, exists in the real world of analog channels. A normal channel is an analog link, whether designed primarily for analog signals or even optimized for digital ones. It is still an analog medium from the physics and basic electronics perspectives. Even a digital data network makes use of channels which are inherently analog, subject to all the characteristics of analog systems and circuits, although the signals sent on the system are digital only.

There are many characteristics of the transmission channel that affect its performance for digital signals in the real world. In order to test and maintain a communications system properly, the test technician must often go one level below the line monitors and similar digitally oriented test equipment and get down to the basics of the ways signals are affected by the transmission link. The instruments that do this, called *transmission impairment measurement sets* (TIMS), allow for full measurement and characterization of the fundamental characteristics of the link. Many TIMS are battery-powered because they are often used in remote locations, away from a regular power source. Also, battery power means that the set can be hooked up without any danger of shock to the operator, especially if there are unknown grounding problems in the wires or connections being checked.

The function of the TIMS is to provide the basic signal readings which help predict the performance of the channel with any signal, analog or digital. There are also tests which are used to determine how well, and at what rates, digital signals will perform on the link. A TIMS unit is used, for example, when everything at both ends of the system seems to be performing to specification by itself, but the system nevertheless cannot achieve the desired BER. Then, a more detailed study of the channel is needed. There may be some aspect of the channel—noise, distortion, bandwidth—that prevents the system from reaching its performance goals.

Many TIMS also have the necessary interface circuitry for connecting into the phone system (two and four wires) and looking like a telephone, modem, or similar telecommunications equipment. These TIMS are used by phone companies or anyone who uses telephone lines as part of an overall data communications system. The TIMS electrically looks like the telephone instrument and also provides the same kind of timing and sequencing that a phone unit has (that is, receive ringing tone, go off hook, send actual signals, go back on hook).

TIMS Tests

The types of test run by the TIMS fall into three general categories of basic, intermediate, and advanced.

Basic Tests   Basic tests include measuring the noise in the channel and signal-to-noise ratio and determining the bandwidth. They may also require looking at the type of noise—is it relatively constant with time, or are there impulses of noise, such as those caused by machinery nearby?

The Intermediate Tests   Intermediate tests involve looking in more detail at the variations in amplitude versus frequency. These cause distortion in any signals sent through the channel, so that the shape of the received signal differs considerably from its shape at the transmitter. There is also *envelope delay,* which means that some frequencies are delayed in time by different amounts than others. (Recall that a delay in time is the same as a phase shift, so this is also called *phase distortion.*) Finally, there can be *intermodulation distortion,* caused by nonlinear characteristics of the channel. The result of intermodulation distortion is essentially similar to the amplitude modulation (AM) process—two frequencies enter the system, and they modulate each other so that their sum and difference frequencies come out as well. The intermodulation process can also produce other frequencies, depending on the exact nature of the nonlinearities.

The intermediate tests are *steady-state* tests that can be performed with steady tones and signals, with the measurements made slowly and carefully. These are important to define the performance, but not always sufficient for digital systems.

Advanced Tests Advanced tests make measurements that are especially critical for data. These include measuring the *phase jitter* (sudden, small phase shifts that are random, often caused by noise) and the *amplitude jitter* (sudden, random small variations in the amplitude received even with a constant amplitude signal transmitted). Both the maximum and average jitter values are needed. These jitters can affect the shape and steadiness of the received digital signals. A sudden drop in amplitude caused by noise, interference with the radio path, and circuit problems, may cause the received signal to be lost momentarily. This signal loss is called *dropout*. It causes an error in the received data and may also result in loss of synchronization. Sync loss is a problem because all subsequent bits are lost, until synchronization can be reestablished.

The reasons for measuring the characteristics of the channel with TIMS are the following:

To understand better what kinds of data performance can be achieved, in terms of bit rate and error rate.

To determine where some type of equalization may help. If the channel has low noise, but poor amplitude versus frequency and therefore severe distortion, the signal at the transmitting end can be modified. As an alternative, the receiver may use special filters and signal shaping circuitry to compensate for this distortion before it tries to synchronize; look at the bits for headers, addresses, and data; or make the decision whether a 1 or 0 was sent by the transmitting end. The automatic equalization that some advanced modems offer is usually preceded in the modem by some specific TIMS-type tests, to determine what kind of equalization will give the greatest benefit in maintaining a low BER at the bit rate to be used.

QUESTIONS FOR SECTION 13-6

1. Why is the analog nature of the channel critical to performance, even in a digital system?
2. What does a TIMS do? Why is it often battery-powered?
3. What does a TIMS do if a phone connection is needed as part of the test?
4. What are the basic tests that are run?
5. What are the intermediate level tests that are made with the TIMS? What are some of the distortions that they check?
6. What are the advanced tests? Why are they used especially for digital communications?
7. What is the impact of jitter? What is the effect of dropouts?
8. How are the achievable bit rate and BER related to the measurements that are made with a TIMS unit?

13-7 Time Domain Reflectometry

A common problem in connection is a break, short circuit, or other damage to the wire or cable carrying the data signal. Imagine trying to find the short circuit in miles of buried cable or undersea cable. Digging or pulling it up is very time-consuming and costly and may cause further damage. On the other hand, a simple problem in a cable can cause the whole communications system to be useless. There are even instances in which it is cheaper and quicker to run a new cable than try to find the problem with the old one.

An elegant way to find the exact location of cable problems is *time domain reflectometry* (TDR). The concept of TDR is very old and was known in the days of the telegraph. TDR is a perfect example of a simple idea that is very useful, but could not be actually implemented until modern high-performance circuitry became available. Today's TDR instruments can find the location of a cable problem to within less than an inch over a range of a few hundred feet, or several feet in the range up to 25,000 to 50,000 ft. This can be done by working entirely from the end of the cable.

The principle of TDR is based on a simple fact. Whenever there is any type of discontinuity in a cable, such as a short circuit to another wire or an open break, this discontinuity causes some of the electromagnetic energy propagating through the cable to reflect back toward the source of the energy. Instead of the cable acting as a near-perfect path for the electromagnetic energy, the cable short circuit or break acts as a big "bump in the road." Some of the energy does not get past the bump, but instead is returned (Fig. 13-12). In TDR, a short pulse is applied to one end of the cable. This pulse propagates along the cable at the characteristic speed for that type of cable (typically, 50 to 80 percent of the speed of light). When it reaches the short circuit or break, most of the energy continues, but some comes back to the source, also at the characteristic speed. If there were a way to measure precisely the round-trip time between sending the pulse and seeing the reflection, and the propagation speed were known exactly, then the distance to the fault could be figured out simply from

Distance = speed × (time/2)

Note: The factor 2 is used because the measured time is for a round-trip, and the one-way distance is desired.

Although this is very simple in principle, there are several technical issues which make TDR difficult:

The pulse sent through the cable must be of very short time duration. Otherwise the pulse is still "on" when the reflection echo returns, and the two signals interfere. A good average value for propagation is 1 foot/nanosecond (1 ft/ns), so a fault 1000 ft away returns a reflection in 2000 ns = 2 ms.

The reflection is usually very small, about 1 to 5 percent of the initial energy wave that is sent. Therefore, the receiver must be very sensitive. It also reinforces the problem of the receiver being overloaded by the large initial transmitted pulse rather than the small reflection.

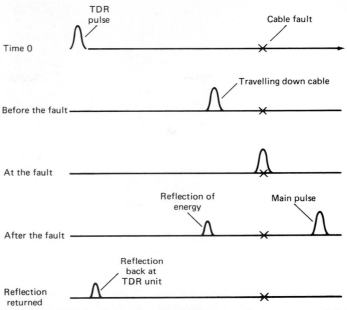

Fig. 13-12 The principle of time domain reflectometry. Some amount of electromagnetic energy is reflected back to the source whenever there is a discontinuity (open or short) in the wire. The time between the original pulse and the reflection is used to calculate the distance to the fault discontinuity. The figure shows a series of "snapshots" of the pulse signal energy on the cable, from the time the pulse is generated to the time the reflection returns.

Since the energy that is reflected is small, the original signal should be as large as possible. This requires circuits that can generate fast pulses of large value.

The rise time of the pulse must be very short, and the pulse must be very sharp. Since the purpose of TDR is to locate a fault precisely, a sharp leading edge results in a sharper reflection. If the leading edge has a longer rise time, then the reflection is less sharp, and there is more ambiguity in the received reflected signal time.

The TDR system must be able to measure time itself with high accuracy.

All of the preceding factors mean that design of the TDR instrument requires a large value, fast rise time, short time duration pulse with a very sensitive receiver that will not be damaged while the initial pulse is transmitted, and precise time measurement.

TDR Instrument

Actual operation of a TDR instrument is relatively simple. The unit looks something like an oscilloscope (Fig. 13-13). The TDR unit sends the pulse down the wire or cable, and the reflection echo is displayed on the screen. The time for the reflection is the time of the horizontal axis of the screen, from the left edge where the initial waveform was sent to the point where the reflection appears. All

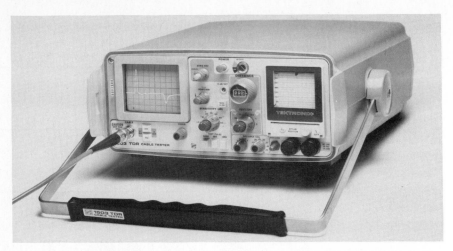

Fig. 13-13 A typical TDR instrument looks like an oscilloscope and is relatively easy to use. The CRT screen helps in interpreting the nature of the fault or faults on the line (photo courtesy of Tektronix, Inc.)

the user has to do is perform a simple calculation using the propagation speed of the particular type of wire in use (which must be supplied to high accuracy from the wire manufacturer or cable company).

Some TDR units do not have a screen and contain circuitry which can electronically detect when the reflection arrives. These units then show the distance to the fault on a digital display, based on the propagation speed number keyed in by the user. For many applications, though, the screen of the unit is necessary. Unfortunately, the reflection is never as sharp and well defined as the original pulse. The bandwidth, inductance, capacitance, and resistance of the wire cause the reflection to look like a distorted version of the first pulse (Fig. 13-14). Judgment has to be applied to determine where the major part of the reflection energy is located. Also, many cable faults have more than one problem. A small break or partial short circuit may be located just in front of or behind a larger one. The TDR screen then shows a series of reflections, with the larger reflection showing a larger fault. The operator studies this and decides the best place to dig. Many TDR instruments have the capability to provide a permanent paper record of what is on the screen, for this reason.

A few sample calculations show the way the TDR process works and pulse width affects operation. Suppose the cable being checked is 1000 ft long and has a propagation factor of 50 percent of the speed of light. A fault 650 ft away returns a reflection in 2.647 μs. If the initial pulse is longer than this, the fault reflection is obscured. For the same fault, but with a cable propagation at 75 percent the speed of light, the reflection time is 1.764 μs. A shorter pulse must be used in this second situation.

There is a conflict in setting the ideal pulse width. A wider pulse has more initial energy and thus more reflected energy and so is easier to see on the screen. It is, however, usable only at longer distances. It also has poorer resolution, since the pulse energy is spread out by the cable impedance, and the exact time

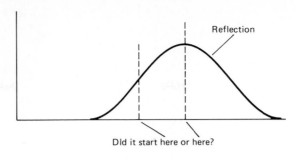

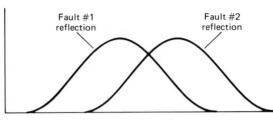

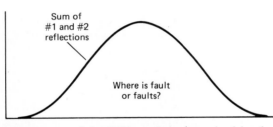

Fig. 13-14 The accuracy of the TDR result is determined by the sharpness of the reflected pulse. The screen may need some interpretation, especially when two line faults are close together and the reflections overlap.

point of the reflection is more difficult to define. To overcome these conflicting characteristics, many TDR instruments allow the operator to look at the screen and set the pulse width to select the best match versus the distance to the fault, the propagation of the cable, and the amount of energy reflected. In addition, not all applications need resolution to several feet over thousands of feet of cable. If the cable is on poles, it may be necessary to know only between which poles the fault lies; if the failure is in an open field it may not matter whether a 1-ft section or a 10-ft section must be dug up by the backhoe.

QUESTIONS FOR SECTION 13-7

1. Why is there a need for locating cable faults?
2. What is the concept of TDR? How does it work?
3. What must be known or measured for a TDR system to show where the cable fault lies?
4. Why must the initial pulse be of short duration?

5. How small is the typical reflection?
6. What happens if a large energy pulse is present when the small echo is expected?
7. Why must the rise time of the pulse be short?
8. Why must the time be measured with high accuracy?
9. Why is TDR simple in theory, but difficult to implement without modern circuitry and electronics?
10. What are the uses of the screen and operator judgment in a TDR system? Why is it difficult to automate the process completely?
11. Why is a pulse width adjustment often used? What are the technical conflicts involved in pulse width?

PROBLEMS FOR SECTION 13-7

1. A TDR system is being used to locate faults in a buried cable, with a propagation value of 68 percent the speed of light. The fault is located 1000 ft away. How long does the reflection take to return to the source?
2. A TDR test is locating a fault in an undersea cable. The cable propagates the applied pulse at 50 percent of the speed of light. The initial pulse is 10 μs long. What is the closest fault that can be found?
3. In the case of Prob. 2, the fault is located at a distance of 2000 m. How much time elapses between the end of the initial pulse and the beginning of the reflection?
4. Two faults are located 50 m apart in a cable with a propagation value of 80 percent the speed of light. What is the time difference in their reflection?
5. Explain what happens if the pulse width used in Prob. 4 is longer than the time between the reflections from the two faults. Give a numerical example to explain.

13-8 Testing Fiber-Optic Systems

The goal of a communications system that uses fiber-optic links is the same as for a system which uses wire or radio—to provide high-quality signals at the necessary rate. In many ways the optical fiber is the ideal link medium. It is immune to electrical interference, has very wide bandwidth, provides electrical isolation, and cannot be "tapped" easily. Testing a fiber-optic-based system requires special instrumentation if the link itself must be checked. This is because fiber optics exists in a world of light energy, glass or plastic fibers, and special connectors and alignment, as compared to the voltages, currents, wire, and more common connectors and measurements of traditional links.

The most common tests run on fiber-optic links are continuity, cable loss and spectral response, optical time domain reflectometry, and bit error rate. The equipment needed to perform these tests includes a calibrated light source, a light power meter, an optical attenuator, and optical transmitters/receivers. Each of these instruments is analogous to an electronic instrument, but is designed for use with optical signals and their special requirements.

The Calibrated Light Source The calibrated light source is the equivalent of a signal generator. It must generate light energy signals of known power levels. These light signals come out through an LED, laser LED, or other electrical-to-light energy device. The other requirement may be the need to vary the wavelength (frequency) of the light wave, just as a signal generator allows the user to vary the frequency. The reason for this is that optical fibers perform best at specific light wavelengths only (depending on the exact type of glass or plastic used) and the light supplied to the optical fiber must have the matching wavelength for that fiber type.

The Light Power Meter The light power meter measures the power of any light signal, which typically ranges from 1 nanowatt (nW) to 2 milliwatts (mW). This is similar to measuring the voltage or power of an electrical signal. The power meter must also be calibrated carefully, since its readings form the basis for determining the amount of light energy that is present. The power meter uses a precise light-to-electrical-energy transducer and then measures its electrical output.

The Optical Attenuator The optical attenuator is similar to a simple potentiometer or circuit used to reduce a signal level. The attenuator is used whenever performance tests must be run, for example, to see how the bit error rate is affected by varying the signal level in the link. The optical equivalent is much more complex than the one used in electronic circuitry. It can be implemented in two ways. One way is to have a precise mechanical setup in which the optical signal passes through a glass plate with differing amounts of darkness and then back to the optical fiber, as shown in Fig. 13-15. The glass plate has gray density ranging from 0 percent at one end to 100 percent at the other end. As the plate is moved across the gap, more or less light energy is allowed to pass. This type of

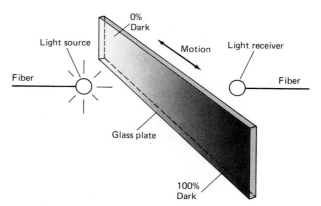

Fig. 13-15 An attenuator for fiber-optic signals can be built from a light source and receiver. In between these is a special glass plate with darkness from 0 to 100 percent, which can reduce the optical signal by any amount.

attenuator is very precise, and can handle any light wavelength (since the plate attenuates any light energy by the same amount, regardless of wavelength), but it is mechanically expensive and not rugged. The second approach is to convert the light energy into an electrical signal, attenuate it electrically, and then convert it back to a light signal (Fig. 13-16). This is less expensive and more rugged than the glass plate method, but it is more wavelength specific, since the light-to-electrical-energy and electrical-energy-to-light transducers do not work equally well at all light wavelengths.

Optical Transmitters/Receivers Optical transmitters/receivers are basically energy converters, such as are used to convert an electrical signal to light energy or vice versa. One side of the transmitter (or receiver) is an electrical connector, and the other side is a fiber-optic connector. They are used whenever some electronic instrument, such as a pattern generator or BER unit, must be connected to a fiber-optic link. They also include the necessary circuitry to convert the signal levels and modulate the light beam, if necessary. A typical transmitter may be designed to work with 0- to 1-V signals. It takes the signal and uses it with circuitry that is designed for properly driving the light source, which may require certain specific current or voltage levels. This level conversion is transparent to the user, so that many types of regular test equipment can be used with fiber optics.

Fiber-Optic versus Electronic Testing

Two key differences between fiber-optic test equipment and purely electronic test equipment are the type of connector and the light wavelengths used. Connecting an electronic test unit to wires or cables of the link is usually fairly simple, if the appropriate connector is used: the connector is attached and mated. In fiber optics, in contrast, the connectors must provide precise mechanical alignment. If a fiber-optic cable must be cut so a signal can be placed on the cable, special tools must be used to cut the cable, polish the end, and hold it properly while it is placed into the connector. (Other schemes are now being designed which may simplify this process.) If fiber-optic cable without connectors is used, special mechanical fittings must be employed so that the light energy

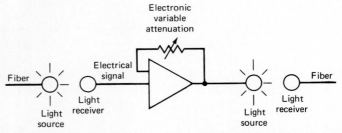

Fig. 13-16 An optical attenuator can also be built using electronic attenuation, in which the signal strength is reduced by a circuit rather than a glass plate. This is less expensive and more rugged, but cannot handle the wide wavelength range of the mechanical attenuator.

is coupled properly into and out of the cable. There is no equivalent to temporarily wrapping some bare wire around the contact or terminal post for electrical signals. Similarly, the frequency of the transmitter and receiver must also be matched to the cable in use, or else a large portion of the light energy may not pass through the cable. The transmitter/receiver must be chosen for a specific type of cable, unless a more expensive set that can provide an adjustable wavelength is used.

Once the instruments for fiber-optic testing are ready, the actual tests can be performed. The simplest is the *continuity test,* to find out whether there is any break in the fiber. It requires only a light source and optical power meter. Some light energy is coupled into the cable, and some should come out the other end. The test is a good/bad test, which doesn't indicate anything about the cable loss or spectral response. The light source and power meter do not have to be calibrated or accurate for this simple test.

(Sometimes, a simple flashlight is used as the light source, and the light coming out can be seen by eye. This is not a recommended practice because in a multifiber bundle, the fiber the technician looks into may not be the one with the flashlight at the other end. Instead, it may be an active fiber that has a laser LED as its source, and eye damage can result. The flashlight and eye should only be used for a single strand of fiber, when the entire length of fiber is visible, and there can be no mistakes. This is typical of a fiber-optic link used between a printer and computer over a short distance, where both ends of this short link can be disconnected.)

The tests for cable loss and spectral response are related, though often only the cable loss test is run. To check *cable loss,* a measured amount of light energy at a specific wavelength is coupled into the cable, and the amount of output light power is measured. The cable loss is specified in decibels and is typically anywhere from 1 to 20 dB for a kilometer of cable. Since the amount of energy coupled into the cable must be known, the light source must be calibrated, or the power meter must first be used to measure the output of the source directly and then the output through the cable using the same source. The light source and optical power meter are sometimes packaged together in a single instrument called an *optical loss set.* The optical loss set has some internal circuitry that allows it to measure automatically the optical power at both points simultaneously, so the resulting cable loss can be displayed in decibels immediately.

The *spectral response test* is an enhancement of the loss test. It requires a light source that can be varied in wavelength through some range, typically from 1 to 1.5 micrometers (μm) (1000 to 1500 nanometers [nm]). The measured output shows the attenuation of energy at each of these wavelengths and results in a graph such as Fig. 13-17. This measurement is important because the light source in use in the data communications system may have drifted from its original wavelength as a result of time, temperature, or voltage supply changes. The performance of the system may be much poorer if the cable in use has maximum transmission (lowest loss) at one wavelength, such as 1.4 μm, and the light source has changed to 1.3 μm. The result can be a 25-dB attenuation, with the data signal carried by the optical cable probably received too weakly to produce an error-free system.

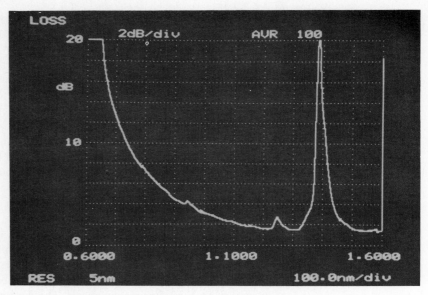

Fig. 13-17 The spectral response of the optical fiber is a critical item. The photo shows the typical loss in a fiber versus a range of optical wavelengths. Note the approximately 20-dB variation across the range and the very high attenuation that occurs at one specific wavelength (photo courtesy of Intelco Corp.).

Optical TDR

Just as TDR is a powerful tool for finding problems with wire cables, *optical TDR* (OTDR) is very useful for finding breaks and poor connections in fiber-optic cables. The concept is the same: any discontinuity causes some of the incident energy to be reflected back toward the source. In OTDR, the source signal is a very sharp light pulse, and a light detector senses the reflection. OTDR is especially useful because of possible problems with the connection used in fiber optics. A typical wire cable and connector have almost no loss and are electrically exactly like the wire itself, with perhaps 0.01 to 0.05 dB of signal lost. A fiber-optic connection, in comparison, may have a 0.5- to 2-dB loss, depending on how well the connection has been made and whether it has been strained mechanically. In a fiber-optic system, these connection losses may accumulate and result in a very weak signal at the receiving end, even if the cable itself is perfect and has little loss. The OTDR display shows the reflections that occur from connections, splices, and cable breaks (Fig. 13-18). The magnitude of the reflection shows the extent of loss. A complete break reflects the maximum amount of energy, but some is reflected from connections as well.

The bandwidth of a fiber-optic cable is defined by the same rule as for a traditional cable. The light wave coupled to the cable is on/off-modulated at some bit rate. The amount of received energy is compared at a high modulation rate to the value at a very low rate. As the modulation frequency increases, the amplitude of the received signal decreases. The bandwidth is the frequency at which the received signal energy is one half (-3 dB) the low frequency value

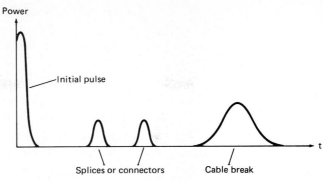

Fig. 13-18 Optical TDR is a powerful technique for checking the condition of optical fiber splices, which have a small loss. In contrast, wire connections rarely have any significant loss.

and typically has a value of several hundred megahertz. The bandwidth test requires that the light source be modulated from the low frequency to the several hundred megahertz value and that the received power be measured at the same time.

Of course, the actual data communications system user is most interested in measuring the performance of the system, not merely characterizing the fiber-optic cable. The *BER test* is very useful here. The same type of equipment is used as is needed for checking nonfiber systems, but the output of the pattern generator must go through the optical transmitter, and the received light energy must be converted at the optical receiver back to an electrical signal for the BER analysis. The BER test is often performed with an optical attenuator for two reasons. The amount of light energy from the test source must not be so great that it overloads the transducer at the receiver, as this can cause malfunction and false results. The attenuator is used to reduce the received energy to the proper range for the receiver in use.

Second, the attenuator is used to characterize the system sensitivity to signal loss. This determination is necessary because the light source and light receiver characteristics may change with time, temperature, and voltage. A system which is working very well at installation time may degrade over months or years of operation. The system designer has to measure the amount of degradation that can be tolerated and the increase in BER as the system components vary. The optical attenuator is used to reduce the signal level (which is in effect a summary of the many effects of change and degradations) and measure the BER that can be achieved at various signal levels. This information is then used to plan the amounts of component change that will be allowed while providing acceptable BER performance to the system users.

QUESTIONS FOR SECTION 13-8

1. Why does a fiber-optic link system require special test equipment and techniques?
2. What is the role of a calibrated light source? Why is a variable wavelength often needed as well?

3. How is the received energy measured? What is the typical range?
4. What is an optical attenuator? How is it implemented mechanically? What is the electronic implementation? What are the advantages and disadvantages of each?
5. What are optical transmitters and receivers? Where are they used?
6. Why is the physical connector critical in fiber optics? Compare this to electronic connectors for wire and cable.
7. Why is the choice of wavelength critical?
8. What is the purpose of a continuity test? Why should it never be performed with a flashlight and eye, except under certain specific circumstances?
9. How is cable loss checked? Why must the output power of the light source be known?
10. How is spectral response checked? What is the typical wavelength range?
11. Why is spectral response important?
12. Why is OTDR especially important for optical fiber systems?
13. How is bandwidth measured? How is the optical signal modulated? What are some typical bandwidth values for optical fiber?
14. How is the fiber-optic BER test performed?
15. What are the two common reasons for using an attenuator when testing BER?

SUMMARY

The voltmeter and oscilloscope of electronics testing and diagnosis are used in the preliminary stages of checking out a data communications system. They are used mainly to verify the power supply voltage values and look at the power signals to determine whether noise is present. The oscilloscope is also used to examine the data signals and find out whether their rise time and general shape are in line with what is expected. The type of noise present—general wideband noise or impulse noise—is also seen on the scope.

In addition, the testing of a communications system uses specialized instrumentation. The simplest instrument is the breakout box, which takes all wires of an interface out to test points and allows each signal wire to be observed with a probe and disconnected from the other side to the interface. Jumpers can then be used to rearrange the interface connections as needed. After the breakout box is used to establish basic communications, a line monitor is employed to observe passively the bits and characters on the data lines. The monitor shows the flow of data. A monitor designed for data communications testing also has the ability to show the full-duplex flow of data, as well as the normally nonprintable control characters. Some monitors can also show the timing relationship between various signal and handshake lines.

Loopbacks allow data to be sent and received at the same end of the system. These loopbacks are installed at various points of the system to diagnose different parts of the end-to-end path. The loopback can be made with a special cable and some internal switches or by external remote command.

Pattern generators are useful wherever a known or repetitive pattern of data bits must be put into the system. The patterns may be all 1s, all 0s, or some mix. The repetition allows an observer to recognize where the data is and what happens to it as it is modified by formatting, encoding, and modulation. The pattern generator can be used as part of a bit error rate test set, which compares the known transmitted data pattern to the received pattern. The difference between these two patterns is the errors. There are several legitimate and useful ways to measure the bit error rate, to indicate the nature and clustering of the errors.

A protocol analyzer is an advanced tool which can actually monitor communications system data flow and help determine the meaning of the bit patterns. Depending on the protocol, the analyzer can separate the addresses, headers, checksums, and other parts of the overall bit group. The analyzer is a valuable tool when advanced protocols, with many rules and timing relationships, must be studied. It can also switch from a passive observer to an active node in a network and put user-specified data patterns and errors into the system, to check the way the system reacts.

Even in the digital world of data communications, the analog physical characteristics of the channel are critical to the performance that can be achieved. Transmission impairment measurement sets can measure the line noise, bandwidth, frequency response, distortions, and amplitude and phase jitters. All of these can contribute to errors and reduced performance. Some of these can be partially corrected by system compensation circuitry which reverses some of their effects.

The site of a buried cable fault can be located with high accuracy, working from one end only, with time domain reflectometry. The TDR technique is simple in principle, but demands short, sharp pulses and precise time measurement. Today's test equipment can fulfill these requirements and allow faults to be located without random digging.

The use of fiber-optic cables for communications links requires test equipment that is different from that used for purely electronic systems. Optical power sources, optical power meters, and optical attenuators are the essential parts of the test kit. These allow measurement of continuity, loss in the optical fiber, spectral response for various wavelengths, and other characteristics. All optical measurements must be made with special concern for the fiber-optic connection, which can itself cause significant optical power loss. The wavelength of the optical source is also critical, since optical fibers are best suited to transmitting only specific wavelengths, on the basis of the glass or plastic of which they are made. If the wrong wavelength is used, most of the optical energy does not reach the other end of the cable.

END-OF-CHAPTER QUESTIONS

1. What is the role of the voltmeter and oscilloscope in testing data communications systems? Why is their role relatively limited?
2. Why are noise spikes a serious problem for system performance?
3. How is the eye pattern used to observe signal quality and the effect of adjustments?
4. What does the breakout box do to help make the relatively simple interconnection of two devices easier?
5. How are the open/close switches of the breakout box useful? What about bringing signals to test/jumper points?
6. What is the role of the line monitor? How can it help beyond what the breakout box can accomplish?
7. How does a properly designed line monitor show duplex data flow? How does it show special control characters?
8. Why is a line monitor which can show signal timing especially useful?
9. What are some of the characteristics a good line monitor cannot show?
10. Where are loopbacks usually implemented? What does a loopback allow the user to test?
11. Compare cable, internally switched, and externally controlled loopbacks in terms of technical advantages and disadvantages.
12. What is the role of the pattern generator? What can it do?
13. How is a pattern generator used to examine signal quality?
14. How can the bit error rate of a system be determined? What must be known?
15. What are three common measures of BER? What does each one indicate?
16. What are the functions of a protocol analyzer which is acting as a passive monitor? What can it show the test technician or engineer? What types of events can it be set to capture and display?
17. What can the protocol analyzer do when it becomes an active unit or node? How is it used to further measure system performance?
18. Why are the analog characteristics of the link critical even for an all-digital system?
19. What types of parameters does a test set for analog characteristics measure? What are the three groupings of tests?
20. How does a TIMS help with telephone-connection-related testing?
21. Why are distortion and jitter critical to determining the potential bit rate and BER of a system? What are these distortions and jitters?
22. What is TDR? How does it work? What must be measured to locate the cable fault? What must be known in advance?
23. Why are the width and shape of the TDR pulse critical? What are some of the technical tradeoffs involved?
24. How does the operator actually use a TDR unit?

25. Explain why a fully automated TDR system is not always feasible.
26. Why is testing fiber-optic links different from testing wire?
27. What instruments are needed for fiber-optic testing?
28. What types of tests are run on fiber-optic links? What do these tests show?
29. What is the critical issue regarding connectors and splices in a fiber-optic system?
30. Why is the selected wavelength used for both testing and communication critical?
31. What are typical values for optical power? For fiber-optic bandwidth? For cable attenuation at optimum wavelength?
32. Why is an attenuator needed when performing BER tests? What does it help determine?

END-OF-CHAPTER PROBLEMS

1. A BER test is being performed on a system which sends data at 20,000 bits/s, divided into frames of 200 bits each. Errors on 1 bit occur in each of the first 25 frames of a 2-s transmission. What is the BER? What are the results for the percentage of error-free seconds and percentage of error-free frames?
2. Repeat Prob. 1 for the case in which all the errors are 4 bits long instead of 1 bit.
3. Repeat Prob. 1 for the case in which all errors are 1 bit only per frame, but occur in each of the first 100 frames.
4. A BER test is performed on an ASCII character-based link between a computer and nearby printer. All characters are represented by one start bit, seven character bits, one parity bit, and one stop bit. The data rate is 1200 baud. A single-bit error occurs in exactly half the characters sent. What is the BER? The percentage of frames without error? (Consider the character to be a frame.)
5. Repeat Prob. 4 for the case in which a 2-bit error occurs in every fourth character. (This shows why the BER may be less meaningful to the user than the percentage of frames without error, since any error [1 or more bits] within the frame invalidates the entire frame.)
6. TDR is being considered for locating a cable fault. The TDR unit under consideration puts out a 0.1-ms pulse. The fault is believed to be about 10,000 ft away, and the cable has a propagation factor of 75 percent the speed of light. Can this TDR unit be used?
7. A TDR instrument is looking for a cable fault that is 10,000 m away. What is the time for the reflection to arrive, if the propagation speed is 85 percent the speed of light? Would a 10-μs pulse work? Would a 100-μs pulse work?
8. Two faults exist in a cable that is propagating signals at 65 percent the speed of light. These faults are 100 ft apart. What is the time difference for the reflections?

Appendix A

CHART OF FREQUENCY ALLOCATIONS IN THE RADIO SPECTRUM

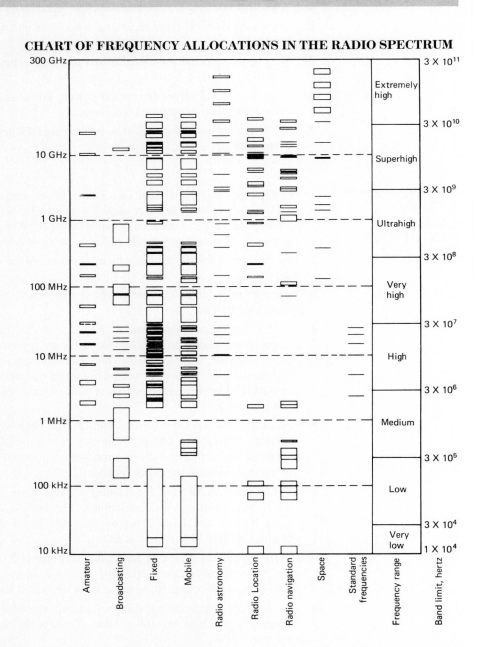

Appendix B

THE ASCII CHARACTER SET (in logical groupings)

b_7 →				0	0	0	0	1	1	1	1
b_6 →				0	0	1	1	0	0	1	1
b_5 →				0	1	0	1	0	1	0	1
Bits b_4	b_3	b_2	b_1								
0	0	0	0	NUL	DLE	SP	0	@	P	\	p
0	0	0	1	SOH	DC1	!	1	A	Q	a	q
0	0	1	0	STX	DC2	"	2	B	R	b	r
0	0	1	1	ETX	DC3	#	3	C	S	c	s
0	1	0	0	EOT	DC4	$	4	D	T	d	t
0	1	0	1	ENQ	NAK	%	5	E	U	e	u
0	1	1	0	ACK	SYN	&	6	F	V	f	v
0	1	1	1	BEL	ETB	/	7	G	W	g	w
1	0	0	0	BS	CAN	(	8	H	X	h	x
1	0	0	1	HT	EM	)	9	I	Y	i	y
1	0	1	0	LF	SUB	*	:	J	Z	j	z
1	0	1	1	VT	ESC	+	;	K	[	k	{
1	1	0	0	FF	FS	,	<	L	\	l	|
1	1	0	1	CR	GS	—	=	M	]	m	}
1	1	1	0	SO	RS	.	>	N		n	~
1	1	1	1	SI	US	/	?	O	_	o	DEL

Control — Numbers, Symbols — Uppercase — lowercase

Character Representation and Code Identification

The standard 7-bit character representation, with b_7 the high-order bit (MSB) and b_1 the low-order bit (LSB), is shown below.

Example:

The bit representation for the character K.

b_7	b_6	b_5	b_4	b_3	b_2	b_1
1	0	0	1	0	1	1

Control Characters

Mnemonic and Meaning		Mnemonic and Meaning	
NUL	Null	DLE	Data line escape (CC)
SOH	Start of heading (CC)	DC1	Device control 1
STX	Start of text (CC)	DC2	Device control 2
ETX	End of text (CC)	DC3	Device control 3
EOT	End of transmission (CC)	DC4	Device control 4
ENQ	Enquiry (CC)	NAK	Negative acknowledge (CC)
ACK	Acknowledge (CC)	SYN	Synchronous idle (CC)
BEL	Bell	ETB	End of transmission block (CC)
BS	Backspace (FE)	CAN	Cancel
HT	Horizontal tabulation (FE)	EM	End of medium
LF	Line feed (FE)	SUB	Substitute
VT	Vertical tabulation (FE)	ESC	Escape
FF	Form feed (FE)	FS	File separator (IS)
CR	Carriage return (FE)	GS	Group separator (IS)
SO	Shift out	RS	Record separator (IS)
SI	Shift in	US	Unit separator (IS)
		DEL	Delete

TRANSMISSION CONTROL CHARACTERS

SYN	Synchronous idle	Used by synchronous transmission systems to provide message and character framing and synchronization
SOH	Start of header	Used at the beginning of a sequence of characters to indicate address or routing information. Such a term is referred to as a *heading*. A STX character terminates a heading
STX	Start of text	Used at the beginning of a sequence of characters that is to be treated as an entity to reach the ultimate destination
ETX	End of text	Used to terminate a sequence of characters started with STX
ETB	End of transmission block	Used to indicate the end of a sequence of characters started with SOH or STX
EOT	End of transmission	Used to indicate a termination of transmission. A transmission may include one or more records and their associated headings
ACK	Acknowledge	A character sent by the receiving station to the transmitting station to indicate unsuccessful reception of a message
NAK	Negative acknowledge	A character sent by the receiving station to the transmitting station to indicate unsuccessful reception of a message
ENQ	Enquiry	A character used to request a response from a remote station. The response that a remote station generates is predefined.

An alternative to showing the ASCII symbols by logical groupings is to list them in numerical order, from 0000000 to 1111111.

ASCII Character	Meaning	(MSB) B7	B6	ASCII Bit Pattern B5	B4	B3	B2	(LSB) B1
NUL	Null	0	0	0	0	0	0	0
SOH	Start of heading	0	0	0	0	0	0	1
STX	Start of text	0	0	0	0	0	1	0
ETX	End of text	0	0	0	0	0	1	1
EOT	End of transmission	0	0	0	0	1	0	0
ENQ	Enquiry	0	0	0	0	1	0	1
ACK	Acknowledge	0	0	0	0	1	1	0
BEL	Bell	0	0	0	0	1	1	1
BS	Backspace	0	0	0	1	0	0	0
HT	Horizontal tabulation	0	0	0	1	0	0	1
NL	New line	0	0	0	1	0	1	0
VT	Vertical tabulation	0	0	0	1	0	1	1
FF	Form feed	0	0	0	1	1	0	0
RT	Return	0	0	0	1	1	0	1
SO	Shift out	0	0	0	1	1	1	0
SI	Shift in	0	0	0	1	1	1	1
DLE	Data line escape	0	0	1	0	0	0	0
DC1	Device control 1	0	0	1	0	0	0	1
DC2	Device control 2	0	0	1	0	0	1	0
DC3	Device control 3	0	0	1	0	0	1	1
DC4	Device control 4	0	0	1	0	1	0	0
NAK	Negative acknowledgment	0	0	1	0	1	0	1
SYN	Synchronous idle	0	0	1	0	1	1	0
ETB	End of transmission block	0	0	1	0	1	1	1
CAN	Cancel	0	0	1	1	0	0	0
EM	End of medium	0	0	1	1	0	0	1
SUB	Substitute	0	0	1	1	0	1	0
ESC	Escape	0	0	1	1	0	1	1
FS	File separator	0	0	1	1	1	0	0
GS	Group separator	0	0	1	1	1	0	1
RS	Record separator	0	0	1	1	1	1	0
US	Unit separator	0	0	1	1	1	1	1
SP	Space	0	1	0	0	0	0	0
!	Exclamation point	0	1	0	0	0	0	1
"	Quotation mark	0	1	0	0	0	1	0
#	Number sign	0	1	0	0	0	1	1
$	Dollar sign	0	1	0	0	1	0	0
%	Percentage sign	0	1	0	0	1	0	1
&	Ampersand	0	1	0	0	1	1	0
'	Apostrophe	0	1	0	0	1	1	1
(	Opening parenthesis	0	1	0	1	0	0	0
)	Closing parenthesis	0	1	0	1	0	0	1
*	Asterisk	0	1	0	1	0	1	0
+	Plus	0	1	0	1	0	1	1
,	Comma	0	1	0	1	1	0	0
−	Hyphen (minus)	0	1	0	1	1	0	1
.	Period (decimal)	0	1	0	1	1	1	0

ASCII Character	Meaning	(MSB) B7	B6	B5	B4	B3	B2	(LSB) B1
/	Slant	0	1	0	1	1	1	1
0	Zero	0	1	1	0	0	0	0
1	One	0	1	1	0	0	0	1
2	Two	0	1	1	0	0	1	0
3	Three	0	1	1	0	0	1	1
4	Four	0	1	1	0	1	0	0
5	Five	0	1	1	0	1	0	1
6	Six	0	1	1	0	1	1	0
7	Seven	0	1	1	0	1	1	1
8	Eight	0	1	1	1	0	0	0
9	Nine	0	1	1	1	0	0	1
:	Colon	0	1	1	1	0	1	0
;	Semicolon	0	1	1	1	0	1	1
<	Less than	0	1	1	1	1	0	0
=	Equals	0	1	1	1	1	0	1
>	Greater than	0	1	1	1	1	1	0
?	Question mark	0	1	1	1	1	1	1
@	Commercial "at"	1	0	0	0	0	0	0
A	Uppercase A	1	0	0	0	0	0	1
B	Uppercase B	1	0	0	0	0	1	0
C	Uppercase C	1	0	0	0	0	1	1
D	Uppercase D	1	0	0	0	1	0	0
E	Uppercase E	1	0	0	0	1	0	1
F	Uppercase F	1	0	0	0	1	1	0
G	Uppercase G	1	0	0	0	1	1	1
H	Uppercase H	1	0	0	1	0	0	0
I	Uppercase I	1	0	0	1	0	0	1
J	Uppercase J	1	0	0	1	0	1	0
K	Uppercase K	1	0	0	1	0	1	1
L	Uppercase L	1	0	0	1	1	0	0
M	Uppercase M	1	0	0	1	1	0	1
N	Uppercase N	1	0	0	1	1	1	0
O	Uppercase O	1	0	0	1	1	1	1
P	Uppercase P	1	0	1	0	0	0	0
Q	Uppercase Q	1	0	1	0	0	0	1
R	Uppercase R	1	0	1	0	0	1	0
S	Uppercase S	1	0	1	0	0	1	1
T	Uppercase T	1	0	1	0	1	0	0
U	Uppercase U	1	0	1	0	1	0	1
V	Uppercase V	1	0	1	0	1	1	0
W	Uppercase W	1	0	1	0	1	1	1
X	Uppercase X	1	0	1	1	0	0	0
Y	Uppercase Y	1	0	1	1	0	0	1
Z	Uppercase Z	1	0	1	1	0	1	0
[	Opening bracket	1	0	1	1	0	1	1
\	Reverse slant	1	0	1	1	1	0	0
]	Closing bracket	1	0	1	1	1	0	1
^	Circumflex	1	0	1	1	1	1	0
_	Underscore	1	0	1	1	1	1	1
`	Grave accent	1	1	0	0	0	0	0

ASCII Character	Meaning	(MSB) B7	B6	B5	B4	B3	B2	(LSB) B1
a	Lowercase a	1	1	0	0	0	0	1
b	Lowercase b	1	1	0	0	0	1	0
c	Lowercase c	1	1	0	0	0	1	1
d	Lowercase d	1	1	0	0	1	0	0
e	Lowercase e	1	1	0	0	1	0	1
f	Lowercase f	1	1	0	0	1	1	0
g	Lowercase g	1	1	0	0	1	1	1
h	Lowercase h	1	1	0	1	0	0	0
i	Lowercase i	1	1	0	1	0	0	1
j	Lowercase j	1	1	0	1	0	1	0
k	Lowercase k	1	1	0	1	0	1	1
l	Lowercase l	1	1	0	1	1	0	0
m	Lowercase m	1	1	0	1	1	0	1
n	Lowercase n	1	1	0	1	1	1	0
o	Lowercase o	1	1	0	1	1	1	1
p	Lowercase p	1	1	1	0	0	0	0
q	Lowercase q	1	1	1	0	0	0	1
r	Lowercase r	1	1	1	0	0	1	0
s	Lowercase s	1	1	1	0	0	1	1
t	Lowercase t	1	1	1	0	1	0	0
u	Lowercase u	1	1	1	0	1	0	1
v	Lowercase v	1	1	1	0	1	1	0
w	Lowercase w	1	1	1	0	1	1	1
x	Lowercase x	1	1	1	1	0	0	0
y	Lowercase y	1	1	1	1	0	0	1
z	Lowercase z	1	1	1	1	0	1	0
{	Opening brace	1	1	1	1	0	1	1
|	Vertical line	1	1	1	1	1	0	0
}	Closing brace	1	1	1	1	1	0	1
~	Tilde	1	1	1	1	1	1	0
DEL	Delete	1	1	1	1	1	1	1

Appendix C

EBCDIC CHARACTER SET (in numerical order)

Character	Bit Pattern	Character	Bit Pattern	Character	Bit Pattern	Character	Bit Pattern
NUL	0000 0000	SP	0100 0000		1000 0000	[	1100 0000
SOH	0000 0001		0100 0001	a	1000 0001	A	1100 0001
STX	0000 0010		0100 0010	b	1000 0010	B	1100 0010
ETX	0000 0011		0100 0011	c	1000 0011	C	1100 0011
PF	0000 0100		0100 0100	d	1000 0100	D	1100 0100
HT	0000 0101		0100 0101	e	1000 0101	E	1100 0101
LC	0000 0110		0100 0110	f	1000 0110	F	1100 0110
DEL	0000 0111		0100 0111	g	1000 0111	G	1100 0111
	0000 1000		0100 1000	h	1000 1000	H	1100 1000
RLF	0000 1001		0100 1001	i	1000 1001	I	1100 1001
SMM	0000 1010	¢	0100 1010		1000 1010		1100 1010
VT	0000 1011	.	0100 1011		1000 1011		1100 1011
FF	0000 1100	<	0100 1100		1000 1100	⌐	1100 1100
CR	0000 1101	(	0100 1101		1000 1101		1100 1101
SO	0000 1110	+	0100 1110		1000 1110	⌡	1100 1110
SI	0000 1111	|	0100 1111		1000 1111		1100 1111
DLE	0001 0000	&	0101 0000		1001 0000	]	1101 0000
DC1	0001 0001		0101 0001	j	1001 0001	J	1101 0001
DC2	0001 0010		0101 0010	k	1001 0010	K	1101 0010
DC3	0001 0011		0101 0011	l	1001 0011	L	1101 0011
RES	0001 0100		0101 0100	m	1001 0100	M	1101 0100
NL	0001 0101		0101 0101	n	1001 0101	N	1101 0101
BS	0001 0110		0101 0110	o	1001 0110	O	1101 0110
IDL	0001 0111	⌐	0101 0111	p	1001 0111	P	1101 0111
CAN	0001 1000		0101 1000	q	1001 1000	Q	1101 1000
EM	0001 1001		0101 1001	r	1001 1001	R	1101 1001
CC	0001 1010	!	0101 1010		1001 1010		1101 1010
CU1	0001 1011	$	0101 1011		1001 1011		1101 1011
IFS	0001 1100	*	0101 1100		1001 1100		1101 1100
IGS	0001 1101	)	0101 1101		1001 1101		1101 1101
IRS	0001 1110	;	0101 1110		1001 1110		1101 1110
IUS	0001 1111	¬	0101 1111		1001 1111		1101 1111
DS	0010 0000	—	0110 0000		1010 0000	\	1110 0000
SOS	0010 0001	/	0110 0001	~	1010 0001		1110 0001
FS	0010 0010		0110 0010	s	1010 0010	S	1110 0010
	0010 0011		0110 0011	t	1010 0011	T	1110 0011
BYP	0010 0100		0110 0100	u	1010 0100	U	1110 0100
LF	0010 0101		0110 0101	v	1010 0101	V	1110 0101
EOB/ETB	0010 0110		0110 0110	w	1010 0110	W	1110 0110

Character	Bit Pattern	Character	Bit Pattern	Character	Bit Pattern	Character	Bit Pattern
PRE/ESC	0010 0111		0110 0111	x	1010 0111	X	1110 0111
	0010 1000		0110 1000	y	1010 1000	Y	1110 1000
	0010 1001		0110 1001	z	1010 1001	Z	1110 1001
SM	0010 1010	:	0110 1010		1010 1010		1110 1010
CU2	0010 1011	.	0110 1011		1010 1011		1110 1011
	0010 1100	%	0110 1100		1010 1100		1110 1100
ENQ	0010 1101	—	0110 1101		1010 1101		1110 1101
ACK	0010 1110	>	0110 1110		1010 1110		1110 1110
BEL	0010 1111	?	0110 1111		1010 1111		1110 1111
	0011 0000		0111 0000		1011 0000	0	1111 0000
	0011 0001		0111 0001		1011 0001	1	1111 0001
SYN	0011 0010		0111 0010		1011 0010	2	1111 0010
	0011 0011		0111 0011		1011 0011	3	1111 0011
PN	0011 0100		0111 0100		1011 0100	4	1111 0100
RS	0011 0101		0111 0101		1011 0101	5	1111 0101
UC	0011 0110		0111 0110		1011 0110	6	1111 0110
EOT	0011 0111		0111 0111		1011 0111	7	1111 0111
	0011 1000		0111 1000		1011 1000	8	1111 1000
	0011 1001		0111 1001		1011 1001	9	1111 1001
	0011 1010	:	0111 1010		1011 1010		1111 1010
CU3	0011 1011	#	0111 1011		1011 1011		1111 1011
DC4	0011 1100	@	0111 1100		1011 1100		1111 1100
NAK	0011 1101	'	0111 1101		1011 1101		1111 1101
	0011 1110	=	0111 1110		1011 1110		1111 1110
SUB	0011 1111	"	0111 1111		1011 1111		1111 1111

Glossary

absorption   the loss of energy as a signal travels through a medium due to the energy's being absorbed by medium or impurities in the medium (water vapor in air, for example)

ACK   an acknowledgment handshake message, which indicates "message received OK"

acoustic coupling   coupling of the tones between a modem and a telephone handset microphone and speaker by using sound energy, rather than electrical signals into the circuit of the telephone loop

active current loop   a current loop user which provides the necessary current (usually 20 mA) for the loop

adaptive equalization   a scheme by which the equalization (correction) of the performance characteristics of a channel is continually changed to conform to changes in the channel characteristics

adjacent channel   the channel whose frequency and spectrum lie directly above or below the channel of interest

aliasing   the misleading digitized values that result from sampling an analog signal at a rate less than twice the Nyquist rate

amplitude modulation   modulation whereby the information signal varies the amplitude of the carrier

analog   a signal which can take on any value within the overall allowable range

analog to digital (A/D)   the process by which an analog signal is converted to a digital one

answer   a modem which uses the tone pairs assigned to the answer (as opposed to originate) mode

antenna   a device which takes electromagnetic energy from a circuit or wire and radiates it into space instead of confining it

AppleTalk   a network standard for personal computers and peripherals developed by Apple Computer, Inc.

application level   the highest layer, layer 7, of the OSI network structure, which allows the person sending or receiving the message to interact and interface with the system

ASCII   a standard pattern for encoding alphanumeric characters into 7- (or 8-) bit patterns; acronym for *American Standard Code for Information Interchange*

asynchronous   communication in which the time period between parts of the message may be of unequal or unknown length, and which uses start and stop signals to identify the beginning and end of the bit stream

attenuation   any loss in signal strength, due to resistance, absorption, capacitance, or any characteristic of the medium or design of the system

bandwidth   the span, in hertz, that the information-bearing signal occupies or requires, or the difference, in hertz, between the lowest and highest frequencies of a band

base band   a modulated signal which has a carrier of 0 Hz

baud   a measurement unit which defines the number of bits per second (or signaling units per second when several bits represent a signaling unit)

binary   a digital system in which only two distinct signal values are allowed, often called *0* and *1* or *low* and *high*

bit   a binary digit (0 or 1)

bit error rate   the rate, as percentage or as bits per million, showing the number of bits in error compared to the data bits really sent

bit field   a cluster of bits of fixed size, reserved for specific message types

bit rate   the rate at which bits are transmitted

BORSHT   the basic functions of a telephone loop interface—battery, overvoltage protection, ring trip, signaling, supervision, hybrid, and test

breakout box   communications test equipment which brings out all the signal lines of a cable and connector so they can be monitored and reconfigured

buffer   a group of memory locations that serve to store data bits temporarily until the system processor can accept and handle them

burst rate   the highest momentary "sprint" rate of bits per second that a channel or system can support

bus topology   a network topology in which all the user nodes share a common pathway called the *bus* and are connected to this bus; similar in concept to a power line into which many line cords can be plugged anywhere along the power line

c   the speed of light in a vacuum, approximately 3×10^8 m/s

capacity   the maximum rate at which a channel can handle data over a sustained period

carrier   a continuous frequency signal which can be modulated or impressed with data

CCITT   acronym for an international consultative committee that sets standards for communications systems

cellular network   a network in which the overall area to be covered is divided into small cells, each with its own transmitter/receiver, and in which each cell coordinates its activities with those of adjacent cells to allow more users within an available bandwidth

central office   the local telephone exchange that handles all phone lines that have the same three-digit prefix and so handles phone lines -0000 through -9999

channel   the path for the communications signal; includes both the electronics equipment at each end and the medium through which the energy passes

character   a group of bits which represent a symbol such as a letter or number

checksum   a group of bits used to check whether the data bits have any errors

clear to send (CTS)   a handshake line indicating that it is OK to send data

clock   a repetitive signal which paces and synchronizes the communications of data bits

coaxial cable   a type of wire in which the inner conductor is surrounded by an outer conductor which acts as an electrical shield

code   a predefined grouping of bits which corresponds to the desired symbols, such as letters, numbers, and control characters

collision   the situation that exists when two users try to send a signal over the same medium at the same time, using the same frequencies and time periods

command/response   a protocol in which a slave unit only responds when it is given a command to speak by the master and never initiates a message on its own

common mode voltage   a voltage that exists between the high-signal wire and ground and equally between the low-signal wire and ground

compliance voltage   the amount of voltage that results from having current in a current loop interface flow through the various voltage drops in the loop

concentrator   a multiplexing unit which combines many users into fewer links and assigns users to unused links as links that were in use become available

contention   the situation in which two transmitters begin to send messages at the same time and collide

control characters   special characters which cause nonprintable operations, such as line feed, carriage return and bell ring

CRC code cyclic redundancy checksum code, based on mathematical theories, allows a small number of checking bits to show whether the overall stream of bits has any errors in it

crosstalk noise and interference caused by electromagnetic signals radiating from one wire into an adjacent wire

CSMA/CD carrier sense multiple access/collision detect, a network protocol that allows collisions among various transmitters and resolves the conflict so each transmitter does get a turn, in time

current loop an interface which uses the presence or absence of current (instead of voltage) to represent 1 and 0

daisy chain an interconnection topology in which any new user node can simply connect to the last user node

data communications communications intended for transmitting and receiving information in the form of digital signals, or data

data communications equipment (DCE) a communications unit designed as the interface between data terminal equipment, such as a computer or terminal, and the data link

data encryption standard (DES) a standardized formula for encrypting data that is designed to be nearly impossible to break even with modern computers; it takes groups of bits and encodes them with a user-supplied key

data link layer (level) level 2 of the seven-layer OSI network model, responsible for synchronization and error control of the link

data set ready (DSR) a handshake line indicating that the data set is ready to accept new data bits

data terminal equipment (DTE) a communications end user, such as a terminal or computer

data terminal ready (DTR) a handshake line indicating that the data terminal is ready to accept new data bits

decibel a unit of relative signal value, which relates a measured signal to some reference signal using a logarithmic scale

dedicated line a reserved, conditioned line between the user terminal/modem and the telephone central office

demodulation the process of extracting the information-bearing modulation from the modulated signal

dial-up line a standard, unreserved, unconditioned telephone line, for example, when an ordinary, dialed phone is connected to the telephone central office

differential a signal existing as the difference potential between two wires, not with respect to ground

digital information which exists with only a specific group of allowed values, in contrast to analog, in which any value can exist within the range

direct connect modem a modem which directly wires into the telephone system, instead of acoustically through the telephone handset

dish antenna a large, parabolic antenna shaped like a dish, which focuses the transmitted and received energy

distributed system a system which distributes small computers around the entire facility, rather than having one larger processor handle all tasks

drift any undesired change in time or temperature of critical parameters such as frequency, voltage, or signal value

DTMF dual tone multifrequency, which uses pairs of sine wave tones to represent numbers being dialed by the phone system

duplex a communications system which is capable of communicating in both directions; can be half duplex or full duplex

EBCDIC   extended binary code for data interchange, a form of binary code for letters, numbers, symbols, and control characters, originated by IBM

echo   a reflection, back to the transmitter, of some of the transmitted signal energy

echo cancellation   a circuit which cancels echo by taking some of the original signal, inverting it, and then adding it to the signal plus echo to produce only the signal

echo suppressor   a circuit which reduces the effect of echoes by disconnecting the reverse direction channel through which the echo returns

efficiency   the number of data bits per second that the channel actually handles as compared to its maximum capacity, taking into account overhead such as start and stop bits, preambles, and time delays

electromagnetic waves   the electrical and magnetic energy used in transmitting information and data in most communications systems

encoding   the process of taking data bits or characters and putting them into some other bit pattern more suitable for the application

encryption   encoding using a secret scheme or key so that only a desired receiver can make sense of the encoded bits

envelope   the outer shape or border of an amplitude-modulated carrier signal, with the carrier amplitude shaped by the information-bearing signal

envelope delay   a time delay (or phase shift) that results when the channel delays various frequenciy components in the signal by various amounts of time

equalization   using circuitry or filters into a channel or correct for frequency distortion or envelope delay

error correction   a communications protocol and circuitry which allows errors to be corrected by the receiver once the receiver determines that an error exists

error detection   a communications protocol and circuitry which allows the receiver to determine that the received data bits contain one or more errors

Ethernet   a network protocol for CSMA/CD

exchange   the local phone office which brings in local loops for connection to any other loop or to a trunk which connects to other phone exchanges

extremely low frequency (ELF)   the frequency band from 3 to 10 kHz

eye pattern   an oscilloscope pattern that results from overlaying many data bits on one another; shows the characteristics of the actual bit signals—noise, jitter, amplitude variations, difference between high and low—and can be used to adjust the circuitry to improve the bit patterns

facsimile   a method of sending printed information by scanning across the page and converting the written lines to digital signals, with a 1 for line and a 0 for blank on the page

fiber optics   the use of thin glass or plastic fibers, light sources, and light detectors as a medium for the transmission of data signals in the form of light waves, instead of using current or voltage on copper wire or radio waves

flat cable   a multiconductor cable with many wires parallel to each other, very easy to connect in a single operation to a single connector which brings out all the wires to pins

format   the specific way that the bits of a message are organized and grouped together, with headers, preambles, postambles, and checking fields

forward error correction   a system in which the transmitted data bits contain extra bits which allow the receiver to detect and correct errors by using this bit stream alone, without going back to receiver for retransmission

Fourier analysis   the mathematical theory which shows how signals expressed as a function of time (the time domain) can be equivalently and equally expressed as functions of combinations of frequencies (the frequency domain)

frame a collection of data bits preceded by a preamble and followed by a postamble, which form a complete block of bits which can be transmitted and understood at the receiver

framing error an error which occurs when the received frame does not fit into the expected pattern of a normal frame and therefore cannot be interpreted

frequency the number of complete cycles per second of a signal which is varying in time repetitively

frequency division multiplexing a scheme which combines several signals so they can share a common medium without interference by shifting each signal to different carrier frequency

frequency domain expressing a signal as a function of sine waves of differing frequencies and amplitudes, as opposed to expressing it as a function of time

frequency modulation (FM) modulation in which the information-bearing signal causes the frequency of the carrier to vary as a function of the information signal

full duplex a communications system that allows simultaneous communications in both directions

fundamental the lowest frequency of a signal which contains many frequencies and multiples of this lowest frequency (harmonics)

gateway a network node that serves as the interface between two different networks

general-purpose interface bus (GPIB) see **IEEE-488**

generator polynomial a mathematical function used to develop a remainder for CRC checksums which can be implemented with simple circuitry

gigahertz (GHz) a frequency of 1000 MHz = 1,000,000,000 Hz

go back n an error correction protocol in which the transmitter automatically goes back n blocks and begins retransmission from that point, when it is told by the receiver that data bits have been received in error

guard band unused bandwidth on each side of the band in use, to provide a border between adjacent assigned frequencies

half duplex a communications system that allows communication in both directions, but only one direction at a time

Hamming codes a series of codes for defining bits to add after the original data bits, which then allow the receiver to detect and correct any of certain types of errors in the received data bits

handshaking a protocol plan in which the sender and receiver signal to each other when there is new data to be sent, and whether and when the new data can be accepted at the receiver

harmonic an integer multiple of the fundamental

header a preamble of a block of data bits, which may contain information such as message number, length, or type, as well as intended receiver

hertz unit of frequency, 1 cycle/s

high frequency (HF) the range of frequencies from 3 to 30 MHz

hybrid transformer a transformer used in communications circuits to convert a two-wire loop to a four-wire signal, with one pair of wires for each direction of signal

IEEE-488 a network standard used primarily for interconnection of test equipment, which uses 24 parallel wires for high speed but is limited to shorter distances

impulse noise electrical noise with sudden, sharp bursts, such as from a motor starting up

information anything new, that was not previously known by the receiver

integrated services digital network (ISDN) a standard for an all-digital telephone network, in which the local loops, central office, and trunks are designed for digital signals only

International Standards Organization (ISO) an international organization which sets standards for use in data communications, especially among the more complex aspects such as networks

interrupt-driven a protocol in which any user node can signal to the master that it has a message and wants to interrupt and have its turn

intersymbol interference the effect of distortion and bandwidth limitations which cause clean digital signals to ''smear'' in time and begin to overlap with the next signal, and so cause confusion and errors in determining what digital value is really sent

isolation the ability to allow information to pass from one subsystem to another without the necessity of having a direct electrical connection

jamming deliberately transmitting an interfering signal on the frequency of another signal, so the intended receiver cannot get the desired signal clearly

jitter undesired, small variations in signal amplitude or timing caused by noise, circuit instability, or component imperfection

key a group of bits used as the starting point in encrypting a message or decoding a message sent with that same key

kilohertz (kHz) 1000 Hz

leased line a dedicated, specially conditioned telephone line used for communications

limiter a circuit used in FM demodulation which evens out the amplitude variations, which are of no interest and contain no information

line driver an IC designed to take a low-level signal and boost it to levels and types suitable for driving long wires

line monitor an instrument which allows the flow of data on a communications link to be monitored without causing any disruptions

line receiver an IC designed to receive signals that have been modified to make them more suitable for long wires and convert them back to the voltages which are normally used with a circuit; the complement of line driver

link any physical path between two subsystems or stages in a communications system

local area network a network designed to link a group of users in a small physical area, up to approximately 3 to 16 km, who share some common application and need to pass data to each other

loop the pair of wires from the local phone central office or communications center to the individual telephone unit

loopback a communications test and diagnosis tool which causes the signal as received to be turned back to the sender, at various stages in the system, so the sender can determine whether the message as received agrees with what was sent

loss a decline in signal strength due to resistance, capacitance, energy absorption, or any other factor between the sending point and the receiving point

low frequency (LF) the frequency band from 30 to 300 kHz

lower sideband (LSB) the difference frequency (carrier modulation) that results from the amplitude-modulation process

Manchester encoding an encoding scheme which ensures that every bit, whether a 1 or a 0, contains one transition between 1 and 0 and so can be synchronized by the receiver circuit

manufacturing automation protocol (MAP) a network standard using token passing that is designed for use in factories for linking automated machinery

mark the signal value in serial communications which represents the idle state or binary 1

media/medium the physical path for the energy of the communications signal; may be air, vacuum, wire, optical fiber

megahertz (MHz) 1,000,000 Hz

metallic modem a modem which is intended to be used where there is a direct wire path, without any blocking elements, between the two ends of the link

modem a modulator/demodulator, which transforms the 1 and 0 binary voltages of the user system into signals compatible with the medium

modulation the act of impressing a data- or information-bearing signal on a carrier signal for transmission over the medium

multidrop a communications link which allows multiple receivers and transmitters to share the same pathway

multiplexing combining, through modulation, various narrower-bandwidth signals into a single larger-bandwidth signal to make use of the bandwidth of the medium

multiplier a circuit which takes two input signals and produces an output equal to the product of these two inputs at any instant of time

NAK a negative acknowledgment signal from the receiver to the transmitter, indicating that data is not received in time or is received with error

network a communications system which interconnects many users who have something in common, either with respect to the type of data being transferred or to the geographic areas that the users cover

network layer (level) a level of a network design, which takes data bits from one level, performs some specific operation on these data bits, and then passes them on to the next level; or level 3 of the OSI network model, responsible for routing messages through the various communications systems resources and layers

node a connection point for users into the network

noise any undesired signal that corrupts the original signal; may be generated within the circuitry or by outside sources

nonreturn to zero (NRZ) an encoding scheme in which one voltage represents a binary 1 and another represents a binary 0 for the entire bit period

null a character which marks a place but has no meaning at all, often used to fill in idle time or character positions in a protocol

null modem a cable which interchanges wire positions, so that a DTE device looks like a DCE, or vice versa

Nyquist rate the rate at which a signal must be sampled so that the original analog signal can be perfectly reconstructed from the sample values alone; the rate is equal to twice the bandwidth of the signal

off-hook the condition when a phone is in use and the phone loop is a completed circuit

on-hook the condition when a phone is hung up and not in use, so the phone loop is an open circuit

one-time key a key for encrypting data which consists of random characters, with only the sender and receiver having the key, which leads to an unbreakable code since knowledge of any part of the message or key gives no information about other characters

optical attenuator a test instrument which reduces the strength of the light energy in a fiber-optic system by precise amounts

optical fiber a hair-thin glass or plastic fiber which acts as a light pipe and lets light energy be used (instead of current or voltage in wires) for communications; capable of very high bandwidths, no interference, and no susceptibility to noise

optical power meter a test instrument used to measure the strength of the optical energy in a fiber-optic system

optical time domain reflectometry a technique for locating faults or poor connections in a fiber-optic system, using the time for a pulse to go from a source to the fault and return

optoisolation a technique for providing electrical isolation between parts of a system while passing data and energy, by using a light-emitting diode and phototransistor to convert the electrical signal to light and then back to electrical energy

originate the role of a modem which is assigned as the starter of the full-duplex data flow and is assigned the originate tones

OSI open system interconnect, a framework or model for the seven layers of a complete network structure

overrun error an error that occurs when a new character arrives at the receiver port before the previous character is removed

packet switching a communications protocol used over large distances, in which messages are grouped into packets of fixed size, and each packet is sent by the best way available at that instant; at the receiving end the packets are disassembled and sent to the correct users

parallel link a link which uses many paths, in parallel, to send data bits and so is much faster than using a single pathway

parity an error-detection scheme which uses an extra checking bit, called the *parity bit*, to allow the receiver to determine whether there has been an error in the received data bits

parity error an error indicator that is signaled when the actual parity of the received data does not agree with the parity value indicated by the parity bit

phase the relative point in the sine wave cycle with respect to some reference waveform

phase-locked loop a circuit which has the ability to lock and synchronize a receiver's oscillator phase and frequency onto the phase and frequency of an incoming signal, used for demodulation of FM signals and for extracting a clock signal from incoming data

phase modulation (PM) modulation in which the information signal varies the phase of the carrier

phase shift a shift in the phase of a signal from its original value, usually resulting from PM with a digital signal

phototransistor a transistor which is sensitive to light and so can allow current flow or stop it, depending on the amount of light received; used in optoisolators

physical layer (level) level 1 of the seven-layer ISO network model, which corresponds to the physical connection to the network media

plaintext the original, unencrypted data bits that are to be encrypted or have been recovered from encrypted data and which anyone can read

postamble the fields of a format which follow the data bits and contain information such as the error-checking bits

POTS plain old telephone service, the most basic type of telephone unit and line

preamble fields which precede the data bits in a format and contain information such as the message number, message length, and intended receiver number

presentation layer (level) level 6 of the OSI network model, makes sure that the information is delivered in a form that the receiving system can understand

private automatic branch exchange (PABX)   a small-scale, automatic telephone exchange, used to connect telephone loops to each other and to trunks to the telephone company central office

private branch exchange   similar to a PABX, but run manually instead of automatically

propagation   the way an electrical signal spreads out in wire, air, or vacuum and is passed, with losses, through the medium

protocol   the set of rules for interchanging data between a sender and receiver that both must follow

pseudorandom sequence   a sequence of data bits used for encrypting that appears to be truly random but is actually generated by a circuit and can be generated repeatedly by the same or an identical circuit

pulse dialing   phone dialing, in which the loop is opened/closed as many times as the number being dialed

quantizing error   the inherent error that occurs when an analog value is converted to the nearest digital value

random sequence   a sequence of data bits that is absolutely random, so that looking at any past bits gives no clue as to what the next bit will be

receiver   the person or equipment that intercepts and makes use of the transmitter's data

reflection   some of the signal energy that is bounced back from various layers of the atmosphere or echoed back to the sender

regeneration   the act of taking a distorted or noisy digital signal and cleaning it up and restoring it to its original digital values

remainder   the value that remains after data bits have passed through a generator polynomial; can serve as the CRC checksum in error-detection schemes

request to send (RTS)   a handshake line in which the device with data asks to know whether the other end of the link can accept new data

resolution   the "fineness" of the division of an overall scale into subsections; usually used in reference to the number of bits that represent an analog signal in analog to digital conversion

return to zero (RZ)   an encoding scheme which returns every binary 1 to 0 in the middle of the bit period, to provide some synchronization for the system clock

reverse channel   a low-speed channel from the receiver to transmitter by which the receiver can send handshake and control information back to the sending end of the system

ring topology   a network interconnection layout in which each node is on a ring and connects to one other node on either side

ring trip   the indication that the phone which is being rung has been picked up and the ringing signal should be stopped

ringing signal   a special signal sent on the phone loop to activate the bell ringer in the phone

root mean square   a statistical analysis of data which involves squaring all values, taking the mean of the sum of the squares, and then the square root of the mean; prevents negative values from canceling the effects of the positive ones and is typically used for analyzing noise values

RS-232/422/423/485   standards for serial communications, which define the voltages, currents, data rates, and other factors about the signals to be used, as well as single-ended, differential, and multidrop operation

S interface the interface of the ISDN system which is closest to the user of the system

sampling theory the theory that shows that sampled values of a continuous signal can be used to reconstruct the original signal perfectly, if the sampling rate is at least equal to twice the bandwidth of the sampled signal

serial link a data link which uses a single path and sends bits one after the other, in sequence

session layer (level) level 5 of the OSI network model, responsible for managing and synchronizing conversations between different applications

Shannon's theory the theory and formula that relate the attainable bit rate in a channel to the system bandwidth, signal power, and noise power

shield a braid around a wire or metal in a chassis which keeps undesired electromagnetic energy out and desired energy from radiating out

short haul modem a modem designed to be used with dedicated wires over a short distance, rather than with telephone system wires and unknown links

sidebands the sum and difference frequencies that result from the AM process

signal-to-noise ratio a ratio of the signal power to the noise power, usually expressed in decibels, which gives an indication of how relatively well the data can be recovered despite the noise, or how many bits per second can be passed by a given bandwidth

simplex a communications system that can transmit in one direction only

single-ended a link in which the system ground also serves as the ground for the data lines

single sideband (SSB) an AM system which suppresses the carrier and one sideband, so as to make better use of available spectrum and transmitter power

slew rate the rate at which a signal makes the transition from one voltage to another, expressed as volts per second

space the signal level in serial communications which represents the nonidle state, the start of a bit stream, or binary 0

space division multiplexing (SDM) a multiplexing scheme in which each user has an individual physical link and uses the same frequencies as all other users, such as one between the standard telephone and the central office

spectrum the range of frequencies of electromagnetic energy waves

spectrum analyzer an instrument which shows the amount of energy that exists at each frequency of the spectrum

spread spectrum a technique for encrypting a transmitted signal by making its carrier frequency constantly change under the control of a PRSQ

star topology a network layout in which there is a central point to which all user nodes connect

start bit the first bit in asynchronous communications, in which the mark value goes to the space value

state transition diagram a diagram which shows all the different conditions (states) of a system, the paths between the various conditions, and the way these various paths can be taken

statistical multiplexer a multiplexer which allows many user channels to take up a smaller bandwidth than the total, by constantly assigning unused multiplex slots to users who need access and then reassigning them when a user is finished

stop and wait an error-correcting scheme in which the transmitter stops between data blocks and waits for an ACK or NAK from the receiver

stop bit the last bit(s) of an asynchronous transmission, in which the signal goes from space to mark values

subscriber loop interface circuit (SLIC) the interface circuitry at the phone office between the exchange equipment and the telephone loop, which must provide the BORSHT functions to the loop

super high frequency (SHF) the range of frequencies from 3 to 30 GHz

superposition the fact that the Fourier transform of a signal which comprises several other signals is equal to the sum of the transforms of the individual subparts

synchronous a communications system in which the data flows continuously, without starts and stops, and the receiver must synchronize to the timing of the transmitter

synchronous data-link control (SDLC) an advanced protocol for synchronous control and management of the data link

syntax the rules for the order of placement of the various parts of the message, equivalent to the location of nouns, verbs, and adjectives in a sentence

T interface an interface for ISDN which takes data from many S interfaces and multiplexes them, and interconnects to the U interface, which then goes to a trunk to the central office

T1 a multiplexed communications system for digitized signals used in telephone systems which supports 24 users with 7-bit digitization, with a 1.544-MHz overall bit rate

telecommunications the integration of computer data and long-distance communications systems

terminal a user communications device which consists of a keyboard and screen (CRT) on which a user can enter messages and see incoming messages

three-state a type of digital driver circuit which can be either at binary 0 or 1, or at a third, high-impedance state in which it effectively appears not electrically connected at all

throughput the rate at which bits or characters can be sent through a system and can be expressed as bits per second, characters per second, or messages per second

time division multiplexing (TDM) a method of multiplexing several user data signals by assigning each one a unique time slot and then sequentially going through all the time slots in order repeatedly

time domain the representation of a signal as a function of signal value versus time, the complement to the frequency domain

time domain reflectometry (TDR) a technique for locating faults in wires and cables by sending a sharp pulse down the line and measuring the time it takes for some of the pulse energy to reflect back from the fault; the time is then related to the fault distance when the propagation speed for waves in the line is known

timing diagram a diagram which shows the sequence of events and the relationship between these events on communications data and handshake lines versus time

token passing a network protocol in which each user node can transmit only when it has the token message, which is passed from one node to the next in sequence

tone dialing a technique for dialing and signaling telephone numbers, using precise tone pairs, which is faster and more computer-compatible than pulse dialing; also called *DTMF dialing*

topology the layout and interconnection of user nodes in a network

total internal reflection the situation in which a light wave in a glass or plastic fiber is completely reflected by the fiber walls, and so no light energy escapes to the outside

transmission impairment measurement set (TIMS) a test instrument which measures the characteristics and performance of the communications channel in areas of noise, bandwidth, amplitude distortion, envelope delay, and jitter

transmitter a device which sends out data and messages

transport layer (level) level 4 of the OSI network model; multiplexes groups of independent messages over a single connection and also divides data groups into smaller ones

trunk a communications link between two central offices in a telephone system, in effect connecting star networks together

twisted pair communications cable made of two wires twisted together to minimize effects of external noise

U interface the ISDN interface between the T interface and the central office trunk

ultrahigh frequency (UHF) the spectrum from 300 to 3000 MHz

universal asynchronous receiver/transmitter (UART) an IC which acts as the interface between a parallel computer bus and the serial data link and provides start/stop bits, parity, and timing to the original data bits (or removes them in the case of received messages)

upper sideband (USB) the sum frequency (carrier + modulation) that results from the amplitude-modulation process

velocity the speed at which the electromagnetic energy propagates through the media

vertical parity a second type of parity, in which a parity bit is formed that is based on the first bit of each character, and another bit is based on the second bit of each character, and so on, as opposed to horizontal parity on the bits within a single character

very high frequency (VHF) the range of frequencies from 30 to 300 MHz

very low frequency (VLF) the range of frequencies from 10 to 30 kHz

voltage-controlled oscillator (VCO) a sine wave oscillator whose output frequency can be controlled by a dc voltage, used in phase-lock loops

waveguide a metal tube which is used to direct and confine electromagnetic energy at UHF frequencies and above

wavelength the distance between successive crests of a sine wave

white noise noise which is wideband and has the same amplitude value (usually rms) at any frequency throughout the bandwidth

wide area network a network designed to transfer data over long distances, for example, among various offices and plants of a large company

wideband noise noise whose bandwidth spans a wide range of frequencies

XOFF a handshake character which indicates to the transmitter that the receiver cannot accept any more characters at this time

XON a handshake character that indicates to the transmitter that the receiver can now accept new characters

zero crossings the points in time where a signal crosses from + to −, especially significant in FM demodulation since the time between crossings indicates the frequency at that instant, which is related to the modulating signal information

Answers for Odd-Numbered Problems

SECTION 2-2
1 0. 537 s, 0.01 s.
3 In air, it travels 1500 m; in wire, 750 m. Since velocity in air is 3×10^8 m/s, in wire it is 1.5×10^8 m/s.

SECTION 2-3
1 3 m.
3 2:3.

SECTION 2-4
1 150,000 kHz; 0.15 GHz.
3 3000 MHz; 3,000,000 kHz.
5 1 m to 0.1 m; 0.67 m to 0.067 m.

SECTION 2-5
1 90.1, 90.3, 90.5, 90.7, 90.9, 91.1, 91.3, 91.5, 91.7, 91.9.
3 50/200 kHz = 25 percent.
5 Less than half; 4500.

END OF CHAPTER 2
1 3.3×10^{-6} s; 1.33×10^{-5} s.
3 0.707 m/c = 2.36×10^{-9} s.
5 $(186,000 \times 3600)/700$ = about 1,000,000 to 1.
7 12,400,000 Hz = 12.4 MHz.
9 0.0255 MHz; 0.0000255 GHz.
11 10,000 m.
13 Channels: 103.5; 110.5; every 7 kHz up to 159.5 kHz. See Fig. 1.

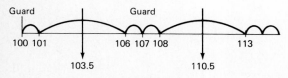

Fig. 1

15 441 kbits/s.
17 a 20 kbits/s.
    b 3.51 kHz.
    c 10 kbits/s.
    d 38 Hz.

SECTION 3-2
1 See Fig. 2.

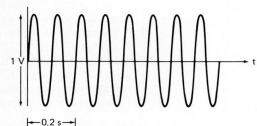

Fig. 2

3 See Fig. 3.

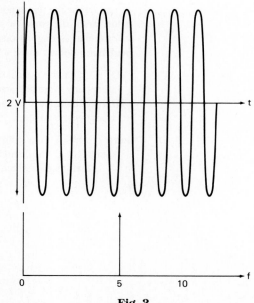

Fig. 3

5 See Fig. 4.

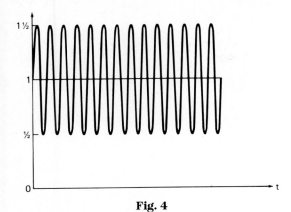

Fig. 4

SECTION 3-4
1 See Fig. 5.

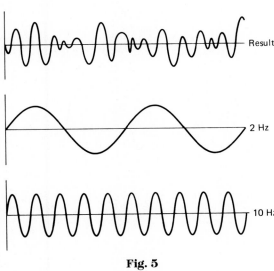

Fig. 5

3 Upper SB: 105 to 110 kHz: lower SB: 90 to 95 kHz.
5 See Fig. 6.

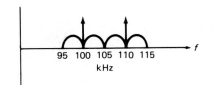

Fig. 6

7 See Fig. 7.

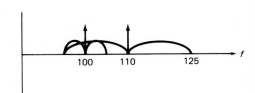

Fig. 7

9 a   20 kHz.
    b   100 kHz.
    c   2 MHz.
    d   4 MHz.

SECTION 3-5
1 50 to 150 kHz.
3 See Fig. 8. Noise of up to 1 V can be tolerated.

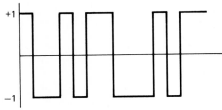

Fig. 8

SECTION 3-6

1 See Fig. 9.

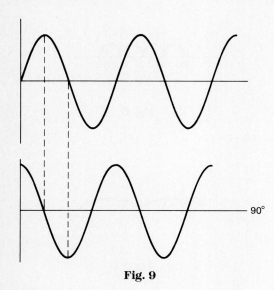

Fig. 9

3 See Fig. 10.

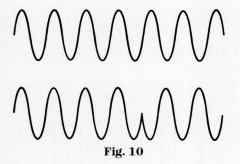

Fig. 10

SECTION 3-7

1 See Fig. 11.

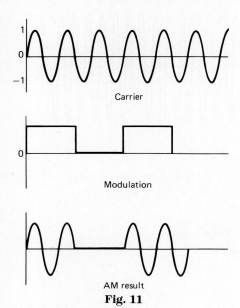

Fig. 11

3 See Fig. 12.

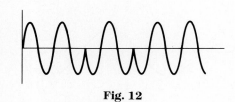

Fig. 12

5 0.05 V.

7 See Fig. 13.

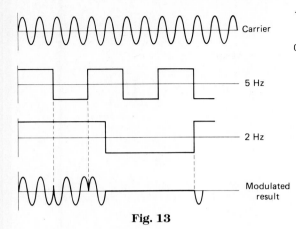

Fig. 13

END OF CHAPTER 3
1 See Fig. 14.

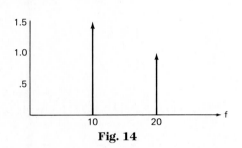

Fig. 14

3 Carriers: 103, 111, 119, 127, 135, 143, 151, 159, 167, 175 kHz.

5 See Fig. 15.

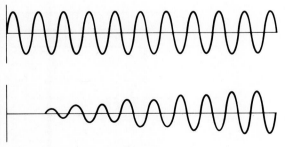

Fig. 15

7 See Fig. 16.

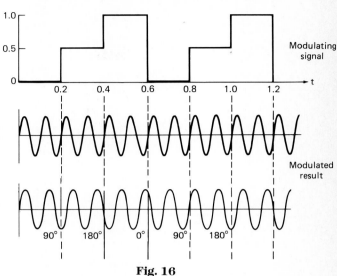

Fig. 16

9 See Fig. 17.

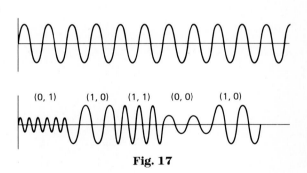

Fig. 17

SECTION 4-4
1 See Fig. 18.

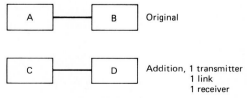

Fig. 18

SECTION 4-5

1 See Fig. 19.

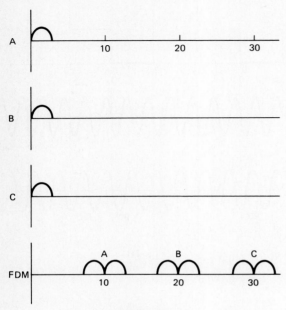

Fig. 19

3 a $(2 + 10 + 5) \times 6$ MHz = 102 MHz.
b $(2 \times 6) + (10 \times 0.2) + (5 \times 0.003) =$ 14.015 MHz.
c See Fig. 20.

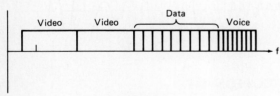

Fig. 20

SECTION 4-6

1 See Fig. 21.

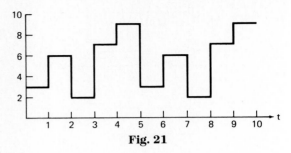

Fig. 21

3 a Once every $5 \times (20 + 20) = 200$ ms.
b Access is $5 \times (10 + 20) = 150$ ms, or 6.6 times/s.
c See Fig. 22.

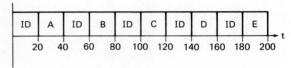

Fig. 22

5 See Fig. 23.

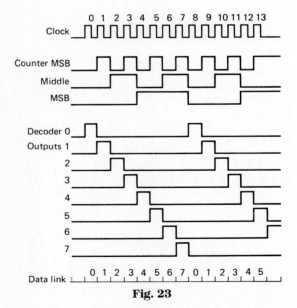

Fig. 23

7 500 times/s for 0.5 ms.

SECTION 4-7

1 a See Fig. 24.

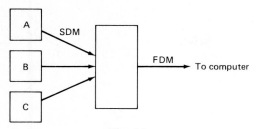

Fig. 24

b See Fig. 25.

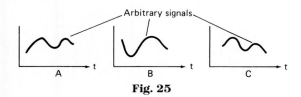

Fig. 25

c See Fig. 26.

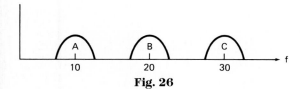

Fig. 26

3 a See Fig. 27.

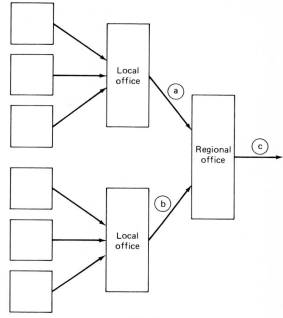

Fig. 27

b See Fig. 28.

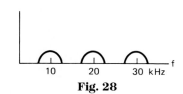

Fig. 28

c See Fig. 29.

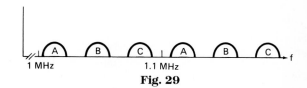

Fig. 29

Answers for Odd-Numbered Problems

SECTION 4-8

1. a 0.5 V.
   b 4.11.
   c See Fig. 30.

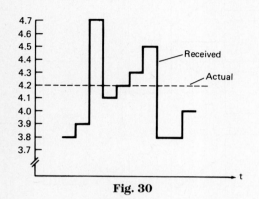

Fig. 30

3. 0.05 V; 0.001; no, 100 mV is greater than 1 percent of any value from 1 to 10 V.

END OF CHAPTER 4

1. a See Fig. 31.

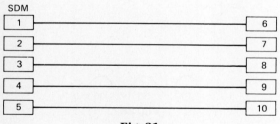

Fig. 31

   b See Fig. 32.

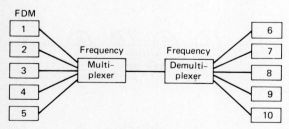

Fig. 32

   c See Fig. 33.

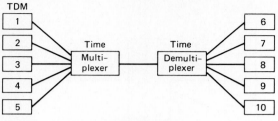

Fig. 33

3. See Fig. 34.

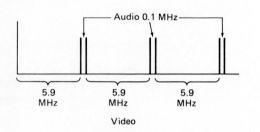

Fig. 34

5. a See Fig. 35.

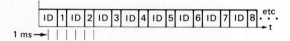

Fig. 35

   b Identifiers are 2/3 = 67 percent of time, independent of number of users.
   c System takes 3 ms/user, or 330 users/s. For 25 users, this is 330/25 times/s, or a little over 13 times/s.

7

		SDM	FDM	TDM
a	Use of capacity	Low	Medium	High
b	Initial cost	Low	Medium	High
c	Complexity for additional users	Low	Medium	Low
d	Cost for additional users	High	Medium	Low
e	Sensitivity to link problems	Low	Medium	High
f	Sensitivity to other user problems	Low	Medium	High
g	Technical complexity	Low	Medium	High

9 a Error is 0 V.
    b Maximum error is ± 0.2 V.

SECTION 5-2

1 Errors will be caused by noise of 1.3, 2, -1.2 V (any value greater than -1 or 1 V).
3 Either 0 V or 20 V.
5 0.0001 percent; 0.0003 percent.

SECTION 5-3

1 200 kHz; 110 kHz; 20 Hz; 2 MHz; 12 MHz; 6 kHz.
3 300 kHz; 165 kHz; 30 Hz; 3 MHz; 18 MHz; 9 kHz for 50 percent greater; 600 kHz; 330 kHz; 60 Hz; 6 MHz; 36 MHz; 18 kHz for three times greater.
5 (3 kHz $\times$ 2) $\times$ 150 percent $\times$ 18 = 9 kHz $\times$ 18 = 162 kHz.

SECTION 5-4

1 $5/2^2 = 1.25$ V; $5/2^4 = 0.3125$; $5/2^6 = 0.078$; $5/2^8 = 0.0195$; $5/2^{10} = 0.00097$.
3 ± 0.0019 V.
5 1.2 V = 001; 2.7 V = 100; 3.2 V = 101; 4.5 V = 111.
7 (5 kHz $\times$ 2) $\times$ 1.5 $\times$ 6 bits = 90 kbits/s.

SECTION 5-5

1 Average value: 0, 0, 0.33, 0.5, 0.6, 0.67, 0.71, 0.63, 0.67, 0.6, 0.55, 0.58.
3 0.33.
5 See Fig. 36.

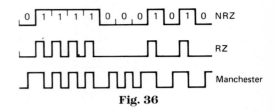

Fig. 36

7 See Fig. 37.

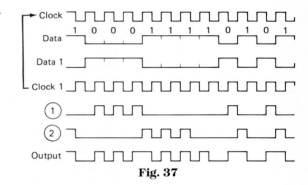

Fig. 37

SECTION 5-6

1 40 kHz.
3 a 40 kHz $\times$ 7 bits = 280 kbits/s.
    b 2.8 MHz.
    c 40 kHz $\times$ 8 bits $\times$ 10 = 3.2 MHz.
5 $10/[10 + (5 \times 8)] = 10/50 = 20$ percent.

END OF CHAPTER 5

1 Either 8 or 12 V.
3 Noise tolerance is ± 0.5 V. Errors from noise of -0.55, -1.1, $+0.75$ V.
5 20 kHz; 180 kHz; 12 MHz.
7 30 kHz; 100 kHz.
9 $150{,}000/(10$ kHz $\times 2 \times 1.5) = 5$ users.
$150{,}000/(10$ kHz $\times 2 \times 2) = 3$ users.
11 The difference is 2 bits, so increased resolution is a factor of $2 \times 2 = 4$.

13 $-1 = 000$; $-0.3 = 001$; $+1 = 111$; $+0.4 = 101$.
15 3 kHz $\times$ 2 $\times$ 8 $\times$ 10 = 480 kbits/s.
17 1, 0, 0.33, 0.5, 0.2, 0, 0.14, 0, 0.11, 0.2, 0.27, 0.5, 0.23, 0.14, 0.067, 0.
19 See Fig. 38.

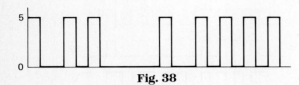
Fig. 38

21 See Fig. 39.

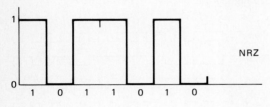
NRZ

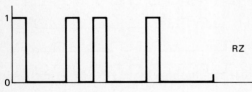
RZ

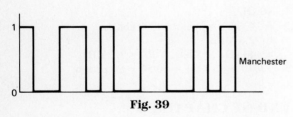
Manchester
Fig. 39

23 a   3 kHz $\times$ 2 $\times$ 2 = 12 kHz.
    b   12 $\times$ 10 = 120 kbits/s.
    c   120 $\times$ 8 = 960 kHz.
25 a   10/[10 + (25 $\times$ 8)] = 2.4 percent.
    b   20/(20 + 400) = 4.7 percent.
    c   Use only 4 framing bits or have 40 users.

SECTION 6-6

1 Average = 0; rms = 0.283.
3 20, 23.5, 41.9, -60, -3.1.
5 10.8, 1.3, 21, 4.8.
7 0.63, 1.37, 1.69, 1.56.
9 0.63, 27.8, 0.35, 111, 16, 1.8.
11 7, 7.5, 4.45, 15.8.
13 1.25, 2.5, 50, 0.45, 0.548.
15 a   20 + 30 + 40 = 90 dB.
    b   100, 1000, 10,000.
    c 10^9 or 1,000,000,000.
    d   92 dB; 100, 1000, 15, 848.9; 1,584,893,192.
17 2, 4, ½, ⅛.

SECTION 6-7

1 1000 Hz—3-ms delay.
2000 Hz—2-ms delay.
4000 Hz—no delay.
3 1 kHz—no gain; 0 dB.
10 kHz—+2 dB.
20 kHz—+3 dB.
See Fig. 40.

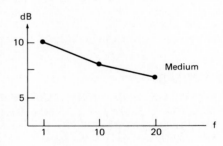

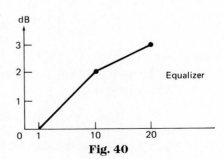
Fig. 40

SECTION 6-8
1. 0.00038 V.
3. 28 dB.
5. Factor of 2; additional 6 dB.

END OF CHAPTER 6
1. A = 0; rms = 4.9.
3. 28; 34; 37.5; 54; 0; −8.5; −20.
5. 9.5; 9.5; −10.5; 7.95.
7. 5.23; −5.23; −3; −6.
9. 1, 70, −2.2, 2.2, 18.9, −1.25, 14.
11. a From 12 to 18 dB.
    b 2; 1.58, 10, 0.5 to 2.
    c 15.8 to 63.2
13. 10 kHz: 13 ms; short by 3.
    20 kHz: 18 ms; short by 8.
    40 kHz: 6 ms.
    50 kHz: 5 ms.
    100 kHz: no delay.
15. 0 kHz: +1 dB.
    5 kHz: +3 dB.
    10 kHz: +4 dB.
    15 kHz: −2 dB.
    20 kHz: −3 dB.
    25 kHz: −5 dB.
17. Max CMV allowed is 5 V. CMRR = 31.6 dB.

SECTION 7-2
1. A = 1000001; a = 1100001; Q = 1010001;
   T = 1110100; 3 = 0110011; 9 = 0111001;
   @ = 1000000.
3. P = 11010111; p = 10010111; 7 = 11110111;
   1 = 11110001; : = 01111010.
5. 245 = 0110010, 0110100, 0110101 (ASCII);
   11110101 (binary); 21 bits versus 8 bits = 62 percent saving.
7. $2^9 = 512$; $2^{10} = 1024$.

SECTION 7-3
1. 2000.
3. 1 s.
5. Number = 0000 0100; character count = 0000 1111. (Spaces are characters.)
7. Message = 1010; receiver number = 0111; character count = 1100.

SECTION 7-4
1. 2/7 = 29 percent overhead (2 bits/character) regardless of number of characters.
3. Overhead = 5 × 8 = 40 bits; 40/(7 × 50) = 11.4 percent overhead; overhead = 2 × 5 × 8 = 80 bits; 80/(7 × 100) = 11.4 percent overhead.
5. Synchronous: 30 characters/10 characters/s = 3 s.
   Asynchronous: 9 bits at 1000 bits/s = 0.009 s.

SECTION 7-5
1. 4; 4; 16; 32.
3. At 300 baud, Throughput is 30 characters/s. It is the same at 1200 baud.
5. Burst rate is 4800 baud = 480 characters/s = 0.002 s/character.
   Throughput is 10 characters in 0.02 s + 1 s = 10 characters/1.02 s.
7. Throughput = 1/(0.01 + 0.05) = 16 characters/s at 1700 baud.
   = 1/(0.0025 + 0.05) = 19 characters/s at 4800 baud.
   = 1(0.33 + 0.05) = 2.6 characters/s at 300 baud.

SECTION 7-6
1. HERE IS OUR REPORT BEL CR LF LF LF.
3. See Fig. 41.

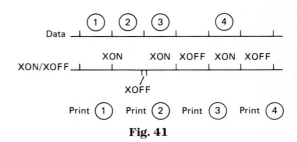

Fig. 41

5. # 1509:;
7. See Fig. 42.

Fig. 42

SECTION 7-7

1. a 350 μs/bit = 2857 bits/s.
   b 2857/9 = 317 characters/s.
3. a <u>01001110</u> 1 0 <u>1001111</u> 1
          N   O
   b <u>0110</u>   <u>0010</u>   <u>10011101</u> <u>1001111</u> <u>0100011</u>
      Message Character   N   O   #
         #   Count
5. ASCII to 16-bit binary.
   75: 0110111 0110101 to 00000000 01001011
   125: 0110000 0110010 0110101 to 00000000 01111101
   257: 0110010 0110101 0110111 00000001 00000001

END OF CHAPTER 7

1.   q   1110001
     #   0100011
     LF   0001010
     BEL   0000111
     DC2   0010010
3. 298 = 0110010 0111001 0111000.
   100 = 0110001 0110000 0110000.
   35 = 0110011 0110101.
5. 4 bits = 16 characters; 12 bits = 4096.
7. Use 0.1 s resolution:
   As soon as buffer fills, no new characters can be taken. As soon as one character is taken, a new one comes in, since there is always a character wanting to come in. See Fig. 43.

Fig. 43

9. 25 × (8 + 2) = 250.
11. For Prob. 9: 50 × (8 + 2) = 500.
    For Prob. 10: 8 + 4 + 8 + (50 × 8) = 420.
13. 1200/9 = 133 characters/s.
    19,200/9 = 2133 characters/s.
15. Burst is 9600/9 = 1066 characters/s.
    Throughput is 10 characters/s.
    Efficiency is 1 percent (approximately).
17. a 0000111 0001010 0001010 1001111 1001011
         BEL   LF   LF   O   K
    b Add 0100011 0100011 in front, 0100001 0100001 at end.

19. See Fig. 44.

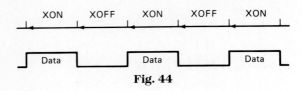

Fig. 44

21. See table below.

Binary	ASCII
0000 0000 01100100	0110001 0110000 0110000
0000 0000 11111110	0110010 0110101 0110100
0000 0000 10000000	0110001 0110010 0111000

SECTION 8-2

1. Noise margin is +7, −7 V.
3. +22, −22 V.
5. See Fig. 45.

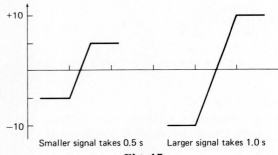

Fig. 45

7. Need to go 10 V in (1/1000) × 10 percent s = 10 V/0.1 ms = 100 V/ms.

SECTION 8-3

1. Q: 2 stop bits, no parity. See Fig. 46.

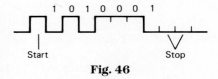

Fig. 46

3 %: See Fig. 47.

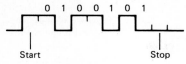

Fig. 47

5 See Fig. 48.

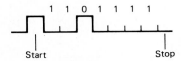

Fig. 48

7 $2 \times (1 + 8 + 1 + 2) = 24$ bits $= 24/1200 = 0.02$ s.

SECTION 8-5
1 See Fig. 49.

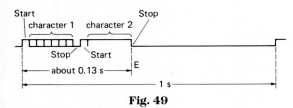

Fig. 49

3 See Fig. 50.

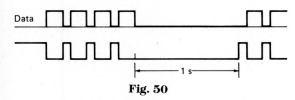

Fig. 50

5 Same as Prob. 3, but when handshake line goes high, now XON is sent; when handshake line goes low, XOFF is sent.

SECTION 8-6
1 See Fig. 51.

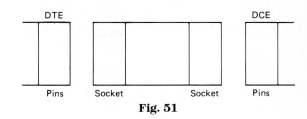

Fig. 51

3 See Fig. 52.

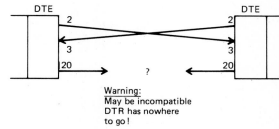

Warning:
May be incompatible
DTR has nowhere
to go!

Fig. 52

SECTION 8-7
1 01110110.
3 10111011. Note: Framing and overrun error almost certainly means that a parity error also is assumed.

END OF CHAPTER 8
1 $+6, -9$ V.
3 10 V/ms takes one-half the time.
5 See Fig. 53.

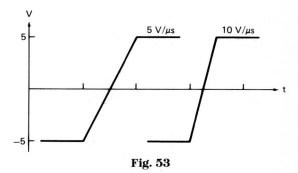

Fig. 53

7 (10 V)/(25 V/μs) = 0.4 μs; (0.4/0.05) = 8 μs; (1/8 μs) = 125,000 bits/s.

9 See Fig. 54.

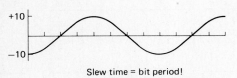

Fig. 54

11 See Fig. 55.

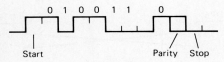

Fig. 55

13 0.019 s; 0.002 s.

15 At 2400 baud, characters arrive every 1/240 s (0.004), so after each character the handshake line says ''not OK'' for 0.001 s.

17 See Fig. 56.

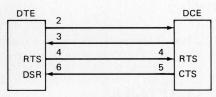

Fig. 56

19 01011010. (Use 16 × clock.)

21 10001101.

SECTION 9-3

1 50/100 = 0.5 ms.

3 RS-232: 0 to 22 V; RS-423: 3.4 to 5.8 V; RS-422: 1.8 to 5.8 V; RS-485: 1.3 to 5.8 V.

5 Binary 1 = −1.5 to −6 V. Decision point is −0.2 V. Worst case: −1.5 + 1.5 = 0 V error! No error at −1.7 + 1.5 = −0.2 V, so binary 1 must be −1.7, at least.

SECTION 9-4

1 See Fig. 57.

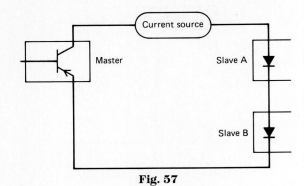

Fig. 57

3 V = (20Ω × 0.020) + 1½ + 1½ = 3.4 V.

5 V = (10Ω × 0.020) + (3 × 2) + 3 × (500 × 0.020) = 16.2 V.

END OF CHAPTER 9

1 Same as text Fig. 9-4, with one more receiver.

3 See Fig. 58.

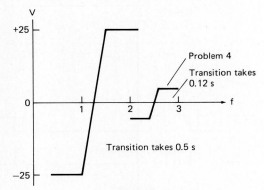

Fig. 58

5 See Fig. 59.

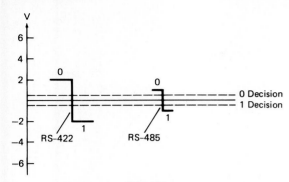

Fig. 59

7 10 × (250 × 0.020) = 50 V
9 R = (10 V/0.020) = 500 Ω minimum.

SECTION 10-2

1 163: 0.1 + 0.5 + 0.6 + 0.5 + 0.3 = 2.0 s.
784: 0.7 + 0.5 + 0.8 + 0.5 + 0.4 = 2.9 s.
223: 0.2 + 0.5 + 0.2 + 0.5 + 0.3 = 1.7 s.
3 1-413-223-3230 = 3.9 s; 1-916-769-8790 = 7.8 s.
5 0: 941, 1336 Hz; 2: 697,1336 Hz; 7: 852,1209 Hz; 7: 852,1447 Hz.
7 1: 512 Hz; 1816 Hz; 6: 677 Hz; 2217 Hz.
9 1.25 s for each.
11 4: Divide by 1298 or 1299; divide by 827 or 828.
9: Divide by 1173 or 1174; divide by 691 or 692.

SECTION 10-3

1 2000 mi × 2/(c × 0.75) = 0.029 s. (Echo time is based on round-trip distance.)
3 10 × 2/(c × 0.60) = 0.000045 s.

SECTION 10-5

1 See Fig. 60.

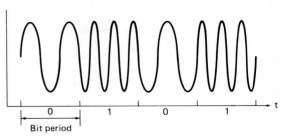

Fig. 60

3 a 1000, 2000, 2000, 2000, 1000, 2000, 1000, 2000 Hz.
b 1000, 1000, 1000, 1000, 2000, 2000, 2000, 2000 Hz.
c 2000, 2000, 1000, 1000, 2000, 2000, 1000, 2000 Hz.

SECTION 10-6

1 Bit pattern is 0100010111: 1070, 1270, 1070, 1070, 1070, 1270, 1070, 1270, 1270, 1270 Hz.
3 Bit pattern is 000011101: 1070, 1070, 1070, 1070, 1270, 1270, 1270, 1070, 1270 Hz.
5 1200 Hz; dibits are 10, 10, 01, 00, so phases are 180°, 180°, 0°, 90°.
7 Dibits are 00, 11, 00, 10, 10. (Note: last bit is "padded.")
Phases are: 90°, 270°, 90°, 180°, 180°.

END OF CHAPTER 10

1 123: 1.6 s; 111: 1.3 s; 777: 3.1 s; 789: 3.4 s.
3 All 1s: 6.1 s.
5 Same for all numbers: 5.25 s.
7 2: 1394, 2672 Hz; 3: 1394, 2894 Hz; 6: 1540, 2894 Hz; 7: 1704, 2418 Hz.
9 1: 827 or 828; 1434 or 1435; 2: 748 or 749; 1434 or 1435; 5: 748 or 749; 1298 or 1299; 8: 748 or 749; 1173 or 1174.
11 Talker echo = (22,500) × 2/(c × 0.97) = 0.25 s.
Listener echo is 2 × talker echo = 0.5 s.

13 Echo distance is between points marked x
$(1500 + 3000)(c \times 0.70) = 0.034$ s. See Fig. 61.

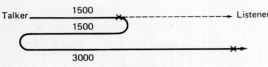

Fig. 61

15 See Fig. 62.

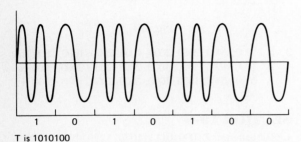

T is 1010100

Fig. 62

17 See Fig. 63.

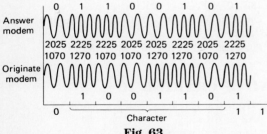

Fig. 63

19 Bit pattern is 0 1010010 11. Dibits are 01, 01, 00, 10, 11. Phase shifts are 0°, 0°, 90°, 180°, 270°.

SECTION 11-2

1 1770.
3 25 to 26: 25 new links. 25 to 30: 135 new links.
5 Two new connectors.
7 12×1 dB = 12 dB.

9 See Fig. 64.

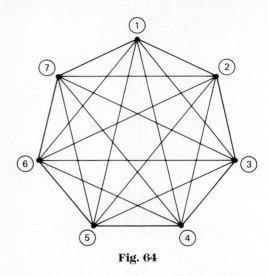

Fig. 64

11 See Fig. 65.

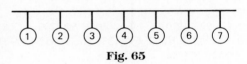

Fig. 65

13 Each user point-to-point: 21. Star: 7. Bus: 1 bus + 7 links. Ring: 7 links.

SECTION 11-3

1 Master: "Slave 3, give me your data."
Slave 3: "Master, here is data: _____."
Master: "Slave 4, give me your data."
Slave 4: "Master, here is data: _____."

3 Master: "Slave 4, give me your data."
Slave 4: "Master, here is data for slave 3 and slave 5: _____."
Master: "Slave 3, here is data: _____."
Slave 3: "Slave 3 has received data."
Master: "Slave 5, here is data: _____."
Slave 5: "Slave 5 has received data."

552 Answers for Odd-Numbered Problems

5 Node 3: 7 ms; node 8: 17 ms.
7 2→3→4→5 means 0.3 ms; to get back to 15 means 1 ms more.
9 2→3→4→5 still means 0.3 ms; to get back now means 1.5 ms. Shortest: 0.1 ms; longest: 1.9 ms.
11 D gets the message out in 3+ ms, compared to 4 ms in the previous case. See Fig. 66.

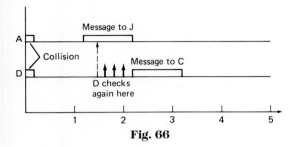

Fig. 66

13 See Fig. 67.

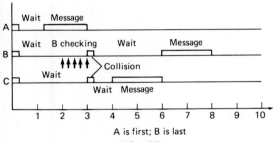

A is first; B is last
Fig. 67

SECTION 11-8

1 Same as Fig. 11-22. LAN uses three layers.
3 Translation is made at layer 6 at sending side (computer describing the position to machine tool control).

5 See Fig. 68.

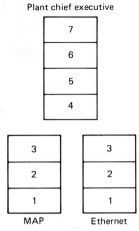

Fig. 68

SECTION 11-9

1 See Fig. 69.

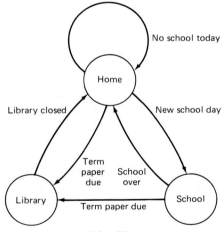

Fig. 69

3 See Fig. 70.

Fig. 70

SECTION 11-10

1 Max = 20 × 200 = 4000 users; 20 percent = 3200 users.

3 See Fig. 71.

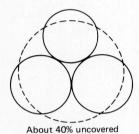

About 40% uncovered

Fig. 71

5 See table below.

Cells, Number	Overlap, Percentage	Total Cost, Dollars	Number of users (theory)	Number of users (actual)	Cost/User, Dollars
1	0	3.00	500	500	6K
7	15	5.95	3,500	2,975	2K
15	25	12.75	7,500	5,625	2.26K
30	35	25.5	15,000	9,750	2.6K

END OF CHAPTER 11

1 For 60: 1770, for 60→65: 310 new links; 65→66: 65 new links.

3 Star: another 500 ft; bus: another 5 ft; ring: 0 ft

5 $\underbrace{3 \times 0.5}_{\text{User 1 to master to user 2}} + \underbrace{2 \times 0.5}_{\text{User 1}} + \underbrace{3 \times 0.5}_{\text{User 2 to master to user 3}} + \underbrace{2 \times 0.5}_{\text{User 1}} +$

$\underbrace{2 \times 0.5}_{\text{User 2}} + \underbrace{3 \times 0.5}_{\text{User 3 to master to user 4}} = 7.5$ ms.

7 Response time for both would be longer, as master had to check with units 1 through 6.

9 For either user 7 or user 85, it would take 200 × 0.3 = 60 ms.

11 See Fig. 72.

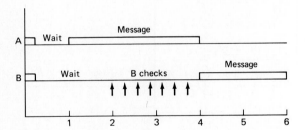

A transmits at 1 ms
B transmits at 4 ms

Fig. 72

13 See Fig. 73.

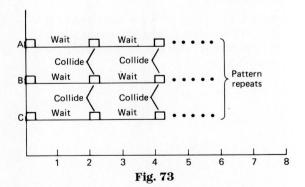

Fig. 73

15 See Fig. 74.

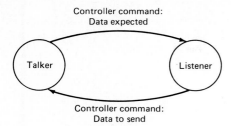

Fig. 74

17 See Fig. 75.

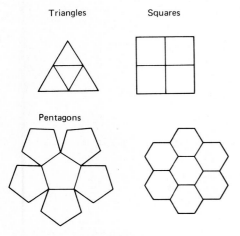

These patterns are called tesselations.
Fig. 75

SECTION 12-2

1 0, 0, 1, 0, 0, 1.
3 0, 1, 1.
5 10100101 10010011.
7 0-0: multiple-bit error 3 bits.
0-0: multiple-bit error 4 bits.
1-1: multiple-bit error 8 bits.
1-0: 1-bit error.

9 See Fig. 76.

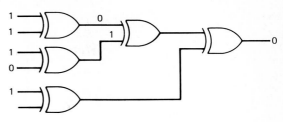

Fig. 76

11 01011000→1→1
11110000→0↗
Conclusion: parity of parity bits of subgroups = parity of larger group.

SECTION 12-3

1 $x^3 + x; x^3 + x^2 + x^1 + 1; x.$
3 1011; 10100; 10101100.
5 $x^3 + x^2 + 1; x^2 + 1; x^5 + x^4 + x^2 + 1.$
7 See Fig. 77.

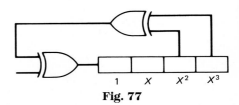

Fig. 77

9 CRC: 16/128 = 12.5 percent.
Parity: 1/8 = 12.5 percent.
11 0010.
13 0011 ← Same data as Prob. 12, different CRC checksum.

SECTION 12-4

1 (200 + 50) ms block = 4 blocks/s = 20 blocks in 5 s.
Time lost = 20 × 50 = 1000 ms = 1 s.
Percentage = 50/250 = 20 percent.
3 Forward   200→   200→
Reverse   50←   50←
5 Forward 1 2 3 4 5 6 7 8 9 10
Reverse   1 OK 2 OK 3 OK 4 OK 5 OK 6 OK 7 OK 8 OK 9 OK

SECTION 12-5

1 See table below.

Data Transmit	Checksum	Data Receive	Checksum	Error Syndrome	Error
0110 1001	11010	0010 1001	10001	10001	Bit 6
0010 1111	01001	0001 1111	10001	00111	2 bits
1001 1001	01110	1011 1001	00011	10010	Bit 5
0111 1100	10000	0000 1100	01000	00111	2 bits
0011 1100	10000	1101 1100	10001	11110	3 bits but system is fooled

3 See table below.

Data Transmit	Checksum	Data Receive	Checksum	Error Syndrome	Error
1100 0001	10100	—	01001	00010	Many errors
1100 0011	10001	—	11101	10011	2 bits
0011 1100	10111	—	10000	11000	Bit 7
0011 1100	01100	—	10000	00011	Unknown

5

	SEF	DEF
a	1	1
b	1	1
c	1	0
d	1	0

7

Data Transmit	Checksum	Data Receive	Checksum	Syndrome
0100 0000	10100	1111 1110	01010	00001

SECTION 12-7

1 FRPH DW WKUHH.
3 PMU OGLD YFW.
5 1000 1100 1110 1100.
7 1100 1100 0110 1100.

9
0011 ←1st	0111
1001	0011
1100	1001
1110	1100
1111	1110 ←16th
0111	1111
0011	0111
1001	0011
1100	1001 ←20th
1110	↑ PRSQ bits = LSBs
1111	

11
0011	1000 ←1st
1001	1100
0100	1110
1010	0111
0101	0011
1010	1001
1101	0100
0110	1010
0011	0101
1001	1010 ←10th

↑ PRSQ bits = LSBs.

13	1010	1010	1010	1010	Sync data
	1001	1010	0011	0001	PRSQ
	0011	0000	1001	1011	Result-encrypted
	0011	0100	0110	001X	PRSQ shifted left
	0000	0100	1111	100X	Result
	X100	1101	0001	1000	PRSQ shifted right
	X111	1101	1000	0011	Result

SECTION 12-9

1. Data can be 1, 2, 3, . . ., 100: 100 possibilities. With two digits, data can be 1, 2, 3, . . ., 99.
3. Eavesdropper sees 10 and thinks data can be 1, 2, 3, . . ., 99. The possibilities are correct, but the answer is not correct.

END OF CHAPTER 12

1. 1, 1, 1, 1.
3. 0000.
5. a   1--1: misses errors.
   b   1--1: misses errors.
   c   1--0: sees some error.
   d   1--0: sees some error.
7. See Fig. 78.

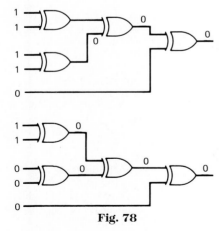

Fig. 78

9. $x^7 + x^5 + x^3 + x^i$; $x^5 + x^4 + x + 1$; $x^7 + x^6 + x^4 + 1$; $x^3 + 1$.
11. $x^5 + x^3 + x^2$; $x^7 + x^4 + 1$; $x^7 + x^5 + 1$; $x^7 + x^3 + x^2 + x$.

13. 8/32 = CRC = 25 percent; 1/8 = parity = 12.5 percent.
15. $2^{32} = 4.294 \times 10^9$ for 32 bits; $2^8 = 256$ for 8 bits.
17. 1011.
19. Full-duplex: 100 ms/(200 + 100) = 50 percent; half-duplex: 100/(150 + 100 + 150) = 25 percent.
21. Transmit 1, 2, 3, 2, 3, 4, 5, 6, 7, 6, 7, 8, 9, 10, 9, 10
    ARQ OK OK OK Not OK OK OK OK OK Not OK OK OK OK OK Not OK OK OK
23. First three cases unchanged; last case is 0000 0001; check bits 00101; syndrome 00101; error in bit 0.
25. CZMFDQ RSZX ZVZX.
27. FWOFCR RQZB AYBX.

29	1010	0110	0011	1001	Data
	0110	1010	0011	1001	PRSQ
	1100	1100	0000	0000	Result
31	1001	0110	Received
	0101	1100	PRSQ
	1100	1010	Original data

33. PRSQ is 1, 0, 0, 0, 1, 0, 0, 1, 1, 0, 1, 0, 1, 1, 1, 1
35. PRSQ is 1, 0, 0, 1, 1, 0, 1, 0, 0, 1

SECTION 13-4

1. BER is one bit error/8000 bits = 0.0125 percent. Percentage EFS = 25 percent (only 3.00 to 4.00 is error-free). Percentage EF frames = 2.5 percent (five frames have errors).
3. First case (one error per frame): BER = 1 percent; percentage EFS = 0 percent; percentage EFF = 0 percent. Second case (all errors in first frame): BER = 1 percent; percentage EFS = 90 percent; percentage EFF = 99 percent.

SECTION 13-7

1. 3.0 μs.
3. Reflection takes 26 μs, so interim time is 16 μs.
5. For pulse widths longer than the reflection gap, the two reflections overlap and cannot be distinguished.

END OF CHAPTER 13

1. BER = 0.0625 percent; percentage EFS = 50 percent; percentage EFF = 87.5 percent.
3. BER = 0.25 percent; percentage EFS = 50 percent; percentage EFF = 50 percent.
5. BER = 5 percent; percentage EFS = 0 percent; percentage EFF = 75 percent.
7. Reflection time is 78 μs; so 10-μs pulse is OK; 100-μs pulse is not OK.

Index

Absorption, wave, 25
Achievable bit rate, 302
Acknowledgment (ACK), 444–446
Acoustic coupling, 345–346
Adaptive equalization, 191
ADC (analog to digital converters), 133–139
Adjacent channels, 28
Air as medium, 167–170
Algebra theory, 436–437
Aliasing errors, 131
AM (*see* Amplitude modulation)
American Standard Code for Information Interchange (*see* ASCII *entries*)
Amplitude, wave, 15
Amplitude jitter, 503
Amplitude-modulated (AM) broadcast radio band, 19
Amplitude modulation (AM), 53–58
　advantages and disadvantages of, 56–57
　baseband, 57–58
　demodulator of, 72
Amplitude response, 330
Amplitude-shift keying (ASK), 71
Analog communications systems, 80–81
　functions within, 81–85
　shortcomings of, 109–111
Analog modulation, 67–72
Analog signals, 9
　TDM for, 129–130
Analog to digital converters (ADC), 133–139
Answer modems, 344–345
Antenna, 82
　dish, 167, 168
　wire, 167, 168
Antialiasing filters, 132
AppleTalk, 389–391
ARQ (automatic request for repeat), 444–445
ASCII (American Standard Code for Information Interchange), 210–214
ASCII character set, 519–523
ASK (amplitude-shift keying), 71
Asynchronous modulation and demodulation, 73–76
Asynchronous systems, 222–227
Autodial feature, 323
Automatic request for repeat (ARQ), 444–445
Averaging signal values, 109

Bandpass, 44
Bands, 19
　frequency, 22–23
　guard, 28
Bandwidth, 26–28
　channel capacity and, 29–33
　distance and, 34–36
　effective, 184
　filter, 45
　of medium, 162-163
　modulated, 59–61
　relative, 27
Bandwidth compression, 31–33
Bandwidth limitation, 184–191
Bandwidth reduction, 31
Baseband AM, 57–58
Basic rate service, 417–418
Baud, 228–229
Baud value, 255–256
Bell 103 modem, 348–349
Bell 202 modem, 350
Bell 212 modem, 349–350
BER (*see* Bit error rate)
Binary coding, straight, 213
Binary system, 9, 67
　random sequences and, 457–458

Bipolar encoding, 496
Bit(s), 121
　data, 257–258
　least significant (LSB), 138
　most significant (MSB), 138
　parity, 430–432
　poll/final (P/F), 238
　per signaling unit, 229
　start, 222
　stop, 223
Bit error rate (BER), 125, 180, 427, 482
　test, 513
Bit error rate analyzers, 495–498
Bit rate, 229
　achievable, 302
BORSHT functions, 319, 321–322
Break detect circuitry, 259
Break indication, 258–259
Breakout boxes, 485–487
Broadband coaxial cable, 383–384
Broadcast mode, 294
Buffers, 125, 218–219
　memory, 138–139
　receiving, 231
Burst aspect of noise, 426
Burst rate, 231
Bus topology, 367–369
Byte, 121

c (speed of light), 18
Cable, 164–167
　coaxial (*see* Coaxial cable)
　multiconductor flat, 165–166
Cable loss, 511
Cable TV systems, 20
Calibrated light source, 509
Carrier envelope, 54–55
Carrier sense multiple access/collision detect (CSMA/CD), 377–380

Carriers, 40
  single-sideband suppressed (SSB), 57
Cathode-ray tube (CRT) terminal, 487–489
CBC (chain block cipher), 468–469
CCITT (International Telegraph and Telephone Consultative Committee), 417–418
CD (collision detect), 377
Cellular networks, 412–416
CFB (cipher feedback), 469–470
Chain block cipher (CBC), 468–469
Channel capacity, bandwidth and, 29–33
Channel efficiency, 120
Channel identification, 97–99
Channel number, 97
Channels:
  adjacent, 28
  communication (*see* Communications channels)
  term, 13–14
  TV, 20
Character Set
  ASCII, 519–523
  EBCDIC, 524–525
Checksum, 432
  CRC, 437–442
Cipher feedback (CFB), 469–470
Clock frequency, 75
Closed networks, 363
CMRR (common mode rejection ratio), 195
CMV (common mode voltage), 192–197
Coaxial cable, 166–167
  broadband, 383–384
Code conversion, 213–214
Coherent communications, 76
Collision detect (CD), 377
Combined modulation, 71–72, 153
Combined modulation systems, 105–108
Command register, 283
Command/response protocol, 297
Command/response systems, 371–373
Common mode rejection ratio (CMRR), 195
Common mode voltage (CMV), 192–197
Communications, 1–3
  coherent, 76
  data (*see* Data communications)
  digital, 118–156

Communications (*continued*)
  digital systems, (*see* Digital systems)
  multidrop, 293–298
  pathways for, 3
  point-to-point, 173
  uses of, 3–5
Communications channels, 13–16
  total, 16
Communications hierarchy, 206, 209
Communications interfaces, 292–313
Communications systems:
  analog (*see* Analog communications systems)
  data communications and, 8–10
  issues in, 205–209
  protocols of, 215–221
  requirements of, 205–246
  spread spectrum, 463–465
  structure and types of, 5–8
  T-1 digital carrier, 151–152
Compliance voltage, 308
Compression, bandwidth, 31–33
Computer-to-computer interface, 269–271
Computer-to-plotter protocol, 236–237
Computer-to-printer interface, 267–268
Computer-to-printer protocols, 235–236
Computer-to-terminal interface, 268–269
Concentrator modems, 354–355
Connection of medium, 163–164
Connections, physical, 384–385
Connectors, 273–274
Contention problem, 294
Continuity tests, 511
Control characters, 210, 213
Control function table, 452
Controllers, 4–5, 393
Cost of medium, 163
Coupling:
  acoustic, 345–346
  transformer, 390
Crashes, system, 240
CRCs (*see* Cyclic redundancy codes)
Crosstalk noise, 176–177
CRT (cathode-ray tube) terminal, 487–489
CSMA/CD (carrier sense multiple access/collision detect), 377–380

Current loops, 302, 305–310
  handshaking on, 309
  implementation of, 306
  installation issues in, 309-311
  multidrop operations on, 307–309
Current source, 305
Cutoff frequency, 184
Cyclic redundancy codes (CRCs), 436–442
  checksum, 437–442

Daisy chain approach, 385
Data, 121
Data bits, 257–258
Data communications, 1–2
  communications systems and, 8–10
Data communications equipment (DCE), 261–265
Data encryption standards (*see* DES entries)
Data rates, 228–234
  throughput and, 230–232
Data security, 428
  nature of, 455–456
Data sets, 265
Data terminal equipment (DTE), 261–265
Data terminals, 265
dB (*see* Decibels)
dc (direct current), term, 52
dc component of noise, 178
dc shift, 140–142
DCE (data communications equipment), 261–265
DE (driver enable) line, 303–304
Decibels (dB), 179
  applications of, 181–182
  calculation of, 179–181
Delay characteristics, 330
Demodulation, 41–42
  AM, 72
  FM, 61
  synchronous and asynchronous, 73–76
DES (data encryption standards), 466–473
  ICs for, 470–473
DES concept, 466–468
DES modes, 468–470
Dial-up network, 331
Dialing, telephone, 323–328
  pulse, 323–324
  tone, 324–327
Difference signals, 54–55
Differential line drivers, 195

Index 559

Differential signals, 194–197
Digital communications, 118–156
Digital converters, analog to (ADC), 133–139
Digital gates, 433
Digital modems, 354
Digital modulation, 67–72, 152–154
Digital signals, 9–10, 69, 81, 119
   encoding of, 140–145
   modulation of, 147–155
   multiplexing of, 147–155
   TDM for, 129–130, 148–153
Digital systems:
   advantages of, 122–127
   description of, 118–121
Digitization and modulation, 138–139
Direct connect modems, 345–347, 353
Direct current (see dc entries)
Disk antenna, 167, 168
Distance:
   bandwidth and, 34–36
   loss, 163
Distortion:
   intermodulation, 502
   phase, 502
   signal, 35
Distributed systems, 4–5
Drift, frequency, 92–93
Driver enable (DE) line, 303–304
Dropout loss, 503
DTE (data terminal equipment), 261–265
Dual tone multifrequency (DTMF), 325–328
Duplex system, 6–7

EBCDIC (Extended Binary Code for Data Interchange), 212–214
EBCDIC character set, 524–525
Echo, telephone, 333–335
Echo cancellation, 335
EDC (see Error-detection and -correction system)
Efficiency, channel, 120
EIA (see Electronics Industry Association)
Electrical noise (see Noise, electrical)
Electromagnetic energy, 167–168
Electromagnetic spectrum, 19, 22–26
   chart of, 518
Electromagnetic waves, 15–18

Electronic code book algorithm, 468
Electronic mail systems, 362
Electronic testing, 510–511
Electronics Industry Association (EIA), 250
   standards, 298–304
Encoding schemes, 142–144
Encryption, 456
   with PRSQ, 464–465
Encryption standards, data (see DES entries)
Envelope, carrier, 54–55
Envelope delays, 330, 502
Equalization, 189–191
   adaptive, 191
Error correction, 428
   forward (FEC), 446–453
Error detection, 425–476
Error-detection and -correction (EDC) system, 448–453
   ICs for, 451–453
Error function table, 452
Error rate, bit (BER), 125, 180, 427, 482
Error syndrome bits, 449
Errors:
   aliasing, 131
   dealing with, 444–446
   nature of, 426–428
   quantization, 136
   quantizing, 136
   RS-232, 259–260
Ethernet, 394
Ethernet ICs, 410–412
Even parity, 429
Exchanges, 331
Exclusive OR (XOR) gates, 438–439
Extended Binary Code for Data Interchange (see EBCDIC entries)
Eye patterns, 187–189

Facsimile machine, 32–33
Fading phenomenon, 169
FEC (forward error correction), 446–453
Federal Information Processing Data Encryption Standard, 46, 466
Feedback taps, 439
Fiber-optic modems, 352–353
Fiber optics, 171–174, 384, 482
   testing, 508–513
FIFO (first in, first out) buffer, 218

Filter bandwidth, 45
Filters:
   antialiasing, 132
   low-pass, 62
First in, first out (FIFO) buffer, 218
Flat cable, multiconductor, 165–166
FM (see Frequency modulation)
FMD (frequency-division multiplexing), 86, 89–93
Forward error correction (FEC), 446–453
Four-wire systems, 344–345
Fourier analysis, 42–52
   examples of, 45–51
Frame format, 237-238
Framing errors, 259–260
Framing synchronization, 150
Frequency, 14, 17, 19–21
   clock, 75
   cutoff, 184
   resonant, 52
Frequency bands, 22–23
Frequency-division multiplexing (FDM), 86, 89–93
Frequency domain description, 43
Frequency drift, 92–93
Frequency-modulated (FM) radio stations, 27–28
Frequency modulation (FM), 59–63
   demodulation of, 61
   phase coherent, 395
Frequency response, 330
Frequency-shift keying (FSK), 71
Frequency-specific noise, 175–177
Full-duplex modems, 343–345
Full-duplex system, 6–7

Gateways, 398–399
Gender changers, 485
General-purpose interface bus (GPIB), 392
Generator polynomial, 437
Geostationary orbit, 24
GPIB (general-purpose interface bus), 392
Ground potential difference, 193
Guard bands, 28

Half-duplex SDLC example, 239–240
Half-duplex system, 7–8
Hamming EDC circuit, 448–451
Handshake messages, 217

Handshaking, 215, 217–218
  on current loops, 309
  hardware, 216–217
  RS-232 interface, 276–278
Hardware, software versus, 241–243
Hardware conversion, 213
Hardware handshaking, 216–217
Harmonics, 48
Hertz (Hz), 19
Hierarchy, communications, 206, 209
High-impedance mode, 294–295
High lead, 194–195
High performance, term, 226
Horizontal parity bits, 432–433
HP4971S, 500–501
Hybrid circuits, 319, 320
Hz (hertz), 19

ICs (*see* Integrated circuits)
Idle value, 222
IEEE (Institute of Electrical and Electronic Engineering), 388
IEEE-488 standard, 391–394
  ICs for, 408–410
Impulse noise, 175, 176
Information mode, 238
Institute of Electrical and Electronic Engineering (*see* IEEE *entries*)
Integrated circuits (ICs), 61–62
  for DES, 470–473
  for EDC, 451–453
  Ethernet, 410–412
  for IEEE-488 standard, 408–410
  for integral modems, 350–351
  network, 407–412
  physical interface, 279
  for RS-232, 279-286
  SN75176, 303–304
  TDM circuit using, 101–103
  tone generation, 327–328
  VLSI, 470
Integrated Services Digital Network (ISDN), 417–418
Intel 8251, 281–285
Intel 82501 Ethernet Serial Interface, 410–412
Interface needs, additional, 292–293
Interface standards, 293
Intermodulation distortion, 502
Internal reflection, total, 171–172
International Organization for Standardization (ISO), 399

International Telegraph and Telephone Consultative Committee (CCITT), 417–418
Interrupt-driven protocols, 373–375
Interrupts, 394
Intersymbol interference (ISI), 187–188
Intracomputer communication, 5
ISDN (Integrated Services Digital Network), 417–418
ISI (intersymbol interference), 187–188
ISO (International Organization for Standardization), 399
Isolation, 196

Jamming problem, 169
Jitter, 74–75
  amplitude, 503
  phase, 503
  sampling, 130

LAN (local area network), 386–388
  OSI and, 405–406
  in wide area networks, 398
LAN coprocessors, 410–411
LAN protocol analyzers, 500–501
Least significant bit (LSB), 138
Light, speed of, 18
Light-emitting diode (LED), 196
Light power meters, 509
Light source, calibrated, 509
Line drivers, 285–286
Line monitors, 487–490
Line receivers, 286
Lines, telephone (*see* Telephone lines)
Links:
  microwave, 168
  radio, 162
  transparent, 107
Listener echo, 333–334
Listeners, 393
Local area network (*see* LAN *entries*)
Long-distance telephone link, 105
Loopbacks, 490–494
  implementing, 492–494
  locations for, 491–492
Loss distance, 163
Low lead, 194–195
Low-pass filter, 62
Lower sidebands, 54–55
LSB (least significant bit), 138

Mail systems, electronic, 362
Manchester encoding, 143–144
Manchester encoding circuit, 146
Manufacturing automation protocol (MAP), 389, 394–395
Mark, binary, 252
Master, network, 371–373
Master stations, 294
Mating pairs, 274
Medium, 161
  air and vacuum as, 167–170
  choosing, 382–384
  role of, 161–164
Memory buffers, 138–139
Message code, 207
Message polynomial, 437
Message protocols, 220–221
Message translation, levels of, 208
Metallic modems, 353
Microwave link, 168
Mixing process, 56
Mobile phones, 413
Modems, 92
  answer, 344–345
  concentrator, 354–355
  digital, 354
  direct connect, 345–347, 353
  fiber-optic, 352–353
  full-duplex, 343–345
  functions of, 340–342
  integral, ICs for, 350–351
  metallic, 353
  null, 274
  operations of, 342–343
  originate, 343–344
  role of, 339–347
  short haul, 353
  specialized, 352–355
  specific, 348–351
Modulated bandwidth, 59–61
Modulation, 40–77
  amplitude (*see* Amplitude modulation)
  analog versus digital, 67–72
  combined, 71–72, 153
  digital, 152–154
  of digital signals, 147–155
  digitization and, 138–139
  frequency (*see* Frequency modulation)
  phase (PM), 64–66
  pulse-code (PCM), 139
  synchronous and asynchronous, 73–76
  types of, 53–54

Index 561

Most significant bit (MSB), 138
Motorola MC14410, 327–328
MSB (most significant bit), 138
Multiconductor flat cable, 165–166
Multidrop communications, 293–298
Multidrop operation on current loop, 307–309
Multiplexers:
  concentrator, 354–355
  statistical, 355
Multiplexing, 30–31, 70, 85–86
  of digital signals, 147–155
Multiplier, 62

Negative acknowledgment (NAK), 444–446
Network ICs, 407–412
Network layers, 399–406
Network protocols, 371–380
Networks, 361–420
  cellular, 412–416
  closed, 363
  defined, 361–364
  dial-up, 331
  examples of, 389–395
  local area (see LAN entries)
  open, 363
  packet switched, 397–398
  private, 363
  standards and, 388
  switched, 331
  topology of (see Topology)
  wide area, 396–399
Nodes, 364
Noise, electrical, 25, 174–177
  burst aspect of, 426
  crosstalk, 176–177
  dc component of, 178
  effects of, 122–123
  frequency-specific, 175–177
  impulse, 175, 176
  measurements of, 178–182
  wideband, 174–175
Nonlinearity, 84
Nonprintable characters, 210, 213
Nonreturn to zero (NRZ) encoding, 142–145
Nonsequenced mode, 238
NRZ (nonreturn to zero) encoding, 142–145
Null modems, 274
Nyquist sampling theory, 128–130

Odd parity, 429
One-time key, 458–459

Open networks, 363
Open standards, 388
Open systems interconnection (OSI), 399–406
  LANs and, 405–406
Optical attenuators, 509–510
Optical loss set, 511
Optical time domain reflectometry (OTDR), 512–513
Optical transmitters/receivers, 510
Optics, fiber (see Fiber optic entries)
Optoisolators, 196–197, 306
Orbit, geostationary, 24
Originate modems, 343–344
Oscilloscopes, 483–484
OSI (see Open systems interconnection)
OTDR (optical time domain reflectometry), 512–513
Overrun errors, 259–260
Overvoltage protection, 319

PABX (Private Automatic Branch Exchange), 336–338
Packet switched networks, 397–398
Parallel systems, 207, 232–233
Parity, 230, 429–434
  circuitry for, 432–434
  weakness of, 431–432
Parity bit, 430–432
Parity check, 429
Parity errors, 259–260
Pattern generators, 495
PBX (private branch exchange), 336
PCM (pulse-code modulation), 139
P/F (poll/final) bit, 238
Phase, 64
Phase coherent frequency modulation, 395
Phase distortion, 502
Phase jitter, 503
Phase-locked loop (PLL), 61–63
Phase modulation (PM), 64–66
Phase-shift keying (PSK), 71
Photons, 17
Physical interface ICs, 279
Physical security, 428
Pinout diagrams, 274–275
Plain Old Telephone Service (POTS), 316–317
Plaintext, 456
PLL (phase-locked loop), 61–63
Plotter, protocol from computer to, 236–237
Plug connectors, 273–274

PM (phase modulation), 64–66
Point-to-point communications, 173
Poll/final (P/F) bit, 238
Polling, 373
Postamble, message, 220
Potential difference, ground, 193
POTS (Plain Old Telephone Service), 316–317
Preamble, message, 220
Precompensation, 189
Primary rate service, 417–418
Printer, protocols from computer to, 235–236
Private Automatic Branch Exchange (PABX), 336–338
Private branch exchange (PBX), 336
Private networks, 363
Propagation, wave, 25
Proprietary standards, 388
Protocol analyzers, 499–501
Protocol converters, 243–244
Protocols, 215–221, 241
  examples of, 235–240
  interrupt-driven, 373–375
  message, 220–221
  network, 371–380
Pseudorandom sequences (PRSQs), 459–461
  encrypting with, 464–465
PSH (phase-shift keying), 71
Pulse-code modulation (PCM), 139
Pulse dialing, 323–324

Quantization errors, 136
Quantization uncertainty, 135
Quantizing errors, 136

Radio-based telephones, 413
Radio links, 162
Random sequences, 457–461
  binary systems and, 457–458
Range of medium, 162
Receive enable (RE) pin, 304
Receiving buffers, 231
Recommended standard (see RS entries)
Reduction, bandwidth, 31
Redundancy codes, cyclic (see Cyclic redundancy codes)
Reference signal, 66
Reference waveform, 64
Reflection:
  total internal, 171–172
  wave, 25
Regeneration, signal, 123–124
Resonant frequency, 52

Return to zero (RZ) encoding, 143–145
Ring topology, 369–370
Root-mean-square (rms) value, 178
RS (recommended standard), 251
RS-232 interface standard, 250-288
   error conditions of, 259
   examples of, 266–271
   handshaking in, 276–278
   integrated circuits for, 279–286
   interconnections with, 272–278
   line designation for, 262
   line types and sources, 263–265
   signals for, 261–265
   voltages of, 252–256
RS-422 standard, 300–301
RS-423 standard, 298–300
RS-485 standard, 301–302
RZ (return to zero) encoding, 143–145

Sampling jitter, 130
Sampling theory, 128–132
SDLC (*see* Synchronous data-link control)
SDM (space-division multiplexing), 86–89
Security:
   data (*see* Data security)
   of medium, 163
   physical, 428
Self-clocking encoding method, 144
Self-handshaking, 277
Serial bit stream, 136
Serial systems, 94, 207, 232–233
Shannon's theorem, 29–30
Shielding, wire, 21
Shift keying, 71
Shift register, 438
Short haul modems, 353
Sidebands, 54–55
Signal distortion, 35
Signal regeneration, 123–124
Signal-to-noise ratio (SNR), 179
Signal values, averaging, 109
Simplex system, 6
Sine waves, 45
Single-ended systems, 195
Single-sideband suppressed carrier, (SSB), 57
Slave, network, 371–373
Slew rate, 254–255, 299
SLIC (*see* Subscriber Loop Interface Circuit)
SN75176 IC, 303–304
SNR (signal-to-noise ratio), 179

Socket connectors, 273–274
Software, hardware versus, 241–243
Software conversion, 213
Space, binary, 252
Space-division multiplexing (SDM), 86–89
Spectral response test, 511
Spectrum analyzer, 43–45
   screen of, 44
Speed of light, 18
Spread spectrum data communications systems, 463–465
SSB (single-sideband suppressed carrier), 57
Standards:
   data encryption (*see* DES *entries*)
   networks and, 388
   open, 388
   proprietary, 388
Star topology, 366–367
Start bit, 222
State diagrams, 407–412
State machines, 407
Statistical multiplexers, 355
Status register, 284
Steady-state tests, 502–503
Stop bit, 223
Store and forward systems, 397
Straight binary coding, 213
Subscriber Loop Interface Circuit (SLIC), 319
   actual circuit, 321–322
Sum signals, 54–55
Superposition principle, 52
Supervisory mode, 238
Switched networks, 331
Sync characters, 74
Sync loss, 503
Sync signal, 97–99
Synchronization, 97
   framing, 150
Synchronous communications, TDM and, 99–100
Synchronous data-link control (SDLC), 237–239
   half-duplex example, 239–240
Synchronous modulation and demodulation, 73–76
Synchronous systems, 222–227

T-1 digital carrier communications systems, 151–152
Talker echo, 333–334
Talkers, 393

TDM (*see* Time-division multiplexing)
TDR (*see* Time domain reflectometry)
Telecommunications, 316
Telecommunications system, 10
Telephone exchange, standard, 87–89
Telephone lines, 329–335
   echoes on, 333–335
   standard, 330–333
Telephone office functions, 319–321
Telephone service, basic, 315–322
Telephone systems, 2, 315–357
Telephones:
   basic, 316–319
   dialing (*see* Dialing, telephone)
   radio-based, 413
Television (TV), 20
Temperature meter application, 266
Test techniques, 481–515
Testing:
   electronics, 510–511
   fiber-optic systems, 508–513
   goal of, 481–482
Tests:
   basic, 481–485
   steady-state, 502–503
Thermocouples, 83
3-dB bandwidth point, 184–187
Three-state output, 294
Throughput:
   actual, 231
   data rates and, 230–232
TI75LS636, 451–453
Time delay phase shift, 330
Time-division multiplexing (TDM), 86, 94–103, 126
   advantages and disadvantages of, 100–101
   for analog and digital signals, 129
   circuit using ICs, 101–103
   for digital signals, 129–130, 148–153
   synchronous communications and, 99–100
Time domain description, 43
Time domain reflectometry (TDR), 504–507
   optical (OTDR), 512–513
Timing diagrams, 216–218
TIMS (transmission impairment measurement sets), 501–503
TMS9914A, 408–409
Token passing, 375–377

Tone dialing, 324–327
Tone generation ICs, 327–328
Topology, 364–370
    bus, 367–369
    ring, 369–370
    star, 366–367
Transducers, 83
Transformer coupling, 390
Transistor-transistor logic (TTL), 426
Transmission control characters, 520
Transmission impairment measurement sets (TIMS), 501–503
Transmitting system, complete, 82
Transparent links, 107
Trunks, 331
    number needed, 337–338
TTL (transistor-transistor logic), 426
TV channels, 20
TV systems, cable, 20

20-mA interface, 305
Twisted wire pairs, 164–165, 383

Ultra high frequency (UHF), 22–25
Uncertainty, quantization, 135
Universal Synchronous/Asynchronous Receiver/Transmitter (USART), 280
Upper sidebands, 54–55
User numbers, 296

Vacuum as medium, 167–170
VCO (voltage-controlled oscillator), 62
Vertical parity bits, 432–433
Very large scale integration (VLSI), 470
Voice communication, 3
Voice grade conditioning, 330–331
Voltage, 42, 193
    common mode (CMV), 192–197
    compliance, 308

Voltage-controlled oscillator (VCO), 62
Voltmeters, 482–484

WANGNET, 389
Watchdog timer, 411
Wave propagation, 25
Waveguides, 169–170
Wavelength, 17–21
WD2001 and WD2002, 470–473
Wide area networks, 396–399
Wideband noise, 174–175
Wire, 164–167
Wire antenna, 167, 168
Wire pairs, twisted, 164–165, 383

X.25 protocol, 397–398
XON/XOFF method, 269–270
XOR (exclusive OR) gates, 438–439

Zero crossings, 61